S

Future
Crimes

EVERYTHING IS **CONNECTED**

FUTURE

EVERYONE IS **VULNERABLE**

CRIMES

AND **WHAT WE CAN DO** ABOUT IT

Marc Goodman

DOUBLEDAY CANADA

Doubleday Canada and colophon are registered trademarks of Random House of Canada Limited.

Library and Archives Canada Cataloguing in Publication

Goodman, Marc, author
Future crimes : everything is connected, everyone is vulnerable and what we can do about it / Marc Goodman.

Includes bibliographical references and index.
Issued in print and electronic formats.
ISBN 978-0-385-68256-5 (bound) ISBN 978-0-385-68257-2 (epub)

1. Computer crimes. 2. Technology--Social aspects. I. Title.

HV6773.G65 2015   364.16'8   C2014-907395-X
                                            C2014-907396-8

Jacket design by Pete Garceau
Jacket art: camera lens © PixelEmbargo/iStock/Thinkstock
phone © yganko/iStock/Thinkstock

Printed and bound in the USA

Published in Canada by Doubleday Canada,
A division of Random House of Canada Limited,
A Penguin Random House Company

www.penguinrandomhouse.ca

10 9 8 7 6 5 4 3 2 1

Penguin
Random House
DOUBLEDAY CANADA

To all my teachers,
who have taught me so much

CONTENTS

Future
Crimes

The Irrational Optimist:
How I Got This Way

My entrée into the world of high-tech crime began innocuously in 1995 while working as a twenty-eight-year-old investigator and sergeant at the LAPD's famed Parker Center police headquarters. One day, my lieutenant bellowed my name across the crowded and bustling detective squad room: "Goooooooodmaaaan, get your ass over here!" I presumed that I was in trouble, but instead the lieutenant asked me the question that would change my life: "Do you know how to spell-check in WordPerfect?"

"Sure, boss, just hit Ctrl+F2," I replied.

He grinned and said, "I knew you were the right guy for this case."

Thus began my career in high-tech policing with my very first computer crime case. Knowing how to spell-check in WordPerfect made me among the techno-elite of cops in the early 1990s. Since that case, I have been a keen observer and student not just of technology but of its illicit use. Though I recognize the harm and destruction wrought by the misapplication of technology, I continue to be fascinated by the clever and innovative methods criminals use to achieve their objectives.

Criminals perpetually update their techniques to incorporate the very latest emerging technologies into their modi operandi. They have evolved well beyond the days when they were the first on the street carrying pagers and using five-pound cell phones to send coded messages to one another. Today, they are building their own nationwide encrypted cellular radio telecommunications systems, like those employed by the narco-cartels of Mexico. Consider for a moment the sophistication required to establish such a fully functioning encrypted nationwide communications network—an amazing feat, especially because many Americans still can't get a decent mobile phone signal most of the time.

Organized crime groups have established themselves as early adopters of technology. Criminals embraced the online world long before the police ever contemplated it, and they have outpaced authorities ever since. News headlines are replete with stories of 100 million accounts hacked here and $50 million stolen online there. The progression of these crimes is striking, and they are accelerating in very much the wrong direction.

The subject of this book isn't just what was going on yesterday or even what is happening today. Nor is its focus how long your password should be. It is about where we're going tomorrow. In my own research and investigations, first with the LAPD and later working with federal and international law enforcement organizations, I have uncovered criminals who have progressed well beyond today's cyber crime into new and emerging fields of technology such as robotics, virtual reality, artificial intelligence, 3-D printing, and synthetic biology. In most instances, my law enforcement and government colleagues around the world are unaware of these looming technological developments, let alone their growing exploitation by both organized crime and terrorist organizations. As somebody who has dedicated his life to public safety and public service, I am deeply concerned by the trends I observe all around me.

Though some may accuse me of fearmongering or being a hard-core pessimist, I am neither. Rather, I am optimistic—perhaps "irrationally optimistic"—given what I've seen about our future. To be clear, I am no neo-Luddite. Nor am I foolish enough to suggest technology is the source of all the ills in our world. Quite the opposite: I believe in the tremendous power of technology to be a driving force for good. It should also be noted that there are many ways it can and has been used to protect individuals and society. But technology has always been a double-edged sword. My real-world experiences with criminals and terrorists on six continents has made it clear to me that the forces of evil will not hesitate to exploit these emerging technologies and deploy them against the masses. Though the evidence and my gut tell me there are significant bumps in the road ahead—ones that government and industry are not dedicating sufficient resources to address or combat—I want to believe in the techno-utopia promised to us by Silicon Valley.

This book is the story of the society we are building with our technological tools and how these very same implements can be used against us. The more we plug our devices and our lives into the global information grid—whether via mobile phones, social networks, elevators, or self-driving cars—the more vulnerable we become to those who know how the underlying technologies work and how to exploit them to their advantage and to the detriment of the common man. Simply stated, when everything is connected, everyone is vulnerable. The technology we routinely accept into our lives with little or no self-reflection or examination may very well come back to bite us.

By shedding light on the very latest in criminal and terrorist tradecraft, I hope to kick off a vibrant and long-overdue discussion among my friends and colleagues in policing and national security. Though most are already overburdened with enough traditional crime, they must confront, sooner rather than later, the exponentially advancing technologies that will arrive as a tsunami of threats capable of destabilizing our common global security.

More important, as somebody who long ago swore to "protect and serve" others, I want to ensure that members of the general public are armed with the facts needed to protect themselves, their families, their companies, and their communities against the horde of emerging threats that will be here much more quickly than anticipated. Limiting this knowledge to "insiders" working in government, security, and Silicon Valley simply won't do.

Throughout my tenure in public service working with organizations that include the LAPD, the FBI, the U.S. Secret Service, and Interpol, it became increasingly obvious to me that criminals and terrorists were out-innovating police forces around the world and that the "good guys" were rapidly falling further and further behind. On a quest for deeper impact against the growing legions of criminals abusing cutting-edge technologies, I left government and moved to Silicon Valley to further educate myself on what would come next.

In California, I immersed myself in a community of technological innovators in order to decipher how their latest scientific discoveries would affect the common man. I visited the scions of Silicon Valley and made friends within the highly talented San Francisco Bay Area start-up community. I was invited to join the faculty of Singularity University, an amazing institution housed on the campus of NASA's Ames Research Center, where I worked with a brilliant array of astronauts, roboticists, data scientists, computer engineers, and synthetic biologists. These pioneering men and women have the ability to see beyond today's world, unlocking the tremendous potential of technology to confront the grandest challenges facing humanity.

But many of these Silicon Valley entrepreneurs hard at work creating our technological future pay precious little attention to the public policy, legal, ethical, and security risks that their creations pose to the rest of society. Yet my own experience putting handcuffs on criminals and working with police forces in more than seventy countries has yielded a different outlook on the potential abuses of the emerging technologies that innocent people everywhere welcome into their daily lives—generally without question.

To that end, I founded the Future Crimes Institute. The goal was to use my own experiences as a street cop, investigator, international counterterrorism analyst, and, most recently, Silicon Valley insider to catalyze a community of like-minded experts to address the negative as well as the positive implications of rapidly evolving technology.

As I look toward the future, I am increasingly concerned about the ubiquity of computing in our lives and how our utter dependence on it is leaving us vulnerable in ways that very few of us can even begin to comprehend. The current systemic complexities and interdependencies are great and growing all the time. Yet there are individuals and groups who are rapidly making sense of them and innovating in real time, to the detriment of us all.

This is their story—the story of organized criminals, hackers, rogue governments, substate actors, and terrorists, all competing to control the latest technologies for their own benefit.

The techno-utopia promised by Silicon Valley may be possible, but it will not magically appear on its own. It will take tremendous intention, effort, and struggle on the part of citizens, governments, corporations, and NGOs to ensure that it comes to fruition. A new battle has begun between those who will leverage technology to benefit humanity and those who prefer to subvert these tools, regardless of the harm caused to others. This is the battle for the soul of technology and its future. It rages on in the background, mostly sub rosa, heretofore well hidden from the average citizen.

Beyond merely cataloging the latest in criminal innovation and technical vulnerabilities, this book offers a path forward to defeat the myriad threats that await us. If we use foresight, I believe it is possible to anticipate and prevent tomorrow's crimes today, before we reach the point of no return. Future generations will look back and judge our efforts to curb these security threats and defend the soul of technology to ensure that it inures to humanity's ultimate benefit.

A friendly warning: if you proceed in reading the pages that follow, you will never look at your car, smart phone, or vacuum cleaner the same way again.

This is your last chance. After this, there is no turning back.
You take the blue pill—the story ends, you wake up in your bed and
believe whatever you want to believe. You take the red pill—you stay in
Wonderland, and I show you how deep the rabbit hole goes.
Remember, all I'm offering is the truth—nothing more.
MORPHEUS'S WARNING TO NEO, *THE MATRIX*

A
Gathering
Storm

CHAPTER 1

Connected, Dependent, and Vulnerable

**Technology . . . is a queer thing; it brings you great gifts with one hand
and it stabs you in the back with the other.**
CHARLES PERCY SNOW

Mat Honan's life looked pretty good on-screen: in one tab of his browser were pictures of his new baby girl; in another streamed the tweets from his thousands of Twitter followers. As a reporter for *Wired* magazine in San Francisco, he was living an urbane and connected life and was as up-to-date on technology as anyone. Still, he had no idea his entire digital world could be erased in just a few keystrokes. Then, one August day, it was. His photographs, e-mails, and much more all fell into the hands of a hacker. Stolen in just minutes by a teenager halfway around the world. Honan was an easy target. We all are.

Honan recalls the afternoon when everything fell apart. He was playing on the floor with his infant daughter when suddenly his iPhone powered down. Perhaps the battery had died. He was expecting an important call, so he plugged the phone into the outlet and rebooted. Rather than the usual start-up screen and apps, he saw a large white Apple logo and a multilingual welcome screen inviting him to set up his new phone. How odd.

Honan wasn't especially worried: he backed up his iPhone every night. His next step was perfectly obvious—log in to iCloud and restore the phone and its data. Upon logging in to his Apple account, he was informed that his password, the one he was sure was correct, had been deemed wrong by the iCloud gods. Honan, an astute reporter for the world's preeminent technology magazine, had yet another trick up his sleeve. He would merely connect the iPhone to his laptop and restore his data from the hard drive on his local computer. What happened next made his heart sink.

As Honan powered up his Mac, he was greeted with a message from Apple's calendar program advising him his Gmail password was incorrect. Immediately thereafter, the face of his laptop—its beautiful screen—turned ashen gray and quit, as if it had died. The only thing visible on the screen was a prompt: please enter your four-digit password. Honan knew he had never set a password.

Honan ultimately learned that a hacker had gained access to his iCloud account, then used Apple's handy "find my phone" feature to locate all of the electronic devices in Honan's world. One by one, they were nuked. The hacker issued the "remote wipe" command, thereby erasing all of the data Honan had spent a lifetime accumulating. The first to fall was his iPhone, then his iPad. Last, but certainly not least, was his MacBook. In an instant, all of his data, including every baby picture he had taken during his daughter's first year of life, were destroyed. Gone too were the priceless photographic memories of his relatives who had long since died, vanquished into the ether by parties unknown.

Next to be obliterated was Honan's Google account. In the blink of an eye, the eight years of carefully curated Gmail messages were lost. Work conversations, notes, reminders, and memories wiped away with a click of a mouse. Finally, the hacker turned his intention to his ultimate target: Honan's Twitter handle, @Mat. Not only was the account taken over, but the attacker used it to send racist and homophobic rants in Honan's name to his thousands of followers.

In the aftermath of the online onslaught, Honan used his skills as an investigative reporter to piece together what had happened. He phoned Apple tech support in an effort to reclaim his iCloud account. After more than ninety minutes on the phone, Honan learned that "he" had just called thirty minutes prior to request his password be reset. As it turns out, the only information anybody needed to change Honan's password was his billing address and the last four digits of his credit card number. Honan's address was readily available on the Whois Internet domain record he had created when he built his personal Web site. Even if it hadn't been, dozens of online services such as WhitePages.com and Spokeo would have provided it for free.

To ascertain the last four digits of Honan's credit card, the hacker guessed that Honan (like most of us) had an account on Amazon.com. He was correct. Armed with Honan's full name and his e-mail and mailing addresses, the culprit contacted Amazon and successfully manipulated a customer service rep so as to gain access to the required last four credit card digits. Those simple steps and nothing more turned Honan's life upside down. Although it didn't happen in this case, the hacker could have just as easily used the very same information to access and pilfer Honan's online bank and brokerage accounts.

The teenager who eventually came forward to take credit for the attack—Phobia, as he was known in hacking circles—claimed he was out to expose the vast security vulnerabilities of the Internet services we've come to rely on every day. Point made. Honan created a new Twitter account to communicate with his attacker. Phobia, using the @Mat account, agreed to follow Honan's new account, and now the two could direct message each other. Honan asked Phobia the single question that was burning on his mind: Why? Why would you do this to me? As it turns out, the near decade of lost data and memories was merely collateral damage.

Phobia's reply was chilling: "I honestly didn't have any heat towards you . . . I just liked your [Twitter] username." That was it. That's all it was ever about—a prized three-letter Twitter handle. A hacker thousands of miles away liked it and simply wanted it for himself.

The thought that somebody with no "heat" toward you can obliterate your digital life in a few keystrokes is absurd. When Honan's story appeared on the cover of *Wired* in December 2012, it garnered considerable attention . . . for a minute or two. A debate on how to better secure our everyday technologies ensued but, like so many Internet discussions, ultimately flamed out. Precious little has changed since Honan's trials and tribulations. We are still every bit as vulnerable as Honan was then—and even more so as we ratchet up our dependency on hackable mobile and cloud-based applications.

As with most of us, Honan's various accounts were linked to one another in a self-referential web of purported digital trust: the same credit card number on an Apple profile and an Amazon account; an iCloud e-mail address that points back to Gmail. Each had information in common, including log-on credentials, credit card numbers, and passwords with all the data connected back to the same person. Honan's security protections amounted to nothing more than a digital Maginot Line—an overlapping house of cards that came tumbling down with the slightest pressure. All or most of the information needed to destroy his digital life, or yours, is readily available online to anybody who is the least bit devious or creative.

Progress and Peril in a Connected World

In a few years' time, with very little self-reflection, we've sprinted headlong from merely searching Google to relying on it for directions, calendars, address books, video, entertainment, voice mail, and telephone calls. One billion of us have posted our most intimate details on Facebook and willingly provided social networking graphs of our friends, family, and co-workers. We've downloaded billions of apps, and we rely on them to help us accomplish everything from banking and cooking to archiving baby pictures. We connect to the Internet via our laptops, mobile phones, iPads, TiVos, cable boxes, PS3s, Blu-rays, Nintendos, HDTVs, Rokus, Xboxes, and Apple TVs.

The positive aspects of this technological evolution are manifest. Over the past hundred years, rapid advances in medical science mean that the average human life span has more than doubled and child mortality has plummeted by a factor of ten. Average per capita income adjusted for inflation around the world has tripled. Access to a high-quality education, so elusive to many for so long, is free today via Web sites such as the Khan Academy. And the mobile phone is singularly credited with leading to billions upon billions of dollars in direct economic development in nations around the globe.

The interconnectivity the Internet provides through its fundamental architecture means that disparate peoples from around the world can be brought together as never before. A woman in Chicago can play *Words with Friends* with a total stranger in the Netherlands. A physician in Bangalore, India, can remotely read and interpret the X-ray results of a patient in Boca Raton, Florida. A farmer in South Africa can use his mobile phone to access the same crop data as a PhD candidate at MIT. This interconnectedness is one of the Internet's greatest strengths, and as it grows in size, so too does the global network's power and utility. There is much to celebrate in our modern technological world.

While the advantages of the online world are well documented and frequently highlighted by those in the tech industry, there is also a downside to all of this interconnectivity.

Our electrical grids, air traffic control networks, fire department dispatch systems, and even the elevators at work are all critically dependent on computers. Each day, we plug more and more of our daily lives into the global information grid without pausing to ask what it all means. Mat Honan found out the hard way, as have thousands of others. But what should happen if and when the technological trappings of our modern society—the foundational tools upon which we are utterly dependent—all go away? What is humanity's backup plan? In fact, none exists.

The World Is Flat (and Wide Open)

For centuries, the Westphalian system of sovereign nation-states has prevailed in our world. It meant that countries were to be sovereign in their territory, with no role for outside authorities to meddle in a nation's domestic affairs. The Westphalian structure was preserved through a system of borders, armies, guards, gates, and guns. Controls could be implemented to limit both immigration and emigration of people from a national territory. Moreover, customs and inspection structures would be established to control the flow of goods across national boundaries. Yet as prescient as the signatories to the Treaty of Westphalia were in 1648, none of them foresaw Snapchat.

Though physical borders still matter, such divisions are much less clear in an online world. Bits and bytes flow freely from one country to the next without any border guards, immigration controls, or customs declarations to slow their transit. The traditional transnational barriers to crime faced by former generations of thieves, thugs, and convicts have been demolished in the online world, allowing unsavory individuals to freely enter and exit any virtual location they please.

Think about that and its implications for our security. Once upon a time, if criminals attempted to rob a bank in New York's Times Square, several things were considered to be self-evident. First and foremost, it was assumed that the perpetrators had entered a physical location within the boundaries of the NYPD's Midtown South Precinct. The bank robbery would have breached both New York State and U.S. federal law, and the NYPD and the FBI would share joint jurisdiction to investigate the matter. The victim (in this case the bank) was also colocated within the physical jurisdiction of the concerned law enforcement authorities, greatly simplifying their investigation. Attempts to solve the case would have been bolstered by physical evidence likely left behind at the scene of the crime by the bank robber, including fingerprints on a note handed to a teller and DNA on the counter he jumped over, and perhaps via the images of his own face visible on the bank's security camera system. In addition, the crime itself was subject to certain physical limitations. The dollar bills stolen would have had weight and mass, and only so many could be carried away. The piles of cash might also have had an embedded exploding dye pack to flag the suspect to the police. But in today's world, long-established, tried-and-true investigative givens, such as jurisdictional commonality and physical evidence—fundamental tools for authorities to solve crimes—often no longer exist.

Compare the Times Square heist scenario above with the infamous 1994 Internet bank robbery carried out by Vladimir Levin from his apartment in St. Petersburg, Russia. Levin, a computer programmer, was accused of hacking the accounts of several of Citibank's large corporate customers and making away with $10.7 million. Collaborating with accomplices around the world, Levin transferred large sums of cash to accounts in Finland, the United States, the Netherlands, Germany, and Israel.

Who had jurisdiction for this matter? Was it the police in the United States, where the victim (Citibank) was located? Was it the cops in St. Petersburg, where the suspect carried out the alleged offense? Or was jurisdiction held in Israel or Finland, perhaps, where the ill-gotten funds were electronically delivered to fraudulent accounts? Levin never physically entered the United States to commit the crime. He left no fingerprints or DNA and was never marked by an exploding dye pack. Importantly, he never needed to physically carry the thousands of pounds of cash out of the bank; it was all accomplished with a mouse and a keyboard. No need for a mask or sawed-off shotgun either; Levin merely hid behind his computer screen and used a circuitous virtual route to cover his digital tracks.

The nature of the Internet means that we are increasingly living in a borderless world. Today anybody, with good or ill intent, can virtually travel at the speed of light halfway around the planet. For criminals, this technology has been a boon, as they hop from one country to the next virtually hacking their way across the globe in an effort to frustrate police. Criminals have also learned how to protect themselves from being tracked online. A smart hacker would never directly initiate an attack against a bank in Brazil from his own apartment in France. Instead, he would route his attack from one compromised network to another, from France, to Turkey, to Saudi Arabia, toward his ultimate target in Brazil. This ability to country-hop, one of the Internet's greatest strengths, creates enormous jurisdictional and administrative problems for the police and is one of the main reasons why cyber-crime investigation is so challenging and often feckless. A police officer in Paris has no authority to make an arrest in São Paulo.

The Good Old Days of Cyber Crime

The nature of the cyber threat has changed dramatically over the past twenty-five years. In the early days of the personal computer, hackers were mostly motivated by the "lulz," or laughs. They hacked computer systems just to prove that they could do it or to make a point. One of the very first computer viruses to infect IBM PCs was the Brain virus, created in 1986 by the brothers Amjad Farooq Alvi and Basit Farooq Alvi, aged twenty-four and seventeen, of Lahore, Pakistan. Their virus was intended to be innocuous, to stop others from pirating the software the brothers had spent years developing. Brain worked by infecting the boot sector of a floppy disk as a means of preventing its copying and allowed the brothers to track illegal copies of their own software. The brothers, upset that others were pirating their software without paying for it, included an ominous warning that appeared on infected users' screens:

```
Welcome to the Dungeon © 1986 Brain & Amjads
(pvt). BRAIN COMPUTER SERVICES 730 NIZAM BLOCK
  ALLAMA IQBAL TOWN LAHORE-PAKISTAN PHONE:
 430791,443248,280530. Beware of this VIRUS . . .
       Contact us for vaccination . . .
```

Their message is notable for several reasons. First, the brothers claimed to have copyrighted their virus, a ballsy move indeed. Even more strange was the fact that they included their address and phone numbers for users to contact the virus originators for "vaccination" or removal of the virus. Their reasoning for creating the virus seemed logical enough to Basit and Amjad, but what they hadn't realized was their creation had the capacity to replicate and spread and did so the old-fashioned way, through human beings' carrying around 5.25-inch floppy disks from computer to computer. Eventually, Brain had traveled the globe, introducing Basit and Amjad to the rest of the world.

Over time, hackers grew more ambitious—and more malicious. Our interconnection to one another via computer bulletin board services meant that digital viruses no longer needed to travel via a "sneakernet," carried by human beings on floppy disks, but rather could spread via modem over telephone lines through early online services such as CompuServe, Prodigy, EarthLink, and AOL. Newer viruses and Trojans such as Melissa (1999), ILOVEYOU (2000), Code Red (2001), Slammer (2003), and Sasser (2004) could now infect Microsoft Windows computers around the world with ease, destroying the term papers, recipes, love letters, and company spreadsheets we had saved to our hard drives. Suddenly we all had skin in this game.

Computer "malware," a portmanteau combining the words "malicious" and "software," now comes in many forms, but all seek to damage, disrupt, steal from, or inflict some illegitimate or unauthorized action on a data system or network:

- Computer viruses propagate by inserting a copy of themselves into another program, just as a real-world virus infects an available biological host.
- Computer worms also cause damage, but do so as stand-alone software and do not require a host program to replicate.
- Trojans, named after the mythical wooden horse used by the Greeks to infiltrate Troy, often masquerade as legitimate pieces of software and are activated when a user is tricked into loading and executing the files on a targeted system. Trojans frequently create "back doors" that allow hackers to maintain persistent access to an infected system. Trojans do not reproduce by infecting other files per se, but rather spread by deceiving users into clicking on a file or opening an infected e-mail attachment.

Virus writers today recognize that the public is slowly (very slowly) beginning to get the message not to click on files sent by strangers. As a result, criminals have updated their tactics to create so-called drive-by downloads, which use malware to exploit vulnerabilities in computer scripting languages such as Java and ActiveX, languages commonly used by Internet Web browsers. The world has moved online, and hacking tools such as Internet Explorer, Firefox, and Safari makes sense for criminals, though the new modus operandi comes with a heavy cost for unsuspecting users. Researchers at Palo Alto Networks discovered that as much as 90 percent of modern malware is now spread via previously hacked popular Web sites that serve up the computer infection the moment an unsuspecting visitor stops by the site. Many large companies, including Yahoo!, a major destination portal around the world, have had their Web sites hijacked by criminals and thus unknowingly poisoned their own customers who innocently stopped by to check sports scores or the latest stock market returns.

The Malware Explosion

Now it's not just about the "lulz" but for want of money, information, and power that hackers ply their trade. In the early twenty-first century, as criminals figured out ways to monetize their malicious software through identity theft and other techniques, the number of new viruses began to soar. By 2015, the volume had become astonishing. In 2010, the German research institute AV-Test had assessed that there were forty-nine million strains of computer malware in the wild. By 2011, the antivirus company McAfee reported it was identifying two million new pieces of malware every month. In the summer of 2013, the cyber-security firm Kaspersky Lab reported it identified and isolated nearly 200,000 new malware samples every single day.

Taking a cynical approach to these statistics and presuming that antivirus companies might be incentivized to overstate the problem that they were established to combat, one could be inclined to deflate the numbers dramatically, say by 50 or even 75 percent. Even so, that would still mean that fifty thousand new viruses were generated each and every day. Think about the tremendous research-and-development effort that would be required on a global scale to create that volume of uniquely coded malware.

As any business owner knows, R&D is an expensive proposition. As such, the return on investment (ROI) required to support international organized crime's ongoing illicit computer programming efforts must be vast. An independent study by the trusted Consumers Union, publisher of *Consumer Reports* magazine, seems to confirm the mounting impact of computer malware. A survey of its members revealed that one-third of the households in the United States had experienced a malicious software infection in the previous year, costing consumers a whopping $2.3 billion annually. And that's just the people who realize they've been attacked.

The Security Illusion

Each year, consumers and businesses around the world put their faith in the computer security software industry to protect them from the burgeoning threat of computer malware. According to a study by the Gartner group, worldwide spending on security software totaled nearly $20 billion in 2012 and is forecast to skyrocket to $94 billion spent annually on cyber security by 2017.

Ask most individuals what to do about computer viruses, and their very first answer would be to use an antivirus product from a company like Symantec, McAfee, or Trend Micro. The response is instinctual from a public that has been trained well. While such tools might have proven useful in the past, they are rapidly losing their efficacy, and the statistics are deeply revealing. In December 2012, Researchers at Imperva, a data security research firm in Redwood Shores, California, and students at the Technion–Israel Institute of Technology decided to put the standard antivirus tools to the test. They collected eighty-two new computer viruses and ran the malware against the threat-detection engines of more than forty of the world's largest antivirus companies, including Microsoft, Symantec, McAfee, and Kaspersky Lab. The results: the initial threat-detection rate was only 5 percent, meaning that 95 percent of the malware went completely undetected. That also means the antivirus software you are running on your own computer is likely only catching 5 percent of the emerging threats targeting your machine. If your body's own immune system had a batting average like that, you would be dead in a matter of hours.

Months later, the behemoths of the security software industry eventually update their software, but of course by then it's often too late. The fact of the matter is that criminals and virus writers are completely out-innovating and outmaneuvering the antivirus industry established to protect us against these threats. Worse, the "time-to-detection rate"—that is, the amount of time it takes from the initial introduction of a piece of malware "in the wild" to be uncovered—is growing. For example, in 2012 researchers at Kaspersky Lab in Moscow uncovered a highly complex piece of malware known as Flame that had been pilfering data from information systems around the world for more than five years before it was detected. Mikko Hypponen, the well-respected chief research officer at the computer security firm F-Secure, called Flame a failure for the antivirus industry and noted he and his colleagues may be "out of their leagues in their own game." Though millions around the world rely on these tools, it's pretty clear the antivirus era is over.

One of the reasons it is proving difficult to counter the wide variety of technological threats in our lives today is that there has been a burgeoning increase in the number of so-called zero-day attacks. A zero-day exploit takes advantage of a previously unknown vulnerability in a computer application that developers and security staff have not had time to address. Rather than proactively looking for these vulnerabilities themselves, antivirus software companies generally only consider known data points. They'll block a malicious bit of code if it's just like the other malicious bits of code they have seen previously. It's essentially like putting up a wanted poster for Bonnie and Clyde because we know they have robbed banks previously. Bank tellers would know to be on the lookout for the couple, but as long as no one fitting that description materialized, they might let their guard down—until a different bank robber struck, that is. These zero days are increasingly being generated for a wide variety of techno-products commonly used in our lives, affecting everything from Microsoft Windows to Linksys routers to Adobe's ubiquitous PDF Reader and Flash Player.

Eventually, hackers figured out that the more noise they made breaking into your systems, the more quickly you would fix the problem and kick them out. Now it's all about stealth and clandestinity, like having a sleeper cell in your computer. You might think the abysmal 5 percent computer virus detection rate revealed in the Imperva study applied only to average citizens using personal security software in their homes. Surely businesses with their massive budgets for information technology and security would fare much better against hackers? Not so much. Tens of thousands of successful hacks against major corporations, NGOs, and governments around the world reveal that enterprises, for all they spend, are not much better in protecting their own information.

According to Verizon's *2013 Data Breach Investigations Report*, most businesses have proven simply incapable of detecting when a hacker has breached their information systems. The landmark survey carried out by Verizon business services, working in conjunction with the U.S. Secret Service, the Dutch National Police, and the U.K. Police Central E-crimes Unit, reported that on average 62 percent of the intrusions against business took at least two months to detect. A similar study by Trustwave Holdings revealed that the average time from the initial breach of a company's network until discovery of the intrusion was an alarming 210 days. That's nearly seven months for an attacker—whether organized crime, the competition, or a foreign government—to creep around unfettered in a corporate network stealing secrets, gaining competitive intelligence, breaching financial systems, and pilfering customers' personally identifiable information, such as their credit card numbers.

When businesses do eventually notice that they have a digital spy in their midst and that their vital information systems have been compromised, an appalling 92 percent of the time it is not the company's chief information officer, security team, or system administrator who discovers the breach. Rather, law enforcement, an angry customer, or a contractor notifies the victim of the problem. If the world's biggest companies, firms that collectively spend millions on cyber defense and have whole departments of professionals working 24/7 to protect their networks, can be so readily penetrated by hackers, the prospects for home users protecting their information look grim indeed.

How hard is it to break into the average computer system? Laughably easy. According to the Verizon study, once hackers set their sights on your network, 75 percent of the time they can successfully penetrate your defenses within minutes. The same study notes that only 15 percent of the time does it take more than a few hours to breach a system. The implications of these findings are profound. From the time an attacker decides to target your world, 75 percent of the time the game is over in minutes. You'll be punched, knocked out, and on the floor before you ever knew what hit you. In today's world, hackers are living unfettered and free inside your very own data systems for months and months, watching, waiting, lurking, and pillaging everything from your passwords to work projects to old selfies. You are easy marks and sitting ducks. How odd that we as a society tolerate this. If any of us noticed a burglar in our home watching over us as we slept or filming us in the shower, we would immediately dial 911 (or alternatively scream or reach for a gun). In cyberspace, this is a daily occurrence, yet most of us remain calmly, even blissfully unaware of the threat, despite our deep vulnerabilities and the bad guys looming over us as we sleep.

The cost of our universal cyber insecurity continues to mount. Though businesses around the world may be on track to spend nearly $100 billion by 2017 for a variety of software and hardware security measures, that price is merely a starting point when considering the full economic impact of our technological fragility. Take, for example, the 2007 cyber strike against TJX, the parent company of the retail stores T.J. Maxx and Marshalls in the United States and T.K. Maxx throughout Europe.

In that case, hackers stole the credit card details of over forty-five million customers, making it the largest retail store hacking case of its time. Later in court filings, it was revealed the actual number of victims was closer to ninety-four million. Though TJX reached a settlement with Visa, MasterCard, and its customers in the amount of $256 million, many analysts believe the true costs could easily have been closer to $1 billion. One of the most authoritative sources for research on the cost of data breaches comes from the Ponemon Institute, which conducts independent research on data protection and information security policy. In calculating cybersecurity breaches, it notes it is important to extend the loss analysis well beyond direct consumer theft amounts.

For example, the victim company targeted in the attacks, such as TJX, must spend handsomely on detecting the breach, containing the attackers, investigating the matter, identifying the perpetrators, and repairing and recovering its computer network. Moreover, there are often heavy sales declines as a wary public shies away from using the services of a company perceived to be unsafe and insecure. Add to that the price of credit card replacement fees (currently estimated to be $5.10 per card), consumer credit-monitoring services that need to be purchased by the victim company to prevent ongoing credit card fraud against its customers, and rising cyber-insurance premiums, and it is easy to see how quickly the costs of these losses can escalate. No wonder most companies are loath to admit that they have been hacked and many attempt to deny the breach as long as possible.

There are yet even greater costs to be considered, including how the stock market punishes victim firms via precipitous drops in their stock prices after a cyber intrusion. In one case, Global Payments saw its market valuation slashed by 9 percent in just one day until the New York Stock Exchange halted trading of its shares. Adding to the financial headaches in these cases are the ensuing class-action lawsuits from a firm's customers, shareholders, and regulators. All told, the Ponemon Institute estimates that companies face nearly $188 in costs for each and every record stolen. Multiply that amount by the nearly 100 million account records stolen at TJX, and it's easy to see how quickly the cost of these breaches escalates and grows exponentially.

All told, between the sums spent on mostly ineffective prevention measures and retroactively closing the cyber barn door after the horses are out (and the hackers are in), we pay dearly as a society for our technological insecurity. Worse, our growing connectedness to the networked world and our concomitant radical dependence on wholly penetrable technologies can bite in ways that hurt much more than our collective wallets.

The Internet has lost its innocence. Our interconnected world is becoming an increasingly more dangerous place, and the more we incorporate assailable technologies into our lives, the more vulnerable we become. The next Industrial Revolution, the information revolution, is well under way, with massive yet unrealized implications for our personal and global security. Yet as daunting as the threats to individuals, organizations, and even our critical infrastructures seem today, there is a proverbial technological train leaving the station, one that is rapidly and exponentially picking up speed. There are signs of it everywhere, if one knows where to look.

Just over the horizon are newly emerging technologies, including robotics, artificial intelligence, genetics, synthetic biology, nanotechnology, 3-D manufacturing, brain science, and virtual reality, that will have major impacts on our world and pose a panoply of security threats that will make today's common cyber crime seem like child's play. These innovations will play essential roles in our daily lives in just a few years, yet no in-depth, broad-based study has been completed to help us understand the unintended attendant risks they pose.

The depth and extent of this transformation and its concomitant risks have gone unnoticed by most, yet before we know it, our global society will connect as many as one trillion new devices to the Internet—devices that will permeate every aspect of our lives. These persistent connections will bind us to both man and machine, across the planet, for good and for ill, and will be woven into the entire realm of our exponentially expanding common sentience. As a result, technology will no longer be just about machines; it will become the story of life itself. Those who know how these underlying technologies work will be increasingly well positioned to exploit them to their advantage and, as we have seen, to the detriment of the common man. The cornucopia of technology that we are accepting into our lives, with little or no self-reflection or thoughtful examination, may very well come back and bite us. These risks portend the new normal— a future for which we are wholly unprepared. This is a book about man and machine and how the slave may become master.

CHAPTER 2

System Crash

**If we continue to develop our technology without wisdom or prudence,
our servant may prove to be our executioner.**
OMAR N. BRADLEY

Something had to be wrong with the signals. It was a Tuesday in early January 2008 when a tram in Lodz, Poland, suddenly veered to the left. This in itself wasn't so strange—except for the fact that the conductor had been attempting to turn his train right. Moments later, the rear cars skidded off the rails before crashing into another tram and coming to a screeching halt.

Amazingly, given the size of the collision, no one was killed, but more than a dozen passengers were injured, and many others were left scratching their heads. What had gone wrong? Rather than a circuit failure or human error on the part of the conductor, rail engineers quickly suspected foul play. They were right, but for reasons they would probably never have wagered.

It turns out that a fourteen-year-old computer whiz kid had created an infrared remote transmitter capable of controlling all the junctions on the transit line. The boy spent months studying the city's rail system to determine the best places to redirect trains in order to cause the most havoc and hacked the switches throughout the town to redirect trains on his command.

In other words, the teen was able to use the city's tram system as his "personal toy train set" by hacking and electronically commandeering the city's transit infrastructure. The teen was believed to have utilized the device on numerous occasions, and when he was caught and arrested, he admitted, like Mat Honan's hacker, that he had carried out his act "just for the lulz" of it all.

But this prank resulted in the derailment of four trains and could easily have led to passenger deaths, an important distinction that left many security analysts fuming that more wasn't being done to ensure the safety of the city's critical infrastructure. They fittingly reasoned that if a fourteen-year-old boy, acting alone, could hack the transit system's network and wreak this sort of havoc for his own amusement, what was to stop criminals, terrorists, or a warring nation from doing the very same?

A Vulnerable Global Information Grid

We've seen how easy it is to hack into the majority of computer systems and how quickly it can be done. Mat Honan's experience proved that our digital lives can be struck from the record in an instant. T.J. Maxx and Citibank both learned the hard way what can happen when criminals thousands of miles away bring you into their sights. Given the obvious dangers, one would have thought a measure of prudence would be in order before adding everything that plugs into the wall or has a battery to the global information grid, yet we proceed full steam ahead in our growing love affair with all things tech.

As a result, we are increasingly connected to computer systems in ways we don't understand. Moreover, these connections are wholly untrustworthy and vulnerable—a poor foundation upon which to build the information society of the twenty-first century. Yet that is exactly what we are doing. Not only are our personal and work computers deeply enmeshed with the Internet, but so too are all of the critical infrastructures upon which our modern society depends. The electricity grid, gas pipelines, 911 dispatch systems, air traffic control, the stock market, our drinking water, streetlights, hospitals, and sanitation systems are all dependent on technology and the Internet to function. In this brave new world, we've taken the human being out of the loop and have entrusted the backbone of civilization to machines.

The credit card transactions, point-of-sale payment terminals, and ATMs that keep the global flow of commerce and capitalism humming would come to a screeching halt without the computers to run the network. Computers decide how, where, and when to route electricity to ensure the power grid's stability. Computer-aided dispatch systems keep track of police cars, ambulances, and fire trucks so dispatchers know who is available and closest to respond in case of emergency. For a sneak peek of what this dystopian world without computers and electricity looks like, one need only turn on the television for a taste of techno-Armageddon–cum–

zombie apocalypse from shows such as *The Walking Dead* or from films like *Planet of the Apes* and *Live Free or Die Hard.* Hollywood machinations aside, our computer-based critical information infrastructures are increasingly under attack and deeply vulnerable to systemic failure—the impact from which could be truly catastrophic.

Much of the world's critical infrastructures utilize supervisory control and data acquisition (SCADA) systems to function. SCADA systems "automatically monitor and adjust switching, manufacturing, and other process control activities, based on digitized feedback data gathered by sensors." These are specialized, and often older, computer systems that control physical pieces of equipment that do everything from route trains along their tracks to distribute power throughout a city. Increasingly, SCADA systems are being connected to the broader Internet, with significant implications for our common security. Unfortunately, these systems were not designed with security in mind and were not engineered to be resistant to an Internet-connected world. The problem is worse than you think: in a July 2014 study of critical infrastructure companies across multiple sectors, nearly 70 percent of them had suffered at least one security breach that led to the loss of confidential information or the actual disruption of operations during the preceding twelve months.

What might a hacker do with access to these systems? Take, for example, the complex information technology systems regulating the local water treatment facility. The SCADA system consistently measures and adjusts the appropriate mix of chemicals to clean our water and make it safe to drink. But what if such a system were hacked? Could the wrong amount of chemicals be mixed in, poisoning our water rather than purifying it? It may sound like fantasy, but hackers have already reportedly carried out an attack against the South Houston, Texas, Water and Sewer Department, according to a BBC report from 2011. The Internet protocol address of the attacker was traced to Russia, and the hackers involved were said to have repeatedly turned on and off a pump, quickly causing it to fail. While nobody fell ill in the attack, the proof of concept has been demonstrated.

What other infrastructure hacks might be possible? As it turns out, the sky's the limit, as the Federal Aviation Administration's control tower in Worcester, Massachusetts, found out all the way back in 1998. There a local teenager used his knowledge of computing to sever communications between inbound aircraft and the tower and even turned off the runway lights for approaching aircraft. While nobody was killed in the incident, the potential for disaster is obvious. Of course there have been many more attacks against critical information infrastructures around the world. One of the earliest occurred in Maroochy Shire, Queensland, Australia, in 2001 wherein a hacker attacked a sewage treatment plant. He gained control of the facility's industrial control systems and "caused millions of litres of raw sewage to spill out into local parks, rivers and even the grounds of a

Hyatt Regency hotel." The attack destroyed significant amounts of local marine life and flora alike, to say nothing of the environmental threat to local residents.

Perhaps one of the most critical systems vulnerable to attack is our national electricity grid. Without electricity, all the fineries of our modern world cease functioning:—no lights, elevators, ATMs, traffic control, subways, garage doors opening, refrigerators, and pumping gas. And when the backup batteries and emergency generators inevitably die, no cell phones and no Internet. Despite our vital dependence on electricity as the technological infrastructure most central to our contemporary lives, the former U.S. secretary of defense Leon Panetta noted, "the next Pearl Harbor that we confront could very well be a cyberattack that cripples" our power systems and our grid.

Panetta's concerns seemed validated and further bolstered by a report by the U.S. Department of Energy which noted that the American energy grid—often called the most complex machine in the world—connects fifty-eight hundred individual power plants and has more than 450,000 miles of high-voltage transmission lines. Yet 70 percent of the grid's key components are more than twenty-five years old. Each of these components uses much older SCADA technologies that are readily attackable and persistently targeted.

An investigation by the House Energy and Commerce Committee revealed that "more than a dozen American utility companies reported 'daily,' 'constant,' or 'frequent' attempted cyber-attacks ranging from phishing to malware infection to unfriendly probes. One utility reported that it had been the target of more than 10,000 attempted cyber attacks each month." The report concluded that foreign governments, criminals, and random hackers were all hard at work either planning or attempting to take down the grid. The findings build upon prior statements by intelligence officials to the *Wall Street Journal* confirming that cyber spies had "penetrated the U.S. electrical grid and left behind software programs that could be used to disrupt the system." The same officials went on to say that spies from Russia and China have allegedly mapped the American grid so that in times of crisis or war with the United States its entire electrical network "could be taken out."

Terrorists also have designs on digitally attacking America's infrastructure. In the summer of 2012, a video from al-Qaeda's As-Sahab media wing was uncovered by the FBI. In the video, the terrorist organization called for its " 'covert mujahidin' to carry out waves of cyber attacks against the U.S. networks of both government and critical infrastructures, including the electric grid." Earlier FBI investigations revealed numerous instances of al-Qaeda's conducting online target research and surveillance on emergency telephone systems, electric generation plants, water distribution facilities, nuclear power plants, and gas storage networks in the United States.

The terrorist organization had even completed elaborate targeting packages on potential critical infrastructures to be attacked, including photographs of intended targets, detailed notes, and online research.

Hackers too are working to understand, expose, and exploit vulnerabilities of SCADA and other critical information infrastructures. At the Chaos Communication Congress, an annual hacker gathering held in Germany, analysts at Positive Research demonstrated how to get full control of industrial infrastructures in the gas, chemical, oil, and energy industries. Equally troubling is the fact that hackers share this information with each other and have even created fully searchable public databases of known exploits that can be used to commandeer critical infrastructures. One well-known hacker database, Shodan, provides tips on how to exploit everything from power plants to wind turbines, searchable by country, company, or device, providing detailed how-tos and greatly lowering the technical bar and knowledge for any rogue individual to hack our critical infrastructures. In effect, for attackers interested in gaining control of our connected world, Shodan has become their Google—one that is near impossible to shut down because Shodan is hosted on multiple servers in foreign countries around the world and publishing vulnerabilities is not currently a crime in most of these places.

Organized crime groups are also turning their attention to infrastructure attacks as a logical means of extorting money from utility companies and governments. Several such incidents reportedly occurred in Brazil between 2005 and 2007, when a wave of cyber attacks was carried out north of Rio de Janeiro and in the state of Espírito Santo. In that incident, nearly three million people were stuck in the dark when the local electricity provider failed to meet the extortion demands of a local crime syndicate. As a result, the city of Vitória, one of the world's largest iron ore producers, had numerous plants forced off-line, costing the company nearly $7 million. The attacks were confirmed by U.S. intelligence officials, security researchers, and even obliquely by President Obama when he noted, "We know that cyber intruders . . . in other countries . . . have plunged entire cities into darkness."

WHOIS It?

The famous Chinese general Sun Tzu, author of *The Art of War,* cleverly observed twenty-five hundred years before the creation of the Internet that "if you know thy enemy and know thyself, you will not be imperiled in a hundred battles." Accordingly, in order to understand the vast technological threats before us, we must understand our enemies. Each has different means and motives, but what they have in common is the risk each poses to our profoundly interconnected world.

The cast of characters responsible for cyber malfeasance is vast and includes nation-states, neighborhood thugs, transnational organized crime groups, foreign intelligence services, hacktivists, military personnel, cyber warriors, state-sponsored proxy fighters, script kiddies, garden-variety hackers, phreakers, carders, crackers, disgruntled insiders, and industrial spies. Each plays its role in what the U.S. military has declared to be the "fifth domain" of battle: cyberspace (following the militarization of land, sea, air, and space of generations past).

These parties frequently use similar tactics, albeit with varying degrees of sophistication. Yet all attackers benefit from the asymmetric nature of the technology: the defender must build a perfect wall to keep out all intruders, while the offense need find only one chink in the armor through which to attack. Among the factions battling in the cyber underground, there is cooperation, witting and unwitting, as the players often learn from and imitate the operational success of one another. For example, transnational organized crime groups use highly sophisticated reconnaissance operations to plan their attacks but often rely on neighborhood thugs to carry out elements of their plots, such as the placement of ATM skimmers, money laundering, or the fencing of stolen goods on eBay. Terrorist organizations learn from cyber criminals and hack for financial gain to fund real-world operations. Patriotic bands of citizens are often formed into cyber posses by their state-sponsored benefactors in nations such as China, Russia, and Iran and receive tacit approval, funding, and training. In doing so, they share some of the same techniques and tools with their government patrons. There is a symbiosis in the cyber underground and a commonality of methodologies across the full spectrum of threat actors.

Perhaps the first vision to come to mind when thinking of a hacker is the stereotypical image of a teenage male living in his mom's basement, pounding away at the keyboard, surrounded by empty bags of Fritos and discarded Coke cans, trying to change his grades on his high school's computers (as did Matthew Broderick in the 1983 film *WarGames*). In the early days of hacking, it was the telephone system that was the target of hackers' attention as so-called phone phreaks manipulated the network to avoid the sky-high costs of long-distance calls. Let's not forget two hackers who spent part of their youth back in 1971 building "blue boxes," devices capable of hacking the phone network and making free calls: Steve Wozniak and Steve Jobs. The pair sold blue boxes to students at UC Berkeley as a means of making money that would effectively help fund their other small start-up, the Apple computer company.

As time passed, other notable hackers emerged, such as Kevin Mitnick and Kevin Poulsen. Mitnick famously broke into the Digital Equipment Corporation's computers at the age of sixteen and went on to a string of such cyber intrusions, earning him the FBI's ire and the distinction of being "America's most wanted hacker." Poulsen's ingenious 1990 hack allowed him to commandeer all of the telephone lines of a local Los Angeles radio station to ensure he would be the 102nd caller, thereby securing the top prize of a $50,000 Porsche 944 S2.

These hacks from the 1970s, 1980s, and 1990s would seem benign by today's standards. In the intervening years, hackers have gone on to become highly organized and have formed global online crime syndicates. They commit identity theft, credit card fraud, health-care fraud, welfare fraud, and tax fraud. Organized crime groups are now going after bigger and more sophisticated targets as well, including the vast amounts of intellectual property created by businesses around the world, from a company's product plans to its computer source code. For example, in October 2013, criminal hackers targeted Silicon Valley's Adobe Systems, stealing thirty-eight million account log-ins and passwords as well as millions of credit card numbers (nothing new there). But what changed in that attack was that the criminals also stole more than forty gigabytes of computer source code for Adobe's flagship products, including Photoshop, ColdFusion, and Acrobat.

As a result, not only can criminals now freely sell Adobe products, but they could also alter the code and insert untold numbers of hidden back doors, malware, and additional exploits into the product, causing Adobe's legitimate and unsuspecting customers to suffer widespread hacking attacks and identity theft—a troubling development indeed given Adobe's massive global footprint among computer users. Even Symantec, the maker of pcAnywhere and Norton AntiVirus, has had its source code stolen. Yep, the company that is selling you antivirus software to protect you from being hacked was itself compromised when a hacker stole 1.27 gigabytes of its security software source code and demanded a relatively paltry $50,000 in exchange for not posting the data on the well-known hacker Web site The Pirate Bay.

Traditional organized crime groups, such as the Italian Mafia, Japanese Yakuza, Chinese triads, and Colombian drug cartels, have all diverted efforts and resources from their usual criminal activities to take advantage of the easy profits, greater anonymity, and limited police scrutiny afforded in cyberspace. Moreover, they do not have to worry about the draconian sentencing minimums often associated with their former economic activities, such as narcotics smuggling and trafficking in human beings. Organized crime groups in cyberspace are responsible for spam, phishing, fake pharmaceutical ads, the dissemination of child sexual abuse images, denial-of-service attacks, and extortion, to name but a few of their favored activities.

In addition to the stalwart old guard of organized crime, a more nimble class of pure-play cyber-crime organizations are exploding onto the scene. These newly emerging and professionally organized criminal hacking groups are highly profitable and truly global, with heavy concentrations in

China, Indonesia, the United States, Taiwan, Russia, Romania, Bulgaria, Brazil, India, and Ukraine. New syndicates, such as the Russian Business Network (RBN) in St. Petersburg, have even made names for themselves as multi-product-line, full-service cyber-crime organizations.

RBN famously provides "bulletproof" Web site hosting services for all manner of other criminal enterprises and takes a completely hands-off approach to hosted content, freely welcoming anything from child pornography to malware exploit exchanges on its servers. Other professional criminal hacking groups, such as the ShadowCrew, offer online havens for "carders," who specialize in the murky world of stolen personally identifiable information, including forged passports and driver's licenses, and stolen credit cards—key ingredients to the world's growing identity-theft economy. ShadowCrew operated the now-defunct Web site CarderPlanet .com, where over four thousand criminals from around the world could freely gather to buy and sell stolen and hacked identities, documents, and account numbers. Founded by the notorious criminal hacker Albert Gonzalez, ShadowCrew offered fellow criminals tutorials on everything from cryptography to card-cloning techniques, and Gonzalez's organization was reportedly responsible for stealing and reselling more than 180 million credit and ATM cards. The number and reach of these highly profitable transnational organized cyber-crime rings have grown, and the security intelligence firm CrowdStrike was actively tracking more than fifty such major organizations globally.

Besides transnational organized crime syndicates, hacktivists— politically motivated cyber attackers—represent one of the most influential and powerful groups in cyberspace. Anonymous, LulzSec, AntiSec, WikiLeaks, and the Syrian Electronic Army fall into this group and launch their attacks in retaliation for perceived injustices. Personalities such as Julian Assange, Chelsea (Bradley) Manning, and Edward Snowden have become household names for challenging some of the world's most powerful institutions and for releasing data that others would most certainly have preferred remain hidden. While Assange, Manning, and Snowden have been propelled onto the covers of newspapers around the world, other hacktivist groups prefer that their individual members remain discreetly hidden in subordination to the organization itself and its broader agenda. One such notable example is Anonymous, a self-described leaderless organization whose members have become recognizable in public for wearing Guy Fawkes masks.

The group's motto, "We are Anonymous. We are legion. We do not forgive. We do not forget. Expect us," manifests its organizational ethos: "The corrupt fear us. The honest support us. The heroic join us." When MasterCard, Visa, and PayPal all agreed to stop funneling donations to Julian Assange's WikiLeaks organization, Anonymous responded by launching a series of effective cyber attacks against the financial firms.

Anonymous is strongly against what it perceives to be rigid antipiracy laws, and it took credit for an earlier attack against the Sony PlayStation Network in response to Sony's support of U.S. antipiracy legislation known as the Stop Online Piracy Act.

Anonymous views itself as hacking for good and has taken on a wide variety of social causes, including its support of activists throughout the Middle East during the Arab Spring. Even some of the group's most ardent critics might find themselves supporting some of Anonymous's lesser-known activities in combating criminal organizations and injustice. For example, during an attack dubbed Operation Darknet, members of Anonymous targeted child pornography Web sites hosting vile images of young children being sexually abused. The hacker collective knocked the sites off-line and released the names of fifteen hundred pedophiles using the services. Whether one supports or detests the actions taken by Anonymous and other hacktivist organizations, one thing is clear: they are a force to be reckoned with among the wide tapestry of threat actors in our über-connected world.

Hacktivists are capable of targeting any individual or corporation and can even have geopolitical impact the world over, as was seen during the Arab Spring. In recognition of their growing power, *Time* magazine ranked Anonymous as one of the hundred most influential people in the world in 2012. Their burgeoning influence and capabilities have not gone unnoticed by government, and it was recently revealed that the Government Communications Headquarters (GCHQ), the British equivalent of the American National Security Agency, had launched its own series of denial-of-service attacks against Anonymous and its membership in an effort to disrupt their activities. This dramatic response by the state against a non-state actor and hacktivist group demonstrates the impact Anonymous is having in the world.

Meanwhile, terrorist organizations too are increasingly using the Internet and other technologies to plan, support, and execute their murderous activities. Technology helps terrorists recruit new members in underground chat rooms, finance operations (through cyber crime or online fund-raising), communicate clandestinely, and disseminate propaganda, such as the gruesome beheading videos produced by ISIS (the Islamic State). ISIS is tech savvy and in its latest recruitment videos even edited in scenes from the video game *Grand Theft Auto V* for effect. In its online video production, the reviled terror group offered new recruits the opportunity to "do the things you do in games, in real life on the battle-field . . . like attack a military convoy or kill police officers." The video is plastered with the ISIS logo.

Internet reconnaissance and research by terrorists are commonplace, and on more than one occasion officials have found Google Earth images of intended targets, including a 2007 planned attempt by terrorists to blow up fuel tanks at New York's John F. Kennedy International Airport. Terrorists have been early adopters of technology, particularly in their use of data encryption to secure their communications. For instance, "Ramzi Yousef, the convicted mastermind of the first World Trade Center Bombing in 1993, used encrypted files to hide details of his plot to destroy 11 U.S. airliners." In the Yousef case, it took law enforcement authorities more than a year to break the encryption algorithm used by the terrorist, and in doing so, police officials were fortuitously able to prevent the plot against the airlines.

Some counterterrorism experts have referred to the Internet as a "terrorist university," a place where terrorists can learn new techniques and skills to make them more effective in their attack methodologies. Widely available online are documents such as *The Mujahideen Poisons Handbook*, which contains various "recipes" for homemade poisons and poisonous gases. The six-hundred-page *Encyclopedia of Jihad* is also widely available online and includes chapters such as "How to Kill," "Explosive Devices," "Manufacturing Detonators," and "Assassination with Mines." In a striking example of how dangerous such online education can be, Dzhokhar Tsarnaev, the terrorist suspect arrested for his role in the April 2013 Boston Marathon bombings, admitted to authorities he and his brother learned how to make the pressure-cooker bomb used in the attack after reading step-by-step instructions published in al-Qaeda's online magazine, *Inspire*, in an article titled "Make a Bomb in the Kitchen of Your Mom."

Not only are terrorists using the Internet for operational support and planning, but they have turned to both hacking and cyber crime as means of funding and carrying out their real-world terrorist operations. In June 2007, a U.K. terrorist cyber cell was disrupted by police when three British residents, Tariq al-Daour, Waseem Mughal, and Younes Tsouli, were charged with using the Internet to incite murder. Evidence presented showed that the men had used hacked credit card accounts to purchase goods to assist fellow jihadists—items such as night-vision goggles, global positioning devices, airplane tickets, and prepaid mobile phone cards—for the purpose of providing direct tactical support for terrorist operations. "The trio reportedly made fraudulent charges totaling more than 3.5 million U.S. dollars and was in possession of a database containing nearly 40,000 stolen credit card accounts."

Even the infamous 2002 Bali bombing mastermind, Imam Samudra from the al-Qaeda-linked terrorist group Jamaah Islamiyah, funded his attack in which more than 200 people were murdered with the $150,000 he obtained by hacking into Western bank accounts and credit lines. Samudra was technologically savvy and while in prison wrote an autobiographical manifesto containing a chapter titled "Hacking, Why Not?" In the book, Samudra shared his hacking and "carding" techniques with his disciples, encouraging them "to take the holy war into cyberspace by attacking U.S.

computers, with the particular aim of committing credit card fraud, called 'carding,'" to fund operations. Terrorists seem to be getting the message, and both the 2004 Madrid bombings at the Atocha train station, in which 190 people were killed and nearly 2,000 wounded, and the 7/7 London bombings, in which 52 civilians were slain and over 700 injured, were funded in whole or in part through hacking and credit card fraud.

As the technical hacking skills of terrorist organizations increase, so too does the amount of ill-gotten gains they are capable of generating online. For example, in late 2011, police in the Philippines working with the FBI uncovered a telephone hacking scam against AT&T that defrauded the company and its business customers of $2 million. The Filipino hacking cell was working with Jamaah Islamiyah and funneled the millions back to a Saudi-based terror group that in turn funded the Pakistani-based Lashkar-e-Taiba, the terrorist group responsible for the deadly 2008 bombing siege that gripped the city of Mumbai, India, and killed and maimed hundreds.

It is clear that criminals, hacktivists, and terrorists use our interconnectivity against us, whether for profit, politics, or massacre. They have schooled themselves in science and technology and have proven a formidable force in exploiting the fundamentally insecure nature of our twenty-first-century technological skin. Yet thieves, hackers, activists, and terrorists are not the sole inhabitants of the digital underground. They are accompanied by a phalanx of nation-states, cyber warriors, and foreign intelligence services, each handily playing in the so-called fifth domain, fully leveraging for their own purposes the insecurity of the underlying digital infrastructure that unifies the planet.

Though the average Internet user today may be busily updating his Facebook status or playing *Angry Birds*, it is important to recall that today's Internet was born of the Defense Advanced Research Projects Agency (DARPA), a U.S. Department of Defense invention created to ensure redundancy in military communications in the event of nuclear attack. The Internet is a military creation with significant corollary geopolitical ramifications.

It is when governments turn their attention (and budgets) to offensive cyber operations that we can see the full range of vulnerabilities in the hardware and software upon which we depend and that our common technological frailty is fully exposed. Though a $50,000 criminal extortion demand against Symantec or even a $1 billion hacking loss at Target remains noteworthy and surely merits our attention, it is chump change compared with the computer spying breach of the Pentagon's $300 billion F-35 Joint Strike Fighter project—the most expensive Defense Department weapons program in history.

In May 2013, the U.S. government specifically named China as being responsible for a series of hacks against vital American defense and government systems, including the F-35. Over the years, it has been reported that many other defense blueprints and technologies have been stolen, including those from an advanced Patriot missile system known as PAC-3, the navy's Aegis Ballistic Missile Defense system, the F/A-18 fighter, the V-22 Osprey, the Black Hawk helicopter, and the littoral combat ship. According to an FBI report, China has secretly developed an army of 180,000 cyber spies and warriors, mounting an incredible ninety thousand computer attacks a year against the U.S. Defense Department networks alone. The totality of the thefts and their impact on American national security are breathtaking.

China's purported cyber-hacking activities provide it with significant strategic advantages, including an immediate tactical and operational edge in any conflict with the United States. Having the blueprints to so many U.S. defense systems provides key details on how they work and importantly how to defeat them in times of crisis. Moreover, this great "brain robbery" saves China billions in research dollars from its own military development costs (and decades of work) by merely appropriating and building upon work paid for by American taxpayers.

Of course it is not just the American military's technology that is being targeted by China, but rather a litany of Washington institutions, including law firms, think tanks, human rights groups, contractors, congressional offices, embassies, and any number of federal agencies. Moreover, a 2009 report by Canadian researchers at the Infowar Monitor, the SecDev Group, and the Citizen Lab at the University of Toronto uncovered the so-called GhostNet, "a vast global cyber espionage network" extending to 103 countries that was controlled by servers in China and targeted the Tibetan government in exile and the Dalai Lama himself.

China has also been accused of hacking numerous media outlets, including most famously the *New York Times* in early 2013 after the paper reported that the relatives of China's prime minister, Wen Jiabao, had accumulated billions of dollars of wealth through their business dealings since Wen had entered office. The breach gave the perpetrators access to any computer on the *New York Times*'s network, and it was believed the Chinese were working to uncover sources and contacts that might damage the reputations of China's leaders. The *Times* hired the private cyber-security firm Mandiant, which investigated the incident and in a fascinating report tied the attack back to Unit 61398 of the People's Liberation Army. The unit's headquarters, on Datong Road in the Pudong district of Shanghai, is a 130,000-square-foot, twelve-story building in which thousands of employees go to work every single day hacking governments, companies, and individuals around the world.

Those technological thefts that are not directly carried out by the Chinese state itself are often sponsored by the state and performed by appointed proxies, with profound implications and deep costs for businesses around the world. In 2012, *Bloomberg Businessweek* covered China's ongoing theft of global intellectual property in a cover story that screamed in full-page letters, "Hey China! Stop Stealing Our Stuff." The piece featured the tale of Dan McGahn, the CEO of the Massachusetts-based American Super-conductor (AMSC), a green energy technology company specializing in the design of power systems and the software that runs large wind turbines. In March 2011, AMSC's largest customer—China's formerly state-owned Sinovel Wind Group—suddenly began refusing shipments at its assembly plant in Liaoning province and canceled more than $700 million of pending orders from AMSC. The market's response to AMSC's canceled orders was brutal: a 40 percent drop in valuation in a single day and an 84 percent decline by September of that year.

An investigation into the matter revealed that Sinovel's own turbines "appeared to be running a stolen version of AMSC's software" and that the Chinese firm had snatched a complete copy of the American company's proprietary computer source code. Because Sinovel had in its possession all of AMSC's intellectual property, it no longer needed AMSC or its products and could merely produce AMSC's products itself. As a result, the Chinese company canceled its existing supply contracts worth more than $700 million with the Massachusetts firm.

All told, between thefts of commercial, governmental, and military intellectual property, China's hacking efforts have netted the nation the greatest transfer of wealth in human history. According to Akamai's *State of the Internet* report, China is the source of a shocking 41 percent of all of the cyber attacks in the world. Of course China vehemently and routinely denies involvement in any global hacking activities. When allegations arise, responses typically come from a Chinese embassy spokesman in the capital city of the hacked country in question, whether Paris, Berlin, or New Delhi. A message issued by a Chinese embassy official named Wang Baodong in Washington, D.C., is typical of their response: "China is firmly against international hacking activities and is ready to work with other countries to secure the cyberspace." Wang's denial is hardly the first time such a response has been issued: a Google search for the phrase "China denies hacking" yields a mere thirty-five million such denials.

Though China is the most populous nation on earth, it is not the only country to engage in cyber operations. According to the former FBI director Robert Mueller, there were at least 108 nations with dedicated cyber-attack units going after industrial secrets and critical infrastructure alike, including Iran. In late 2012, a previously unknown hacker group called Cutting Sword of Justice took responsibility for carrying out the most destructive computer sabotage against a company to date when it targeted the oil and gas giant Saudi Aramco. The offensive took place on the eve of one of the holiest nights in the Islamic calendar, Lailat al-Qadr, the day Muhammad is said to have revealed the Koran to his followers and when Aramco's fifty-five thousand employees stayed at home to celebrate with family and friends. At stake, 260 billion gallons of oil, valued at over $8 trillion (fourteen times the market cap of Apple Inc.).

During the incident, an unknown insider with access to the facility inserted an infected USB thumb drive into a single PC connected to the company's computer network. Within minutes, the drive's viral payload, known as Shamoon, was spreading like wildfire across all of Aramco's thirty thousand corporate computers. Though its goal was to disrupt oil and gas production at Aramco's facilities, good security practices meant that the virus "only" destroyed corporate data. The toll? Shamoon erased 75 percent of the company's thirty thousand corporate hard drives, wiping out "documents, spreadsheets, e-mails, files—replacing all of it with an image of a burning American flag."

The Cutting Sword of Justice claimed its attack was in response to Saudi "crimes and atrocities" in Syria and Bahrain against Shiite protesters. American intelligence officials suspect that the Cutting Sword of Justice is nothing more than a front for Iran, which they believe was to blame for sponsoring the attack. The shocking capabilities demonstrated in the Aramco attack predated several other successful attacks by the Iranian government, including a series of distributed-denial-of-service (DDoS) disruptions in early 2013 targeting the American financial services industry. Numerous marquee banks, including JPMorgan Chase, Bank of America, Wells Fargo, BB&T, HSBC, and Citigroup, were affected by the attack, which made their corporate networks and public Web sites inaccessible for extended periods of time and prevented customers from accessing their money. A hacker group calling itself Izz ad-Din al-Qassam Cyber Fighters claimed responsibility for this cyber blitz, but American officials say the group is merely a proxy for Iran.

The widespread denial-of-service attack against America's financial industry by Iran was shocking for its size and scope and for the massive volume of data generated by the perpetrators. "Some banks were hit with a sustained flood of traffic that peaked at 70 gigabits" per second. To put that volume of DDoS traffic in perspective, it is as if 1 billion people simultaneously phoned your bank, hung up, and immediately dialed back one second later. In order for your call (or visit to its Web site) to get connected, you would be number 1,000,000,001 on the list to get through. In other words, for all intents and purposes, you would never reach the bank.

Strikingly, the Iranian-backed attack against the financial services sector was reported to be multiple times larger than the infamous 2007 attack against the nation of Estonia by Russian-based hackers—an attack that nearly knocked the small Baltic nation entirely off-line. The incident is widely believed to have been carried out with the direct support of the Russian government via nationalist proxy hackers after Estonia decided to move a Soviet-era grave marker from its long-standing position in Tallinn's city center to the outskirts of town—a move that deeply offended Moscow. Many security experts called the all-out digital assault against Estonia the "world's first cyber war" because of its size and scale. Given that Iran had just bested that attack, one security researcher noted the Islamic Republic's technical bombardments have graduated from being the equivalent of "a few yapping Chihuahuas into a pack of fire-breathing Godzillas."

Of course, there have also been widespread allegations of hacking by the United States against the rest of the world, based largely on the numerous classified documents stolen and unilaterally released by the former National Security Agency (NSA) contractor Edward Snowden beginning in June 2013. Snowden detailed at great length the global technical surveillance apparatus run by the National Security Agency and provided documentary evidence to support his claims in discussions with the journalists Glenn Greenwald and Laura Poitras. Programs such as PRISM and XKeyscore subsequently came to light, as did the NSA's purported ability to track billions of e-mails, phone messages, chat sessions, and SMS texts each and every day.

While living in Moscow, Russia, under temporary asylum status, Snowden continued to catalog U.S. offensive cyber and technical operations, including the tapping of the personal mobile phones belonging to world leaders ranging from the German chancellor Angela Merkel to the Brazilian president Dilma Rousseff. Moreover, Snowden divulged that millions of citizens in ally nations such as France and Germany were having their communications recorded, to the tune of 120 billion calls a month around the world. Snowden's leaks also served to limit international sympathy for U.S. complaints regarding intensive cyber operations by the People's Republic of China, especially when he revealed that America too had launched cyber operations against Chinese targets, including China Mobile and the prestigious Tsinghua University. Depending on one's political beliefs and point of view, Snowden is either an enemy of the state, a hero, a whistle-blower, a dissident, a traitor, or a patriot. Most feel strongly one way or another. Regardless of how history judges Snowden, his disclosures, if true, paint a highly detailed portrait of how governments are entering the cyber fray.

An analysis of the threat-actor landscape in cyberspace reveals hacktivists, criminals, proxy warriors, terrorists, and rogue governments, all fully capable of exploiting the insecurity of the technological infrastructure of our world. Our financial data, identity, children's baby pictures, and power grids are all vulnerable and at risk, easy targets for the picking. Yet for as ubiquitous as technology seems in our lives today, the exponential rate of growth means that just over the horizon is a tidal wave of technological advances that will leave our heads spinning. Not only will the breadth and depth of our connection to the global information grid vastly expand, but new technologies heretofore relegated to the realm of science fiction will soon emerge as science fact. In short, we ain't seen nothing yet.

CHAPTER 3

Moore's Outlaws

The future is already here.
It's just not evenly distributed yet.
WILLIAM GIBSON, *NEUROMANCER*

To learn the mathematical power of exponents and exponential curves, schoolchildren in France were asked to imagine a pond with a small water lily leaf growing on it. The leaf, they were told, would double in size every day and would take thirty days to cover the entire pond. If the lily did cover the pond, it would smother and kill all other forms of life in the water. The question then posed to the children was, on what day will the lily cover half the pond?

At first, there wasn't much to worry about. The lily grew at a rate that was barely noticeable, reaching only one-tenth of 1 percent of the pond covered by day 20. Just 0.1 percent. Five days later, it had reached 3 percent, but, still relatively unconcerned, the children let the lily continue to grow. Until suddenly, at day 29, the lily covered half the pond. By then, there was preciously little time to save the pond, which was strangled by the lily the very next day. The twenty-ninth day can often seem just like any other day, but given the nature of exponentials, the pond is already half-choked to death.

The lessons of the pond are that the magical nature of exponential growth can sneak up on us very, very quickly and that our continued linear thinking may come at our own peril.

The World of Exponentials

In his book *The Singularity Is Near*, the futurist Ray Kurzweil describes the exponential nature of the technological world around us and introduces the concept of what he calls the "knee of an exponential curve." The knee of the curve is an inflection point in time at which an exponential trend becomes truly noticeable. Shortly thereafter, however, the trend line becomes explosive and appears essentially vertical as the mathematical impact of an exponential growth curve is felt. Malcolm Gladwell might describe this phenomenon as a "tipping point" wherein the sum of many small things drives toward making a notable massive difference in outcomes. Given the exponential nature of technology and its omnipresence in our life, there is overwhelming evidence that we are rapidly approaching such an inflection point. The question is, will we be tipping good or breaking bad?

According to the International Telecommunication Union, there were a mere 360 million people online in the year 2000. Though it took nearly forty years to develop, by 2005 the global community that is the Internet reached its first 1 billion members. The second billion were added just six years later, achieving the milestone in March 2011. The greatest growth has been in the developing world, with Asia and Africa experiencing an 841 percent and a whopping 3,606 percent rocketing climb, respectively, since 2000. And while half the world regrettably does not yet have access to the Internet today, Google's executive chairman, Eric Schmidt, has boldly predicted that by 2020 everybody in the world will be online.

The relentless pace of these changes and the ever-expanding presence of technology in our lives have been catalyzed through an axiom of technology known as Moore's law. The concept was named after Gordon Moore, the former chairman of the Intel Corporation, who in 1965 famously predicted that the number of transistors per square inch on an integrated circuit would double every year into the future. This principle, later revised to a doubling every eighteen months to two years, is referred to as Moore's law and is now commonly applied more broadly to the power and capabilities of all circuit-based technologies. As a result, an increasing spectrum of emerging scientific discovery, everything from biotechnology to robotics, is governed by Moore's law and its consequences. Moore's law also has implications beyond science, ranging from geopolitics to economics as every domain of human existence is increasingly touched by technology. Importantly, the implications of Moore's law can have both positive and negative impacts on our world.

It is the persistent doubling of computer processing power stipulated in Moore's law that makes it so deeply significant. It means that all computer-based technologies are exponential in their growth curves—not linear. In other words, these technologies benefit not from the power of mere addition but from multiplication. It is the difference between 1, 2, 3, 4, 5, 6, 7 and 2, 4, 8, 16, 32, 64, 128. The more the linear versus exponential trend line continues, the more stark and shocking the results. To put the concept into perspective, taking thirty steps linearly, one might walk across the living room. But taking thirty steps exponentially—doubling the distance with each successive step—would be the equivalent of traveling the distance from earth to the moon. The fact that today's technologies are exponential in their growth curves, not linear, is absolutely fundamental to understanding the next phase of human evolution. We are now living in exponential times.

As information technologies continue to double in their price performance, capacity, and bandwidth, amazing things become possible. Take, for example, the iPhone that hundreds of millions of users carry in their pockets today. Incredibly, it literally has more computer processing power than that which was available to all of NASA during the *Apollo 11* moon landing forty years ago. The modern smart phone is more than "a million times cheaper and a thousand times faster than a supercomputer of the '70s." As a result of mathematical repercussions of exponentials and Moore's law, "we won't experience a hundred years of progress in the twenty-first century; it will be more like twenty thousand years of progress (at today's rate)."

Given the exponential pace of change in computer processing power and sophistication, it should be obvious that in the very near future computers will become profoundly capable. Ray Kurzweil describes the constant doubling of computing's price performance and power in his "law of accelerating returns." He predicts a point in time where a technological singularity will take place—that is, a moment in time where computing progress is so rapid it outpaces mankind's ability to comprehend it and machine intelligence will exceed human intelligence. Whether or not that day eventually comes (Kurzweil predicts the year to be 2045), one thing is clear: computing power is growing exponentially, and our ability to understand the global information grid and map its vast interconnections is waning.

It's not just your imagination, technology is indeed progressing faster than most of us can keep up, and it's not your fault. Human beings heretofore have developed evolutionally to think in a linear fashion; it's been coded into our brain since the dawn of mankind. From our days on the plains of the Serengeti, we've intuitively done linear calculations in our heads to determine the best path of escape from a charging lion. But that is not the world in which we live today. Kurzweil believes that the coming years will bring a "technological change so rapid and profound it represents a rupture in the fabric of human history." Given this ever-accelerating rate of change and our journey from building-sized computers to iPhones in the past forty years, what might the next forty years bring? Much more good, and potentially much more evil, than most of us could possibly imagine.

Ours is not a simple binary story of whether technology is good or evil but rather one of accelerating returns. How can we remain safe and secure in a world that is moving so quickly? We are building a civilization that is deeply interconnected yet technologically insecure at the same time. In other words, we are constructing a world that is wired for crime and a panoply of other security threats. Mounting evidence demonstrates these dangers and introduces us to a newly emerging class of elite criminals, terrorists, and foreign governments that can exploit these technologies at will. The result? We now find ourselves increasingly connected, dependent, and vulnerable.

The Crime Singularity

In the criminal days of yore, crime was a simple affair. Any would-be criminal need only buy a knife or gun, hide in a dark alley, and then leap out at an approaching victim and demand, "Give me your money." Apart from the unsavory morality issue, robbery was a great entrepreneurial business model that had survived for millennia. The start-up costs were low, and criminals could set and work their own hours and schedules. Of course like all entrepreneurs, criminals struggled with an obvious problem: how to scale and grow their businesses. Even a very good robber could only steal from so many people a day, perhaps five or six a day, if lucky.

Fortunately, however, technology provided an answer for would-be criminals on how to surmount the scalability issues their illicit businesses faced, and the solution came from an unlikely place: the locomotive. Of course when trains were invented, nobody ever envisioned that they might become subjected to train robberies. Criminals, however, foresaw the opportunity and lost no time in taking advantage of the new technology. Now, rather than robbing one person at a time, thanks to the locomotive, armed gunmen could rob two hundred or three hundred people simultaneously, thereby vastly expanding their business opportunities and their profits.

Early criminal entrepreneurs such as Bill Miner, Jesse James, and Butch Cassidy in the mid- to late nineteenth century made their fortunes robbing trains of their cargo and passengers of their cash and jewelry. Attacks against trains remained a viable form of criminal employment for more than one hundred years, culminating in the U.K.'s great train robbery of 1963, wherein a band of robbers commandeered a Royal Mail train headed from Glasgow to London. Their carefully planned heist netted the crew £2.6 million, the equivalent of £46 million today ($7.28 million and $76 million, respectively).

Fast-forward to today, and we see that crime too can benefit greatly from the exponential nature of technology. Using the Internet, thieves have gone from robbing individuals and hundreds of people at a time to stealing from thousands and now even millions of individuals. As a result, we are witnessing a fundamental paradigm shift in the nature of crime and how it is committed. With technology, crime scales, and it scales exponentially.

As noted previously, the 2007 T.J. Maxx hack was the largest retail crime of its kind at the time, initially affecting forty-five million customers and their financial data. But news headlines have made it abundantly clear TJX was not an isolated incident. In June 2011, attackers compromised the Sony PlayStation gaming network and gained access to more than seventy-seven million online accounts, including victims' credit card numbers, names, addresses, dates of birth, and gaming log-in credentials. The incident kept the PlayStation Network off-line for days and affected customers around the world. Criminals have lost no time in taking advantage of all the technological conveniences in our lives, including our gaming consoles. In the end, financial analysts estimated that the repair bill for the Sony PlayStation hacking incident cost the company in excess of $1 billion from lost business, outside consultants, and various lawsuits.

Later, in 2013, Target stores across the United States admitted that they too had become victims of a cyber attack against their point-of-sale credit and debit card terminals. The episode could not have come at a worse time for the retailer, at the very height of the Christmas shopping season. In that incident, data from more than 110 million accounts were stolen, in an attack apparently masterminded by a seventeen-year-old hacker in Russia.

Think about the scale and the enormity of the loss. Nearly one-third of the American population was simultaneously robbed. Never before in the history of humanity has it been possible for any one person to steal 110 million of anything, let alone concurrently rob more than 100 million people.

As incredible as the Target hack was for its size and scope, just over a year later, in August 2014, that number was surpassed by a Russian hacking group that gathered 1.2 billion user names, passwords, and other confidential data from 420,000 Web sites, according to Hold Security. Crime too has entered the age of Moore's law, and it has exponential consequences for us all.

Control the Code, Control the World

**Technological progress is like an axe
in the hands of a pathological criminal.**
ALBERT EINSTEIN

As the entire human race drives itself toward ubiquitous connection to the Internet, we are transforming both ourselves and our world. From this global interconnectivity will flow tremendous good. Man grows omniscient as every fact or thought ever recorded becomes available in real time regardless of its source or location. From the chemical formula for photosynthesis, to the current temperature in Baku, to who won an English county cricket match in 1901, to the latest shenanigans of Justin Bieber, all is becoming knowable as we plug ourselves into the global brain that is the Internet.

At the same time, man is also growing omnipotent as the world's objects go online. You can activate your DVR from the freeway and start your car from the living room. 3-D printers churn out auto parts, clothing, and construction materials. Diabetic pumps, pacemakers, and implantable cardiac defibrillators are all connected to the Internet and transmit lifesaving digital data to your doctor in real time. Physicians can even perform transatlantic surgeries via tele-connected robotic surrogates, projecting surgeons into villages where none had ever traveled. Human beings now have the capacity to control things on the other side of the planet in ways that would previously have been both unimaginable and impossible.

While there are obvious cost, efficiency, and capability advantages to these transformations, they add tremendous complexity to our world. One very rough approximation by which to examine these complexities is by considering the number of lines of computer code (LOC) required to make a particular piece of software or system function. For example, the 1969 *Apollo 11* Guidance Computer that safely guided astronauts the 356,000 kilometers from earth to the moon and back only contained 145,000 LOC, a ridiculously paltry sum and a remarkable achievement by today's standards. By the early 1980s, when the space shuttle became operational, its primary flight software had grown to a relatively slim 400,000 LOC.

By comparison, Microsoft Office 2013 is 45 million LOC, slightly fewer than the 50 million lines of code required to run the Large Hadron Collider located at the European Organization for Nuclear Research. Today, the software required to run the average modern automobile clocks in at a remarkable 100 million LOC, many fewer than the unprecedented reported 500 million LOC that ran the much maligned U.S. HealthCare .gov Web site. Though direct comparisons are difficult, HealthCare.gov was roughly thirty-five hundred times more complex than the guidance system that brought *Apollo 11* to the moon and back. Is it any wonder the Web site crashed and burned?

The growing complexity of computer software has direct implications for our global safety and security, particularly as the physical objects upon which we depend—things like cars, airplanes, bridges, tunnels, and implantable medical devices—transform themselves into computer code. Physical things are increasingly becoming information technologies. Cars are "computers we ride in," and airplanes are nothing more than "flying Solaris boxes attached to bucketfuls of industrial control systems." As all this code grows in size and complexity, so too do the number of errors and software bugs. According to a study by Carnegie Mellon University, commercial software typically has twenty to thirty bugs for every thousand lines of code—fifty million lines of code means 1 million to 1.5 million potential errors to be exploited. This is the basis for all malware attacks that take advantage of these computer bugs to get the code to do something it was not originally intended to do. As computer code grows more elaborate, software bugs flourish and security suffers, with increasing consequences for society at large.

Growing system complexities, even when not intentionally exploited by bad actors, can still pose significant safety risks. Take, for example, the 2003 Northeast blackout that left fifty-five million people in Canada and the United States in the dark for days. A labyrinthine electrical grid, an operator error, and a software bug led to the largest blackout in North American history. Computer failures also played a role in the 2010 *Deepwater Horizon* disaster that killed eleven workers and created the largest environmental catastrophe in American history, leaking 4.9 million barrels of oil into the Gulf of Mexico. At a government hearing into the disaster, Michael Williams, the chief electronics technician aboard the *Deepwater Horizon*, testified that crucial drill monitoring and control systems were crippled by frequent software crashes and a "blue screen of death" on the oil rig's computer prior to the explosion that sank the rig.

Though the 2003 Northeast blackout and the *Deepwater Horizon* disaster were by all accounts accidents, they provide useful insights into the tremendous harm that can stem from computer system malfunctions. Whether a computer system fails because of accidental or criminal action, however, is only a question of intent. Given the large number of bugs in modern computer code, what might come to pass if malicious intent were applied? The very same technology that can save the world and enable globalization can be used by radicals, criminals, terrorists, and governments to destroy it.

Unfortunately, once a cyber weapon is released into the wild, it does not die. It can be repurposed. Unlike conventional bombs, which explode into a million parts when dropped on their targets, weaponized malware can be used over and over again. Though military and intelligence officials may spend millions of dollars secretly developing a particular weapon, computer code is easy to copy. Once released, it becomes available to hacktivists, crime groups, and terrorists to exploit for their own purposes, enabling new forms of cyber-weapons proliferation.

Think of it as a virtual Molotov cocktail that, once lobbed their way, can now be thrown back over the fences at us. We've already seen this happen as criminal organizations and rogue governments have copied the code designs initially used against them and repurposed them for their own attacks. As computer code continues to be weaponized, attacks like these will become both more common and more sophisticated.

It is unsettling but true that to date no computer system has been created that cannot be hacked—a sobering fact given our categorical reliance on these machines for everything from communication to transportation to health care. Not only are the passwords and system checks that left Mat Honan so vulnerable a farce, but so is the software we use to run the world. Plainly stated, when everything is connected, everyone is vulnerable.

The power of Moore's law applies not just to the positive aspects of technology but to its negatives as well. With Moore's law come Moore's outlaws—criminals, terrorists, hacktivists, and state actors who exploit technology at will. They are keenly aware of how to take advantage of system complexities and poorly written software to wring what they want from our rapidly developing technology-based civilization. With all objects transforming themselves into computers and all computers run by code, these powerful new illicit actors clearly understand that if you control the code, you control the world.

But it's not just criminals and rogue governments that we should be worried about. Often, the very companies and organizations that we count on for protection, advice, and entertainment are leaving us incredibly vulnerable, for they too control the code that runs our lives.

CHAPTER 4

You're Not the Customer, You're the Product

The truth will set you free, but first it will piss you off.
GLORIA STEINEM

Parkinson's, relapsing remitting multiple sclerosis, necrotizing fasciitis, acute lymphocytic leukemia, childhood-onset diabetes, HIV, amyotrophic lateral sclerosis—a diagnosis of any one of these diseases would undoubtedly strike fear in the hearts of most patients receiving such life-altering news. In years past, individuals with such illnesses would have found themselves depressed and alone, unable to discuss their plight with others who knew exactly what they were going through. Moreover, the paucity of comprehensible medical information written for actual human beings would have further isolated these patients from friends and family. That is why Jamie and Ben Heywood (whose brother was diagnosed with Lou Gehrig's disease) founded the Internet site PatientsLikeMe.com—to allow visitors to share their stories and connect with others going through the exact same health trials and tribulations. Since the site's founding in 2004, it has grown to a global community of more than 200,000 patients diagnosed with fifteen hundred unique diseases. For thousands of people, PatientsLikeMe.com has been both a figurative and a literal lifesaver, as users learned more about their illnesses and exchanged survival strategies and treatment protocols in a variety of online discussion forums.

It was that opportunity to connect with others that first drew Bilal Ahmed, a thirty-three-year-old businessman from Sydney, Australia, to the site. Ahmed had been suffering from anxiety and depression since the death of his mother, and he found it difficult to discuss his condition with friends and family. Ahmed created a pseudonym account on Patients-LikeMe and joined its Mood Forum, where users share intimate details about emotional disorders such as bipolar disease, PTSD, bulimia, addiction, OCD, and thoughts of suicide. On the Mood Forum, he dutifully listed his symptoms, test results, and all the drugs he had been prescribed to treat his depression. There he connected with other patients around the world, formed friendships, and shared the most intimate of details regarding his illness on the password-protected site—receiving exactly the type of support he had yearned for.

It was for that reason that Ahmed felt so violated after having been notified by PatientsLikeMe of "unauthorized activity" on its Mood Forum discussions board. At 1:00 a.m. on May 7, 2010, system administrators noticed suspicious activity coming from several new accounts that were "scraping," that is, copying every single message off the private online forum and downloading the information to a third-party site. PatientsLikeMe eventually identified the intruder responsible for the break-in: the Nielsen company, the same advertising giant known for TV's Nielsen ratings. A Nielsen subsidiary known as BuzzMetrics admitted taking the data, which it added to its online collection of information purloined from the other 130 million blogs, eight thousand message boards, Twitter, Facebook, and other social media sites it tracked. Nielsen sells these data to advertisers, marketers, and, in this case, major pharmaceutical houses as raw material in a multibillion-dollar global data-mining industry.

Nielsen's egregious activity, though ethically repugnant, was technically legal under current federal law, and on May 18, 2010, PatientsLikeMe disclosed the incident to its entire user community. The company also took the opportunity to remind users of its own privacy policy terms and conditions:

> We take the information patients like you share about your experience with the disease and sell it to our partners (i.e., companies that are developing or selling products to patients). These products may include drugs, devices, equipment, insurance, and medical services . . . [Y]ou should expect that every piece of information you submit (even if it is not currently displayed) may be shared.

Wait, what? The note disclosing the Nielsen break-in was bad enough, but the follow-up e-mail detailing the Web site's privacy policy was a massive wake-up call. For most users of PatientsLikeMe, it was the first time they realized that all the medical information that would previously have remained securely locked in the filing cabinet of their doctors' offices—their conditions, diagnosis dates, family histories, symptoms, CD4 counts, viral loads, lab results, biographical information, gender, age, photographs, and even entire genetic sequences—was now being sold by PatientsLikeMe, the very place these desperate patients had entrusted to help them and to protect their information.

Though PatientsLikeMe claimed it only sold de-identified/anonymized data on its patients, new and emerging data companies such as PeekYou LLC of New York had long ago worked out a variety of patentable techniques to match people's real names to the pseudonyms they used on blogs, chat forms, and Twitter. In other words, any pharmaceutical company or health insurer that wanted the information from PatientsLikeMe need only hire PeekYou to reverse engineer your user name or pseudonym en masse to fully identify you. For Bilal Ahmed, this meant that all of the personal data he had entrusted to PatientsLikeMe were now owned by Nielsen/BuzzMetrics. In a public interview after his identification, Ahmed noted he felt totally violated by the incident and immediately deleted all his posts from the site as well as the list of drugs he had been prescribed, but by then of course it was too late. Every time he and other patients had posted highly detailed accounts of their illnesses and symptoms on PatientsLikeMe, there were companies like Nielsen lurking in the background scraping up all the data he shared. Those that weren't siphoned off by third parties were freely sold by PatientsLikeMe itself, disclosed in the fine-print privacy policy that Ahmed and so many others failed to ever read when they created their accounts.

As Ahmed discovered, social networks are the new public records. All that you share, wittingly or not, is being scraped, sorted, and warehoused by newly emerging global data behemoths and sold to advertisers, governments, and third-party data brokers, each with an increasingly voracious appetite to know the most intimate details of your life. These data can be used to determine if you have any preexisting conditions, if you should pay higher life insurance rates, or if you should be denied a job or promotion. While sharing may be caring, it also could mean higher insurance rates. As a result, the hundreds of thousands of people using PatientsLikeMe learned a valuable if painful lesson: they were not in fact the Web site's customers; they were its product, sold to the highest bidder in an effort to drive the company's own bottom line.

Our Growing Digital World—What They Never Told You

By 2013, Americans were spending more than five hours a day online with their digital devices. We read the news on Web sites run by CNN, the *New York Times,* and ESPN. We check our bank balances at Citibank and Wells Fargo. We shop at Amazon and Macy's. We pay our ConEd and Comcast bills, make appointments with our doctors, and check our health insurance with Blue Cross. We watch *House of Cards* on Netflix and *Downton Abbey* on Hulu. And that's just the beginning. Take a moment to think about how you used your smart phone today. Eighty percent of us check our mobile phones for messages within fifteen minutes of waking up. Did you provide a quick status update today to your friends on Facebook? You'll probably get a "Like" or two or maybe a funny comment from a friend. And what about those selfies you sent your boyfriend? The Internet has become a vast and free treasure trove of information and entertainment, and so we dutifully gorge ourselves at the trough. And at every step of the way, we are collectively leaving behind a daily digital exhaust trail big enough to fill the Library of Congress many times over. How all these data are created, stored, analyzed, and sold are details that most of us readily gloss over, but do so at our own peril.

There is no denying the power of social media. In a mere ten years since its creation in 2004, Facebook sprinted from zero to 1.3 billion members around the globe. Each day, more than 350 million photographs are uploaded, while the omnipresent Like button is pressed approximately six billion times. Social media chronicle our dates, graduations, home purchases, childbirths, new pets, marriages, and divorces. They also can be instruments for geopolitical change, as was seen during the 2010 Arab Spring, when a Google executive named Wael Ghonim created a Facebook page to highlight the slaughter of a young Egyptian protester at the hands of Hosni Mubarak's internal security forces. "Two minutes after he started his Facebook page, 300 people had joined. Three months later, that number had grown to more than 250,000." Similarly, Twitter, Google, and other services were credited with helping to drive change in Tunisia, Iran, and Libya. While history will judge the role social media played in the Arab Spring, there is no doubt these services can be a force for good.

And the allure of these tools is clear. After all, most of us spend our lives trolling the Web for music, recipes, investment advice, news, directions, business opportunities, celebrity gossip, and sports scores. When we're not checking e-mail, we're playing *Temple Run* or *Fruit Ninja*. And it's all entirely free of charge. Even the fees we once paid to travel agents, newspapers, and music companies have all but disappeared—eliminated thanks to the generous people who brought us the World Wide Web. But did you ever stop and wonder why Google never sends you a bill?

Ask the average person why Google, Facebook, Twitter, YouTube, and LinkedIn are free, and he or she may be a little fuzzy on the details. Many think it has to do with advertising—that is, those annoying banner ads or pop-up screens with which we are constantly barraged. Perhaps, but that is only a small part of the story. Others may believe the trade-off to be pretty simple. These companies give us valuable services for free, such as e-mail, news, videos, and a place to post pictures, and in return we give them a little bit of information about ourselves. Occasionally, we need to watch an advertisement that has been specifically designed to suit our needs, but privacy settings put us in the driver's seat and nobody gets hurt, right? If only it were that straightforward. The reality of the bargain we've made is much more disconcerting.

Take Google as an example, a company founded in 1998 by two Stanford PhD students, Larry Page and Sergey Brin, in a friend's garage in Menlo Park, California. The pair invented a groundbreaking algorithm that vastly improved search results on the nascent World Wide Web and attracted a loyal following drawn in by their simple interface and high-quality search results. In 2000, they began selling ad keywords for particular products aligned with any given search phrase. For example, if your query were "Paris, France," you would be served up sidebar ads for Air France, travel insurance companies, or Hilton Hotels. Companies looking for new customers could now market via Google's keyword ads with previously unknown precision, getting a much better return for their advertising dollars. What started as a humble idea by two Stanford students in 1998 had by 2015 grown into a global juggernaut.

Over the years, Google has introduced dozens of products that make our lives simpler and more productive. When it launched Gmail in 2004, it offered an amazing one gigabyte of data, vastly outmatching the paltry two megabytes offered by the dominant player of the day, Microsoft's Hotmail. As the young organization hit its stride, other fantastic products emerged, and eventually we were introduced to Google Calendar, Google Contacts, Google Maps, Google Earth, Google Voice, Google Docs, Google Street View, Google Translate, Google Drive, Google Photos (Picasa), Google Video (YouTube), Google Chrome, Google+, and Google Android, to name but a few. One by one, services such as phone calls, translation, maps, and word processing—services for which we would previously have paid hundreds of dollars (think Microsoft's Office)—were now suddenly free.

The most benevolent interpretation of this bounty would be that Google was merely providing products the public wanted, satisfying our ever-growing technological needs (and those of advertisers). A less altruistic explanation might be that each and every one of the aforementioned products was created with the specific intent to trick, cajole, and coax users to reveal an ever-increasing volume of data about themselves and their lives ad infinitum. People might balk if they fully understood the true nature of the exchange. So, to paraphrase Otto von Bismarck, it's best for Google's customers if they don't see or know quite how the sausage is made. But pulling back the curtain and studying the sausage factory are fundamental in order to understand an ever-growing mountain of data security risks facing our world today.

The gradual siphoning of your data begins innocently enough when you first start using Google to search the Web. You search, and it tracks and records the queries, not to mention every link you click on. From that initial search product, the carefully orchestrated acquisition of your personal information is carried out with artful precision. Eventually, a search engine wasn't enough, and Google craved additional ways to gain further insights into you, your hopes, dreams, and desires. The result? Gmail. By providing a vast amount of storage space and a wonderfully seamless experience, Google gained access to both your personal and your professional e-mails. Now Google could understand not just your searches but everything you were writing and to whom. Google scanned and electronically read your messages and found new insights it could offer to advertisers, increasing its fees along the way as it refined its profile of you. If you wrote an e-mail to your mom telling her you were sad over a recent breakup, Google might suggest an antidepressant, a comedy club, or a Caribbean vacation. As long as you remained logged in to Gmail, it could track all of your searches to your unique identifier with the company; as a result, Google's profile of you grew richer, as did the company.

When Google offered you the opportunity to store your contacts online, it in turn could evaluate the size, strength, and purchasing power of your social network. When Google introduced its Maps program and provided gratis GPS and driving directions, it could now track the places you went. Google wondered whom you were calling and created Google Voice to find out. Not only could it now track every one of your telephone calls, but it would transcribe your voice-mail messages using voice-recognition and voice-transcription software. A very cool technical feat at the time, it also allowed Google to understand what you and your interlocutors were discussing. If somebody left you a voice-mail message suggesting Italian food for dinner, Google would sell that information to its advertisers, and suddenly ads for pizza would show up across your Google world. For further precision, Google created the Android operating system (OS) and gave it away for free. In exchange, Google could now track you anywhere and everywhere you took your smart phone.

Of course if Google told you all this, you might be freaked out, so instead a pretty ruse was created, a fig leaf of sorts. When Google was founded, it projected itself as the underdog, the little guy battling evil Microsoft. In fact, Google would tell its users that it was so benevolent that it decided to make "Don't be evil" its official company motto. To allay any lingering doubts, Google's icons and graphics, like its childlike multicolored logo and the adorable little green Android guy, were created to be so cute and nonthreatening that surely they could be trusted. Google Doodles, drawings on its home page celebrating everyone from Martin Luther King to Gandhi, further reassured the public that these were the good guys. Besides, Google had all these privacy policies too, which protect me, right? Not so fast.

Viewed with a skeptical eye, Google creates products not to give you free e-mail but to get more data out of you. Like a pusher holding that first dime bag of heroin over a soon-to-be junkie, Google gave you something "on the house," and only later might you realize the implications of the bargain you were making. By then, it was too late. This became clear when in early 2012 Google announced it was merging its data across all of its seventy products and services. The result: a unified, profound, and unprecedented view of you and your world. Previously, your searches in Google, what you did on your Android phone, and the videos you watched on YouTube were data that *in theory* Google held separately. Not anymore; now Google has a single unified, highly detailed picture of you and everything you do across its Googleverse. Many have even argued that Google knows you better than you know yourself. It is precisely because it has all of these data that it can command top dollar from advertisers for your information.

If it wasn't abundantly clear previously, you are not Google's customer; you are its product. That's why you don't get a bill. That's why there's no 800 number for technical support. Those items are reserved for its real customers: the advertisers who are purchasing all the data you litter along Google's information superhighway. You are the thing Google sells to other people; that's the deal it never quite made truly clear to you, and whether or not you realize it, you are completely complicit in the process.

Google to its credit provides mostly wonderful products that serve the needs of its users, and the company is filled with loads of extremely talented and dedicated employees. But make no mistake, its loyalty will always be first and foremost to its advertisers, who pay the bills, and to its shareholders, to whom it has a fiduciary obligation to extract from you (its product and supply chain) the maximum value possible. That is why Google stores every search you have ever conducted on the site indefinitely: "the Ohio State college Republicans" you asked about ten years ago, the "symptoms of gonorrhea" after that one-night stand, the "Girls Gone Wild videos" while traveling for work at your hotel, "is my husband gay?" when feeling dejected and lonely.

Google does not forget, and Google does not delete. Each of the above queries is used to profile you, categorize you, and sell you to advertisers and data miners who make further assumptions about you based on your searches, e-mails, voice mails, photographs, videos, and locations as cataloged by Google. How many data is Google processing every day? you might wonder. About 24 petabytes' worth (that's 1 million gigabytes or 1,000 terabytes—a measure used to describe a volume of data). To put that in perspective, it takes approximately "1 gigabyte of data to store 10 yards of books on a shelf." If all the data Google processed on a daily basis were printed and those books were stacked on top of each other, the pile of books would reach halfway from earth to the moon. That's how much information Google is storing on users—every day!

With all of these data come immense insights and tremendous power, but as the old saying goes, power corrupts. Around the world, Google has been repeatedly sued for privacy violations, security breaches, mishandling user data, theft of intellectual property, tax evasion, and contraventions of antitrust laws. After a lawsuit by thirty-eight American state attorneys general in 2013, Google admitted that its bizarre-looking Street View cars, those outfitted with high-tech 360-degree roof cameras, were not just taking photographs for its Street View mapping product as they drove down the streets of our neighborhoods but also pilfering data from computers inside our homes and offices, including passwords, e-mails, photographs, chat messages, and other personal information from unsuspecting computer users.

In October 2013, a federal judge refused to dismiss a class-action lawsuit against Google claiming its practice of reading and scanning users' Gmail accounts violated U.S. laws against unlawful interception and wiretap. Before that, in 2012, Google was fined a record $22.5 million by the Federal Trade Commission when it was revealed it routinely circumvented privacy settings on Apple computers and for those using Apple's Safari Web browser to track users across the Web against their clearly stated wishes.

Of course Google is a highly innovative company, and in its quest to coax ever more data out of you for its real customers (advertisers), Google has planned a host of new products that may make past privacy concerns pale by comparison to future ones. One such product is Google Glass—a wearable computer in the shape of a pair of eyeglasses sporting an "optical head-mounted display" that connects to the Internet and is capable of projecting visual information onto a screen embedded in the glass. The device runs on the Android operating system and can take photographs and video and stream them in real time via its built-in camera and microphone.

In early 2014, Google Glass was the subject of a *Simpsons* episode titled "Specs and the City" in which all of Mr. Burns's employees were given a pair of "Oogle Goggles." On the show, Homer Simpson and his colleagues use the glasses to see new information about the things and people around them. Ominously, perhaps presciently, Mr. Burns, while seated in his office command center, is able to access the glasses of all his employees and see what they are viewing and doing in real time (in an effort to reduce theft of office supplies).

Even the former head of the Department of Homeland Security (DHS) Michael Chertoff has raised privacy and public policy concerns about Google Glass. He rightfully asked who owned the users' video data and whether the entire video database would be mined and analyzed for commercial purposes. One could also legitimately ask about government access to these data, either retrospectively or in real time, for reasons ranging from crime fighting to "national security." Consider the implications for a moment: By using Google Glass, are you granting the company the right to capture all the live-streaming moments of your daily life, everything you see and hear, so that it can sell these data to advertisers? For example, if, while you were wearing the glasses making your morning coffee in your bathrobe, the Google Glass vision algorithm recognized the object in your field of view as a coffeepot (entirely possible), might you start seeing coupons for Starbucks on your eyeglass screens? Given the aforementioned privacy violations we've seen from the search giant, what else might it be capable of as we enter the age of wearable surveillance?

The Social Network and Its Inventory—You

Of course Google is not alone in its business model of selling you to its advertisers, and there are thousands of companies around the world that do the exact same thing, including most notably Facebook. Founded by Mark Zuckerberg in his dorm room at Harvard in 2004, Facebook is the iconic Silicon Valley success story. With more than 1.2 billion monthly active users, Facebook is by far the largest social network in the world. It has succeeded by getting people to talk about themselves in ways never previously imagined. Sexual orientation, relationship status, schools attended, family tree, lists of friends, age, gender, e-mail addresses, place of birth, news interests, work history, catalogs of favorite things, religion, political affiliation, purchases, photographs, and videos—Facebook is a marketer's dream. Advertisers know every last intimate detail about a Facebook user's life and can thus market to him or her with extreme precision based upon the social graph Facebook has generated.

Moreover, Facebook created a variety of innovations that allow it to track users across the entirety of the Web, including via its omnipresent Like button. You've been trained to click on the cute little blue thumbs-up button to express your support for a particular idea, status update, or photograph; after all, it's the polite thing to do. Your friends see that you support their message, but what neither of you see is what happens with the data generated with each and every Like—data that are captured, dissected, and sold to marketers and data brokers around the world. When you use Facebook's ubiquitous log-in credentials to visit other sites on the Web, such as Spotify and Pandora, Facebook's data-mining engine is crunching your preferences for Lady Gaga over Blake Shelton, just as it is tracking all the Web sites you visit with the Facebook icon on them (even if you don't log in).

In case you aren't sharing enough, Facebook is happy to create new rules and regulations to force you to share more, as it did in 2012 when it instituted its mandatory timeline "feature." The change provided advertisers a dynamic, ever-updating window into your life's interests at any moment in time and more fodder for Facebook to sell to advertisers. Facebook, like Google, has been broadly criticized on issues including privacy, child safety, and hate speech. It has been sued repeatedly around the world, most recently in U.S. federal court in San Jose, California, for regularly and "systematically intercepting users' private messages . . . and sharing the data with advertisers and marketers."

Of course, Google and Facebook are by no means alone in getting you to reveal personal data about yourself and selling them: so too are Twitter, Instagram, Pinterest, and hundreds of other firms. For example, did you realize every time you speak a query into Apple's Siri artificial intelligence agent, your voice recording is analyzed and stored by the company for at least two years? The question, however, is not who is storing your data—everybody seems to be these days—but what are they doing with the information? If the Faustian bargain were as simple as we provide you cool services "for free" in exchange for a few data, all would be fine in the world. But things are not so simple, and as you will soon see, keeping and storing these massive volumes of data in a world that is so connected, dependent, and vulnerable puts you at risk in ways you might never have imagined.

You're Leaking—How They Do It

Every time you visit a modern Web site, the site places invisible digital marking files known as cookies on your computer or phone's hard drive. With these tiny computer files, it becomes possible to track you and your activities across the Web. In addition, all your digital devices have their own unique fingerprints that allow you to be traced, shadowed, and cataloged. Unique identifiers, such as the Internet protocol (IP) address for the computer network you use to access the Internet, the media access control number (MAC address) from your Wi-Fi network cards, and the IMEI or IMSI number from your mobile phone, all allow online companies to know exactly what devices (and users) are utilizing their services.

All of these data are tracked, unified, and exploited to give Internet companies and their advertisers a clear and persistent look into you and your online activities. According to a 2012 *Wall Street Journal* study, one of the fastest-growing businesses today is spying on Internet users. In its report, it highlighted fifty of the most popular Web sites and discovered that on average each left more than 64 cookie tracking files for advertisers to trace and surveil your online activities. The Web site with the most tracking software was Dictionary.com with a total of 234 tracking files implanted on your computer with every visit. All of these tracking beacons and cookies are combined with your Likes, pokes, and tweets to paint an eerily detailed picture of the digital you. Accordingly, your computer's cookies transform themselves into cookie monsters, revealing data you never intended to become public.

Not only are you leaking data as a result of your own social networking activities, but your friends and family are leaking data about you as well. Every time a colleague puts your name and address into his Google Contacts or iPhone, he is providing Google and Apple with your personal details. Record a birthday in Microsoft's Outlook calendar for your nephew, girlfriend, or business colleague, and Microsoft now knows the person's date of birth. When your friends tag you at a party on Facebook (after you called in sick to work), they have shared your location with marketers and potentially the rest of the world, including your boss. Social media and Internet companies love when users do this work for them; it's like having free code monkeys completing survey after survey feeding their big-data machine.

This even happens when your friend uses a particular Internet site or service that you don't. For example, for those who do not have a Gmail account but whose friends do, sending an e-mail to any one of Gmail's 425 million users means that Google has now become a party to your conversation. So if you use your university or work e-mail address to send your sister a message on her Gmail account, and even though you never opened a Google account yourself, Google will still read, scan, and search the message for words of interest that it can sell to advertisers—a practice for which it is now being sued in federal court. In its defense motion for the case filed with Judge Lucy Koh, Google shockingly claimed "a person has no legitimate expectation of privacy in information he voluntarily turns over to third parties." In other words, Google's argument is that by e-mailing any Gmail user, you have automatically waived any privacy rights and consented to its seizure and sale of your e-mail message and its contents, even if you intended the message to be private and don't have a Gmail account yourself.

It's not just your friends who are leaking data about you to third parties such as Google but also your children. In fact, sites aimed at children install more tracking technologies on computers than those for adults. Though a federal law titled the Children's Online Privacy Protection Act limits the information online marketers can collect on kids under thirteen, the rule is blatantly violated on a routine basis. Children are repeatedly presented with requests to sign up for contests, play games, and fill out online questionnaires in an effort to elicit more and more data about them, in contravention of federal law. Well-known companies such as McDonald's, General Mills, Viacom, Turner Broadcasting System, and Subway have all been fined for coaxing data out of young kids on their highly cartoonified Web sites such as HappyMeal.com, ReesesPuffs.com, Nick.com, and SubwayKids.com. In another case, Sony BMG Music asked minors to enter their street addresses and phone numbers on fan pages for their favorite bands, information that Sony then sold to data brokers on at least thirty thousand occasions, without first obtaining parental consent as required by federal law. But why would venerable companies like McDonald's, Google, Facebook, General Mills, and Sony commit acts such as these? Simply stated, there are vast sums of money at play, and you and your data are worth the risk to them, given the exorbitant rewards.

The Most Expensive Things in Life Are Free

The business proposition that most Internet users do not understand is that they are indeed paying for all of the so-called free services they receive online—and paying dearly. That sucking sound you hear is your privacy, your data, and all of the details that make up your unique identity being inhaled by the giant Internet vacuuming system. The details of your searches—stuff you wouldn't dream of sharing with your closest friends and family—are being filtered into a big computer algorithm in the sky, aggregated into petabytes, and sold for billions. That is why you have free search and Google has a $400 billion valuation. Because of you—its product. That's the deal you've made, whether you realize it or not.

Google's consolidated revenue for 2013 was more than $59 billion. That amount is the difference between how much your privacy is worth to Google's advertisers and how much you are not being paid. Google gets $59 billion, and you get free search and e-mail. A study published by the *Wall Street Journal* in advance of Facebook's initial public offering estimated the value of each long-term Facebook user to be $80.95 to the company. Your friendships were worth sixty-two cents each and your profile page $1,800. A business Web page and its associated ad revenue were worth approximately $3.1 million to the social network.

Viewed another way, Facebook's billion-plus users, each dutifully typing in status updates, detailing his biography, and uploading photograph after photograph, have become the largest unpaid workforce in history. As a result of their free labor, Facebook has a market cap of $182 billion, and its founder, Mark Zuckerberg, has a personal net worth of $33 billion. What did you get out of the deal? As the computer scientist Jaron Lanier reminds us, a company such as Instagram—which Facebook bought in 2012—was not valued at $1 billion because its thirteen employees were so "extraordinary. Instead, its value comes from the millions of users who contribute to the network without being paid for it." Its inventory is personal data—yours and mine—which it sells over and over again to parties unknown around the world. In short, you're a cheap date. You've happily provided Internet companies everything you know, everything you do, and everywhere you go in exchange for a modicum of convenience or entertainment.

As if that cash-for-data deal weren't stark enough, Google has decided to grow its $400 billion valuation by using you and your photographs in its advertisements. In October 2013, the company announced a new feature, called shared endorsements, that began to appear in search, map, and

Google Play store results. So if, for example, you rated a song five stars in Google Play's music store or gave a +1 thumbs-up of support to your local bar or bakery, Google now granted itself the right to sell your likeness, name, and endorsement to ad firms and data brokers. This way when your friends Charlie and Juanita searched for a bar or a song in Google, they would see your shining face endorsing the product next to their search results. George Clooney and Angelina Jolie get paid for their celebrity endorsements; do you?

Google introduced "shared endorsements" after a highly controversial similar program called Sponsored Stories was established by Facebook in which the company used your Likes to endorse its true customers—advertisers and the products they represented. After a class-action lawsuit was filed against Facebook, the social network ended the controversial feature, but not before earning $230 million in the eighteen months that the program ran. In the end, Facebook settled the lawsuit for $20 million, or about two cents for each of its users. Right about now you may be asking yourself, "But how can they get away with this?" The answer is simple: you said they could.

Terms and Conditions Apply (Against You)

"I have read and agree to the Terms of Service"
is the biggest lie on the Web.
TERMS OF SERVICE: DIDN'T READ, HTTP://TOSDR.ORG

We've all seen them. Those impossibly long, fifty-page disclaimers written in four-point font, single-spaced, with one-eighth-of-an-inch margins. They are designed to be impossible to read—so we don't. We don't read them, we don't understand them, and for this we pay a heavy price. In today's world, terms of service (ToS) are attached to every Internet site, mobile phone contract, cable TV subscription, and credit card. They delineate how all your private data can be sucked away from you and used in ways unimaginable—including many that we would object to, if only we could actually understand the agreement we were making.

According to a study at Carnegie Mellon University, the average American encounters 1,462 privacy policies a year, each with an average length of 2,518 words. If one were to read each and every one of those policies, it would take seventy-six full workdays, at eight hours a day, from our lives. In the aggregate, that works out to 53.8 billion hours for all Americans, at an estimated national opportunity cost of $781 billion of lost productivity every year because of the nightmare and disgrace that are the ToS.

Of course if this were just a story about lost productivity, it might not be so egregious, but these policies directly affect your wallet as well.

A study in the *Wall Street Journal* estimated that the completely one-sided language in the ToS policies costs each American household $2,000 a year—for a total of $250 billion annually—money that we are cheated out of as a result of a deck of cards that is entirely stacked against us. Though companies call these policies terms of service, for consumers they would be much more aptly described as "terms of abuse."

Let's just take one example of what it is costing you to use a social media site, in this case, LinkedIn, whose privacy policy states,

> You grant LinkedIn a nonexclusive, irrevocable, worldwide, perpetual, unlimited, assignable, sublicenseable, fully paid up and royalty-free right to us to copy, prepare derivative works of, improve, distribute, publish, remove, retain, add, process, analyze, use and commercialize, in any way now known or in the future discovered, any information you provide, directly or indirectly to LinkedIn, including, but not limited to, any user generated content, ideas, concepts, techniques and/or data to the services, you submit to LinkedIn, without any further consent, notice and/or compensation to you or to any third parties. Any information you submit to us is at your own risk of loss.

So by using LinkedIn, you are granting it irrevocable and perpetual access (for free) to any information you have ever listed on the site; there's no take backs, no do overs. Once it has your data, social network graph, work history, skills, and education, it can sell them now or in the future in anyway it desires, including in ways not yet discovered (for example, to own perpetual holographic rights to use your images in advertising?). To demonstrate how ridiculous privacy policies have become of late, the British retailer GameStation ran an experiment to see if anybody ever read its ToS. The company amended its privacy policy to read,

> By placing an order via this GameStation Web site on the first day of the fourth month of the year 2010 Anno Domini, you agree to grant us a non transferable option to claim, for now and for ever more, your immortal soul. Should We wish to exercise this option, you agree to surrender your immortal soul, and any claim you may have on it, within 5 (five) working days of receiving written notification from gamesation.co.uk or one of its duly authorised minions.

That's right, seventy-five hundred GameStation customers who purchased something on its site for the one day it ran the experiment irrevocably granted their immortal souls to the U.K. online retailer. While the point was irreverently made, ToS are no laughing matter, and courts around the world have found your click to be legally binding, with significant financial, privacy, and security implications for you.

Nearly all Internet companies have similarly draconian policies that work against you. While most stop just short of claiming rights to your immortal soul, many come pretty close, and the more words they use, the worse it is for you. Facebook's privacy policy has grown from 1,004 words in 2005 to 9,300 words by 2014 (not counting links to various sub-policies, terms, and conditions). To put that in perspective, Facebook's privacy policy is more than twice as long as the U.S. Constitution. Meanwhile, PayPal's privacy policy and amendments are the longest in the industry at 36,275 words. By comparison, Shakespeare's "*Hamlet* has 30,066 words, including the famous 'To be, or not to be' soliloquy and the moving 'Good night, sweet prince' speech at the end." Complicating matters further, Facebook and others grant themselves full rights to change their privacy policies at will and do so frequently.

Worse, many companies make the privacy settings nearly impossible to access and understand—Facebook alone has fifty different privacy settings with 170 options, well beyond the understanding of the average human being—and that's the point. Moreover, even if you did spend the hours it might take to customize the privacy options to your liking, any updates by Facebook to its ToS automatically put all users back to the default settings, which are the maximum level of openness (so it can sell more of its product—you—to its actual customers, advertisers). Unless you frequently check the settings—as you should—you will find that Facebook has en masse disregarded your explicit previously established privacy wishes. As a result, Facebook reserves for itself the power to monetize you without interference or cacophony from its assembly line of noisy human products.

Three months after it was purchased by Facebook, Instagram outraged its users when it announced it would sell their names, images, and photographs to advertisers. According to its updated ToS, Instagram argued that parents who had uploaded photographs of their minor children had implicitly consented to the use of the images in advertisements. The picture of your infant you uploaded to share with your parents could now be used to sell baby food because Instagram had granted itself those rights. Your brilliant shot of the sun setting over Manhattan? It could be sold as stock imagery for newspapers and magazines. As a result of its change in ToS, sixteen billion user photographs now became Instagram's intellectual property, explaining why Facebook paid $1 billion for a company with thirteen employees.

Google too has demonstrated its penchant for ridiculous ToS. For example, anybody who uses Google Docs or happens to upload a spreadsheet, PDF, or Word document to Google Drive automatically grants ownership of the document to Google. According to Google's ToS,

> When you upload or otherwise submit content to our services, you give Google (and those we work with) a worldwide license to use, host, store, reproduce, modify and create derivative works, such as those resulting from translations, adaptations or other changes and license to communicate, publish, publicly perform, publicly display and distribute such content.

Think about that. If J. K. Rowling had written Harry Potter in Google Docs instead of Microsoft Word, she would have granted Google the world-wide rights to her work, the right to adapt or dramatize all the Muggles as Google saw fit, to say nothing of the Hogwarts School of Witchcraft and Wizardry. Google would have retained the rights to sell her stories to Hollywood studios and to have them performed on stages around the world, as well as own all the translation rights. Had Rowling written her epic novel in Google Docs, she would have granted Google the rights to her $15 billion Harry Potter empire—all because the ToS say so.

That Facebook, Google, Twitter, and others all store your online data and monetize it to the fullest extent possible should be of little surprise by now. What might surprise, however, is the growing number of platforms from which advertisers can collect and process this information, including the formerly humble telephone. Alexander Graham Bell would be shocked to see how his invention had been transformed into the smart phones and apps of today, each reaching deeper and deeper into our lives, with notable risks to both our privacy and liberties.

Mobile Me

Right, my phone. When these things first appeared, they were so cool.
Only when it was too late did people realize they are as cool as
electronic tags on remand prisoners.
DAVID MITCHELL, *GHOSTWRITTEN*

There are currently more mobile phones on the planet than there are people, and as a result an omnipresent digital skin has begun to envelop the earth, with consequences for all. Today's smart phones are powerful miniature computers that we carry with us 24/7 and have become an indispensable part of our lives. Sixty-three percent of Americans admit to checking their phones every hour, and nearly 10 percent check every five minutes. These devices are taken to the bathroom, to the gym, and to our beds. Mobile phones and tablets have replaced our cameras, computers, calculators, calendars, address books, radios, televisions, and games. In fact, making calls is only the fifth most common activity for which people use their smart phones, after Web browsing, social networking, playing games, and listening to music. These gadgets form an integral part of our lives, and 84 percent of us admit that we could not go a single day without our mobiles. Smart phones serve as our trusted companions and as such have unlimited access to our day-to-day lives. But has this access been granted without sufficient consideration of what it means to be virtually conjoined with a computer we carry on our person round the clock?

For as ubiquitous and useful as mobile phones are, they are also veritable snitches in our pockets, digital spies tracking our every move. That device in your purse or jeans that you think is a cell phone is in reality a beacon, constantly signaling the world and providing an incessant stream of data about you, your location, and your life's activities. In the United States alone, mobile phones generate about 600 billion unique data events every day, including where you were, whom you texted, and the photographs you uploaded. The volume of data we leak via our home and work computers pales in comparison with what we're leaking through our digital pals in our pockets. Mobile phones provide the clearest picture into any one person's habits and preferences, and indeed into his or her life. So who has access to all these data? Many more people than you realize. Knowing your physical location, where you spend your time and money and with whom, provides greater opportunities to mine you ever more deeply for monetizable information. Data brokers, spies, and criminals too have come to view the cell phone as a rich fountain of intelligence for their targeting purposes. As a result, they, like others, are all going after the smart phone with a vengeance.

The opportunity to own all your mobile data was the reason Google created its Android mobile phone operating system and gave it away to developers and users for free. But as we've seen previously, free can be a very expensive proposition. The Android mobile phone software provides Google with your mobile phone number, network information, device storage data, call logs, and contacts lists, plus access to a bevy of new sensors that can detect your motion, location, and even ambient temperatures, humidity, and sound levels.

Given all these new technological hooks into your life, Google lost no time in filing a patent application for what it called "Advertising based on environmental conditions." *Ka-ching!* Now Google can detect if you are in a hot location and accordingly serve up an ad for air-conditioning or ice cream. Using its ambient sound technology, Google could also listen in on your phone calls for background noise and then serve up ads based on those preferences. So if you listen to Usher in the background while talking to Aunt Margaret on your Android phone, Google has the ability to detect that and show you ads for his upcoming concerts the next time you check your Gmail or search the Web.

Facebook too has now added the ability to use your phone's microphone to listen in on you and the ambient sounds nearby—all agreed to in the updated ToS and part of its big push for mobile users. When Facebook revealed in the fourth quarter of 2013 that it had reached 945 million monthly mobile users and that 53 percent of its revenue was now coming from mobile ads, the market showered its love on the company, adding billions of dollars to its valuation in the days following the announcement. By finally creating its own mobile app for users, Facebook devised not only a better user experience but also a new tool for grabbing voluminous amounts of data from a user's mobile device.

Pilfering Your Data? There's an App for That

An Apple iPhone commercial in 2009 famously introduced us to the phrase "there's an app for that" as a means of demonstrating there is an iPhone application for every possible human need. A bold statement at the time, but perhaps Steve Jobs was right. Since its launch in 2008, there have been more than sixty-five billion downloads from Apple's App Store, generating revenues of over $10 billion in 2013 alone. To compete with Apple, Google launched its own app store known as Google Play, and each company hosts more than a million separate applications available for download. The growth rate of these small software programs known as apps has been phenomenal, but what motivates tens of thousands of developers around the world to create apps? Money of course, but how do they make money when the majority of apps are free? Well, as we've seen before, free is a great business model, as long as you are monetizing people by commandeering their personal data in volumes. As it turns out, apps are a brilliant ecosystem for doing just that. It also helps to explain why companies like Rovio (maker of the wildly popular game *Angry Birds*) have grown from near obscurity to a $9 billion valuation in just a few years.

In the same way that cigarettes are nothing more than efficient nicotine-delivery systems, apps are nothing more than highly efficient and streamlined tools for the transmission of your private data to advertisers (though cigarettes are at least regulated). The amount of information siphoned off your mobile phone by apps is staggering. For example, the mere act of downloading the Facebook app to an Android phone automatically shares the phone's number with the social network—even before a user has signed into the service for the first time or even agreed to the ToS. Once Facebook has been downloaded, users agree in the ToS to grant it permission to "take pictures and videos with the camera," a setting that allows Facebook to turn on your mobile phone's camera at any time without your confirmation. The social network's ToS also allows it to read your text messages. More recently, Facebook began to ask hundreds of millions of users of its mobile app to allow its new Photo Syncing option to automatically upload every image taken with your phone to the social network's vast data servers.

Though Facebook maintains it isn't really turning on your camera or reading your texts, it has reserved for itself the right to do so. But frankly, how does any user actually know what data are being taken from the phone: it's all done in the background, hidden in the underbelly of the app, and not meant to be seen by those Facebook is productizing.

The appropriation of your personally identifiable information (PII) occurs the moment you agree to download an app. For example, when you buy an Android app in the Google Play store, Google provides the app company with your complete name, your e-mail address, and where you live. This occurs without any clear warning every single time you download an app. But who exactly are these app companies—thousands of them located around the world—who now have your name, address, and phone number? What are their privacy policies, and what do they do with this information? How securely do they store it, and to whom are they selling it? The fact is, it's the Wild, Wild West out there, and there is little or no regulation that effectively protects you and your data from these third-party information hawkers. By sharing millions of names and contact details with its app vendors, Google increases the likelihood that your data will leak, be stolen, or be misused.

As you probably now suspect, Zynga's wildly popular Facebook games *FarmVille*, *Texas Hold 'Em Poker*, and *Mafia Wars* are free of charge because they, too, have been tapping into your PII, including the names of all your Facebook friends. This information is sold to dozens of advertising and Internet-tracking companies, even if you have your privacy settings set to maximum protection. While putting birds in slingshots and shooting them at egg-thieving pigs can certainly be fun, as attested to by the one billion people who have downloaded the *Angry Birds* app, the game has a ravenous ability to collect its users' personal information, including all locations they travel with their mobile phones. And yet, a study by Carnegie Mellon's Human-Computer Interaction Institute revealed that only 5 percent of *Angry Birds* users knew that the company was storing their locational data in order to track them in the real world for targeted advertising purposes. But *Angry Birds* is far from the only offender. McAfee reported that 82 percent of Android apps track your online activities and an astounding 80 percent collect location information.

Location, Location, Location

For advertisers, there are three key questions: Who will buy their products, what are customers looking for, and where are they? In the online world, Google won the "what" question long ago with its powerful search algorithms. Google knows what you are looking for and will even fill out the search box at the top of your screen before you've finished typing your own query. Facebook owns "who"—it knows you and your social network in greater depth than any other firm. But "where"—where is not yet owned by any single company, and a battle is brewing among existing titans and a breed of new start-ups looking to definitively own where. Serving you up an ad or coupon for frozen yogurt just as you are approaching a Pinkberry is akin to advertising nirvana. Until now, the technology to do this was lacking, but that has changed because of the mobile phone revolution, and thus the gold rush for your locational data is on. McKinsey has estimated the market value of your personal location data to be more than $100 billion to the retail, media, and telecommunications industries over the next ten years.

"Where" is determined by a number of techniques—by your phone's GPS antenna, by triangulating your mobile's location and distance between cell-phone towers, and even by the Wi-Fi networks you connect to. These locational data are increasingly being appended to more and more of your online transactions in so-called file metadata—that is, data that provide data about other information. For example, when most people take a photograph with their mobile phones, their locational data (GPS coordinates, longitude and latitude, and so forth) are embedded in the image file. When you upload those photographs and videos to Craigslist, Flickr, YouTube, Facebook, and hundreds of other services, those revealing locational metadata can be passed right along with the original file. For some apps, the request for your location is perfectly logical, such as with Google Maps or GPS navigation tools. But for others, capturing your locational data is just another way for app makers to sell your data at greater value.

Your postings on Facebook, your tweets, and your searches on Yelp all use locational data. Moreover, there are an increasing number of location-based service (LBS) start-ups that are incorporating your "where" into everything from shopping to real estate. Perhaps one of the fastest-growing niches in LBS apps are those involving romance and love, especially "love" of decidedly limited duration. Apps like Tinder and Grindr have been downloaded millions of times and may be responsible for more than fifty million hookups, according to Tinder's CEO. Ah, love apps . . . a many-splendored thing.

But with all the potential benefits of these new streams of locational data come new risks as well. In 2012, a Russian company launched an app called Girls Around Me, which was approved for inclusion in both the Apple App Store and Google Play. Girls Around Me took advantage of all those public postings, status updates, photographs, and check-ins with locational metadata that women had been posting on services like Facebook and Foursquare. When a user launched the Girls Around Me app on his phone, he need only press a single button to be presented with an interactive map showing the faces of young women in his vicinity and their exact locations. Using the app's "radar mode," anybody could geo-locate these women and see their Facebook profiles.

For example, if a man used Girls Around Me and saw that a young attractive woman had just checked in at the local Starbucks, he could track her, access her Facebook profile, see what high school or college she had attended, learn that she recently vacationed in Las Vegas, and find out the names of her parents, her favorite drink, and the fact that earlier that day she had watched *Orange Is the New Black* on Netflix. Armed with this information, the man, a stranger, could casually walk up to the woman as she stood in line ordering her daily grande extra-hot soy latte and strike up a conversation discussing how much he loved Vegas and *Orange Is the New Black*. A very useful tool for dating, as well as for stalkers and rapists searching for women of interest.

For advertisers, "where" is not just about your current location; it's also about where you were yesterday and last month and where you are likely to be tomorrow. Locational records detail exactly how much time you spend in Macy's versus Best Buy, and the sequence of these movements tell much more. In today's world of location-based advertising, when a woman takes her smart phone and its apps with her to the gynecologist's office, an interesting data point is registered across the mobile advertising ecosystem. But when the same woman three weeks later walks into a Babies "R" Us, a much deeper truth is potentially revealed. By aggregating your locational data over time, advertisers can deduce whether you go to church or synagogue, work out at a gym, drink often at the local bar, see a psychologist, or are cheating on your spouse. But who exactly are these people collecting all of this information, how much of it do they have, and what do they do with it? As you are about to discover, vast amounts, growing at an exponential rate for uses we and they haven't even begun to identify.

CHAPTER 5

The Surveillance Economy

In digital era, privacy must be a priority. Is it just me,
or is secret blanket surveillance obscenely outrageous?

AL GORE

Leigh Van Bryan was looking forward to his first American holiday. A few days before his trip to Los Angeles, the twenty-six-year-old Brit checked in with a friend on Twitter and asked if she was "free this week for quick gossip/prep before I go and destroy America." Van Bryan's use of the word "destroy" would easily have been understood by any of his U.K. mates in their twenties as being British slang for "getting trashed and partying." Unfortunately, Van Bryan did little partying upon his arrival in America.

The Department of Homeland Security had been broadly monitoring social media for threats against America, and now Van Bryan was caught in its web. Upon his arrival in Los Angeles, Van Bryan and his travel companion, twenty-four-year-old Emily Bunting, were greeted by armed Customs and Border Protection agents, handcuffed, and put in a cell with alleged Mexican drug dealers for twelve hours. Though the pair attempted to explain their slang usage of the word "destroy," U.S. officials would have none of it. Federal agents repeatedly searched the couple and their suitcases, inexplicably looking for shovels. It turns out another tweet about "diggin' Marilyn Monroe up"—a nod to an episode from the cartoon *Family Guy*—had also raised serious flags with Homeland Security, which feared for the late starlet's remains. After an uncomfortable night in separate cells, Van Bryan and Bunting were reunited, just in time to be thrown on a plane back to the U.K. The couple had been denied entry to the United States and deported back to Britain. In the end, the only things destroyed were their visas and vacation.

You Thought Hackers Were Bad?
Meet the Data Brokers

Acxiom, Epsilon, Datalogix, RapLeaf, Reed Elsevier, BlueKai, Spokeo, and Flurry—most of us have never heard of these companies, but together they and others are responsible for a rapidly emerging data surveillance industry that is worth $156 billion a year. While citizens around the world reacted with shock at the size and scope of the NSA surveillance operations revealed by Edward Snowden, it's important to note that the $156 billion in annual revenue earned by the data broker industry is twice the size of the U.S. government's intelligence budget. The infrastructure, tools, and techniques employed by these firms rest almost entirely in the private sector, and yet the depth to which they can peer into any citizen's life would make any intelligence agency jealous with envy.

Data brokers get their information from our Internet service providers, credit card issuers, mobile phone companies, banks, credit bureaus, pharmacies, departments of motor vehicles, grocery stores, and increasingly our online activities. All the data we give away on a daily basis for free to our social networks—every Like, poke, and tweet—are tagged, geo-coded, and sorted for resale to advertisers and marketers. Even old-world retailers are realizing that they have a colossal secondary source of income—their customer data—that may be even more valuable than the actual product or service they are selling. As such, companies are rushing to profit from this brand-new revenue stream and transform their data infrastructure from a cost center into a profit center. Though credit bureaus such as Experian, TransUnion, and Equifax have been with us for decades, our increasingly digitally connected online lifestyle enables new firms to capture every drop of data about our lives previously unthinkable and impossible.

Just one company alone, the Acxiom Corporation of Little Rock, Arkansas, operates more than twenty-three thousand computer servers that "are collecting, collating and analyzing" more than 50 trillion unique data transactions every year. Ninety-six percent of American households are represented in its data banks, and Acxiom has amassed profiles on over 700 million consumers worldwide. Each profile contains more than fifteen hundred specific traits per individual, such as race, gender, phone number, type of car driven, education level, number of children, the square footage of his or her home, portfolio size, recent purchases, age, height, weight, marital status, politics, health issues, occupation, and right- or left-handedness, as well as pet ownership and breed.

The goal of Acxiom and other data brokers is to provide what is alternatively called "behavioral targeting," "predictive targeting," or "premium proprietary behavioral insights" on you and your life. In plain English, this means understanding you with extreme precision so that data brokers can sell the information they aggregate at the highest price to advertisers, marketers, and other companies for their decision-making purposes. For example, showing an ad for Pampers to a nineteen-year-old male college student might very well be a waste of an executive's marketing budget, but the same information presented to a thirty-two-year-old pregnant housewife might result in hundreds of dollars of sales. To maximize the value of the digital intelligence they collect, data brokers are forever segmenting us into increasingly specific groupings or profiles. Welcome to the world of dataveillance.

Acxiom sells these consumer profiles to twelve of the top fifteen credit card issuers, seven of the top ten retail banks, eight of the top ten telecommunications companies, and nine of the top ten insurers. To command the billions it charges its advertising customers every year, "Acxiom assigns you a 13-digit code and puts you into one of 70 'clusters' depending on your behavior and demographics." For example, people in cluster 38 "are most likely to be African American or Hispanic, working parents of teenage kids, and lower middle class and shop at discount stores." Someone in cluster 48 is likely to be "Caucasian, high school educated, rural, family oriented, and interested in hunting, fishing and watching NASCAR." These data are also sold to other third-party brokers who apply their own algorithms, further refining the data sets to create category lists of their own such as "Christian families," "compulsive online gamblers," "zero mobility," and "Hispanic Pay Day Loan Responders."

Those in the Christian family category might receive ads for Bibles and ChristianMingle.com, whereas gamblers and those deemed by an algorithm to have "zero mobility" would be targeted with ads for subprime lenders and debt-consolidation schemes. While being listed as a Christian family or as an urban Hispanic college-educated female might on the surface not appear too troubling, some data brokers have sold much more disturbing lists to advertisers and other parties unknown. For example, some brokers offer lists of seniors with dementia and people living with AIDS, while another firm, MEDbase200, has even auctioned off lists naming both victims of domestic violence and survivors of rape.

The depth and extent of the commercial data collection and surveillance economy were highlighted in early 2014 when a grieving father in Lindenhurst, Illinois, received a sales flyer in the mail from the retailer OfficeMax. Printed on the address label were the words "Mike Seay, Daughter Killed in Car Crash," followed by the man's home address. OfficeMax had indeed reached the right guy: Seay's seventeen-year-old daughter had been killed in a car crash with her boyfriend the year prior. When Seay called OfficeMax to complain about the incident, the manager refused to believe him and dismissed the allegation as "impossible." It wasn't until a local NBC reporter in Chicago ran the story that OfficeMax acknowledged the error was "the result of a mailing list rented through a third-party provider." Eventually, Seay received a phone call from a lower-level OfficeMax executive who apologized for the incident but refused Seay's repeated requests to name the data broker responsible for the incident. Nor would the executive reveal whether the company held similar data on other prospective customers. Seay's story is obviously troubling, especially because he was an infrequent OfficeMax shopper who had only on occasion purchased printer paper in the store.

The incident highlights some serious questions about the data broker industry. For example, what other deeply personal data does OfficeMax have on its customers? For the data broker that sold the information in the first place, what else might its massive data banks reveal about you and your family? Brother an alcoholic? Mother diagnosed schizophrenic? Thirteen-year-old daughter with an eating disorder? What regulations exist to limit what data brokers can do with this information, and what can you do if the information they hold on you is incorrect? There are hardly any regulations, as it turns out. It's reminiscent of the plotline from Franz Kafka's famous novel *The Trial*, in which a man is arrested without being informed why and only later learns that a mysterious court has a secret dossier on him, which he cannot access. Today's modern data brokers, unlike credit-reporting agencies, are almost entirely unregulated by the government. There are no laws, such as the Fair Credit Reporting Act, that require them to safeguard a consumer's privacy, correct any factual errors, or even reveal what information is contained within their systems on you and your family.

As a result of Seay's experience and those of thousands more like him, Congress, led by Senator Jay Rockefeller of West Virginia, the Federal Trade Commission, and the Consumer Financial Protection Bureau, have begun to investigate the nature and scope of the multibillion-dollar data broker industry. Any meaningful regulatory changes will be vehemently opposed by the data brokers; there is just too much money to be made. Moreover, once the data is out there, it is virtually impossible to put the proverbial toothpaste back in the tube. In the interim, Acxiom and others continue their stockpiling of your information. In late 2013, Acxiom's CEO, Scott Howe, proudly announced that his firm had collected nearly 1.1 billion third-party cookies and had identified and profiled the mobile devices of more than 200 million customers. "Our digital reach will soon approach nearly every Internet user in the US," Howe affirmed.

By mining public databases and aggregating that knowledge with all the personal information people share, wittingly and unwittingly, about themselves, their friends, and their families on social media, companies such as Acxiom have been able to deploy the most comprehensive intelligence surveillance system that has ever existed into the lives of nearly every American alive today. This technological feat represents the "new normal" of our data surveillance society and is part of what the former vice president Al Gore dubbed the "stalker economy" while speaking at the 2013 South by Southwest interactive festival in Austin, Texas.

Gore is right. As should be obvious by now, surveillance is the business model of the Internet. You create "free" accounts on Web sites such as Snapchat, Facebook, Google, LinkedIn, Foursquare, and PatientsLikeMe and download free apps like *Angry Birds, Candy Crush Saga, Words with Friends,* and *Fruit Ninja,* and in return you, wittingly or not, agree to allow these companies to track all your moves, aggregate them, correlate them, and sell them to as many people as possible at the highest price, unencumbered by regulation, decency, or ethical limitation. Yet so few stop and ask who else has access to all these data detritus and how it might be used against us. Dataveillance is the "new black," and its uses, capabilities, and powers are about to mushroom in ways few consumers, governments, or technologists might have imagined.

Analyzing You

Each of us now leaves a trail of digital exhaust throughout our day—an infinite stream of phone records, text messages, browser histories, GPS data, and e-mail that will live on forever. The analysis of this information allows companies to find prospective customers with much higher degrees of accuracy and at greater value than previously possible. For example, let's say that you're interested in taking a family vacation to Miami Beach. You search for flights on Kayak. Later you go into a store and buy a bathing suit using your credit card. The data retrieved from the purchase of the swimsuit combined with your browsing data reinforce the likelihood that you are interested in booking a hotel room in Miami. As a result of this behavioral analysis, you now have a quantifiable data value to the hotels in Miami, which can outbid each other for your business, in real time, by presenting advertising that reaches you with highly relevant messages and offers based on your intended behavior.

Google Now, which promises "just the right information at just the right time," is another example of deep analysis applied to large data sets. The Google Now app provides consumers with wonderfully convenient information that helps them capture and leverage all of the data invisibly swirling around them. Once users agree to Google's ToS, Google Now will show them when their friends are nearby, provide traffic alerts, determine the quickest travel routes home and to work, automatically furnish the morning's weather report, and keep track of favorite sports teams and update their scores in real time. Google Now automagically tells you when your flight is delayed and when your gate has changed and offers flight alternatives when available. Because Google Now knows where all your appointments are located and monitors traffic jams along all your intended routes in real time, the app will alert you at your current location, advising you to leave early if you hope to make your next appointment on time. Using a technique known as geo-fencing, Google Now will analyze your to-do list and match it against your persistently tracked location in order to alert you as you drive past the grocery store that you need to buy milk. To enjoy this cornucopia of information and bounty of convenience, you just need to provide Google Now with access to your entire online digital footprint, including your Gmail in-box, Web searches, hotel bookings, flight plans, full contact lists, friends' birthdays, restaurant reservations, and calendar appointments, as well as your physical location at all times via the GPS on your mobile phone. From this massive data set, Google (and others) can re-create what intelligence analysts call your pattern of life, knowing and mapping your physical location over time as well as what you are doing and with whom. How terribly convenient.

But what else might Google or any other company that had access to your pattern of life be able to determine? Let's say, for example, your mobile phone was on the nightstand in the same home as your wife's telephone six nights a week. From these data, it would be logical to conclude that the owners of the two cell phones lived together and were likely sleeping together. But what if one night a week your mobile phone was on a nightstand next to another woman's mobile phone? What might that suggest to Google or others about your fidelity? An analysis of your locational data and that of the phones (and apps) around you is an excellent approximation of the strengths and bonds of your personal and professional networks. When your data exhaust patterns are studied over time, many more revelations about your life become possible. For example, researchers in the U.K. studied the past whereabouts of mobile phone users and using basic data analysis techniques were able to determine to within twenty meters of accuracy where a mobile phone user would be at the same time twenty-four hours later, a very useful tool for both advertisers and stalkers. Your phone today knows not only where you've been but also where you are going.

Analysis of your social network and its members can also be highly revealing of your life, politics, and even sexual orientation, as demonstrated in a study carried out at MIT. In an analysis known as Gaydar, researchers studied the Facebook profiles of fifteen hundred students at the university, including those whose profile sexual orientation was either blank or listed as heterosexual. Based on prior research that showed gay men have more friends who are also gay (not surprising), the MIT investigators had a valuable data point to review the friend associations of their fifteen hundred students. As a result, researchers were able to predict with 78 percent accuracy whether or not a student was gay. At least ten individuals who had not previously identified as gay were flagged by the researchers' algorithm and confirmed via in-person interviews with the students. While these findings might not be troubling in liberal Cambridge, Massachusetts, they could prove problematic in the seventy-six countries where homosexuality remains illegal, such as Sudan, Iran, Yemen, Nigeria, and Saudi Arabia, where such an "offense" is punished by death. A study of fifty-eight thousand Facebook users published by the National Academy of Sciences demonstrated that by merely studying their Likes, one could determine intimate details and personality traits with surprising accuracy. The rigorous study carried out in conjunction with the University of Cambridge predicted whether users had a high or low IQ, were emotionally stable, or came from a broken home. The challenge with the data we are leaking is that, as has been shown numerous times, others can pick up our digital bread crumbs and interpret them without our knowledge in ways that can cause us harm.

But I've Got Nothing to Hide

In December 2009, when CNBC's Maria Bartiromo asked Google's own CEO, Eric Schmidt, about privacy concerns resulting from Google's increasing tracking of consumers, Schmidt famously replied, "If you have something that you don't want anybody to know, maybe you shouldn't be doing it in the first place." Schmidt, and others, dismiss privacy concerns by saying that if you haven't done anything wrong, you should not be afraid of people (corporations, governments, or your neighbors) knowing what you are doing.

This sentiment has been echoed by Facebook's CEO, Mark Zuckerberg, who has argued that "privacy is no longer the social norm." While privacy may no longer be the norm—at least for the general public—in his own life, Mr. Zuckerberg seems to treasure privacy quite a bit. In late 2013, it was revealed that the Facebook CEO spent $30 million to buy the four homes surrounding his own property in order to ensure his privacy would remain free from intrusion or disturbance.

Facebook's chief operating officer, Sheryl Sandberg, too has suggested that your assertion of any privacy rights is in contrast with "true authenticity." Sandberg notes that "expressing authentic identity will become even more pervasive in the coming years . . . And yes, this shift to authenticity will take getting used to and it will elicit cries of lost privacy." Convenient for Schmidt, Zuckerberg, and Sandberg that these "naturally occurring shifts" in social norms are tied to their personal and professional bottom lines, which directly benefit from monetizing you and the mountains of information you are leaking to the fullest extent possible, as a result of their highly one-sided ToS.

But "I have nothing to hide" is absolutely the wrong way to think about our new dataveillance society. It is a false dichotomy of choice: either we accept total surveillance, or we are criminals worthy of suspicion. If proponents of the "nothing to hide" argument meant what they said, then they would logically not object to our filming them having sex with their spouses, publishing their tax returns online, and projecting video of their toilet use on the Jumbotron of a crowded stadium, right? After all, they have nothing to hide. The fact is that each of us has private special moments in our lives, made exceptional by limiting with whom we share such intimacies.

For those who believe the fallacy of nothing to hide, perhaps a lesson in something to fear might be appropriate, for all of us have details in our lives we would rather not share. For example, Google Voice, Skype, your mobile phone carrier, and any number of government agencies have records of anyone who has ever phoned an abortion clinic, a suicide hotline, or a local chapter of Alcoholics Anonymous. Data aggregators know who has searched for "slutty cheerleaders," "Viagra," or "Prozac" across any of their electronic devices. While all these behaviors may be perfectly legal, no doubt they have repercussions in our society should the information come to light.

Given that Google and Facebook alone have hundreds of petabytes of data on their users stored in perpetuity, perhaps it is more worthwhile to question not what any of us may have to hide today but what we might wish to keep private in the future—and if Facebook existed in 1950, how might history judge an off-color joke today? What future crime might you be convicted of without ever knowing you were in fact violating the law? Did you drive across the border to New Jersey or Delaware to save on taxes when buying back-to-school clothes for the kids? Your cell-phone and credit card receipts document your tax evasion. That photograph on Twitpic of the family dinner showing your twenty-year-old son drinking a glass of wine—evidence of alcohol furnished to a minor. As the computer security researcher Moxie Marlinspike pointed out, there are "27,000 pages of federal statutes" in the United States and another "10,000 administrative regulations. You probably do have something to hide, you just don't know it yet."

Privacy Risks and Other Unpleasant Surprises

As *Wired*'s Mat Honan and the grieving father Mike Seay discovered, our personal data can end up in the hands of those who we assuredly would prefer did not have access to such information. The combination of our social data commingled with public databases, cookies, beacons, and locations can lead to a series of unintended and even harmful consequences. Put another way, your data are increasingly promiscuous. They flow from one system to the next, from database to database, obscured and distributed across cloud-based networks around the world, shared, processed, and sold. But as we have learned from the real world, promiscuity can often lead to social diseases and other unintended consequences.

In an incident not too dissimilar from the OfficeMax debacle, a Minneapolis man learned his daughter was pregnant, not from her, but from his local Target store. The discovery was made when Target began sending the fifteen-year-old girl coupons for items that did not meet her father's approval. Armed with the coupons and a letter addressed to his daughter, the father furiously marched into Target and began berating the store manager. "My daughter got this in the mail! . . . She's still in high school, and you're sending her coupons for baby clothes and cribs? Are you trying to encourage her to get pregnant?" A few days later, the man phoned the store to apologize, noting, "There have been some activities in my house I haven't been completely aware of. She's due in August. I owe you an apology." But how in the world did Target know the girl was pregnant? Through its pregnancy prediction algorithm of course, which aggregated a customer's entire purchase history with demographic statistics purchased from data brokers. Target reasoned that if it could find those women before their second trimester of pregnancy and hook them as customers, it would receive the lion's share of their purchases, not just for baby wipes, cribs, and diapers, but for toys and clothes as the infants aged through adolescence. After an in-depth study by Target's statisticians, Target noticed that women in the baby registry were "buying larger quantities of unscented lotion at the beginning of their second trimester in addition to vitamin supplements such as calcium, magnesium and zinc." In total, Target was able to identify twenty-five products that, when analyzed together, allowed it to assign each shopper a "pregnancy prediction score." When this model ran against the millions of women in Target's customer databases, thousands and thousands of pregnant women were identified before any other companies had made the connection. Target and the company's marketers were ecstatic with this discovery. Less enthralled was the father of that fifteen-year-old girl in Minneapolis who would learn of his forthcoming grandchild via a corporate coupon mailer. Given Target's 2013 hacking, in which the financial data of 110 million of its customers were compromised, what guarantees do consumers have that the vast additional troves of highly personal data in Target's vaults also won't be stolen? Can customers trust Target or any other large retailer with the volumes of data it collects, stores, and analyzes? Likely not, and therein lies the problem.

The risks to our personal data come not just from hackers but, as more and more people are finding out, from big-data analytics as well. Previously, many of the data aggregated were held in limbo, as our collection abilities well surpassed our ability to make sense of all that had been collected. That is now changing, and the data we leak on social media sites like Facebook are showing up in unexpected ways. One such person affected was Bobbi Duncan, a twenty-two-year-old lesbian student at the University of Texas at Austin. She came from a strict Christian family and worked hard to keep her sexual orientation from her parents. As she began to understand herself better, she joined a number of student groups on campus, including the Queer Chorus, as a means of meeting other gay and lesbian students at her school. When she joined the organization, the president of the Queer Chorus welcomed Bobbi by adding her to the group's Facebook discussion page, which he was able to do without her permission (there is no setting in Facebook to prevent a third party from adding you to his or her group). When he did so, Facebook sent an automatic system notification to Bobbi's entire list of friends—including her father—notifying them that she had joined the Queer Chorus. Two days after receiving the notification, Bobbi's father wrote a reply on his Facebook page: "To all you queers. Go back to your holes and wait for GOD. Hell awaits you perverts. Good luck singing there." Facebook outed a closeted lesbian and caused her parents to disown her. In response to the irreparable harm she suffered, Bobbi was unequivocal in her stance: "I blame Facebook . . . It shouldn't be somebody else's choice what people see of me."

When you are the product of Internet and social media companies, the challenge you face is that data you provide in one context can be used in unexpected ways in another, with notable consequences. Such is the case with the highly popular "free" dating site OkCupid. Users seeking dates are asked to fill out questionnaires on the site, and most presume, wrongly, that the data they provide remain exclusively within the OkCupid system, used solely for the purposes of finding a suitable match for a date. Yeah, right! To allegedly get the best matches, OkCupid asks users a bevy of deeply personal questions about their number of past sexual partners, whether they support abortion rights, whether they own a firearm, if they would sleep with somebody on a first date, if they smoke cigarettes, and if they drink alcohol frequently or use illegal drugs (including which drugs and how often). At least that's what users see on their screens when completing their profiles . . .

What they don't see is the fifty or so companies with whom OkCupid shares this information, including ad firms, data brokers, and marketers. To understand the extent of the data leakage, Ashkan Soltani, a digital privacy specialist who used to work at the Federal Trade Commission, created a dummy account on OkCupid. Using several free privacy browser plug-in tools, including Collusion and mitmproxy, Soltani was able to observe that the answers provided by OkCupid's users were parsed and forwarded to dozens of data brokers in real time. When Soltani completed his test

OkCupid profile and clicked that he frequently used drugs, he was able to observe a cookie file that shared his purported drug usage with a data broker known as Lotame. You think you're just filling out a confidential profile for a "free" online dating service; in reality, you've been had and are instead detailing information that you would never otherwise share with any marketing company or data broker. It's a huge ruse: dating is just the "cover story" for massive data extraction. In an ensuing investigation into Soltani's research by NPR, both OkCupid and Lotame declined to comment on the matter. Such is the state of affairs in the world's unregulated data broker industry. Who else might be willing to pay for OkCupid's archive of your drug use and sexual history? An insurance company, prospective employers, or perhaps the government after that DUI incident you had last June?

Even when you have "nothing to hide," your continually tracked social network graph and location can come back to bite you and even affect your financial status. A handful of tech start-ups have begun to use the quality of the friends in your social network to determine whether or not you are a good credit risk. One such company, Lenddo, determines if you are friends with somebody who is behind on paying back her loans and how often you interact with that person. As a result, your creditworthiness can drop because of whom you've friended on Facebook. If your pals on Google+ and Pinterest are deadbeats, chances are you may be too (according to the big-data gods). Facebook may become the next FICO credit-scoring agency as financial data aggregators take full advantage of your social data feeds to rate your financial stability. So as your mom used to warn you, choose your friends wisely.

The fact is that we are all contributing to our own digital pollution. Just as in the twentieth century people thought nothing of pouring industrial waste into a river or tossing garbage onto the street, so too do we fail to comprehend the long-term consequences of our digital actions today. The current state of affairs stems from our fundamental misunderstanding of the bargain we have made for so-called free online services.

Opening Pandora's Virtual Box

People share their most intimate thoughts and secrets online as if they were having a private conversation with a trusted friend. If only the legal system agreed. In the United States, social networks are considered to be public spaces, not private ones, and any information shared there is covered under the so-called third-party doctrine, which in plain English means that users have no reasonable expectation of privacy in the data their service providers (cell-phone companies, ISPs, cable companies, and Web sites) collect on them.

This noted exception to the Fourth Amendment's prohibition on unreasonable search and seizure means that any data you post online in any format (regardless of your privacy settings) or any data that are collected by the third parties with whom you have an agreed-upon business relationship are not considered private. Nor does it meet the constitutional standard of "private papers" but rather forms part of the business records of the institution in possession of the data. Shocking though this may be, it is the current state of jurisprudence in the United States, with noted and profound impact on all citizens both online and off. As a result, your data leak to places you would never want them to, and you cannot claw them back, no matter how hard you try.

Accordingly, the word "Facebook" appeared in a full one-third of divorce filings in 2011. All of this provides excellent fodder for the 81 percent of divorce attorneys who admit searching social media sites for evidence that can be used against their clients' spouses. For instance, all the data shared on Facebook and Twitter and all the cell-phone call records and GPS locational data that neatly recorded whose cell phone was next to whose and when become fair game in the battle royal that can be divorce proceedings. The pictures innocently taken of you at all those parties over the years, blurry-eyed with drink in hand, now become evidence of unfit parenting, a nugget of gold for opposing counsel during cross-examination. That profile you created on OkCupid indicating you were single (which was shared via your browser's cookies with fifty marketing companies)—perfectly admissible when your wife brings it up during your divorce hearing. When a husband complains that his wife is an inattentive and an unfit mother, he has new powerful evidence to support his claims in the form of subpoenaed records documenting the hundreds of hours she logged on *FarmVille* and in *World of Warcraft*, times coinciding with all of her children's soccer and baseball games that she missed. But the data we're leaking affect us not only during divorce but in our jobs as well.

A survey conducted by Microsoft on the matter of online reputation found that 70 percent of human resource professionals had rejected a job candidate based on information they had uncovered during an online search. Worse, some employers are now demanding the social media passwords of job applicants and even current employees. Want to work for the Norman, Oklahoma, Police Department, the Maryland Department of Public Safety and Correctional Services, the city of Bozeman, Montana, or the Virginia State Police? Applicants in all of these jurisdictions were required to turn over their Facebook and other social media passwords as part of their so-called routine background checks. This includes providing prospective employers access to all your messages, photographs, and timelines, private and public, on Facebook, Google, Yahoo!, YouTube, and Instagram. While some states, including California, have barred such practices against employees, there is no federal law banning such practices, and it remains legal in 80 percent of American states, and so the data leak.

Increasingly, more and more teachers and school districts are demanding this information from students as well, without warrant of course. That's what happened to a twelve-year-old Minnesota middle school girl who was accused of posting "inappropriate comments" on her Facebook account. The student at the Minnewaska Area Middle School had posted that she "hated" a particular school official who was "always mean to her." The girl was summoned to the principal's office, where administrators, a school counselor, and a deputy sheriff were waiting for her and demanded that she divulge her Facebook password so that they could review all of her postings. Yes, of course a lawsuit is pending, but the growing number of egregious cases shows your children are leaking data that can come back and bite them as well.

Even college athletes at schools like the University of North Carolina and the University of Oklahoma are being required to provide their passwords on social media sites to their coaches as a condition of playing sports at the schools. Some college athletes have also been compelled to install monitoring software on their personal computers and phones from companies such as UDiligence, which tracks in real time the students' activities to ensure "that collegiate athletic departments protect against damaging posts made by student-athletes."

Governments are also getting in on the action. A survey by the International Association of Chiefs of Police of more than five hundred law enforcement organizations revealed that 86.1 percent of police departments now routinely include social media searches as part of their criminal investigations. The IRS too began training its investigators on how to use social networks to investigate taxpayers back in 2009, and Homeland Security's Citizenship and Immigration Service instructed its agents in 2010 to use social media sites to "observe the daily life of petitioners and beneficiaries suspected of fraud."

Federal agents can readily access your social data through a variety of means, by serving subpoenas, national security letters, and other administrative orders on your service providers, who under the third-party doctrine exception to the Fourth Amendment needn't even notify you of the request. For example, AT&T revealed that in 2013 it received more than 300,000 requests for data relating to both civil and criminal cases. The demands for information came from state, federal, and local authorities and included nearly "248,000 subpoenas, nearly 37,000 court orders and more than 16,000 search warrants." In 2009, Sprint disclosed that it had even created a law-enforcement-only portal that gave police the ability to "ping" (without warrant) any one of Sprint's mobile phones in order to geo-locate users in real time—a feature that was used more than eight million times by police in a one-year period.

What data of yours the government doesn't subpoena, it just buys. The NSA and other government agencies didn't build their global eavesdropping and data-siphoning network from scratch; they purchased or otherwise obtained a complete copy of what the corporate world was already collecting. It makes perfect sense: Why build what they can just buy? ChoicePoint, now owned by Reed Elsevier, maintains seventeen billion records on businesses and individuals that it resells to its 100,000 clients, including to 7,000 federal, state, and local law enforcement agencies. Revelations from Edward Snowden alleged the Central Intelligence Agency pays AT&T $10 million a year for its call data and suggested Verizon too supplies data to the U.S. government. Commercial data brokers have lost no time in offering their paid subscription services to government agents, serving up the streams of information you have freely provided across your social networks.

A brilliant parody on the comedic Onion News Network lampooned the current state of affairs in a fake evening news report:

> Congress today reauthorized funding for Facebook, the massive
> online surveillance program run by the CIA. According
> to reports, Facebook has replaced almost every other CIA
> information-gathering program since it was launched in 2004.
> [A mock CIA official noted,] "After years of secretly monitoring
> the public, we were astounded so many people would willingly
> publicize where they live, their religious and political views, an
> alphabetized list of all their friends, personal e-mail addresses,
> phone numbers, hundreds of photos of themselves, and even
> status updates about what they were doing moment to moment.
> It is truly a dream come true for the CIA. Much of the credit
> belongs to CIA agent Mark Zuckerberg, who runs the day-to-day
> Facebook operation for the agency."

As hilarious and spot-on as the faux news report was, the leakage of our personal information to both shadowy data brokers and government agencies is no joking matter. The cost of the surveillance economy, owing to great advances in technology, is dropping exponentially. Gone is the need for vast teams of special agents to follow you around, tailing you on foot and in vehicles as you traverse a city. Instead, one study has estimated that by using proxy surveillance technologies such as mobile phones, online activity, social data, GPS information, and financial transactions, the government now spends just "$574 per taxpayer, a paltry 6.5 cents an hour," to track each and every American.

Upon learning the true extent of the NSA's domestic and international spying prowess, the former head of the East German Stasi Wolfgang Schmidt admitted publicly that such a system "would have been a dream come true." Schmidt noted that during his reign as head of the much-feared secret police service of the former German Democratic Republic, the Stasi could tap only forty telephones nationwide at a time, but clearly now technology had made it possible to monitor all calls and Internet data at all times. He cautioned, "It is the height of naivete to think that once collected this information won't be used . . . This is the nature of secret government organizations. The only way to protect the people's privacy is not to allow the government to collect their information in the first place."

Knowledge Is Power, Code Is King, and Orwell Was Right

In George Orwell's dystopian novel *1984*, he depicted an omnipotent government surveillance state controlled by a privileged few elite who persecuted independent thinking as "thought crimes." Though Orwell clearly would have foreseen the NSA debacle, it's less clear he might have predicted Acxiom, Facebook, and Google. To that point, in those cases it wasn't Big Brother government that "did something to us," but rather we who did something to ourselves. We allowed ourselves to become monetized and productized on the cheap, giving away billions of dollars of our personal data to new classes of elite who saw an opportunity and seized it. We accepted all their one-sided ToS without ever reading them, and they maximized their profits, unencumbered by regulation or oversight. To be sure, we got some pretty cool products out of the deal, and *Angry Birds* is really fun. But now that we've given all these data away, we find ourselves at the mercy of powerful data behemoths with near-government-level power who do as they please with our information and our lives.

In his 1999 book *Code and Other Laws of Cyberspace*, the Harvard Law School professor Lawrence Lessig insightfully demonstrated that the instructions encoded in any software program, app, or platform shape and constrain the Internet, just as laws and regulations do. Thus, when Facebook or Google unilaterally changes its terms of service to allow your news feeds to become public or your photographs to be used in advertisements against your will, it is as if a new "law" has been passed. Code, is in effect, law.

Perhaps then the only way to opt out of such a system would be to close one's account or never create one in the first place? Unfortunately, both approaches are problematic and increasingly impossible. A *New York Times* article previously noted that Facebook keeps all your data even after you've closed your account. Even if you chose not to participate in an online social network, your friends would continue to tag you in pictures, the GPS in your car would still track your location, and Target would track all of your purchases.

The unprecedented volumes of data about ourselves that we have entrusted to private companies are up for grabs, and once the genie is out of the bottle, there's no putting it back in. The troika of opportunity created by our online data exhaust, ridiculous terms of service, and little or no regulation means that modern data brokers can surveil us with better-than-government-grade surveillance capabilities, capturing our every thought, photograph, and location and subjecting them to big-data analytics. As Mat Honan, Bilal Ahmed, Mike Seay, Bobbi Duncan, Leigh Van Bryan, and Emily Bunting all learned firsthand, there are social costs and risks associated with our continued data leakage. But privacy implications are just one of the great threats resulting from the exponential growth in data.

Hackers are hard at work stealing all of the social data you have dutifully reported on yourself and are successfully breaking into the computers of data brokers and Internet giants responsible for storing it all. As Sony, Target, and even the Department of Defense have learned, data stored in insecure information systems are data waiting to be taken. As such, all data gathered will eventually leak, with potent implications for our personal and professional lives and even for our safety and security.

The problem with our being the product as opposed to the customer of massive data brokers is that we are not in control of our data and thus not in control of our destiny. The continued aggregation of this information, unregulated and insecure, sits as a ticking time bomb, with our every thought and deed available for the picking by a new and emerging class of bad actors whose intents are far worse than selling us discounted diapers and adjusting our insurance rates. International organized crime groups, rogue governments, and even terrorists are rapidly establishing their own data brokerages and bolstering their analytic capabilities in order to take full advantage of the single largest bonanza that has ever come their way, with frightening implications for us all.

CHAPTER 6

Big Data, Big Risk

**Our technological powers increase,
but the side effects and potential hazards also escalate.**

ALVIN TOFFLER

On the evening of November 26, 2008, a sixty-nine-year-old man checked into room 632 at the luxurious Taj Mahal Palace hotel in Mumbai, India. The guest, K. R. Ramamoorthy, was visiting from Bangalore on a routine business trip. Little did he know that his life was about to change forever.

At about 11:00 p.m., Ramamoorthy heard a brief commotion outside his door, then suddenly a knock. "Room service," a voice said. Ramamoorthy knew he had not ordered any food and sensed something was gravely wrong. He attempted a retreat to the bathroom, accidentally bumping into the door. The noise gave away his presence inside the hotel room. The response was swift: a hail of bullets came flying through the door, obliterating the lock separating the businessman from the world outside.

Two heavily armed men forced their way into Ramamoorthy's room, and in the blink of an eye he was beaten, stripped naked, and tied up in what would become the most terrifying night of his life. The men were from a Pakistani-based al-Qaeda-affiliated terrorist organization known as Lashkar-e-Taiba (LeT), and Ramamoorthy had unfortunately found himself at the center of the deadly 2008 terrorist siege on the city of Mumbai.

"Who are you, and what are you doing here?" his LeT captors demanded of him. "I'm just an innocent schoolteacher," Ramamoorthy replied. Of course, the terrorists knew that no Indian schoolteacher could afford to stay in a suite at the city's most opulent hotel. The terrorists located their hostage's identification card on his bedside table and now had his true name, which they called in to their terrorist commanders on the satellite phone they had brought with them.

The LeT ops center receiving the call resembled any modern military command-and-control facility. From across the border in Pakistan, terrorist cell leaders tracked the progress of their attack on the people of Mumbai. They had carefully selected their targets, including two luxury hotels, a busy railway station, a Jewish community center, a popular tourist café, and even a women and children's hospital. On the ground in Mumbai, terrorist operatives ruthlessly threw hand grenades at innocent people as they sat eating in cafés and gunned down unarmed civilians waiting to catch trains on their way home from work.

As the attacks unfolded, LeT commanders in Pakistan used their war room to carefully monitor the BBC, Al Jazeera, CNN, and local Indian TV stations to learn as much as possible about the progress of their operatives and the response of the Indian government. Regrettably, the terrorists did not limit their information-gathering operations to broadcast media; they also mined the Internet and social media in real time, to deadly effect.

When the terrorists holding Ramamoorthy phoned in his name to their Pakistani base, the ops center deftly conducted an Internet search on their hostage. Moments later, they had his photograph. Then his place of work. They learned Ramamoorthy was not an innocent schoolteacher as he claimed when pleading for his life but rather the chairman of one of India's largest banks, ING Vysya. Based on the image they had found online, the terrorist commanders asked their operatives at the Taj Mahal Palace to compare the man before them with the photograph of the bank chairman located online:

Your hostage, is he heavyset?
Yes.
Is he bald in front?
Yes.
Does he wear glasses?
Yes.

"What shall we do with him?" asked Ramamoorthy's captors. Moments later, the terrorist war room gave its reply. Kill him.

In an instant, a simple Internet search was all the terrorists needed to decide the elderly man's fate. Though we may worry about advertisers and data brokers abusing our privacy settings on Facebook, the fact of the matter is that our openness can be used against us in ways worse than we had ever imagined. When we leak data, it's not just captured by corporations or governments. Criminals and terrorists have access to our social data as well, and they are leveraging it with killer precision. In today's world, a search engine can literally determine who shall live and who shall die.

The men who carried out that attack on Mumbai were armed with AK-47s and RDX explosives. Guns and bombs are nothing new in terrorist operations, but these LeT operatives represented a deeply disturbing new breed of terrorist. They had seen the future and leveraged modern information technologies every step of the way throughout their assault to locate additional victims and slaughter them.

When the attackers set out to sea from Pakistan under cover of darkness, they wore night-vision goggles and navigated to Mumbai using GPS handsets. They carried BlackBerrys containing PDF files of the hotel floor plans and used Google Earth to explore 3-D models of target venues to determine optimal entry and exit points. During the melee, LeT assassins used satellite phones, GSM handsets, and Skype to coordinate with their Pakistan-based command center, which monitored broadcast news, the Internet, and social media to provide real-time tactical direction to its ground assault team.

When a bystander tweeted a photograph of police commandos rappelling from a helicopter onto the roof of the besieged Jewish community building, the terrorist ops center intercepted the photograph, alerted its attackers, and directed them to a stairwell leading to the roof. The police, who had hoped to surprise the terrorists, instead found themselves ambushed inside the stairwell the moment they opened the door. When the BBC mentioned on air that witnesses had reported the terrorists were hiding in room 360 or 361, their war room phoned them immediately and told them to reposition themselves to avoid capture.

At every point during the siege, the LeT attackers exploited readily available technology to gain situational awareness and maintain tactical advantage over police and the government. They monitored the Internet and social media, gathered all available open-source data, and even mounted a sophisticated online counterintelligence operation to protect their operatives. Throughout their assault on Mumbai, the terrorists were so dependent on technology that numerous witnesses reported seeing LeT operatives shooting hostages with the guns in their right hands while simultaneously checking BlackBerry messages with their left.

Not only was technology crucial to the operational success of the siege, but as we learned in chapter 2, criminal abuse of technology also funded the attack. It was a Filipino hacking cell working on behalf of the al-Qaeda affiliate Jamaah Islamiyah that committed extensive cyber crime and online fraud to bankroll the LeT operation in India. The hackers funneled their millions in ill-gotten cyber gains back to their handlers in Saudi Arabia, who in turn laundered the funds and forwarded them to the Lashkar-e-Taiba team responsible for the brutal onslaught against the people of Mumbai.

In the end, it took police sixty-eight hours to end the siege on the city of Mumbai. Counterassault teams eventually killed nine of the terrorists and arrested the tenth. Shockingly, one of the innocents to survive the attack was K. R. Ramamoorthy. At the very moment the LeT command center had given the order to kill him, there was an explosion in the Taj Mahal Palace, which his attackers thought was the police closing in. As the terrorists ran to investigate, they gave Ramamoorthy the brief moment in time he needed to free himself and escape. Not so lucky were the 166 men, women, and children who lost their lives that day, as well as the hundreds more who were gravely wounded as a result of the carnage.

Let us pause for a moment to consider the implications of this terrorist assault. Ten men, armed not just with weapons but with technology, were able to bring a city of twelve million people, the fourth-largest metropolis on earth, to a complete standstill, in an event that was broadcast live around the world. The militants proved fully capable of collecting open-source intelligence mid-attack (traditional media, Internet, mobile, and social data) and using it for synchronous operational decision making. LeT simply processed the data the public was leaking and leveraged them in real time to kill more people and outmaneuver authorities. That was terrorism in the digital age circa 2008. What might terrorists do with the technologies available today? What will they do with the technologies of tomorrow? The lesson of Mumbai is that exponential change applies not just for good but for evil as well.

Data Is the New Oil

Data is constantly being generated by everything around us. Every digital process, sensor, mobile phone, GPS device, car engine, medical lab test, credit card transaction, hotel door lock, report card, and social media exchange produces data. Smart phones are turning human beings into human sensors, generating vast sums of information about us. As a result, children born today will live their entire lives in the shadow of a massive digital footprint, with some 92 percent of infants already having an online presence. From their parents' posting of their first in utero sonogram until the disconnection of their Internet-enabled pacemaker more than a hundred years later, every moment from birth to death will be digitally chronicled and preserved in the cloud in perpetuity. Our data creation cycle never sleeps, and in 2014 each and every minute of every day, we

- sent 204,166,667 e-mail messages
- queried Google's search engine 2 million times
- shared 684,000 pieces of content on Facebook
- sent out 100,000 tweets on Twitter

- downloaded 47,000 apps from the Apple App Store
- uploaded 48 hours of new video on YouTube
- posted 36,000 new photographs on Instagram
- texted 34 million messages on WhatsApp

Put another way, every ten minutes, we created as much information as did the first ten thousand generations of human beings. The cost of storing all these data is dropping exponentially as well. For instance, as of late 2014, a six-terabyte hard drive can be purchased on Amazon.com for just $300 and store all of the music ever recorded anywhere in the world throughout history.

This vast growth in the world's information infrastructure has been dubbed the big-data revolution. The promise of big data is that long-standing complex problems become quantifiable and thus empirically solvable. Consider medicine. As all patient data are cataloged in electronic medical records, it becomes easier for doctors to mine these data sets to identify the most effective treatments, spot deadly drug interactions, and even predict the onset of disease before physical symptoms begin to emerge. Untold lives could be saved.

Across all industries, whether retail, transportation, or pharmaceuticals, there will be tremendous economic value realized as a result of big data, so much so that the World Economic Forum recently dubbed data "the new oil." There is a new-age gold rush afoot in which hundreds of companies such as IBM, Oracle, SAS, Microsoft, SAP, EMC, HP, and Dell are aggressively organizing to maximize their profits from the big-data phenomenon. And if data are the new oil, the modern currency of a digital world, then those in possession of the greatest amounts will have enormous power and influence. Just as the first oil barons, such as John D. Rockefeller and J. Paul Getty, ruled their era, so too will those who possess the largest amounts of data in today's world, as Mark Zuckerberg and Eric Schmidt have demonstrated. Companies such as Facebook, Google, and Acxiom are creating the largest data sets about human behavior ever accumulated in history and can leverage this information for their own purposes, whatever they might be, whether profit, surveillance, medical research, political repression, or blackmail.

But if data are the new oil, then like other long-established natural resources, they must be safeguarded. We don't leave 100 million barrels of oil unprotected, and yet for the most part that is exactly what we do with the vast majority of data created. The protection of our digital information is nowhere near the levels it should be. The 100 million barrels of oil are protected by guards, fences, guns, video cameras, and sensors in the ground and along the pipelines. But what about the 100 million credit cards and customer records stored by retailers such as Target? They, as we have seen, are stored in inherently insecure and poorly defended databases.

When you aggregate such vast troves of valuable data and fail to protect it, what do you think is going to happen? Our ability to capture and store information is greatly outpacing our ability to understand it or its implications. Though the business costs of storing the world's information may be driving toward zero, the social costs may be much higher, posing huge future liabilities for society and our world.

History here can be instructive. Willie Sutton, the famous American bank robber, stole nearly $2 million over his multi-decade career in crime, which began in the 1920s. After his capture by the FBI, a reporter asked Sutton, "Hey, Willie, why do you rob banks?" His answer, oft repeated, was, "Because that is where the money is." Though Sutton could have robbed two million people of a dollar each, he chose a more logical and time-efficient approach, deciding instead to rob the currency aggregators (banks). Thus is it any surprise that criminals are going after Target, Sony, and other data aggregators when the rewards are so high and the risks so low? In today's world, data are where the money is.

Taking a cue from Gordon Moore and his eponymous law, I too decided a dictum was in order to describe the risks associated with the growing mountains of data produced. Thus I present Goodman's law:

The more data you produce and store, the more organized crime is happy to consume.

Eventually, your personal details will fall into the hands of criminal cartels, competitors, and even foreign governments. While big data may be the new oil, our personal data are more like weapons-grade plutonium— dangerous, long lasting, and once they are leaked, there's no getting them back.

Even the federal government is realizing it too can fall victim to this problem. Just look at the 2010 WikiLeaks debacle and the hundreds of thousands of classified diplomatic cables Private Chelsea (Bradley) Manning was able to steal while working as an army intelligence analyst in Iraq. Of course just a few years later, the world would meet Edward Snowden, who used his skills and access as an NSA system administrator to steal millions of highly classified files from America and its allies and share them with journalists for publication online. Some have called this type of mass information theft and disclosure the "civil disobedience of the information age." But if Manning and Snowden could (after purportedly thorough background investigations) amass and steal such vast volumes of sensitive data from the federal government, what might they do if they were working for Target, Citibank, or Apple? The exponential growth in corporate data means that trade secrets, engineering designs, technical know-how, customer lists, employee salary tables, pricing strategies, suppliers, and any other information stored on a digital device can leak. Today, any company, large or small, can have a Snowden in its midst with notable implications for its data security, privacy, and long-term economic viability.

Just one compromised e-mail account on Facebook, Google, or Apple can give hackers access to years of your e-mail messages, calendar appointments, instant messages, photographs, phone calls, purchase histories on Amazon, bank and brokerage accounts, and documents in Dropbox or on Google Drive. It is important to note, however, that the data losses we imagine today will pale in comparison with what becomes possible tomorrow. In this world, our ability to aggregate all information created by both man and machine and store it in perpetuity is far exceeding our understanding of the concomitant risks.

Bad Stewards, Good Victims, or Both?

What I did in my youth is hundreds of times easier today.
Technology breeds crime.
FRANK W. ABAGNALE

When Sony, Target, and T.J. Maxx were hacked, whose fault was it? Were these firms innocent victims of ingeniously inventive cyber attacks perpetrated by sophisticated transnational organized crime groups? Or were they deeply lax with their security precautions, remiss in implementing the most basic of protections for the hundreds of millions of accounts entrusted to them? The answer lies between the two extremes. Not only are retailers doing an ineffective job of protecting their customer data, but so too are legions of Internet start-ups and the behemoths of social media. When you volunteer your data to Facebook, Google, LinkedIn, and others, you need to be aware not only of the numerous privacy ramifications of doing so but of the criminal implications as well. These firms are routinely hacked, and the data taken are yours. How often does this happen? Way more than you might ever imagine.

Facebook's own security department has shockingly acknowledged that over 600,000 accounts are compromised every day. Did you get that? Not 600,000 accounts per year or even per month, but per day. That's one account every 140 milliseconds (a blink of an eye is 300 milliseconds). These data can be used for identity theft, criminal impersonation, tax fraud, health insurance scams, and a host of other criminal offenses. Consider the tremendous volumes of personal data you share on Facebook, and now think what organized criminals might be able to do with them. Mother's maiden name, check. Place of birth, check. Date of birth, check. Photographs of your kids, check.

Compromising your Facebook account is not the end goal; it's just the beginning. Because 75 percent of people use the same password for multiple Internet sites and 30 percent use the same log-in information for all their online activities, once your Facebook account password is compromised, it can potentially be used to access your bank, credit card, and e-mail accounts. In addition, third-party companies are increasingly allowing you to use your Facebook log-on credentials as your passport to the rest of the digital world. While using your Facebook account to shop, listen to music, and play games is greatly convenient, once that single sign-on is compromised, so too are all the other services.

Many social media companies have been breached, including LinkedIn (6.5 million accounts), Snapchat (4.6 million account names and phone numbers), Google, Twitter, and Yahoo! Transnational organized crime groups are responsible for carrying out a full 85 percent of these data breaches, and their goal is to exfiltrate the greatest amount of data possible, with the highest value in the cyber underground. Sometimes organized crime groups don't even need to hack into a computer system; it's already wide open. Just as predators on the plains of the Serengeti won't pass up an already dead animal as a free meal, so too are hackers happy to take advantage of any free data bounty that comes their way. Such was the case, for example, when the mega cloud storage data company Dropbox accidentally turned off the need for any account password whatsoever across its entire network back in 2011. As a result, any person could read any file posted on the Dropbox network.

You might think that if your social media or Internet accounts were compromised in such a manner and you suffered harm, such as identity theft or tens of thousands of dollars stolen from your bank account as a result of somebody else's negligence, you might have recourse to sue those who put your information at risk. Of course, you do not. You waived all of those rights when you clicked "I have read and agree to the terms of service," a caveat that holds these companies completely harmless for such breaches.

And Facebook makes it clear:

> We try to keep Facebook up, bug-free, and safe, but you use it
> at your own risk. We are providing Facebook as is without any
> express or implied warranties . . . We do not guarantee that
> Facebook will always be safe, secure or error-free . . . [Y]ou
> release us, our directors, officers, employees, and agents from any
> claims and damages, known and unknown, arising out of or in
> any way connected with any claim you have.

By the way, it is not just organized crime groups that are going after the massive data repositories you've created with Google, Yahoo!, and Facebook; it is governments, foreign and domestic, as well. For example, in January 2010 Google went public with news of a massive attack across its network and blamed the attack on the Chinese government. Google reported that Chinese authorities were going after the Gmail accounts of activists in the United States, Asia, and Europe who had raised concerns about China's human rights practices. Also targeted in the incident were trade secrets and Google's source code—the very software that runs Google and its products.

Though Google admitted being attacked, the exact extent and nature of what was taken were closely guarded company secrets. Later, however, it was revealed that hackers tied to China's People's Liberation Army (PLA) took the source code for Google's global password management system. The theft of Google's source code could readily have provided the Chinese persistent access to the passwords of millions of Google's customers worldwide and have allowed the PLA to remain hidden within Google's systems on a long-term basis. Have you changed your Google password since 2010? If not, the PLA may have a copy of it. Whether Internet and social data companies are bad stewards of our data, highly targeted victims, or a little bit of both, the fact of the matter is that any data we entrust to the sites and companies could leak to criminals, terrorists, and others.

Data Brokers Are Poor Stewards of Your Data Too

One of the problems with having shadowy and poorly regulated data brokerages amass huge volumes of information on us is that these companies can readily be hacked as well. When firms such as Acxiom store trillions of records on each of us, those records will be targeted by organized crime because, as Willie Sutton reminds us, that's where the money is. This theft of large-scale data sets from data brokers has been going on for many years, and back in 2002–3 more than 1.6 billion customer records were stolen from Acxiom and its clients. According to court documents, the hacker responsible for the theft, Scott Levine, was able to download more than eight gigabytes of Acxiom files, making it one of the largest ever intrusion cases involving the theft of personal data.

More recently, in 2013, the data broker Experian mistakenly sold the personal data of nearly two-thirds of all Americans to an organized crime group in Vietnam. The epic fraud meant that the Social Security numbers of 200 million Americans were now available to the thieves around the world. The data sets obtained were known as "fullz" in the criminal underground because they contain the full set of information required for criminals to apply for credit cards and take out loans in the names of their victims. The massive breach of security occurred because Experian failed to do due diligence on the Vietnamese hacking organization, which had established a front company posing as a U.S. private investigation firm in order to purchase the data for its crime commission. Get that? Experian sold 200 million user data files to an ID theft ring. The data were eventually put up for sale on dozens of hacker Web sites, including SuperSet.info and FindGet.me, selling for just sixteen to twenty-five cents a record, with payment accepted only via untraceable online currencies such as Liberty Reserve and WebMoney. Experian learned of the compromise and its complicity in the affair only after it was contacted by the Secret Service, which discovered the information for sale on the hacker Web sites.

And why on earth would a purportedly reputable firm sell the data without doing due diligence? The answer, as usual, lies in the money. Data brokers make money when they sell data, not when they protect it. In the course of the investigation, it was uncovered that criminals had accessed the Vietnamese data set at least 3.1 million times before it was taken down—but of course by then, the damage had been done.

Given the ready availability of data on any one of us, organized crime groups have now even started their own data brokerages and front companies providing illicitly obtained information on any particular target of interest. An example of this was seen when Russian hackers created a Web site known as Exposed.su to demonstrate their hacking prowess to their fellow criminally inclined buying public—bona fides, if you will. Boasting of their ability to get data on anybody, the hackers freely hosted credit files on a wide number of public figures in politics, law enforcement, and entertainment.

To obtain their ill-gotten goods, the thieves subverted the security systems at Equifax's AnnualCreditReport.com Web site and obtained the full credit reports of all those targeted. Those who fell prey to the attack included a who's who of celebrities, among them Ashton Kutcher, Kim Kardashian, Jay-Z, Bill Gates, Beyoncé, Robert De Niro, Lady Gaga, and Sean Combs. Also breached were the credit reports of a number of extremely high-profile government figures such as First Lady Michelle Obama, Vice President Joe Biden, former president George Bush, the FBI director Robert Mueller, the CIA director John Brennan, and Attorney General Eric Holder, as well as the LAPD chief Charlie Beck.

Once the Exposed.su hacker crew obtained the complete credit reports on those listed above, they posted them in full online in PDF format. There for all the world to see were the Social Security numbers of the victims, their dates of birth, every address they had ever used, personal phone numbers, legal judgments against them, and other personally revealing information such as how many hundreds of thousands of dollars they charged every month on their American Express black cards or how many millions they owed on their mortgages. The credit reports of those affected were viewed nearly one million times before the sites were eventually taken down.

As noted in the last chapter, large data brokers create highly segmented lists of clustered individuals such as "Caucasian, high school educated, rural, family oriented, and interested in hunting, fishing and watching NASCAR." Now it appears as if certain data brokers are also creating lists that are of direct benefit to organized crime groups, who will pay top dollar for such criminal leads. Scammers are a lucrative fountain of revenue for data brokers, and as such the data industry is happy to create lists that cater to criminal customers as well. Though data brokers would disagree or disavow any responsibility for what happens with the lists, clusters of individuals such as " 'gullible' pensioners who 'want to believe their luck can change'" are nothing more than an invitation to defraud senior citizens of their life savings.

For data brokers such as ChoicePoint, Experian, and Equifax, the economic incentives are gravely misaligned from a public risk and security perspective. This is particularly true in the age of big data, when organized criminals now find themselves in the knowledge management business. They are an efficient, effective, and industrious force in the world of big data, and the more we create, the more they are happy to consume.

Social Networking Ills

Social media are great fodder for identity theft, as all the information criminals need to pursue you is freely available online, whether it's your date of birth or your mother's maiden name on your Facebook account. You may be thinking, "Criminals can't see that information . . . I've blocked it in my privacy settings." If only the system worked as advertised. There are many reasons why any information you post to Facebook leaks. First, as noted above, when Facebook updates its ToS, it will often reset your desired privacy settings back to the least private options available, making these data available to anybody, especially its advertisers. Second, with 600,000 Facebook accounts compromised daily, it's only a matter of time before criminals get around to you. Last, given that social data are now "where the money is," criminals have created specialized tools, in the form of targeted viruses and Trojans, to take over your Facebook and other social media accounts without your permission.

At least 40 percent of social media users have been exposed to one form of malware, and more than 20 percent of us have had our e-mail or social networking account compromised or taken over by a third party without our permission. Bad guys trick users into clicking on links in posts and messages that purport to come from friends or colleagues using a technique known as social engineering. Criminals take advantage of the trust we extend to those in our social network by masquerading electronically as these trusted parties, invariably tricking users to click on a link that will ultimately infect a computer with a virus, Trojan, or worm. Moreover, organized crime groups are extremely quick to react to breaking news, which they use to dupe innocent users to click on as a means of infecting them. Whether it was the earthquake in Haiti, Justin Bieber's arrest, or

Miley Cyrus in the nude, the headlines are too good to ignore and thus people click on them. When Malaysia Airlines Flight MH370 went missing over the Indian Ocean, scammers were ready to go with fake photographs of the plane and purported videos showing "MH370 found at sea, shocking video just released by CNN." The messages spread like wildfire over social media to a curious public eager for answers, not realizing they had just infected their machines with viruses. Sometimes curiosity really does kill the cat.

One of the best-known pieces of social media malware was known as Koobface (a variation of "Facebook"), which targeted Facebook users around the world. The malicious social media worm spread by tricking users to click on a Facebook link with an impossibly compelling headline such as "OMG—I just saw this naked video of you!" Who wouldn't click on such a message? Unfortunately, one curious click can lead to a flurry of malware. Once infected, the Koobface worm steals any available log-in credentials it can find on your machine, including those for your Facebook, Skype, Yahoo! Messenger, and Gmail accounts. Koobface can also force your computer to take part in denial-of-service attacks against third parties and hijack your Web search returns and clicks to take you to untrusted Web sites. The malware was designed and disseminated by a hacking group in St. Petersburg, Russia, and though the criminals responsible were individually identified and even publicly named, Russian authorities have refused to extradite them to face trial for their crimes.

Of course these days, social media attack tools have become streamlined, and one need not even be a master hacker to steal information. For example, Firesheep is a simple Firefox browser plug-in that anyone could download to take over the Facebook session of others on the same network and hijack their Facebook accounts. With this, for example, if you checked your Facebook account at a local Starbucks while sharing the network with twenty-five other people in the coffee shop, and if one of them was running Firesheep, the hacker could use the plug-in to log in as you on your own Facebook account. Easy as pie. Once logged in, the crook could see all your personal information, change your account settings, and post anything he or she wanted on your wall or in messages to other users. This technique is sometimes known as session hijacking or "sidejacking" and is ridiculously easy to execute.

Criminals are also targeting users on social media sites via third-party apps and online games, attacks that grant them access to your bank account and can ruin your credit. This was a lesson Lisa Lockwood of Baltimore, Maryland, learned the hard way when the information her seventeen-year-old son had provided a Facebook gaming app came back to bite them both. The game offered the teen extra gaming points in exchange for filling out an account application that asked for his Social Security number. Without thinking, and with visions of more points to "level up" dancing in his head, the teen completed the application, never realizing his Social Security number was about to be used by criminals to complete seven separate car loan applications on his behalf in a matter of days. It was only when the boy's mother received a telephone call from a local Subaru-Volkswagen dealer inquiring about her son's credit application for a new car that she learned of the incident.

Illicit Data: The Lifeblood of Identity Theft

The explosion of data has led to the creation of a brand-new industry for transnational organized crime groups, and mass identity theft is the result. According to the Congressional Research Service, identity fraud cost Americans nearly $21 billion in 2012, and more than 13.1 million Americans are reportedly victims of identity fraud annually. That works out to about one American every two seconds. Furthermore, the theft of this personally identifiable information is a gateway crime that leads to any number of other criminal offenses such as financial fraud, insurance fraud, tax fraud, welfare fraud, illegal immigration, and even terrorist finance. Exponential growth in data is leading to exponential growth in online crime.

Children are the fastest-growing group of victims of identity theft. They are particularly vulnerable because they don't have early-warning systems built in as do adults. If somebody fraudulently charged $500 or $1,000 to your credit card, you would likely notice it on your next billing statement, but children don't get credit card statements. Thieves who pilfer their identities can use them for eighteen years, and only when these young adults apply for credit themselves, such as college student loans, do they learn that their credit history has been destroyed by information thieves.

In the United States alone, 500,000 children are victims of identity theft annually. According to a study of 40,000 children by Carnegie Mellon University's CyLab, kids are shockingly fifty-one times more likely to be victims of identity theft than adults. From toddlers to teenagers, young people are readily targeted because their credit histories are nonexistent and thus a tabula rasa for organized crime groups. Parents don't find out about the crimes and identity thefts until years and years later, when suddenly they are confronted by aggressive bill collectors attempting to collect on their children's unpaid debts. Given the extent to which children and young adults live their lives online, and the aggressive ways in which data brokers and large companies are pursuing them, it is perhaps to be expected that they would face significant threats from identity thieves. If only these financial woes were their biggest problem. As we shall see, the data we all leak can lead to physical dangers as well.

Stalkers, Bullies, and Exes—Oh My!

The volumes of data about you sloshing around online are useful not only for identity thieves but for legions of other criminals as well. Old-world crimes are increasingly enabled by newer technologies, and big data allow traditional criminals to target you with ever-greater precision. Through our persistent 24/7 online lifestyles, we are reachable at all times, even by those who we wish could not reach us. What is odd about this phenomenon is that often we, through our voluntary provision of information or via data leaks, are making it easier for stalkers, harassers, and criminals to find us.

Take, for example, the case of cyber bullying. Though bullying has always been a problem in schools, the Internet provides cyber bullies with instant access to their victims, not just in the school yard, but everywhere, at all times. The threats come online, via e-mail, on social media, on mobiles, and even via messaging apps and games. According to the National Crime Prevention Council, nearly half of all teens are affected by cyber bullying. For young people facing the persistent harassment, it seems as if there is no escape; as a result, a full 20 percent of middle school students admitted to "seriously thinking about attempting suicide" because of online bullying.

Children are not the only ones victimized by cyber bullying; cyber stalking is increasingly affecting the adult population as well. In fact, the ever-expanding flows of data about us and our persistent online presence have helped to transform the Internet into a fertile ground for a new breed of criminal known as the cyber stalker. These offenders use the Internet as "their weapon to harass, threaten and intimidate their prey." Cyber stalkers do this by sending unwanted e-mails, text messages, postings, and tweets and by spreading rumors about the victim online. Using the data that each of us leaks every day or that are commonly available via data brokers, cyber stalkers can easily obtain detailed information about their victims, including their home and work addresses and phone numbers. Often these details are used by cyber stalkers to confront their victims in person.

Facebook has been of particular use to stalkers. With each of us having hundreds of friends, many of whom we've never met, it would be wise to more carefully consider who was actually sending those friend requests. Christopher Dannevig used Facebook to find his victim, Nona Belomesoff, an eighteen-year-old woman from Sydney, Australia, and meticulously studied her profile before contacting her. It was Belomesoff's frequent posting about her love of animals on her Facebook page that gave her stalker the idea of how to persuade her to meet him. Using the social media data the young woman was innocently leaking, Dannevig created a fake profile under the name "James Green" and claimed he worked as an HR recruiter for a well-known local animal rescue group. The woman's stalker used the very details she had posted to con her. After creating the fake pro-file, Dannevig contacted Belomesoff and exchanged a series of messages, eventually befriending her and gaining her trust. Shortly thereafter, he announced there was a job opening at the animal rescue charity for which she would be a perfect fit. Belomesoff agreed to meet him for an interview, and her stalker offered to drive the young woman to the animal shelter located in a secluded area just outside Sydney. Thrilled at the prospect of having found a paying position working with the animals she loved, she agreed to travel with the man. It was there, in the deserted outskirts of Sydney, that Dannevig strangled and murdered the girl.

Though the threat from strangers using our data to find and stalk us is real, it pales in comparison to the perils we face as a result of domes-tic violence and harm by those with whom we once shared an intimate relationship. Facebook makes it easy to keep tabs on a former boyfriend, girlfriend, or spouse out of a normal, though unhealthy, level of prurient curiosity. New friends, life updates, changes in relationship status, travel locations, and vacation plans are all of great interest to former partners. The phenomenon is so common that "Facebook stalking" has even entered the common lexicon.

But for some, the data we leak fuel much more than curiosity on the part of our ex-partners. In those relationships that involved domestic vio-lence in the real world, 45 percent of victims admitted that their abus-ers followed and attacked them online as well, causing many to suffer from PTSD. Social data can also provide details as to their locations, and because abusers often go to great lengths to find their victims, an innocent tweet, check-in, or status update could be as effectively dangerous as a bullet. For instance, Paul Bristol flew from Trinidad to England to stab his ex-girlfriend to death after seeing a post of a photograph with her new boyfriend on Facebook.

Another challenge in the world of big data is that the information we share and intend to keep private leaks out to others. Often we are betrayed by those with whom we have entrusted the most intimate details of our lives, particularly with the photographs we have shared. Sexting, or the sharing of sexually explicit SMS photographs via mobile phones, is a growing phenomenon, and some 67 percent of college-aged students have admitted to engaging in the practice. Unfortunately, photographs shared in such a manner do not just disappear, and these data detritus, like all other forms, often come back to bite their originators in unexpected ways.

Web sites such as MyEx.com allow the jilted to share photographs of their ex-lovers on a single Web site. There are more than seven hundred pages of photographs of naked men and women, with paragraphs of com-plaints about those in the photos—terrible lover, cheated on me with my sister, small penis! Another wildly popular site, IsAnyoneUp.com, was cre-ated by twenty-four-year-old Hunter Moore as a data repository for anyone to submit naked photographs of their exes and enemies and was visited by a quarter of a million people daily. The phenomenon has become so popular it now has its own name: revenge porn. Moore's site by design listed next to each photograph links to the person's Facebook or Twitter account, her full name, and hometown and made this information fully indexical and retrievable in Google so that it would show up in an innocent search by a third party looking for that person. Every naked photograph was accompanied by a comments section that allowed members of the public, and Moore himself, to comment on and ridicule the photographs.

Online Threats to Minors

According to the Pew Research Center, today 95 percent of young people in the United States are online, and 74 percent of teens aged twelve to seventeen are mobile Internet users, often accessing the online world via cell phones and tablets. Moreover, 95 percent of young people aged ten to twenty-three have at least one social media account. Much of this Internet access takes place outside the purview of their parents, 74 percent of whom say they are overwhelmed by modern technology and don't have the energy, time, or ability to monitor what their children do online. That is unfortunate, for although cyber bullying by peers is a major source of stress for young people, they face even greater dangers in our increasingly connected world.

Child predators have used technology to great effect to zero in on children for the purposes of sexual abuse. So common is the practice it even had its own television show highlighting the phenomenon, NBC's *To Catch a Predator*. The challenge for children is that four out of five of them cannot tell when they are talking to an adult posing as a child online. Their new online friend—the other eight-year-old girl one town over—could just as likely be a fifty-year-old man two states away, willing to travel across state lines for the purpose of child abduction.

Given that pedophiles have noted preferences in the children they pursue (age, gender, hair color, height, and so forth), any photograph posted on social media or elsewhere online can be used as a shopping catalog or virtual marketplace for child sex abusers looking for victims to target. Pedophiles make it their business to know the very latest games, messaging services, and virtual worlds of interest to children and will seek out their victims in every possible online forum, using a variety of tools ranging from Xboxes to iPads. Lest you think the demand for such disturbing photographs is limited, law enforcement sources have identified at least twenty-two million such images and videos in the United States alone, and some password-protected child pornography Web sites have up to thirty thousand paying members.

Today the volume of pedophile images is growing, not because an adult has necessarily abducted a child and abused him or her, but because young people are readily targeted via subterfuge and social engineering.

Such was the case with Amanda Todd of British Columbia, Canada, who at the age of twelve was coerced into flashing her breasts on a live video chat site popular with teens known as blogTV. The anonymous person who made the request seemed nice and complimented and flattered young Amanda on how pretty she was. In a moment of teen naïveté, Amanda revealed her breasts, assuming the requester was another teen. As time passed, however, she realized she had encountered a much darker force. A year after her disrobing, Amanda received a message on Facebook from a man under a pseudonym who demanded the young girl reappear nude and perform sex acts on camera for him. If she refused, he threatened to release the original video of her topless. To prove he was serious, the attacker revealed the names of Amanda's friends and family, her address, and the school she attended and said all would be shown her video. The young girl demurred, and the harassment began.

Her tormentor created a fake Facebook profile in Amanda's name and used a photograph of her bare chest as the profile picture. He then began sending friend requests to all of Amanda's friends, family members, and teachers whom he had discovered on her true account. Amanda was unaware of the incident until the police, worried about the implications, came knocking on her family's door at 4:00 a.m. on Christmas Eve. Amanda was horrified. Upon returning to school, she was bullied and harassed relentlessly. For the young teen, the pressure was unbearable. Depression, anxiety, and panic disorders set in. She cried herself to sleep every night and was disowned by her friends, who blamed her for appearing in the video. She ate lunch alone every day and began cutting herself.

To avoid the pain and ridicule, Amanda switched schools and moved to a different town. Sadly for her, the persecution continued. Her stalker followed her activities online and created a fresh Facebook page to advise her new teachers and classmates of her topless video. At the new school, the bullying in class got so bad that a group of girls attacked her on the playground, punching her and pushing her into a muddy ditch. Adding insult to injury, the perpetrators even posted a video of their attack on YouTube. Amanda went home that afternoon and drank from a gallon of bleach in an effort to bring all her pain and suffering to an end. She was rushed by ambulance to a hospital, where her stomach was pumped. Though Amanda survived, the harassment continued. On her Facebook page, other students posted photographs of Clorox containers and encouraged her to "try harder next time." In response, on September 7, 2012, Todd posted a nine-minute YouTube video detailing her struggle with bullying and self-harm. In the deeply emotional video, Amanda shared her experiences of being bullied, while powerfully moving music played in the background. Shortly thereafter, the pain became too much for Amanda to bear, and she committed suicide at the age of fifteen.

Amanda's video went viral after her death and was viewed millions of times. Some police thought that Amanda could have been a victim of "cappers," a disturbing trend in which gangs of online pedophiles revel in coercing kids into stripping on camera and recording them. Worse, the pedophiles then use the videos to blackmail teens into performing more explicit sex acts online and in person. The tragedy of the Amanda Todd affair is multifold. A young girl innocently leaks data about herself online and is stalked on social media and in the real world, leading to her death. But tragic though it may be, this is not an isolated occurrence, and the speed at which the trend is growing is disturbing. Big data bring with them big risks, and even information innocuously shared by an adult can be used by pedophiles.

In 2011, police in Melbourne, Australia, uncovered a number of child molesters targeting lonely single mothers with young daughters by trawling their online profiles and searching references to their children. The goal of the pedophiles was to work their way into the home, usually using a false name and pretense, in an effort to begin a relationship with the child's mother. Once invited into the home and welcomed there, the pedophile would use his time alone to ultimately target the single mom's young daughters. Criminals and predators play by different rules and are happy to use all our data as fodder for a wide variety of unwanted consequences.

Haters Gonna Hate

Your social media profile may also make you vulnerable to another type of attack—a hate crime—one in which bigots, racists, and homophobes target individuals online based on their race, religion, creed, color, gender, or sexual orientation. Such incidents have taken place on Facebook, Instagram, ICQ, Twitter, and numerous other social media services. Facebook was accused of hosting so much violent hate speech that CNN ran a story titled "Facebook/Hatebook?" to document the phenomenon.

Online data allow criminals to locate victims based on the attacker's individual biases. In one case, an assailant in Texas targeted a gay man whom he met on the social media platform MeetMe.com. After arranging a rendezvous with the victim, the attacker kidnapped him, beat him unconscious, bound his wrists, and threw him into the trunk of his car before dumping him on the roadway. Brice Johnson of Fort Worth was charged in the attack and upon his arrest said he just wanted to teach the gay man a lesson but admitted "the joke may have gone too far."

As horrific as the Texas incident was, the volume of hate crimes involving social media in the United States pales in comparison to that in Russia, with thousands of attacks credited to an emerging neo-Nazi youth movement in the country. In a one-hour documentary produced for the U.K.'s Channel 4, reporters documented over fifteen hundred kidnappings in which bands of vigilantes hunted young gay men on the streets and online. The victims, mostly teenagers, are kidnapped, assaulted, and terrorized during their abductions, which are often boldly filmed. The attackers fear no retribution from complicit police in the country and thus have posted videos of their brutal attacks on both Facebook and Instagram in an effort to further humiliate those injured. Despite the volumes of documentary evidence posted online in hundreds of cases, no arrests or prosecutions have ever occurred, even when the victims were murdered or left permanently disabled—an odd incongruity given the ability of Russian police to systemically monitor all Internet activity in the country across all social media channels.

Burglary 2.0

Have you ever casually posted on Facebook that you were going on vacation? A strikingly large percentage of people will talk about their future travel plans online, mentioning how much they are looking forward to their trip to Disney World or their weekend beach getaway. What they don't realize, however, is that criminals are perfectly capable of scraping these data off the Internet and using them for their own purposes (think Goodman's law: *the more data you produce and store, the more organized crime is happy to consume*).

In the old days, if a burglar wanted to target a particular home, he would traditionally look for the telltale signs that the residents were away on vacation—a pile of newspapers in front of the house or a porch light that remained off at night. But even burglars have modernized their tools and are increasingly using technology to find their targets and property to steal. Welcome to the world of burglary 2.0. These criminals are increasingly searching your postings on Facebook, Google+, and Twitter and using the data you leak there for lead generation purposes, just as would any other good salesman or prospector. To highlight this threat, a group of Dutch computer developers concerned by our oversharing created a Web site called PleaseRobMe.com. There they aggregated locational data from people's tweets and Foursquare check-ins and created a searchable database of the information collected. The result: would-be burglars could check by postal code and see who was away and for how long and was thus suitable for burglary. It's criminal target selection at the click of a mouse.

The threat is not purely hypothetical: real-world burglars are indeed monitoring social data. One such example was uncovered in 2010 when a group of local criminals in Nashua, New Hampshire, turned to Facebook to determine when victims were away from their homes. Nashua police discovered that this crime ring checked Facebook updates of their victims before carrying out more than fifty break-ins and stealing nearly $200,000 in property during their burglary spree. These are not your grandfather's burglars, and criminals are rapidly adapting to technologies that help them commit more crime. According to a 2011 study of convicted burglars in the U.K., 78 percent of them admitted monitoring Facebook, Twitter, and Foursquare before pinpointing a specific home to rob. They also admitted using tools such as Google's Street View to scope out the property before-hand and plan escape routes when fleeing from the scene of the crime. The results highlight the ways in which criminals can use the data we are leaking against us.

Another way in which burglars are targeting you is via the locational data embedded in the files you've posted online. As noted previously, these so-called metadata are silently implanted and hidden in the photographs, videos, and status updates you share with others via your mobile devices and reveal the date and time the photograph was taken, the serial number of the phone or camera, and, most important, the longitude and latitude (GPS coordinates) of where the picture was snapped. The metadata con-taining this information, though not immediately obvious when watching a video or viewing a picture, lie there readily accessible by anybody who knows how to download a simple browser plug-in to access them. With any one of hundreds of free tools, suddenly your photographs come to life and magically appear on a Google Map that allows anybody to zoom in on the precise location where the picture was taken. Such is the miracle of cyber casing—using hidden geo-locational data to plan one's crimes.

Those very same metadata are contained in millions of photographs posted to sale and auction sites such as Craigslist and eBay. For example, a photograph of a diamond ring or an iPad posted on Craigslist might have embedded with it the precise location of your home where the photograph was taken. This information allows tech-savvy thieves to use Craigslist as nothing more than a shopping catalog for goods soon to be stolen.

When Keri McMullen and Kurt Pendleton of New Albany, Indiana, decided to sell their plasma TV and stereo system, they posted photographs of the items online. A few days later, the couple mentioned on Facebook they would be attending a concert in nearby Louisville that Saturday night. That's all the information that thieves needed to burglarize the home con-taining the electronic goods they were seeking. The burglars knew they could take their time because the couple would be at the concert for hours. In the end, the couple were robbed of their flat-screen television, two lap-tops, a stereo rack and all its components, and a high-end 35 mm digital camera. This is but one of several ways in which criminals too are engag-ing in e-commerce, cyber casing your home from the inside with the data you are leaking.

Targeted Scams and Targeted Killings

Another way in which criminals take advantage of your online vacation postings and travel status updates is by scamming your grandparents. Yes, criminals monitor your social media and watch you post your vacation photographs online in real time. Once you've done so, con artists analyze your social networks looking for elderly relatives, usually grandparents, whom they notify of your "unfortunate accident." The scam goes something like this: "Hello, Grandma? Yes, I have some terrible news. Your grandson Peter was involved in a terrible accident in Barbados. The hospital won't accept his American insurance and refuses to treat him until we put down $10,000 for the surgery. If you don't help, Grandma, he may not make it." How do criminals get away with this? Because we help them, though unwittingly, through the information we share in the new world of big data. Facebook tells the world—including organized crime—exactly who our grandparents are and how to find good old aunt Margaret to pressure her: "It's looking really bad . . . Peter has gone into a coma . . . Please send the funds immediately!" Hundreds of victims have been defrauded by the scam, and millions of dollars have been sent via Western Union and MoneyGram as a result.

While Internet scammers monitoring your social media accounts may cost you a few thousand dollars, when narcos follow you on Twitter, it may cost you your life. Drug cartels implement a wide variety of sophisticated counterintelligence programs to collect data from social media platforms, blogs, and government tip lines as a means of exposing potential threats.

Comments made online that are perceived to be unfavorable to the cartels are dealt with swiftly. In September 2011, just across the Texas border in Nuevo Laredo, Mexico, residents on their way to work awoke to find two bodies strung up by their arms and legs, hanging in the air from a pedestrian overpass. The victims, a man and a woman in their twenties, had been badly tortured, and the female was fully disemboweled. Above the dangling bodies read an ominous warning on a large posted sign: "This will happen to all the internet snitches . . . Be warned, we've got our eye on you. Signed, Z," a reference to the Zetas, one of Mexico's largest and most violent drug cartels. These cartels are equally savvy with their social media campaigns, uploading photographs and videos of themselves on Facebook and Twitter in the act of decapitating their victims with chain saws and machetes.

Meanwhile, terrorists not only leverage social media for operational purposes, as we saw in Mumbai, but also tweet in real time to drive public opinion and cause more fear among their targets. During the September 2013 assault on the Westgate shopping mall in Nairobi, Kenya, members of the al-Shabab group who carried out the attack live tweeted their slaughter from inside the shopping center. The Somali-based terrorists murdered sixty-three innocent civilians and wounded nearly two hundred more. The group even posted photographs on Twitpic of the carnage inside the Westgate and accused the Kenyan government itself of destroying the shopping center, using the hashtag #Westgate.

Counterintelligence Implications of Leaked Government Data

Organized crime and narcotics organizations are using social media to gather intelligence on government and law enforcement officials. For example, when two Maricopa County, Arizona, sheriff's deputies pulled over a vehicle in 2010 for DUI, a search of the vehicle uncovered a variety of data CDs, including one with the names, photographs, and Facebook profiles of nearly thirty patrol and undercover officers. Non-state actors and hacktivist groups such as Anonymous and LulzSec also go after the social data leaked by government officials.

In a 2012 incident, the hacktivist group LulzSec demonstrated its power to even go after the FBI. Because the hacktivist organization had begun to tap the personal e-mail addresses of individual police officers, especially those working the cyber-crime beat, they were able to intercept an e-mail message notification of a conference call taking place between the FBI, Scotland Yard, and several other global police agencies. The topic of the call? A discussion of "the on-going investigations related to Anonymous, LulzSec, Antisec and other associated splinter groups." Once in possession of the e-mail, the hacktivists merely used the dial-in information and access code to silently participate in the call. As the world's premier law enforcement organizations discussed the case against Anonymous and LulzSec, there the hackers sat on the line listening to police unknowingly brief them on the status of the investigation. The call was even recorded by LulzSec, which then posted it to YouTube, greatly embarrassing the police authorities involved and undermining their investigation.

So No Online Profile Is Better, Right?

Not necessarily. Given all the risks from posting social data online, it may seem as if not participating in Facebook or LinkedIn would be the obvious solution. But a social media boycott brings its own challenges. If you don't own and control your own online persona, it's extremely easy for a criminal to aggregate the known public information about you and create a social media profile for you and use it for a wide variety of criminal activity, ranging from identity theft to espionage. Indeed, there are many such examples of this occurring, particularly for high-profile individuals. For instance, in late 2010, an organized crime group commandeered the identity of the secretary-general of Interpol, Ron Noble, and created a Facebook Web page for him. The criminals took his official photograph from Interpol's own Web site and extracted data from his official biography to build out his fake Facebook profile. The organized crime group began friending other senior law enforcement officials around the world in the persona of Noble and posed operational questions of them via the social media service. In particular, criminals posing as Noble tried to gather intelligence regarding Operation Infra-red, an Interpol global undercover operation to locate and arrest high-priority international fugitives. It is unclear how many fell for the ruse and how much data is shared, but dozens of the world's most senior police officials accepted the purported friend requests.

The Spy Who Liked Me

Industrial espionage too has found a powerful ally in social networks. In the second chapter of this book, we learned about the Massachusetts wind-turbine firm AMSC, which lost nearly a billion dollars of valuation after its computer source code was stolen via a Chinese espionage operation. What wasn't explained, however, is how the attack was carried out.

When the Chinese officials decided to purloin the source code on behalf of Sinovel, a state-owned company whose wind turbines were supplied by AMSC, a simple check of LinkedIn would have provided their agents with access to the roster of employees working at the Massachusetts firm. Once the Chinese had completed a review of all employees and their positions, a list was generated highlighting those targets likely to have best access to AMSC's highly prized source code. One such person identified was a Serbian engineer working for AMSC's office in Austria by the name of Dejan Karabasevic.

The Chinese began monitoring Karabasevic across a variety of social media sites such as LinkedIn, Facebook, and Twitter. They learned he was going through a nasty divorce and had recently been demoted at work—the exact types of vulnerabilities any modern intelligence agency looks for when targeting potential recruits. Through his various postings, the Chinese were able to re-create Karabasevic's "pattern of life"—plotting on a map his favorite coffee shops, gyms, and restaurants, his home and office, his travel times, and his daily routines. They also learned that he had a penchant for Asian women. Armed with all of this information, the Chinese began their recruitment process.

Chinese handlers approached Karabasevic and offered him a "consulting" opportunity to work with them. In the end, they were readily able to persuade Karabasevic to provide the source code (secret sauce) that allowed

Sinovel to build its own wind turbines without AMSC. Most important for Karabasevic, his Chinese handlers established an office in Beijing for him and promised "all the human contact he could want . . . in particular with female co-workers." After the theft occurred, hundreds of Skype chats and e-mail messages were uncovered between Karabasevic and his Chinese handlers. In one note Karabasevic wrote, "All girls need money. I need girls. Sinovel needs me." To allay his financial concerns, meet his companionship needs, and bolster Sinovel's bottom line, the Chinese offered Karabasevic $1.7 million for the source code. The economics of the deal were fascinating and instructive: Karabasevic receives $1.7 million; AMSC loses a billion dollars in valuation and intellectual property, which Sinovel captures by purportedly selling pirated AMSC products around the world. A great return on investment for Sinovel and for those unencumbered by the moral implications of such dealings.

As should be obvious by now, there are numerous risks from the exponentially growing sea of data in which we find ourselves. Not only do we face an onslaught of data mining from Internet companies, marketers, and third-party data brokers, but criminals, terrorists, and rogue governments too have us under constant assault and surveillance, forever aggregating and amassing. Those trails of data detritus, however, are growing exponentially longer thanks to the computers we're now carrying with us at every turn—our mobile phones.

CHAPTER 7

I.T. Phones Home

**Mobile phones are one of the most insecure devices that were
ever available, so they're very easy to trace and they're very easy to tap.**
EVGENY MOROZOV

On March 21, 2002, Milly Dowler, a thirteen-year-old from Surrey, England, phoned her father to say she'd be home soon. Hours later, the teen still hadn't arrived, and calls to her cell phone went unanswered. By the following evening, a massive search of the area was under way, and Milly's disappearance had made national news.

As part of their investigation, the Surrey police accessed the mobile phone voice mail of the missing girl in their search for clues. Ongoing checks with her mobile phone carrier revealed that five days after her disappearance, the voice-mail system had been accessed, and a new message that had arrived that day was played by parties unknown. The discovery gave the Dowlers hope that their daughter was still alive. As the weeks dragged on, additional messages left on Milly's voice mail continued to be retrieved and deleted, leading investigators to question whether the missing girl was in fact a runaway.

Sadly for the Dowlers, Milly was no runaway but was abducted, her body discovered twenty-five miles away from where she was last seen alive six months prior. In an instant, Milly's case was declared no longer that of a missing person but a full-blown murder investigation. But one fact continued to confound police: Who had been repeatedly accessing the young girl's mobile phone long after she was missing and now presumed to have been dead? Was it the killer? A jealous boyfriend? Her parents? For nearly a decade, the haunting question went unanswered until June 2011 when the mystery was finally solved. The culprit was one nobody would ever have guessed.

In a lengthy article published by the *Guardian* newspaper, it was revealed that Milly's phone was among those targeted by Rupert Murdoch's *News of the World* in a scandal dubbed Hackgate by the British press. Milly's phone was hacked not by the killer, her parents, or a boyfriend but by those looking for a scoop for their tabloid pages. Poor Milly and the Dowlers were not the only victims of Hackgate; so were numerous celebrities, politicians, and even members of the British royal family, which might make sense given their high public profiles. But ultimately, it was discovered that reporters and private investigators hired by the *News of the World* had extended their mobile data theft operations well beyond that of high-profile public figures. Shamefully, they had also hacked into the mobiles of the relatives of British soldiers killed in Iraq and Afghanistan, as well as victims of the tragic 7/7 London terrorist bombings. The appalling details of the case led to a global public outcry and the closure of Murdoch's *News of the World* after 168 years of continuous publication. Dozens of employees and contractors of the paper were arrested, including the private investigator it had hired to gain details of the young girl's disappearance.

Of course for two grieving parents, the sanctions and arrests of those involved were of little comfort. For the Dowlers, the news that it was a newspaper that had broken the security of Milly's phone was unfathomable. Had this unlawful mobile phone hacking of a missing thirteen-year-old girl somehow impeded the investigation of their missing daughter's whereabouts? What police resources were wasted trying to get to the bottom of what appeared to be a prominent clue possibly left behind by Milly's suspected murderer—precious time wasted that might have prevented her tragic and untimely death. We will never know, nor will the Dowlers, who have to live with the tragedy and the burning question every day for the rest of their lives.

While the actions are indeed deplorable, sadly they are all too easy to perpetrate. Our mobile phone security—that of the device that most modern citizens hold most dear—is a farce, easily exploited by organized crime, stalkers, terrorists, and even journalists lacking a moral compass or a shred of decency.

Mobile Phone Operating System Insecurity

Mobile phones are becoming our computers of choice. These "snitches in our pockets" serve as constant beacons of our activities and our locations. Just as mobile phones provide a treasure trove of data to advertisers, so too do they for criminals. Even worse, mobile phones may be the most inse-cure of all devices. The software is notoriously easy to subvert, the risks are poorly understood, and the systems for device protection are imma-ture and wholly underdeveloped. As a result, smart phones are among the easiest devices to hack. While law enforcement and security services have been able to target and tap mobile phones for years, now the very same techniques are readily available to criminal enterprises and everyday hack-ers as well.

Today, there are viruses and Trojans specifically designed to give attackers access to your cell's microphone, recording any sounds nearby, even when you are not on a call. Anything you do or any data you store on your mobile phone—your entire text-messaging history, your address book, photographs, call logs, social networking passwords, and account information—can all be intercepted, hacked, and forwarded to criminal organizations for future exploitation.

Mobile phone malware can be used to track your persistent location and allow criminals to see your position in real time, conveniently plotted on a Google map. Even your smart phone's video camera can be turned on (without any warning light) to record you. There are so many YouTube videos, instructional Web sites, and prefabricated criminal software pro-grams for sale that even novices can hack a mobile phone. In fact, it's often as easy as sending an infected SMS message to your target.

One might legitimately ask, how can mobile phones be so readily com-promised? The answer is that it's all about the operating system. Mobile phone operating systems are newer than their long-standing desktop coun-terparts and even more insecure. Criminals fully recognize that the world of big data is going mobile, and that is where they are concentrating their efforts to ensure the largest return on their malware investments. Mobile is the platform of choice. It is intimate and always with us, and criminals are adapting and innovating with alacrity.

By 2014, McAfee had already identified nearly four million distinct pieces of mobile phone malware, a 614 percent increase over the previous year. Moreover, according to a study by Cisco (and widely touted by Apple's senior vice president of worldwide marketing, Phil Schiller), 99 percent of all mobile malware is targeted against Google's Android mobile operating system. The findings are deeply troubling, especially given that 85 percent of smart-phone handsets shipped worldwide as of mid-2014 were Android and that it is expected that one billion additional Android mobile devices will shipped by 2017. To be sure, the open-source nature of Android's operating system is one of the platform's biggest selling points, but with such openness and the ability to customize the free software as one sees fit comes a huge liability—security. The majority of device manufacturers and phone carriers simply implement the software poorly.

So what is it that makes it so easy to pilfer data from Android devices? Simply stated, it is a lack of updates and bug fixes to the mobile phone's operating system. New versions of Android dribble out to users from carriers as a means of forcing upgrades. In addition, carriers and handset manufacturers need to tweak each installation of Android and customize it to work with individual mobile phone models, an expensive and time-consuming process resulting in far fewer updates per device in the Android world. Worse, according to several studies, it is this customization process and the insecure software added by mobile phone companies and handset manufacturers that lead to 60 percent of the security threats in the Android ecosystem. All those annoying apps and skins that come with your phone are known as bloatware, because they take up space on your device, are of dubious value, and serve as little more than marketing gimmicks for your handset manufacturer or wireless carrier. Not only are they bothersome, but their poorly thought-out implementation leads to the majority of security threats on Android devices.

By comparison, Apple controls its entire hardware and software ecosystem. As such, it can ensure that its mobile iPhone operating system (iOS) software works more seamlessly, and carriers are prohibited from fundamentally altering the underlying operating system with their bloatware. A comparison between Android and iOS clearly tells the story: five months after its release in 2013, 82 percent of Apple's 800 million mobile devices were using 7, the most up-to-date version of its mobile operating system. Only about 4 percent of Android users were running the latest version of Google's Android, also released the same year. What is deeply frustrating about these figures is that if all Android users were simply upgraded to the latest version of the mobile phone operating system, a full 77 percent of security threats could be largely eliminated. It is the failure of Google and its partners to make security updates widely available to its user base that allows criminals the time they need to find hole after hole in the Android OS and target it for exploitation.

Mind the App

App makers such as Rovio, Zynga, and Snapchat aren't the only ones creating apps as a means of acquiring and selling your data; organized crime groups have also adopted the practice. Though one might logically presume that any app submitted by a developer to Google's Android or Apple's App Store would have undergone a vigorous security review of its computer code and its developer, all is not as it seems. With more than a million apps in both the Android and the iOS ecosystems, strikingly little if any human verification occurs—a fact known by criminals, who have subverted the mobile app stores on numerous occasions. Instead, computer-automated algorithms do all the heavy lifting in the review process, and the app store creators just hope it all works out.

As a result, mistakes are common, and apps that contain malware are increasingly hosted in what you would presume to be a reputable app site. By 2013, more than forty-two thousand apps in Google's store had been found to contain spyware and information-stealing Trojan programs. The malware in these apps specifically targets the data on your phone, particularly financial information. Just days after the original Android Market app store had launched, criminals had uploaded fake banking apps for major financial institutions around the world. The apps were deeply realistic and used the correct bank logos, fonts, and color schemes to add to their credibility. Tens of thousands of people were tricked into downloading the apps, and when they failed to work, angry customers called their banks, only to find out "we don't have an Android app yet."

Cyber criminals have retooled their operations to create many more fake banking apps. Though only sixty-seven banking Trojans had been identified in 2012, the number had grown to more than thirteen hundred by the end of 2013, according to Kaspersky Lab. To date, mobile malware packages have been uncovered targeting customers of the world's largest banks, including Citibank, ING, Deutsche Bank, HSBC, Barclays, and sixty-six other financial institutions around the world.

And malware is even more rampant in third-party app stores. While there is at least some limited algorithmic security screening in the official Android marketplace, there is often none whatsoever in third-party sites. As a result, more than five hundred such third-party app vendors have been found to be offering Android apps containing malware. Because there are no security reviews in these app stores, apps containing viruses and Trojans can have near-infinite shelf lives, providing lifetime annuities to the criminals who create and upload them.

While much less common, malicious apps have also been found in Apple's App Store. Although the Apple iOS ecosystem is tightly regulated and controlled, many users find the environment too stifling. When they initially purchase their products, iPhone users cannot customize their keyboards, change their default browsers, manage files locally, or add widgets to their home screens—all features standard in Android. To get around these limitations, many users "jailbreak" their iOS devices, using specialized software to hack their own mobiles in order to obtain root administrative access to phones and control over features locked down by Apple. Jailbreaking an iOS device allows users to gain access to thousands of software programs not officially approved by Apple. Nearly ten million iOS devices have been jailbroken, and their users have turned to third-party app stores like Cydia to download their apps. While jailbreaking these devices provides much greater control to users, it also opens iOS mobile devices to the same security threats common in the Android ecosystem, including a variety of financial frauds.

Why Does My Flashlight App
Need Access to My Contacts?

Hundreds of millions of smart phone users around the world have down-loaded that ever-popular and convenient flashlight app. It's so useful when looking for keys in a purse or trying to open the door late at night, and most of us paid absolutely nothing for the privilege. But why does your flashlight app need access to your contacts? Why does it ask for my loca-tion? My location should be obvious: I'm in the dark; that's why I need a flashlight app! As it turns out, the majority of these apps, especially in Android, are just convenient mechanisms to steal your data, download all your contacts, ascertain your persistent location, install keystroke log-gers, and capture your financial information. As a result, we are seeing the "app-fication" of crime, criminal acts reduced to the simplicity of a mobile phone application.

The permissions granted to these apps, especially in the Android eco-system, where there is no way to deny a specific permission to a given app prior to installing it, mean you and your data are at risk. App permissions on mobile devices are akin to terms of service: we all click yes but never really stop to think about the implications of our decision. The reality is that permission means a rogue or criminal app developer now has the authority needed to commit fraud or steal from your bank account via your mobile device.

Criminals are also creating fake apps specifically to commit telecom-munications fraud. Three-quarters of all cell-phone malware exploits loopholes in mobile payment systems by sending fraudulent premium-rate SMS messages, each one generating $10 in immediate profit. Multiplied by hundreds of thousands of fake SMS premium messages, the money gener-ated is huge. In one incident, scammers were able to post fake versions of wildly popular games such as *Angry Birds* and *Assassin's Creed* into an app store. Once the app was downloaded, every time the user opened it, it would send three premium SMS messages without the user's knowledge at $7.50 per message. In just a few hours, thieves generated tens of thousands of dollars of fraudulent charges.

Hijacked mobile phones are increasingly being used to send spam e-mail messages as they join so-called botnet networks. Botnets are a collection of enslaved, malware-infected computers that work in unison, under the control of hackers or criminals, to pump out massive amounts of spam or take part in DDoS attacks, unbeknownst to the device's legitimate owner. While botnets were previously limited to desktops or laptops, now millions of mobile phones have also been commandeered, and these drone devices are under the full control of criminals and hackers who have joined them to their exponentially growing "zombie networks." These massive networks of hacked devices lie in wait, ready to be unleashed against any target at a moment's notice. Given that mobile device shipments are outpacing desktops and laptops by a factor of ten to one, it is clear that the future of computing is mobile. As such, criminals have figured out that the future of data theft, DDoS, and malware is mobile too.

Even legitimate apps can put you and your data at risk if the software is poorly written or contains undetected security vulnerabilities. Such was the case with the highly popular "social photo booth" app known as Snapchat. Snapchat is a service that allows users to send "selfie" photographs (often involving nudity) that purportedly disappear in just a few seconds after arriving on the recipient's phone. More than one billion photographs have been sent via the service, and in late 2013 Facebook unsuccessfully tried to buy the company for $3 billion. In early 2014, it was revealed that Snapchat contained a security flaw that exposed millions of iPhone users to denial-of-service attacks.

The vulnerability meant that hackers could target your phone specifically by sending a thousand Snapchat messages in just five seconds, thereby crashing your phone and making it unavailable for your use until you performed a hard reboot of the device. Moreover, hackers were also able to compromise nearly five million Snapchat user accounts and published a database of user names and phone numbers on a hacker Web site. Worse, it was revealed that Snapchat's foremost feature—the ability to send naked photographs that would self-destruct in ten seconds or less—was also flawed. The images did not self-destruct as promised and could still be retrieved both on the recipient device and on Snapchat's own computer servers. As a result, tens of thousands of Snapchat photographs thought to have been deleted have shown up across the Internet, reposted on Instagram and on numerous revenge-porn sites. The photographs have subsequently been used for the purposes of extortion and other criminal offenses.

Mobile Device and Network Threats

The emerging threats to the data we carry on our mobile devices are not only affecting consumers but also having a major impact on business. In today's modern enterprise, BYOD, or bring your own device, has become the standard and allows employees privileged access to sensitive corporate data and applications from their own personal mobile devices. Today 89 percent of employees are accessing work-related information on their mobile phones, and 41 percent are doing so without permission of their companies.

The phenomenon, now standard practice in the workplace, means more and more corporate information is at risk thanks to point-and-click spyware attacks against mobile devices. Even when a corporate network is locked down and protected, personal mobile phones are an easy place from which to pilfer data. Criminal organizations won't waste their time going to the most secure place where you have stored your information; they will always go after the weakest link in the chain to get what they want.

Criminals are growing increasingly inventive in how they target the information on your mobile device and even the phone networks themselves. For just a few hundred dollars, criminals can purchase and set up a femtocell, a wireless network extender to help people improve mobile phone service in areas with poor network signals. The device is in effect a mini-mobile-phone tower, and criminals can hack it to trick your mobile into believing it is legitimate, when in fact it is merely connecting to a portable mobile phone tower run and operated by criminals. Their goal in doing this? To capture all the data that is sent from your mobile, such as the password you type for your bank account or sensitive e-mails you might be sending. Rogue femtocells can be particularly useful for industrial espionage, and hackers need only set up the device outside the perimeter fences of your corporation to take advantage of the data coming off the mobile devices of hundreds of your employees. Other prime targets would be airports and big conferences where lots of businesspeople congregate. As it turns out, you are not the only one with an affinity for the data on your smart phone.

Hacking Mobile Payments

Of course, today's mobile phones are just at their earliest stages of development, and many new sensors, such as radio-frequency identification (RFID) and near-field communication (NFC), will bring new capabilities to mobile phones, as well as new vulnerabilities. One area in which this will be most clearly seen is in the disappearance of physical currency. The future of money is mobile and virtual, and a bevy of new sensors and apps are on track to replace your wallet and the cash in your pocket. In fact, some mobile phone providers, such as Safaricom in Africa, dominate the overall payment space. In Kenya, for example, 25 percent of the nation's GNP is actually transacted on Safaricom's M-PESA payment system. Mobile money payment systems, which did not even exist at the turn of the last century, are now available in over seventy countries and are used to move billions of dollars every month. In particular, they have been incredibly useful in getting previously "unbanked" populations in the developing world access to the global world of commerce with significant positive impact for local economies.

In the developed world, there has also been a rush to adopt and deploy mobile phone payment systems. MasterCard and Visa have implemented numerous NFC payment programs that allow users to launch an app on their phones and wave or tap the device on a contactless sensor to quickly charge goods and services. From Starbucks, to Best Buy, to parking meters in San Francisco and cabs in New York City, "wave and pay" is increasingly the choice of users for quick checkout and payment. Though Google was an early adopter of NFC payment systems for its Android phones, in September 2014 Apple joined the bandwagon and added swipe-and-pay technology to its latest batch of iPhones. Within the Android ecosystem, Google's Wallet payment system allows users to store their debit and credit card information with Google and launch the Google Wallet app to check out in an increasing number of stores via any PayPass-enabled store checkout terminal. Google Wallet works with the NFC chips on a wide variety of mobile phones from HTC, LG, Motorola, and Samsung.

The money as represented on these mobile devices is nothing more than data—data that are stored in vulnerable applications, controlled by deeply vulnerable mobile operating systems, using insecure sensor technologies and sensor data-transfer protocols. The obvious result? The future of mobile money may also be the future of mobile pick pocketing. The Google Wallet system has already been subverted by criminals on numerous occasions, and apps such as Wallet Cracker allow anybody to see a user's personal identification code (PIN) number for the system on demand. Moreover, if and when a user loses his or her Android phone, any pre-stored money in the user's Google Wallet (data on the device) can readily be spent in a store by the person who happens to steal or find the device. With the rise of NFC applications, and Apple's notable entrée into mobile payments, we will undoubtedly see growing hacker attention targeting these and other sensors embedded in mobile devices, including GPS.

Your Location Becomes the Scene of the Crime

Advertisers and data brokers are not the only people interested in persistently tracking your location. Criminals, fraudsters, and stalkers have also found good use for the GPS chip on your smart phone. Often hackers can merely piggyback on the good work already done by data brokers as a means of subverting the data you are already leaking. Take, for example, the location-based dating application Tinder, which we discussed in chapter 4. Given the volumes of data, salacious photographs, and potential sex partners, it was not surprising that hackers worked to discover a security vulnerability in the app that allowed anybody to uncover the real-time location of any other users within five feet, information that was designed to be kept private. The best-case scenario with this locational data is that they lead to a positive dating experience. The worst-case scenario is that they and locational data generated by rogue apps such as Girls Around Me could prove a tremendous tool for stalkers, rapists, and potential child abusers. In fact, in 2012 police in South Australia warned members of the public that pedophiles were using geo-tagging data embedded in photographs of children posted online to track down potential targets, putting the subjects of these photographs at risk.

Increasingly, mobile data are being used to ill effect in cases of relationship discord and domestic violence. In 2012, the U.S. Department of Justice revealed that there were 3.4 million victims of stalking annually among those hundreds of thousands who were tracked by spyware and GPS hacks. Welcome to the world of point-and-click surveillance. To be clear, using such spyware is considered an unlawful interception under federal law and is illegal, but the tools are widely available to even novice hackers—or exes—with no prior experience. One product, Mobile Spy, will turn any phone into a bugging device, allowing for ambient recording of the surroundings, even when not on a call. The company also makes an "iPad monitor" product, and all of its software includes a "stealth camera" mode that allows third parties to remotely activate and monitor your camera in real time and store any photographs or videos they choose to record from your device on a central server for later download. Another product, Mobistealth, was used in 2011 by the convicted murderer Simon Gittany, a jealous and abusive boyfriend, to monitor the mobile phone activity of his fiancée, Lisa Harnum, in Sydney, Australia. Thus when Harnum sent an SMS message to her girlfriend confiding that she planned on leaving her abusive relationship, Gittany was immediately notified via Mobistealth of her intent on his own mobile phone. Furious with her plan to leave him, Gittany drove to her home and, in the altercation that ensued, threw her off the fifteenth-floor balcony of her apartment.

But in some cases, domestic abusers did not even need to add third-party spyware to a phone: they just activated the AT&T FamilyMap program offered as a service by the wireless carrier, which allows a cell-phone account holder to track all the devices on his plan. Using his wireless carrier's family mapping service, Andre Leteve of Scottsdale, Arizona, was able to locate his estranged wife and their two children, whom he murdered. Today it's no longer even necessary to pay for such a service from carriers like AT&T; the features are already built in to both iOS and Android devices, services with names such as Find My Friends and Find My Phone, which can be turned on to remotely track others. To help combat these threats, domestic violence shelters have learned the moment a new client arrives at their facilities to take away her mobile phone, remove the battery, and disassemble it to avoid acting as a beacon to potential stalkers and violent individuals. It's not just victims of domestic violence who need to worry about inadvertently sharing their locations; even soldiers on the battlefield have cause for concern as terrorists monitor their online activities for potential avenues of attack.

"Is a badge on Foursquare or a check-in worth your life?" That question, now commonly asked by the U.S. Army of its soldiers, is not rhetorical when even terrorists are taking advantage of geo-tagged data. For instance, when American military forces received a new fleet of AH-64 Apache helicopters at their base in Iraq, some deployed soldiers uploaded photographs of themselves in front of their new choppers to Facebook. Unbeknownst to them, their phones had accidentally embedded their GPS coordinates in the photographs. Not only were insurgents monitoring the soldiers' Facebook postings, but they were also downloading the photographs and analyzing them for useful intelligence. The longitude and latitude information embedded in the photos allowed the terrorists to launch a series of precise mortar attacks that directly targeted and destroyed four of the newly arrived Apaches on the compound.

Not only can we be tracked via the data we are leaking from our mobile phones and embedded files, such as photographs and videos, but increasingly we are leaking our locational data in the physical world. GPS bugging devices are cheap to buy online and are even available for sale in the ubiquitous *SkyMall* magazine available on every flight we take. In that catalog, Tracking Key sells a GPS hardware bug that attaches via magnet or Velcro to any car and allows owners of the device to replay on an online map everywhere the targeted vehicle drove, including speed of the vehicle, determined at one-second intervals. "Useful to see whether your teenager was speeding, where your spouse is going or where your employees are driving." Previously, such high-tech gear would only have resided in a spy agency or with the FBI, but now, given the exponential drop in pricing of these technologies, even a neighborhood mom can spy on her kids or potentially cheating spouse.

In the world of big data, we can even leak our physical location without a bugged mobile phone or GPS tracker hidden in our car. A new technology, known as an automatic license plate reader (ALPR), allows both governments and individuals to use video cameras and optical character recognition to record the locations of cars as they pass from one camera point to another, revealing the real-time movement of any vehicle throughout a city or country with great detail. From Minnesota to New Jersey, and from Ankara to Sydney, hundreds of millions of individual license plate records have been stored. As a result, a query can be applied against these massive databases to determine the position of any vehicle over time. Interestingly, those being photographed are not charged with or suspected of any crime in the overwhelming majority of cases, but these data are being stored nonetheless, because they might be useful for a criminal investigation at some point in the future.

ALPR units are also being mounted on police cars and even tow trucks, vastly expanding these databases. Private companies such as Digital Recognition Network of Texas and MVTRAC of Illinois are also building massive databases of ALPR data, which they sell to agents in the vehicle repossession business. This way, if somebody falls behind on his payments, these companies know all the locations the vehicle has been and can send out a tow truck to repossess it. Just as Google Street View cars are driving up and down our city streets recording video of all they see, so too are private ALPR companies. They want to track your car and place its location in front of your home, at your workplace, and at all the places you shop. These data too of course will be monetized, and the practice in 2014 is entirely legal. But as these massive databases of ALPR mushroom, so will the criminal and privacy risks.

If Experian and Acxiom can suffer data breaches or sell their data sets to criminal organizations, why would ALPR vendors be any different? As a result, even victims of domestic violence who had no online presence and did not even carry mobile phones could still be tracked by where they drove in their cars. We've seen abuse of ALPR data in the past, way back in 1998, when a Washington, D.C., police lieutenant used his computer system to identify the owners of vehicles parked in the parking lot of a popular gay bar in town. He then used the data to extort the men, threatening to out them, unless they paid him a bribe. While the nature of the threats to ALPR data may be different today, they no doubt exist. How might this information be used in divorce cases (his car was parked at the home of the other woman) or by health insurers (we see his car parked at the bar five days a week)? There are other risks as well: ALPR systems are not flawless in their reading of license plate data, and errors can lead to grave consequences. In 2009, a forty-seven-year-old woman was pulled over in San Francisco by multiple police cars at gunpoint with six police officers pointing weapons at her—all because the ALPR system misread a single digit on her license plate, flagging her car as stolen, when in fact the woman was just out to buy groceries.

Even retailers have begun to pilfer our locational details in new and unexpected ways. The Nordstrom department store, for example, recently began tracking its customers via their Wi-Fi signals and MAC addresses on their smart phones when shopping in its stores. As you walked through the stores, Nordstrom could digitally follow you to see how much time you were spending in ladies' underwear versus men's shoes. The high-end retailer contracted with Euclid, a company specializing in helping retailers track customer movements via in-store Wi-Fi connections. To date, Euclid has fingerprinted and tracked more than fifty million mobile devices in the four thousand locations using its service, including hundreds of national retailers such as Home Depot; yes, the same company that leaked fifty-six million credit cards because of a data breach in September 2014 wants to collect even more data on you and your location within its stores. Absent any regulation of the phenomenon, shopping under surveillance will undoubtedly become the new norm, and technology is increasingly turning to tracking people off-line in real space.

At Nordstrom, the only notification customers were given about its use of the new tracking technology was a small, well-hidden sign, barely visible at the entrance to stores. The verbiage on the sign made it clear that this was an opt-out model only, meaning that if you did not want to participate, you had only two choices: don't come in the store, or turn off your cell phone. Data obtained from services such as these can and will be stored in perpetuity. As a result, your spouse's divorce attorney will be able to subpoena Nordstrom and Euclid to see if you and your mistress were in the same store together buying intimates. Your boss will be able to contract with a data broker to find out your location that day you called in sick: "If you were sick, why were you (and your cell phone) at the movie theater and Hooters that afternoon?" Worse, criminals will gain access to all this information over time via the digital underground and use it to blackmail, bribe, and stalk targets of their choosing.

Even Disneyland, the "Happiest Place on Earth," is turning to location-based technologies to track its guests using bracelets called MagicBands, RFID-chip-enabled devices that allow Disney to track its guests throughout its parks. Its goal is to use big data to maximize your stay (and spending) in the Magic Kingdom. As goes Disney, others are likely to follow, and you can expect such human-tracking technologies to be deployed at casinos, resorts, and even airports in the future.

Cloudy Weather Ahead

Though massive amounts of data are leaking from our mobile devices, an increasing number of big-data risks come from the world of "cloud computing." The cloud refers to the massive network of computing resources available online and the practice of using these remote servers to store, manage, and process the world's information. The changing paradigm in computing means that less information is stored locally on our machines and is instead being hosted elsewhere on earth. We mostly do not buy software anymore; we just rent it or receive it for free using a new business model known as Software as a Service (SaaS).

On the personal front, cloud computing means Google is storing our mail, Instagram our photographs, and Dropbox our documents—not to mention what mobile phones are automatically uploading to the cloud for us. In the corporate world, enterprise customers not only are using

Dropbox but also have outsourced primary business functions that would have previously been handled inside the company to SaaS providers such as Salesforce.com, Zoho.com, and Box.com. From a crime and security perspective, the aggregation of all these data, exabytes and exabytes of it, means that our most personal of information is no longer likely stored solely on our local hard drives but now aggregated on computer servers around the world. By aggregating everybody's important data, financial and otherwise, on cloud-based computer servers, we've obviated the need for criminals to target everybody's hard drive individually and instead put all the jewels in a single place for criminals and hackers to target—think Willie Sutton and his love of banks.

The cloud is here to stay, and at this point there is no going back. In early 2014, Google decreased the pricing of its cloud storage offerings by nearly 70 percent, to just $0.026 per gigabyte per month (just under three cents versus the $437,000 it cost in 1980). The move sent shock waves through the industry, and a price war ensued with the cloud storage giants Amazon and Microsoft also joining the fray. The availability of such cheap computing resources and a growing array of SaaS offerings will have untold positive impact on personal productivity, entrepreneurship, and innovation, which in turn will only hasten the inevitable transition to cloud computing. But with this move to store all available data in the cloud come additional risks. Think of the largest hacks to date—Target, Heartland Payment Systems, TJX, and Sony PlayStation Network. All of these thefts of hundreds of millions of accounts were made possible because the data were stored in the same virtual location. The cloud is equally convenient for individuals, businesses, and criminals. To deal with these risks, organizations such as the nonprofit Cloud Security Alliance have been formed to promote best practices and improve security in the age of cloud computing.

The virtualization and storage of all of these data are highly complex and raise a wide array of security, public policy, and legal issues. First, where exactly is this magical cloud storing my data? Most users have no idea when they check their status on Facebook or upload a photograph to Pinterest where in the real world this information is actually being stored. That we do not even stop to pose the question is a testament to the great convenience, and opacity, of the system. Yet from a corporate governance and personal risk perspective, whether your data are stored on a computer server in America, Russia, China, or Iceland makes a difference.

The corporate and individual perimeters that used to protect our information internally are disappearing, and the beginning and end of our computer networks are becoming far less well defined. It's making it much harder to see what data are coming and going from a company, and the task is nearly impossible on the personal front. The transition to the cloud is a game changer for security because it completely redefines where data are stored, moved, and accessed, creating sweeping new opportunities for criminal hackers. Moreover, the nonlocal storage of our data raises important questions about our deep dependence on cloud-based information systems. When these services go down or become unavailable via DDoS attack, or you lose your Internet connection, your data become unavailable, and you are out of business.

As Mat Honan discovered, entrusting highly prized personal information, such as photographs of one's child and years of e-mail, to cloud service providers comes with particular risks. All the major cloud service providers have already been remotely targeted by criminal attacks, including Dropbox, Google, and Microsoft, and we can surely expect more in the future. Indeed, several years after Honan was attacked and published an entreaty to "kill the password" for its near-total inefficacy, thousands upon thousands of individuals and businesses continue to have their cloud-based accounts compromised and their data stolen, including a number of high-profile Hollywood actresses. In late 2014, hundreds of photographs—many of them of a deeply private nature and containing nudity—were stolen from celebrities such as Jennifer Lawrence and Kate Upton when hackers successfully subverted the user names, passwords, and security questions protecting their Apple iCloud accounts. Although it may be your cloud service provider that is targeted in the attack, you're the victim, and the data taken are yours. Of course the rights reserved in the ToS mean that companies bear little or no liability for when data breaches occur. These attacks threaten intellectual property, customer data, and even sensitive government information.

In 2008, the top secret design specs for the president's Marine One helicopter were found freely available online, hosted on a peer-to-peer (P2P) network in Iran. These P2P networks allow for easy decentralized file sharing and are most often associated with the distribution of pirated films and music on the digital underground. How did the top secret plans and capabilities of one of the most technologically advanced helicopters in the world end up in the hands of the Iranians? Simple. A military contractor in Bethesda, Maryland, working on the Marine One project decided he wanted to listen to free music on his work laptop. When he downloaded the popular P2P sharing software, he accidentally and unknowingly installed the program in the wrong directory on his computer. As a result, the plans and defensive security features of the military helicopter that shuttles the president from the White House to Air Force One leaked to P2P music-sharing networks around the world, including those in Iran. For the want of free music, a billion-dollar military project was compromised, and the blueprints for the president's Sikorsky VH-3D helicopter ended up on a peer-to-peer network in Iran, hosted next to the pirated songs of both Michael Jackson and Shadmehr Aghili, the undisputed king of Persian pop. The former military contractor, interrogated by both the

FBI and the Department of Defense, admitted his error, but by then the damage had been one. Our global interconnections and never-ending storage of more and more data mean leaks are inevitable. What data might you or your company be leaking to the cloud?

Big Data, Big Brother

Interestingly, governments are not only victims of data leaks but the cause of many as well. Information is the driving force of all intelligence operations, and governments of all sizes are targeting big data with a vengeance. It is not only the Chinese who are hacking the world; so too are the Americans, Brits, Russians, Australians, Canadians, Syrians, Israelis, Egyptians, Iranians, and even Ethiopians. In fact, there are more than a hundred countries with active, offensive computer-hacking programs, though perhaps none quite as extensive as that of the U.S. government and the National Security Agency. Every single day, the NSA is reported to intercept and stockpile more than 1.7 billion e-mails, phone calls, and SMS messages, compiling a database of nearly 20 trillion data transactions just since 9/11. The agency catalogs who calls whom, who e-mails and texts whom, and who wires money to whom.

Given the exponential growth in its big-data holdings, however, the electronic espionage agency is running out of storage space. In response, the government is in the process of building a new massive operations facility deep in the Utah desert that will allow the NSA to cache and process 100,000 times more data than what is currently held by the Library of Congress. But that is just the beginning . . .

The revelations of Edward Snowden have documented the extensive number of data streams pursued by the NSA, including the ever-growing mounds of social and locational details we all generate. Though the complete list of Snowden's disclosures is too long to outline here, a review of the highlights released to date should make it clear that the private sector is not alone in its aggressive pursuit of big data. NSA's PRISM program allowed the government to collect copious volumes of data from companies such as Microsoft, Google, Facebook, Skype, AOL, and Apple, including users' e-mails, videos, photographs, status updates, and locations.

Snowden also revealed that the NSA accessed and downloaded the interpersonal connections of social media users (whom they spoke to, how often, and where they were located), including the social data graphs of U.S. citizens. These network graphs were supplemented by millions of online user contact lists and address books that the agency had also collected. You see, when you choose to use Google Contacts or iCloud to store the personal details of your friends, family members, and business associates, they can be readily targeted and taken by others, including governments.

Not only did the NSA have cooperative relationships with American firms, but it also targeted them when convenient, including Google and Yahoo!, whose data centers the spy agency infiltrated without authorization. Using the same basic techniques employed by hackers and organized crime groups, the NSA infected more than fifty thousand computer networks around the world with malicious software in order to get access to targets of interest. The agency even posed as Facebook in numerous "man in the middle" attacks to pursue individuals across their social networks. The technique caused targets of interest to connect through a replica Facebook site controlled by the government, allowing the agency to install malware on the machines of its marks.

The NSA did not do all this work by itself, but rather cooperated with sister organizations such as Britain's NSA equivalent, the Government Communications Headquarters. Together, the agencies participated in the program Optic Nerve, which intercepted millions of Yahoo! video chats by commandeering the video cameras on users' laptops and snapping photographs every five minutes. Millions of images were stored, including a large number of sexually explicit pictures containing nudity. Shockingly, many of the video chats intercepted were of individuals not specifically targeted for any particular intelligence operation but because it was easier to grab all the chats than to decide on an individual basis which ones should be seized.

The NSA also replicated the already proven techniques of advertisers and marketers and their commercial data-gathering operations. For example, the spy agency created and installed tracking cookies on hard drives and mobile phones to record the locations and online habits of those under surveillance. According to Snowden, the NSA was even tapping smartphone apps, such as Rovio's *Angry Birds*. The spy agency recognized that *Angry Birds* was already doing such an excellent job of pilfering data that the NSA needn't bother duplicating its efforts. Instead, the NSA merely intercepted the colossal sums of data already forwarded to Rovio by those who naively thought the app's only true purpose was to sling birds at chortling green pigs for amusement.

Only a tiny percentage of the 1.7 billion *Angry Birds* users understood that their "free" app was sharing data ranging from their persistent location to their sexual orientation with Rovio. None—including the app company itself—however, realized that they were now also providing these data (unwittingly of course) to the NSA. Individual NSA analysts were even using the agency's vast spying tools to target their boyfriends, girlfriends, spouses, and ex-lovers. Numerous violations were documented wherein NSA officers entered the e-mail addresses and phone numbers in order to read their e-mails, track their locations, and listen to their phone calls. The actions of these individual employees raise the proverbial but important question of who watches the watchers?

While the overwhelming majority of the NSA's targets appear to be overseas, dozens of security services around the world use electronic espionage to surveil and repress their domestic populations. In China, Iran, Egypt, Syria, Bahrain, and elsewhere, data stored online are routinely monitored and intercepted for reasons of political intelligence and to ensure the status quo. Most countries do not build these surveillance systems, but rather buy them from companies based in other nations, such as Germany's Gamma International, maker of the FinFisher electronic surveillance suite. FinFisher allows domestic intelligence services to monitor thousands of targets simultaneously across their mobile phones, social media networks, and online activities.

Once these systems of mass data surveillance have been established, they can be used for the common good, such as disrupting an imminent terrorist attack, or to the common detriment, such as repressing and harassing human rights activists and subverting the democratic process. While social media did much to empower dissidents in Egypt and Tunisia during the Arab Spring, a story that received tremendous press worldwide, there was a flip side to the social data coin. The millions of tweets and Facebook postings also provided very useful tools for governments to go after critics. Organizing a protest on Facebook gives the government clear access to what opposition activists are planning, and nearly all governments have the skill set to take advantage of such leaking data.

In the uprising against Bashar al-Assad that began in 2011, the Syrian government, with technical support and assistance from Iran, developed a wide variety of programs to monitor social media sites such as Facebook and Twitter to track communications among opposition figures. Leaders of the anti-Assad movement identified online have been targeted for attack, as have members of their families. In the waning days of the former Ukrainian president Viktor Yanukovich, his government forces demonstrated the power of technology to repress and intimidate opposition forces. As demonstrators gathered on the streets of Kiev, the Ukrainian government detected the locations of all mobile phones in proximity to street clashes between riot police officers and protesters. The mobile phones (and their owners) were identified in real time and received what may be the "most Orwellian text message a government has ever sent: 'Dear subscriber, you are registered as a participant in a mass disturbance.'" The language was carefully chosen as Yanukovich had made such participation illegal days earlier and anyone found in violation was subject to immediate arrest.

The Darker Side of Big Data

The emerging legacy of big data may well be one of persistent surveillance, the abolition of privacy, and a rash of criminal threats never previously envisioned. Social media, smart phones, mobile apps, the cloud, and a host of other technologies mean that not only can Nordstrom, Acxiom, Facebook, and Google find you at will, but so too can the Zetas, Lashkar-e-Taiba, domestic abusers, and stalkers. What most people do not understand, however, is that any data collected will invariably leak. Our current computing systems are too insecure to safely store the volumes of information we are generating.

To date, the major threat to big data has been their theft and leakage. But that was only the beginning. As we move forward, we will encounter dangers that may prove to be even more perilous—unauthorized alteration of the information upon which the world depends to run its daily activities. Although we have placed tremendous trust in the data we are feverishly hoarding away, the underlying accuracy of this information, as we will discover, can easily be subverted, with significant consequences for all. For just as bad actors can steal our data, so too can they change it. This gathering storm will leave us vulnerable and shake the foundations of our faith in a data-dependent world in ways not yet fully appreciated.

CHAPTER 8

In Screen We Trust

The world isn't run by weapons anymore, or energy, or money.
It's run by ones and zeros—little bits of data. It's all electrons. There's a
war out there, a world war. It's not about who has the most bullets.
It's about who controls the information—what we see and hear, how we
work, what we think. It's all about information.
AS STATED BY COSMO (BEN KINGSLEY),
THE VILLAIN IN *SNEAKERS*

All systems check. The five thousand spinning centrifuges in operation at Iran's nuclear enrichment facility at Natanz were humming away, and the Islamic Republic was making good progress on its "peaceful" nuclear energy program. If things continued on track, soon Iran would have enough enriched uranium 235 (U-235) to create its own nuclear power plant or its first atom bomb, depending on whom you asked. Although Iran had always insisted its nuclear activities were purely for civilian energy use, much of the world, including the United States, Europe, Israel, and the United Nations, were less convinced.

In 2005, the UN's International Atomic Energy Agency (IAEA) found Iran in noncompliance with the Nuclear Non-proliferation Treaty it had signed, and the inspection agency reported its concerns to the UN Security Council. In response, the UN demanded Iran suspend its nuclear enrichment activities at Natanz, to which the nation's president at the time, Mahmoud Ahmadinejad, responded with an emphatic no. Senior officials at the IAEA concluded Iran had sufficient information to design and produce a workable atom bomb and UN sanctions were imposed. But would the sanctions prevent Iran from getting the bomb? Given Iran's preeminence on America's "Axis of Evil" hit list, something more had to be done.

For political reasons, an overt military strike was ruled out, but the following year President George W. Bush authorized a covert attack on the nuclear facilities at Natanz and dubbed the top secret program Operation Olympic Games, according to the *New York Times*. The result was the "most significant covert manipulation of the electromagnetic spectrum since World War II, when cryptanalysts broke the Enigma cipher that allowed access to Nazi codes."

The Iranians were no easy target and were smart enough to not connect the most prized information network in the Islamic Republic to the Internet. As a result, the operatives associated with Operation Olympic Games could not just tunnel their way in via a poorly protected road on the information superhighway. A network of human agents, engineers, and maintenance workers—spies and unwitting accomplices alike—would have to be assembled and choreographed with tremendous precision if the plan were to succeed. The weapon of choice for this covert operation? A USB thumb drive.

To sabotage the centrifuges at Natanz, a new class of cyber weapon was created, one that could leap from the virtual world of computers and enter the physical world of industrial control systems. Enter Stuxnet, a highly sophisticated computer worm widely believed to have been created by the United States and Israel to keep a notorious foe in check. The authors of Stuxnet copied the worm onto a simple USB flash drive, now locked and loaded, ready to seek out its quarry. How the drive came to be smuggled into Natanz and who inserted it into the computer network at the facility remain unknown, even today.

What is known, however, is how quickly the malware spread across the IT infrastructure of the plant. The mere insertion of the flash drive into a single computer's USB port infected the machine's Microsoft Windows operating system using a zero-day exploit not previously documented. The worm also used a forged digital security certificate, indicating it was reliable and trustworthy, allowing it to replicate across the IT infrastructure at Natanz with impunity. As the worm propagated from desktop to desktop and network to network, it posed a simple question of each machine it infected: Is this computer connected to an industrial control system manufactured by the German multinational company Siemens?

The Americans and the Israelis had done their homework and knew the centrifuges at Natanz were run by Siemens S7-417 industrial programmable logic controllers (PLC), which monitored the valves and pressure sensors of the centrifuges at the plant. If a computer was not connected to a Siemens PLC, the worm failed to replicate and merely died. If, however, Stuxnet detected that a desktop or network computer system was connected to a Siemens PLC, the cyber weapon assiduously began its work, making its way from the Windows computer onto the industrial control system that managed the Iranian centrifuges.

The perpetrators of the attack knew that refining U-235 was a very tricky business. The IR-1 centrifuges used at Natanz were designed to spin at 100,000 rotations per minute (RPM), an incredible feat in speed and technology. If the centrifuges spun too slowly, the U-235 required for nuclear energy (and bombs) does not separate effectively. If they spin too fast, the centrifuges begin to vibrate and shake uncontrollably until the pressure becomes so severe the motors burn out, requiring the centrifuge to be replaced. The authors of Stuxnet understood that no centrifuges meant no enrichment, thus no bomb and no threat.

The Siemens PLCs were key to the attack, but the authors of Stuxnet were not impetuous cyber warriors with a pillage-and-burn mentality. They were patient, strategic, and cunning in their attack on Natanz. In the first phase of the assault on Natanz, Stuxnet did nothing but observe, sitting there silently, stealthily gathering information to understand how the enrichment centrifuges worked. The worm recorded all of its findings in a masterful preplanned move that would prove crucial to the success of the operation.

It was in phase two, however, that Stuxnet began to show its true powers as the worm established dominion over the industrial control systems at Natanz. Slowly, its puppet masters began manipulating the centrifuge valves and motors responsible for enriching U-235 at the facility. For months, and even years, the centrifuges sped up and slowed down, fluctuating from their designed 100,000 RPM specifications. Centrifuge pressure mounted, rotors failed, and yields of enriched uranium began to plummet.

Meanwhile, inside the highly secure operations control room at Natanz, all systems were in full working order—at least according to the computer screens monitored by the engineers at the facility. Every one of the thousands of centrifuges was represented by a light on a computer screen, and each was carefully monitored for system malfunctions. A green light meant the centrifuge was working as designed; a gray or red light indicated problems. Day after day, engineers dutifully watched their screens for any evidence of failure, as the lights continued to shine bright green on the data safety systems before their eyes. Cascade protection system? Check. Centrifugal pressure? Check. Rotor speed? Check. Screens on the walls, screens on their desktops, screens on their control panels—all information systems inside their operations command center told the Iranians their nuclear ambitions were on track. Yet nothing could be further from the truth.

The damage caused by the Stuxnet worm was designed to be subdued at first. Gradually, some centrifuges began spinning out of control, but the Iranians blamed bad parts or the incompetence of their engineers. Each centrifuge that failed seemed to have a different explanation: this device was too slow, that one was too fast, there was too much pressure in these. The uranium processed was increasingly of poor quality and not usable. Inspection after inspection of the facility was carried out, and researchers continued to closely observe the status of their entire operation from the computers inside their control room. As time passed, dozens and then hundreds of centrifuges began to fail. Iran's nuclear ambitions were now in doubt. What the hell was going on? As it turned out, the Iranians had placed too much trust in the computer screens governing their prized secretive nuclear enrichment site.

The data logging and computer recording of the industrial control systems stealthily perpetrated by the Stuxnet worm in phase one of the attack had a clear, if not immediately obvious, purpose: to fully document what the Siemens PLCs looked like when they were in full, proper working order. Rotors spinning according to plan and pressure at expected levels yielded all systems go, all maintenance lights green. Stuxnet captured all of those data and recorded it on the PLC equivalent of a VCR, carefully saved for posterity. What happened next was straight out of a Hollywood blockbuster, portrayed many times in films such as *Ocean's Eleven* and *National Treasure*. The attackers simply prerecorded video footage of the casino vault or safe room to be targeted and played it back on the screens of the watchers and security staff.

As the uranium enrichment centrifuges spun out of control at Natanz, Stuxnet masterfully intercepted the actual input values from the pressure, rotational, and vibration sensors before they reached the operational control room monitored by the plant's engineers. Rather than presenting the correct real-time data from the Siemens PLCs, Stuxnet merely replayed the prerecorded information it had taken during phase one of the operation, showing all systems in full working order. The brilliant move meant that even though in reality the industrial control systems were melting down and digitally screaming for help, the flashing red danger signs displayed by the system were supplanted by a sea of green calm on the monitors of the Iranians controlling Natanz. As the centrifuges spun out of control and tore themselves apart, the human operators in the digital control room had no idea their own reality had been hacked, hijacked by a computer worm with a funny name sent on a mission to search and destroy.

Life in a Mediated World

Unfortunately, you have more in common with the Iranians than you realize. While you may not be producing U-235, you too depend on screens every day to translate the world around you. Your cell phone tells you who has called, your PC reminds you that you need to update your operating system, and the GPS in your car shows you how to get to your morning meeting. All of this and more transpire long before you finish your second cup of coffee. The result? We no longer live life through our own innate primary human sensory abilities. Rather, we experience it mediated through screens, virtual walls that sever us from our intrinsic senses and define the world for us. Screens interpose themselves between us and the real world, projecting information that is purportedly equal to reality but is at best only ever a rough approximation, one that is easily manipulated.

At our airports, hospitals, banks, and ATMs, screens have become an omnipresent fixture in our lives. But screens today are dumb; they do little more than present the underlying information contained in data systems, systems which are eminently hackable. Those who control the computer code also control our screens and thereby our experiences and our perceptions. Everything from video games to voting machines can be tampered with, and in this brave new world seeing something with your own two eyes and hearing it with your own ears is by no means an indication that it is legitimate, correct, or safe. As a result, the screens we watch can deceive us in ways most have yet to understand.

Whether or not you realize it, your entire experience in the online world and displayed on digital screens is being curated for you. Some of this filtering of course is good. With billions of tweets, Snapchats, status updates, and blog posts, there is no way any of us could consume the volume of data thrown our way on a daily basis. Knowing this, Internet companies go to great lengths to learn what you like and to customize your online experience using a series of computer algorithms. Facebook studies your Web links, images, pokes, messages, events, and Likes to customize what you see on your screen every day. As a result, you do not see most of what's posted by your friends or on the pages you follow, and your friends see perhaps only 10 percent of your own updates on their news feeds. For as much effort as Facebook puts into studying and segmenting you for its advertisers, it works at least as hard to determine which of your friend's posts you would most likely want to see every time you visit his site or launch his app. But why does it do this? Simply stated, Facebook, Google, and other Internet companies know that if they provide you the "right" stuff, you'll spend more time on their sites and click on more links, allowing them to serve you up more ads.

Facebook is by no means alone in this game, and Google too quantifies all your prior searches and, more important, what you've clicked on, in order to customize your online experience. In his book *The Filter Bubble*, the technology researcher Eli Pariser carefully documented the phenomenon. Getting you the "right" results is big business, and millions of computer algorithms are dedicated to the task. Google reportedly has at least fifty-seven separate personalization signals it tracks and considers before answering your questions, potentially to include the type of computer you are on, the browser you are using, the time of day, the resolution of your computer monitor, messages received in Gmail, videos watched on You-Tube, and your physical location. Google alters, in real time, the search results it provides you based on what it knows about you. A search for the word "abortion" returns links to Planned Parenthood for some and Catholic.com for others; if your query is "Egypt," you may receive results on the Arab Spring, while your mom sees info on the pyramids or Nile cruises. Like Pariser, you can run this experiment yourself, and the results will provide an illuminating perspective on how Google sees you.

The fact of the matter is that there is no such thing as "standard Google." Eric Schmidt has publicly acknowledged that "it will be very hard for people to watch or consume something [online] that has not in some sense been tailored for them." While none of this is necessarily malicious, there are important questions to be asked about how this information is being culled, sorted, and curated by others purportedly on your behalf. The challenge, however, is that Google, Facebook, Netflix, and Amazon do not publish their algorithms. In fact, the methods they use to filter the information you see are deeply proprietary and the "secret sauce" that drives each company's profitability. The problem with this invisible "black box" algorithmic approach to information is that we do not know what has been edited out for us and what we are *not* seeing. As a result, our digital lives, mediated through a sea of screens, are being actively manipulated and filtered on a daily basis in ways that are both opaque and indecipher-able. This fundamental shift in the way information flows online shapes not only the way we are informed but the way we view the world. Most of us are living in filter bubbles, and we don't even realize it.

Around the world, nations are increasingly deciding what data citizens should be able to access and what information should be prohibited. Using compelling arguments such as "protecting national security," "ensuring intellectual property rights," "preserving religious values," and the peren-nial favorite, "saving the children," governments are ever expanding their national firewalls for the purpose of Internet censorship. Some of these filtering techniques are disclosed to the general public. For example, in France and Germany, sites promoting Nazism or denying the Holocaust are openly censored. In Syria, YouTube, Facebook, Amazon, Hotmail, and pro-Kurdish sites have been blocked. In Saudi Arabia, 400,000 sites have been restricted, including those that discuss any political, religious, or social issue incompatible with Islam or the personal beliefs of the monarch. In many instances, however, there is no indication that your online infor-mation is being censored; instead, your content simply does not appear. In the United Arab Emirates, the government has even blocked all access to the entire .il domain of Israel, digitally erasing the existence of the Jewish state from the virtual world.

Tech companies have collaborated in national censorship programs and acceded to state demands to filter offending content in real time, as Google did when entering the Chinese market in 2005. But perhaps no other government is as adept at and rigorous with its Internet-filtering programs as China. The "Great Firewall" of China ensures that its billion-plus residents are unable to see politically sensitive topics, such as the Tiananmen Square protests, embarrassing details about the Chinese leadership, or discussions of Tibetan rights, the Dalai Lama, Falun Gong, Taiwanese independence, political reform, or human rights. Internet censorship, however, is not restricted to autocratic regimes or despots. As of 2014, there were more than four billion people living in countries that practice Internet filtering of one sort or another.

Screens tell you not what is really out there but what the government or Facebook thinks you should see. If you searched for something and it wasn't there, how would you know it really was? To paraphrase an old philosophical question, if a tree falls on the Internet and no search engine indexes it, does it make any noise? As we live our lives increasingly mediated through screens, when it doesn't exist online, it doesn't exist. If an event is not listed in Google, it never happened. Conversely, if it does appear in Google, it still might not have happened. Welcome to the world of digital trickery, a virtual hall of mirrors represented as screens where all is magically possible.

The profound risk of life in a technologically mediated world is that it creates mammoth opportunities for information to be manipulated in undetectable ways that most neither expect nor understand. Screens are everywhere, beeping, ringing, and blinking for our attention. But what if these screens were lying? Feeding us false information and misleading us? In today's world, all that we see on screens can be faked and is easily spoofed. Ask anybody who has ever visited an online dating site, and he or she will tell you: what you see is not always what you get.

Does Not Compute

Why, sometimes I've believed as many as six impossible things
before breakfast.
LEWIS CARROLL, *THROUGH THE LOOKING GLASS*

What do hackers, fraudsters, and organized criminals have in common with Facebook, Google, and the NSA? Each is perfectly capable of mediating and controlling the information you see on your computer screens. In a world where information is power, the gatekeepers who control the flow of data to your screen can also control others. We encounter this behavior on a daily basis every time we go online. Most of us would not consider making a major purchase or reserving a table at a new restaurant for a special occasion without first doing our own Internet research. Who better to inform us than our fellow shoppers and diners? Nearly 90 percent of consumers say online reviews influence their buying decisions, and a Nielsen study found that a surprising 70 percent trust the reviews they read online as much as recommendations from a friend. Unfortunately, according to an investigation by the New York State attorney general, 25 percent of the reviews on Yelp, one of the most popular sites of this kind, are completely bogus. Worse, in September 2014, a federal appeals court ruled it was completely legal for Yelp to manipulate its ratings based on which companies advertised on the site; big spenders could legally get five stars, even if all users rated them a one. Reviews on eBay, Amazon, and TripAdvisor are also all easily faked, and many of those five-star postings you see were written by the businesses themselves or by paid proxies. There are even professional companies whose entire business model rests on gaming the online review system. The practice is known as astroturfing and is widespread. One company investigated by the State of New York, known as Zamdel Inc., was accused of writing more than fifteen hundred fake reviews on Yelp and Google Places.

I Thought You Were My Friend

According to Facebook's own 2014 annual report, up to 11.2 percent of its accounts are fake. Considering the world's largest social media company has 1.3 billion users, that means up to 140 million Facebook accounts are fraudulent and these users simply don't exist. With 140 million inhabitants, fake Facebook-land would be the tenth-largest country in the world. Just as Nielsen ratings on television sets determine different advertising rates for *The Walking Dead* versus the Super Bowl, online ad sales are determined by how many eyeballs a Web site or social media service can command—if only the data could be believed.

Want 4,000 followers on Twitter? They can be yours for $5. Want 100,000 fans on Facebook? No problem, you can buy them on SocialMediaCorp.org for a mere $1,500. Have even more cash to burn? How about a million new friends on Instagram? "For you we make special deal," only $3,700. Whether you want favorites, Likes, retweets, up votes, or page views, all are for sale on Web sites like Swenzy, Fiverr, and Craigslist. These fraudulent social media accounts are then used to falsely endorse a product, service, or company, for a small fee of course. Most of the work is carried out in the developing world, in places such as India and Bangladesh, where actual humans may control the accounts. In other locales, such as Russia, Ukraine, and Romania, the entire process has been scripted by computer bots, little programs that will carry out your pre-encoded automated instructions, such as "click the Like button," over and over again using different fake personas.

Just as mythological shape-shifters were able to physically transform themselves from one being into another, these modern screen shifters have their own magical powers, and criminals are eager for a taste of the action, studying their techniques and deploying them against easy marks for massive profit. In fact, many of these clicks are done for the purposes of "click fraud." Businesses pay companies such as Facebook and Google every time a potential customer clicks on one of those banner ads or links you see online, but organized crime groups have figured out how to game the system to drive profits their way via so-called ad networks, which capitalize on all those extra clicks. Stung by the criticism, social media companies have attempted to cut back on the number of fake profiles out there. The results of Facebook's actions were revealing. Rihanna and Shakira lost 22,000 Facebook fans, Lady Gaga had 32,000 of hers removed, and Zynga's *Texas Hold 'Em Poker* had 100,000 purported supporters vanish in thin air.

If Facebook has 140 million fake profiles, there is no way they could have been created manually one by one; there has to be something much more sinister at work, and there is. The practice is called sock puppetry and is a reference to the children's toy puppet created when a hand is inserted into a sock to bring the object to life. In the online world, organized crime groups create sock puppets by combining computer scripting, Web automation, and social networks to create legions of online personas. This can be done easily and cheaply enough to allow those with deceptive intentions to create hundreds of thousands of fake online citizens.

One only needs to consult a readily available online directory of the most common names in any particular country or region. Have your scripted bot merely pick a first name and a last name, then choose a date of birth and let the bot sign up for a free e-mail account. Next, scrape online photo sites such as Picasa, Instagram, Facebook, Google, and Flickr to choose an age-appropriate image to represent your new sock puppet. Armed with an e-mail address, name, date of birth, and photograph, you just need to sign up for an account on Facebook, Twitter, or Instagram. As a final step, you teach your puppets how to talk by scripting them to reach out and send friend requests, repost other people's tweets, and randomly "like" things they see online. Your bots can even communicate and cross post with one another. Before you know it, you have thousands of sock puppets at your disposal for use as you see fit. It is these armies of sock puppets that criminals use as key constituents in their phishing attacks, to fake online reviews, to trick users into downloading spyware, and to commit a wide variety of financial frauds—all based upon misplaced trust.

Fatal System Error

We now live in an "in screen we trust" world. We look first and foremost to computers for guidance and direction. We depend on screens to give us answers, and all too infrequently do we question the results. But if your programming is poor or your primary data incorrect, these errors will be reflected in the results you receive. Garbage in, garbage out is one of the axioms of computer science. In the past, our limited reliance on technology insulated and protected us from many of these mistakes. In the age of big data, however, the calculus has shifted—big-time. We are all affected by database errors in one way or another, and the implications of these inaccuracies are growing every day. According to the Federal Trade Commission, nearly 25 percent of all consumer credit reports contain errors, and data brokers such as Acxiom have admitted that 30 percent of the data they maintain on you may be inaccurate.

When the forty million to fifty million Americans affected by these errors attempt to rent an apartment, buy a car, get a mortgage, or apply for a job, they soon discover that somebody else's mistake has now become their nightmare. If "according to our computer" you are a credit risk, you are, no ifs, ands, or buts. Today, millions of decisions are made every day with faulty, incomplete, or imprecise data, often with no further verification. If the problem were limited to credit reports, it might be tolerable, if only barely. But living in the land of "in screen we trust" means computer errors can affect not only your finances but also your life and your liberty.

As the world of medicine rushes to digitize patient records in an effort to save money, improve efficiency, and yield new big-data insights on disease, there has been an unintended cost: accuracy. Tens of millions of electronic medical records contain incorrect information about patients, and erroneous data on computer screens can quite literally kill. Gary Foster, a twenty-seven-year-old from Essex, England, died at University College Hospital in London when a glitch in the facility's computer system meant the young man received an overdose of cancer drugs during his stay. The hospital staff, following an inaccurately entered prescription order, provided him with lethal doses of chemotherapy for his testicular cancer treatments. Not only can too much faith in computer screens kill, but also it can have a deleterious impact on public safety.

In California, a computer glitch led to the release of 450 dangerous criminals after a system error directed prison guards to set some of the state's most violent offenders free. Gang members, rapists, armed robbers, and inmates classified as having a "high risk for violence" walked out of prisons statewide because officials accepted the information on their screens as the truth. Of course, errors in criminal justice data are common, and these mistakes not only free the guilty but also inculpate the innocent. In Britain, police officials at the national Criminal Records Bureau admitted that more than twenty thousand people had been wrongly branded as criminals because of data errors in their system. The massive blunder meant that thousands of innocent individuals were given criminal records for offenses they never committed. "But, Officer, you've got the wrong person" is a refrain cops are used to hearing; unfortunately for those involved, what is on the screen is the truth, until proven otherwise. Across the U.K., victims of these errors were denied jobs and volunteer positions and had their reputations destroyed, all because of our unyielding faith in our screens.

Today we face a confluence of phenomena, both human and technical, that are coming together like a perfect storm to pose particularly striking dangers for our society. With each successive generation, we grow deeply comfortable, even if only unconsciously, with blindly following the directions provided to us by machines. Garbage in, garbage out has been supplanted by garbage in, gospel out: if the computer says so, it must be true. The problem with such reasoning is that we as a society are relying on incorrect data all the time, a festering problem that will come back to bite us. Filter bubbles, invisible search engine censorship, national firewalls, and faulty data mean we have a fundamental integrity problem with the way we see the world, or more precisely with the way the world is presented to us, mediated through our screens.

When Seeing Ain't Believing

In the preceding chapters, we focused extensively on what happens when your data leak and your information confidentiality is breached. No doubt, criminals are having a field day with all the opportunities they have generated by stealing your data. But there is a much more profound and insidious threat to the world's information—changing it. Criminals, hackers, terrorists, and governments are increasingly breaking into data systems, not to steal information, but to surreptitiously manipulate its presentation on our screens, as we saw in Natanz. As a result, the very integrity of the world's information is under attack. Slowly, imperceptibly, and with great precision, attackers can enter our data systems and covertly modify any and all of the underlying information. When hackers attack, stealing our data may be the best-case scenario compared with altering the information without our knowledge.

In the 1995 movie *The Net*, Sandra Bullock plays a reclusive systems analyst who accidentally uncovers a plot by a diabolical cyber-terrorist organization to take over the world's information systems. The film opens with the undersecretary of defense committing suicide after learning he has tested positive for HIV at the Bethesda Naval Hospital. As it turns out, the official was never HIV positive; rather, hackers changed the results on his medical tests, in retaliation for the undersecretary's pursuit of international cyber villains, information that his doctor dutifully reported based on the data on his computer screen. The shame of the test results was too much for the conservative undersecretary, precipitating his suicide.

This is the world of information warfare, where computer disinformation disseminated through an array of blinking screens carries real-world impact. The events depicted in the film are decidedly possible today. Police data systems have been successfully hacked globally, including in Australia, England, Italy, Memphis, Montreal, Hong Kong, and Honolulu. In 2013, the Danish police national driver's license registry was breached, and it was believed the hackers made changes to the underlying law enforcement data systems. In Philadelphia, also in 2013, a classified database of witnesses to some of the city's most notorious crimes was breached by a local criminal group. As a result, the names, addresses, and photographs of dozens of protected witnesses were posted on Instagram with the tagline "Expose the Rats." Many of the individuals exposed had testified in secret grand jury hearings, and within a few days there were nearly eight thousand followers for the Instagram account user known as rats215. One nineteen-year-old witness who had testified in a homicide case was later targeted for retaliation in a shooting. In what amounted to mass witness intimidation, numerous visitors to the site posted comments such as "exterminate the rats" and "put out a hit on them."

In Massachusetts, a prisoner already serving time for computer hacking was allowed access to the prison's library computer for the purposes of legal research on his own case. Once his fingers touched the keyboard, he was able to tunnel through the computer network at the Department of Correction and obtain access to case files on other inmates, as well as the names, dates of birth, Social Security numbers, home addresses, and phone numbers of the prison's eleven hundred guards. Given the insecurity of criminal justice systems, how many prisoners have been wrongly released, like the 450 violent offenders in California, because the underlying data were falsified and purposefully tampered with? The answer is, we simply do not know, and government officials are loath to discuss the matter.

As open and vulnerable as law enforcement computers are, they are a veritable Fort Knox compared with our electronic medical records. Forget for the moment the millions of accidental errors previously noted; the Department of Health and Human Services (HHS) has determined that at least twenty-one million Americans have had their electronic medical records accessed without authorization since 2009. In fact, HHS has documented more than nine hundred such breaches at hospitals across the United States. But how many more weren't reported? Federal law only mandates reporting if more than five hundred records per incident are targeted. Organized criminals are targeting medical data in a wide variety of ways, ranging from Medicare fraud to extortion. In Virginia, hackers accessed eight million patient records and thirty-five million prescriptions maintained by the state's Department of Health and threatened to post the information online unless Virginia paid a $10 million ransom. Globally, electronic medical data systems are deeply porous, and bad actors are fully capable of leveraging these data with deadly consequences.

Time and time again, doctors, nurses, and technicians will follow the directions presented to them on computer screens, even when the information is incorrect, as we saw previously when describing the fatal hospital system errors that led to the death of Gary Foster. If the screen says you are HIV positive, the hospital will deliver that news to you. Worse, if your blood type is listed as O positive and a hacker, enemy, or adversary switches it in the hospital's database to A negative before you go into surgery, the operation will likely result in death. The same would be true if somebody maliciously erased your allergy to penicillin from your digital chart and a nurse innocuously carried out a medical order directing her to inject five hundred milligrams of the drug into your IV.

The profound consequences of the "in screen we trust" mentality can open the door to an array of new crimes, including new ways to commit murder. In response, criminals have developed a panoply of methodologies to profit from a world that has subsumed human intelligence in favor of the digital and the virtual. Nefarious actors are proving particularly adept at so-called man-in-the-middle attacks, wherein they insert themselves between reality and the data we see on our screens. The result? An all-out assault on the integrity of the information we're stockpiling as a result of the big-data revolution.

Screen of the Crime

For every screen in your life, criminals have developed a plan of attack. One of the most common such scams on the Internet is the phenomenon of phishing—a technique by which criminals masquerade as a legitimate Web site in order to acquire information such as passwords and credit card numbers. The term "phishing" is a hacker spelling of the word "fishing," and the technique involves trying to get an innocent fish to take the bait of a malicious link and bite. The organized crime groups that run phishing cons try to trick users into clicking on a link that takes them to a fake Web site controlled by fraudsters. Phishing messages arrive in our in-boxes, via SMS, tweets, instant messages, and Facebook status updates. They allegedly come from our banks, cable companies, retirement plans, social media outlets, and mobile phone operators and target users around the world, with the greatest number of victims in the United States, the U.K., and Germany.

In the end, all phishing attacks depend on an unsuspecting user clicking on a link or attachment in a message that will either take the unsuspecting party to a fraudulent Web site or install malware on the user's machine. Criminals take advantage of HTML hypertext links and embed their attacks in hidden computer code. Phishing messages arrive as fake e-cards, e-mails from our bank, job offers, coupons, or deals too good to be true on social media. These malicious communiqués, replete with grammatical and spelling errors in years past, have become highly professionalized and are today virtually indistinguishable from the real thing. Criminals know exactly how to subvert the trust you have placed in your screens by visually mimicking the sites they impersonate and tricking your senses with a digital sleight of hand.

A typical message might arrive from an address such as security@bankofamerica.com, informing you that you need to update your profile or that your account has been suspended because of suspicious activity. Uh-oh, seems important, better look into it, you think. What you may not know is that the e-mail address that shows up in your in-box is ridiculously easy to spoof or fake. Anytime you set up a new mail account in any mail software program such as Outlook, Mac Mail, or Thunderbird, you will be asked to enter a name and e-mail address. If a crook types "Bank of America Security Team" as the name in an e-mail program, that's what will show up in your in-box. It's just that simple. Only by examining the message headers might you note that the e-mail address used by the bad guys was actually notifications@security-bankofamerica.com—still close enough to fool the average bear.

In every way, the message looks as if it were from your bank—same font, same color, same logo—but it's not. Though the visible link may say www.bankofamerica.com, you will instead be taken to www.bank0famerica.com (0 instead of *o*) or even bankofamerica.accountupdates.com (accountupdates.com is the actual domain you are visiting, owned by the criminals; Bank of America is just a folder on their site to fool you); www.citibank.com will be supplanted by www.citiibank.com (two *i*'s in the fake address, barely noticeable). Phishing messages all subtly scream what they want you to do by making the embedded link containing the poison pill impossible to miss, written in a large font or having a big colorful button: "To update security settings and protect your account, click here." And now they own you.

The fateful click will take you to the Citiibank.com Web site, where you will be prompted to log in and provide your credentials, and when you do, the thieves capture your log-in name and password, as well as a bevy of other personal information. And now is when the criminals really go to work. Phishing is a gateway crime, a fundamental first step that provides thieves with the data they need to perpetrate step two in their plot against you, including identity theft, financial fraud, tax fraud, and insurance fraud. Just as the nuclear engineers in Natanz were presented with a compelling but completely fictional reality on their screens, you too are beset by criminals knocking at your door every day using similar deception techniques.

For criminals, the costs of pulling off such digital charades are ridiculously low. Fully automated phishing kits sold on the digital underground to send scam messages to 500,000 e-mail addresses cost a mere $65, and as mentioned before, criminals take advantage of "sock puppet" accounts to scale up. As a result, more than 100 million phishing messages arrive in our in-boxes every day. According to a study on the economics of these attacks by Cisco, approximately eight people out of a million will succumb to the ruse, with an average loss of $2,000 per victim. Thus for about $130, crooks can generate $16,000, a 12,000 percent return on investment. With 36 billion phishing messages sent annually, the scale, scope, and profitability of cyber crime comes into focus. Although the returns from bulk phishing attacks are impressive, they pale in comparison to "spear phishing," a technique that does not send bulk fraudulent messages to millions, but rather carefully targets specific individuals or organizations.

Spear-phishing attacks have become the go-to tool of choice for those committing industrial espionage, and the costs in those cases can be gargantuan, as the global beverage giant Coca-Cola discovered. As part of its Asian expansion, Coke had entered into advanced-stage negotiations to purchase China's Huiyuan Juice Group. All was proceeding according to plan with the acquisition, until it inexplicably fell apart. Something fishy was going on, and Coke wanted answers. It conducted a full investigation into the matter and began to examine the deal in detail, including communications between Coke and representatives at the Huiyuan Juice Group. In the end, Coke discovered that the Chinese government was aggressively monitoring the deal and pursuing nonpublic insights into Coke's bidding plans and intentions. How did the Chinese ultimately get the access they needed? By manipulating the screen of Paul Etchells, the deputy president of Coca-Cola's Pacific Group.

Etchells opened an e-mail that had been spoofed to appear as if it had come from a senior executive in Coke's legal department. The subject line was enticing: "Save Power, Save Money—from Coke's CEO." Etchells knew his boss at Coke had been heavily pushing energy savings in the company (as did the Chinese who had infiltrated Coke's corporate information systems). The perpetrators of the attack spoofed reality by making the message appear as if it came from a trusted colleague, on the internal corporate network, with a subject line that was compelling and contextually made sense. When Coke's deputy president innocently clicked on the link, he also silently downloaded malware onto his workstation, including a keystroke logger that captured everything the executive typed. As a result, the Chinese were able to download a barrage of computer files related to the deal. While Coke has refused to publicly comment on "security matters," what is clear is that one single spear-phishing attack against a senior Coke official cost the firm its $2.4 billion acquisition of China's Huiyuan Juice Group. Coke is not alone in having succumbed to spear phishing: it has become the preferred method of attack for online criminals and digital spies, responsible for a full 91 percent of all targeted cyber attacks.

Criminals can now even change what you see on your screen in real time, including your financial statements. But what if your bank account balance were zero, and you didn't know it? There are so many thousands of malware programs that will steal money from your bank account these days; the whole process has become both routine and even automated by the bad guys. Criminals infect your computer or mobile phone, capture your log-in credentials, and then use them to drain your account balance. Of course should you yourself happen to log in and notice the low balance, you may have a shot at notifying your bank's fraud department and stopping the flow of the funds. Banks usually reserve settlement periods of a day or so, particularly internationally, in which they can cancel, stop, or reverse a transfer or wire, but the time frame to do so is incredibly short. To this end, criminals will go to great lengths to ensure that what you see on your computer screen is not what you have in your bank account. Highly specialized Trojan horse software, such as SpyEye and URLZone, not only steals your money but will even offer you false reassurances that it is still in your bank account. The magic of these Trojans is that they give thieves more time to use your banking, debit, and credit card information without your realizing what is happening. Your first clue there is a problem is when you try to use your own bank card to make a withdrawal, only to learn you are overdrawn and there are insufficient funds.

The crimeware (criminal software) involved is so sophisticated it even knows how much it has stolen from each bank account compromised. Thus, if the thieves stole $2,419 from your checking account, an algorithm will add that portion back to what you see on your screen in real time as you view your online account balance. Purchases made by criminals with your credit or debit card are automatically struck from the recent transactions list and the online statement before they appear on your screen. Even PDF copies of your banking and credit card transactions sent to your printer are modified before they come out of your machine. When these thieves own you, they really own you.

These types of man-in-the-middle attacks are powerful reminders that criminal hackers are perfectly capable of intermediating reality for you via the ever-increasing number of screens in your life. Just like the perpetrators of Stuxnet, these criminals recognize that screens are merely a proxy for reality, one that is completely malleable and easily manipulated. Yet not all manipulation of the data we see on our screens is carried out by global cyber-crime cartels or espionage services.

Pedophiles routinely take on the digital personas of children, and those under eighteen cannot tell when they are speaking to an adult 80 percent of the time, making the screens of the young particularly vulnerable to attack. Recall the case of Amanda Todd, the twelve-year-old girl who was tricked into showing her breasts on video camera to a person she thought was a boy her age. Todd was blackmailed and tormented by her virtual attacker, driving the Canadian student to take her own life. For years, the case went unsolved, and Amanda's parents never knew who had tortured their daughter until April 2014, when a break in the case led the Royal Canadian Mounted Police to the Netherlands and a suspect five thousand miles away. Police in Holland identified thirty-five-year-old Aydin Coban and charged him with "one count each of extortion, internet luring, criminal harassment, possession of child pornography for the purpose of distribution." Coban's alleged modus operandi was to establish a fake online persona, gain the trust of underage girls, and then seduce them into performing sex acts in front of a Webcam. Dozens of other victims both in Canada and around the world are believed to have been targeted by the Dutch pedophile.

Even for adults, interpersonal relationships and screen manipulation can be a dangerous mix. Such was the case with Elizabeth Thrasher, who was accused of lashing out at the daughter of her ex-husband's new girlfriend. The jealous woman allegedly copied two photographs from the teen's Myspace account and posted them to Craigslist in the Casual Encounters section of the site. She also listed the innocent girl's home address, phone number, e-mail, and work information, stating that she was looking for sex. Thrasher's prey had no idea about the Craigslist posting until she started to receive phone calls, text messages, and photographs (including nude pictures), along with solicitations for sex. In court, the girl testified that she felt as if she'd been "set up to get killed and raped by somebody."

Stock Screeners

Not only can individuals and companies be manipulated based upon what appears on their screens, but so too can financial markets. While whispers and suppositions have moved markets in the past, the breakneck speed of the Internet means the world often reacts before information has been verified. In August 2000, a hacker by the name of Mark S. Jakob, a twenty-three-year-old community college student from El Segundo, California, created a fake press release and forwarded it to the Internet Wire, a Web-based distributor of corporate announcements. Jakob chose the Emulex Corporation, a Nasdaq-traded communications equipment manufacturer, as his target. The hacker merely copied the stationery and style of previous Emulex press releases, spoofed the company's e-mail address, and forwarded his news story to the Internet Wire. The fictional press release stated that the Securities and Exchange Commission (SEC) had opened an investigation into Emulex, that its quarterly earnings were to be restated, and that the company's CEO, Paul Folino, had resigned in response. The sensational story went viral and was picked up by numerous other newswires, including TheStreet.com, CNBC, Bloomberg, and the Dow Jones Newswires.

The market's response was both predictable and immediate. "In a sixteen-minute period following the republication of the fake press release, 2.3 million shares of Emulex stock were traded, and the price plummeted $61, from $104 to $43, resulting in Emulex's losing $2.2 billion in market capitalization." Exactly the response Jakob had been hoping for when he shorted the stock, earning the young market manipulator a paper profit of $250,000. Immediately, Emulex's CEO appeared on Bloomberg and in other financial press denying the story, but by then the damage had been done. Within six days, the FBI, working with the SEC, had identified Jakob, who was arrested and pleaded guilty to wire and securities fraud. When all was said and done, legitimate investors in the market lost more than $110 million because a kid at a community college manipulated the trust they had in their screens.

Screen manipulation in the financial services sector is commonplace, and so-called pump-and-dump schemes are the meat and potatoes of online securities fraud. The practice involves traders artificially pumping up the price of a stock through false and misleading positive statements posted online and then dumping their overvalued shares before their lies are discovered. The practice has flourished in cyberspace, and the FBI has arrested dozens of criminals for taking part in these scams. Though pump and dump is generally unsophisticated in its approach, both individuals and organized crime groups have earned hundreds of millions of dollars by manipulating the information we all see online.

Sometimes financial screens can manipulate you in ways you don't even realize, by watching you while you watch them. That's what professional traders at Goldman Sachs and JPMorgan learned about the Bloomberg trading terminals they had been using for years. Bloomberg terminals are the lifeblood of Wall Street, and firms pay $20,000 per year per terminal to mine the reams of data they provide in order to make their daily trades. What these traders didn't realize, however, was that reporters from Bloomberg's news division had been granted administrative access that allowed them to monitor client activities as traders used their Bloomberg boxes. In other words, Bloomberg news reporters were keeping tabs on terminal use to facilitate news reporting. Traders who thought they were viewing information privately on a dumb terminal learned the terminal wasn't quite so dumb and was actually watching back.

More than 300,000 of the most influential people in the financial world, including bankers, hedge fund managers, and Treasury Department officials, rely on these Bloomberg boxes to carry out their deeply proprietary research, with each query tied to a specific individual. The scandal came to light when a Bloomberg journalist phoned Goldman Sachs to inquire whether a partner still worked there, noting he hadn't logged in to his terminal for days. The casual observation set off alarm bells at Goldman Sachs, which went public with the story.

It was later revealed that Bloomberg's twenty-four hundred reporters were able to see histories of user log-in information on the company's terminals, as well as the various search functions they used, such as for equities or commodities. Goldman officials complained that Bloomberg reporters eavesdropped on customers using their terminals and used this private information to spy on the activity of individual Goldman partners, information they used to generate Bloomberg news stories. One former Bloomberg reporter noted that "there was always a discussion in the newsroom of how to use the terminals to break news."

Financial screens can also be hacked and manipulated via high-frequency trading (HFT). In his seminal 2014 book, *Flash Boys*, Michael Lewis tells the story of how Wall Street insiders have gamed the entire financial trading system by hacking time. By spending hundreds of millions of dollars on vastly superior technical infrastructure, high-frequency traders were able to shave mere milliseconds off their trading times, providing them with an exploitable advantage over their peers. *Flash Boys* follows Brad Katsuyama, a trader at the Royal Bank of Canada's New York office, and his incredibly complex multiyear investigation into the world of HFT. What he discovered was startling: the stock market as it appeared on the screens on his desk was an illusion.

As it turned out, every time Katsuyama attempted to execute a trade, the stock price moved before his order was completed. How did this happen? High-speed traders had figured out a way to exploit the variable speeds at which trading information moved along fiber-optic cables to the stock exchanges. Though the signals traveled at two-thirds the speed of light, over longer distances minuscule time lags added up and could be taken advantage of. By paying huge sums of money for the fastest cables, the most powerful computers, and the privilege of colocating their data servers on-site with the exchanges themselves, high-speed traders were able to see Katsuyama's intent to purchase a stock at a particular strike price and purchase it out from underneath him at the price displayed on his screen. Katsuyama was not alone; we were all being affected by the same problem; he was just the first to document it. High-frequency traders were front running the market and in return screwing us all, including you and your mutual fund purchases, your 401(k) plan, and even city pension plans.

High-frequency traders had hacked time and screen, in what amounted to a man-in-the-middle attack. They inserted themselves between the purported real-time stock market data projected on Katsuyama's screens and a much faster reality, which they controlled and owned. Their computers were so fast that they could detect other people's orders, cut in line before them, buy the stocks in question, and sell them back to the person who had originally attempted the purchase, albeit at a higher price. A few pennies' difference here and there on millions of trades a day meant high-frequency traders made billions of dollars in cumulative profits out of a five-millisecond trading advantage. To put that level of speed into context, the blink of a human eye takes approximately three hundred to four hundred milliseconds. It's akin to the scene in the movie *The Matrix* when the bad guys start shooting at Neo (Keanu Reeves) and he can see the bullets coming and can move at light speed to avoid them, except in this case it's all about hacking the financial system and none of us little people have Neo's powers.

In the days after *Flash Boys* was released, a series of investigations were launched by the SEC, the FBI, and the New York State attorney general. But their sudden interest begs an important question: How is it this entire system was even able to develop in plain view of the SEC right on the heels of the 2008 global financial crisis? Michael Lewis rightfully pointed out that the "market is rigged," but before this could happen, the firms involved in flash trading had to undermine your screen in order to create the fiction of a transparent and trustworthy marketplace. It is troubling to know that we live in a world in which the screens of hospitals, prisons, police departments, banks, brokerages, and news sites are so readily hackable, but as we shall see, screens are proliferating, threats are mounting, and these attacks can cost us much more than just our money.

CHAPTER 9

Mo' Screens, Mo' Problems

**In a world that daily disconnects further from truth,
more and more people accept the virtual in place of the real,
and all things virtual are also malleable.**

DEAN KOONTZ, *THE GOOD GUY*

Robin Sage was a young and attractive twenty-five-year-old woman working as a cyber-threat analyst at the U.S. Navy's Network Warfare Command. She had degrees from MIT and had interned at the NSA. Like many her age, Robin was a consummate networker on social media, with profiles on Facebook, LinkedIn, and Twitter. Shortly after starting her career with the navy, she began to send out friend requests to other cyber geeks working in the government. In under a month, she had grown her network to more than three hundred connections in the cyber-security world, including military personnel, defense contractors, and staff at various intelligence agencies. Among her new online pals were the chairman of the Joint Chiefs of Staff, the CIO at the NSA, senior intelligence officials at the Marine Corps, the chief of staff for a U.S. congressman, and executives at Lockheed Martin, Northrop Grumman, and Booz Allen Hamilton.

Though some who received her friend requests at first didn't quite remember the young woman, Robin assured them that they had met the previous year at DEF CON, a large hacker gathering frequented by the "hackerati" and government spooks alike. Those with any lingering doubts just looked at Robin's network and saw how many friends they already had in common, allaying any fears about accepting her connection requests. Robin even made friends on Facebook and LinkedIn with those who worked with her in the same building at the Naval Network Warfare Command. As her social media presence and network began to grow, Lockheed Martin and other firms became interested in hiring the young woman to work for them and the job offers began to pour in. There was just one slight problem: Robin Sage did not exist.

Sage was the invention of Thomas Ryan, a security consultant who wanted to test the threats social media posed to professionals working in the national security community. His goal was simple: to see what intelligence he could covertly gather via social media networks using a fictitious persona. In less than a month, his new contacts began openly sharing extensive data with his attractive alter ego, Robin Sage. Using his virtual Mata Hari routine, Ryan duped an army Ranger whom he had friended into sending Sage photographs—pictures that included embedded geo-location data for his secret base in Afghanistan. The Ranger further revealed details about his and other troop movements in Iraq to his new "friend."

Robin Sage's screen presence was so convincing that she even received confidential documents to review as well as offers to speak at several high-profile conferences on cyber warfare and security. How hard was it for Ryan to carry out his ruse against an elite group of seasoned military and intelligence professionals working in America's national security community? Dead easy. Ryan merely plucked a photograph off the Internet and used it to create Sage's social media profiles. In reality, the picture belonged to a lesser-known porn star of limited repute. Even her name, Robin Sage, was actually the name of a large military exercise carried out annually by the army in North Carolina. Robin's address? That of the infamous military security contractor Blackwater. The Robin Sage experiment proves how easily the trust people place in their screens can be undermined. If trained military and intelligence professionals took the bait, what chance might the general public have of protecting themselves from these types of threats? But when everything is connected, computers are far from the only screens you have to worry about.

Call Screening

Given the explosion in mobile devices, it's not surprising that criminals are turning their attention from big screens to little ones—especially because phone software is often less secure than its desktop counterparts. Though we are all used to seeing caller ID screens on our phones, in our offices, and in our homes, like any other screen they are easily hacked. There are any number of software programs and Web sites that have been created to alter the caller identification on outbound phone calls.

Web sites and apps such as SpoofCard.com and SpoofTel.com make it incredibly easy to display a different number on any outbound telephone call. All you need to do is enter the number you want it to appear as if you were calling from and the name of the party calling, and that is what will be displayed in the outbound caller ID to your target. Want to pretend to be the president? No problem, just enter "202-456-1414" and "The White House" in the app, and you are good to go. Phone-spoofing companies offer a variety of packages meant to deceive other senses beyond just the visual. They also offer the ability to change your voice from male to female and even insert background noise into any conversation to convince the other party that you are calling from a busy office, nightclub, traffic jam, or airport. These companies tout their products as a means of "protecting your identity" or "pulling pranks on your friends." Of course, text messages can be altered using the same techniques. While no doubt teenagers would have fun pretending to be everybody from Lady Gaga to the FBI director, there are obviously more nefarious uses, which criminals are all too eager to exploit.

In the case of the epic News Corp phone-hacking debacle, a spoofed caller ID screen allowed reporters to access the voice-mail system of Milly Dowler and others, an attack that could just as easily happen to you. The scam worked because by default many mobile phone carriers do not require a password to enter a voice-mail box. The system just relies on your mobile phone caller ID to play your messages. Because all mobile companies worldwide have a central 800 number you can dial to check your messages when calling from a landline, bad guys just spoof your outbound number when dialing the mobile phone company's voice-mail system and, voilà—they have full access to your messages. Not only were public figures in the U.K. targeted with the technique, but some celebrities seeking gossip have even used it to hack into the voice mail of rivals, such as when Paris Hilton notably used SpoofCard to listen to the messages of Lindsay Lohan.

Back in the real world of noncelebrities, a spoofed caller ID means criminals can listen to your messages at the office and learn valuable information regarding pending business transactions, mergers and acquisitions, and even personal medical data. From a social engineering perspective, telephone spoofing creates a powerful tool for the criminal mind. A spoofed telephone call to a company's IT department requesting a system password reset or the latest Wi-Fi key is much more likely to be successful if the call appears to be emanating from within the company's own internal telephone infrastructure, a perennially successful ruse.

On the personal front, deceiving your mobile phone's screen with a spoofed caller ID is also a go-to tool of bank scammers. Financial institutions such as Bank of America and Chase offer telephone banking, and criminals have routinely spoofed the phone number of an account they want to access. Once the bank's telephone system sees a call coming in from your phone number, all the bad guys need is a little bit of personal information (such as the last four digits of your Social Security number or your mother's maiden name)—information readily available on the digital underground or in your Facebook profile—and they are in. Worse, criminals can spoof your bank's telephone number and call you to elicit information, such as your security questions, and then turn around and call your own bank using your spoofed mobile number and the security data you innocently provided to access your account.

Organized crime groups have even successfully spoofed the federal government in their outbound calls and made millions doing it. In what the Internal Revenue Service described as its largest tax scam to date, fraudsters have spoofed the agency's phone number calling your mobile phone. You see an inbound call from the IRS—oh, crap . . . what's this all about? you wonder. On the other end of the line is an IRS agent who informs you that you are delinquent on your taxes and that you must pay the IRS immediately to avoid further penalties. Victims of the scam are told that "given the severity of the crime and your prior delinquent status, we will only accept payment via bank wire or prepaid debit card." To add credibility to the scam, the purported IRS agents confirm the last four digits of the taxpayer's Social Security number (which leaked in any one of the numerous big-data breaches previously discussed). Those individuals who pose questions of the so-called IRS employee are met with a barrage of threats such as arrest, revocation of their business proceeds or driver's licenses, and even deportation if the mark has a foreign-sounding name.

Scammers bolster the creditability of the claims by hacking other screens and by adding bit players to the ruse. After their phone calls, victims often receive official-looking e-mails on "IRS letterhead" that confirm the call and demand payment. They also receive additional calls from their local police departments with caller ID spoofed (such as Amherst, Massachusetts, Police Department) or purportedly from state DMV officials (such as State of Georgia DMV). These additional "officials" confirm the ruse with scripts such as "This is Detective Smith from the Amherst Police Department. We just received notification from the IRS saying you're delinquent on taxes and now criminally liable for payment. I don't want to have to come out there and arrest you in front of your family. If you can make the payment this week, then the IRS has told me no arrest will be necessary." According to the Treasury Department inspector general, more than twenty thousand individuals have fallen prey to the scam.

Trust in screens may cost you not only your money but potentially your life. In a phenomenon known as swatting, bored hackers have been able to telephone police 911 emergency phone systems with spoofed telephone identities in order to report nonexistent crimes, resulting in the response of heavily armed police SWAT units. Even though the hacker may be in Maine, because he used your phone number in Miami, that's where the cops are heading. The deadly game begins when criminals spoof your phone number and then dial 911. A woman screams into the phone, "My husband shot my mother and baby, and now he's holding me hostage . . . PLEASE come quick . . . He's got a shotgun and an AK-47 . . . Hurry . . . he's crazy!" A recording of gunshots can be played in the background for good measure. Now a deadly trap has been set.

In the meantime, you're sitting at home on the couch eating ice cream with your wife and kids, enjoying the latest episode of *The Big Bang Theory*. The cops think a woman inside is moments away from being murdered, and the police are marshaling all available black-and-whites and the local SWAT unit to come save her. When the two sides meet, there is a deep cognitive dissonance and the encounter proves a dangerous powder keg. The cops have surrounded the house and are yelling for you to put your hands up and come out. Your kids are screaming, and your wife is confused. You're not complying well with the police commands, making them further suspicious something horrible is going on. You don't want to go out of your house and confront a group of maniacs (even if they are cops) pointing rifles at you. To the police, your refusal to cooperate heightens the tension. Their next step is shoot some flash-bang grenades through the windows of your house and see what happens. Perhaps alternatively you were fast asleep when a teenage hacker a few states away swatted you. The cops show up, and you are awoken by noises outside your window. You think it's burglars, and you reach for your gun to go investigate. Now you've walked outside with a gun in hand, and you are met with six members of the local SWAT team pointing little red lasers at your forehead. This scenario can't end well.

The FBI recorded at least four hundred incidents of swatting in 2013 alone, with victims across the country, from Ohio to California. Mostly, it's just hackers doing it for the "lulz," because they can. Pre-Internet days, the big teenage hoax was ordering pizzas and having them sent to the kid you didn't like in school. Now kids are ordering SWAT units with guns to carry out their pranks. For instance, in 2009 a group of teenagers from Massachusetts were convicted of carrying out more than three hundred swatting attacks. In some of the cases, the teens met their victims on social networks or online dating sites and retaliated against them if, for example, they refused to engage in sexual conversations. In fact, swatting is the perfect complement to the fictitious Craigslist sex ads that jealous exes post to exact revenge on former lovers, putting them at even greater risk.

Increasingly, celebrities and high-profile public figures are targeted by the practice. In 2013, a twelve-year-old boy in Los Angeles was prosecuted for swatting Ashton Kutcher's Hollywood home and Justin Bieber's Calabasas, California, estate. He also swatted a local bank and reported a robbery in progress. Other celebrity victims of swatting include Russell Brand, Tom Cruise, Rihanna, Charlie Sheen, and Miley Cyrus. It is only by a miracle that no innocent civilians have been killed as a result of swatting incidents, though several police officers have been injured while risking their lives responding at full speed to bloodcurdling spoofed calls for help to 911.

Another way criminals can subvert the screen on your phone is by attacking its baseband—that is, the inner guts actually running your phone. The baseband handles all communication between what you see on your screen and a bevy of radio antennas that control everything from the text messages you receive to your voice calls and Wi-Fi signals, as well as super-geeky telecommunications protocols such as GSM, UMTS, HSDPA, and LTE. Because the baseband is both proprietary and nonpublic, most telephone handset manufacturers implement these underlying operating systems in an insecure fashion. They believe in security by obscurity: this software is so far down in the weeds nobody can figure it out, and thus we needn't worry about security, goes the logic. Of course they are wrong.

A number of hackers, governments, and security researchers have begun to successfully reverse engineer the baseband chips and code, revealing a wide array of security vulnerabilities that can be used to access and modify data on a phone remotely. In early 2014, such a security flaw and back door were found in the baseband software of Samsung Galaxy phones, allowing hackers access to user data stored on the devices. Because modern smart-phone handsets are nothing more than miniature computers, their screens, like those of their larger brethren, can be manipulated to display an altered reality meant to deceive. The FBI has reportedly used the technique to turn phones into bugging devices by altering the phone's usual interface and having it place a covert call to the FBI to allow for remote monitoring. In other words, even when the device showed nothing more than your home screen of apps, it was actually on a phone call to the Feds listening in on you.

Criminals can manipulate these screens the same way, often using techniques that you might never expect. For example, when you dial a number on your mobile device, you do so by pushing a series of numbers on your screen in order to be connected to your party. But how do you know the number you dialed was the one you were connected to? Simple enough if you dial Mom and she picks up your call. But what about when you dial Citibank, Bank of America, or Wells Fargo? You're not reaching a banker at the local branch the way you might have twenty years ago. Instead, you're being connected to someone you've never spoken to before from a call center, generally in a foreign country manned by people with foreign accents.

By using mobile phone malware, hackers can install a "rootkit" on your mobile phone that gives them control over all features of the device, including its touch screen and number pad. Rootkits are malicious software that hide normal computer processes and functions from a user's view and give hackers administrative or "root" access to any device. Organized crime groups know the 800 numbers for financial institutions around the world.

If your phone is infected with malware, once you dial your bank's customer service number, the rootkit detects one of its targeted institutions is being phoned and can intercept and reroute the call. It is another classic man-in-the-middle attack that allows the criminals to shape and mold the reality you see on your screen and bend it to their desired outcome.

As a result, when you dial 1-800-4MY-BANK, your call is invisibly rerouted to a call center manned and operated by international organized crime. Given the wide use of foreign call centers by financial institutions, who would question an accent on the other end of the line when you spoke to your "bank"? The spoof would be relatively easy to perpetrate. Once you were connected, you would be asked for your account number, mother's maiden name, password, and other security information "just to verify you." Next you would be told, "Oh, I'm sorry, sir or ma'am, our computers have just gone down. Tech support tells us they should be up in the morning. Would you mind calling back?" Nothing in that conversation would seem suspicious to anybody who's had to deal with call center employees in the past. The only difference would be that by the end of the phone call, the criminal would have access to your personal and banking details and would use them in rapid succession to remove all funds from the account. All of this is possible because the screens on our phones show us not reality but a technological approximation of it. Because of this, not only can the caller ID and operating system on a mobile device be hacked, but so too can its other features, including its GPS modules. That's right, even your location can be spoofed.

Lost in Space: GPS Hacks

In the 1997 James Bond movie, *Tomorrow Never Dies*, we find Mr. Bond investigating a spoofing attack on the GPS navigation of a British frigate. In the story line, the navigation of the HMS *Devonshire* is tampered with by an evil genius who uses an "encoding box" to send the ship off course. As a result, the *Devonshire* enters Chinese territorial waters and appears to be sunk by the Chinese navy. To the British, however, the frigate was clearly in international waters, and thus the Chinese have committed an act of war. The actions of the villain have their desired effect: Britain and China are now on a path toward war. Once again Hollywood was prescient in its vision of future evil.

The Global Positioning System (GPS) is a space-based, low-earth-orbiting "constellation of 24 navigational satellites" that provides location and time information anywhere on the planet. It is an "invisible utility" that we rely on to get around town, deliver packages, find the closest Starbucks, coordinate air traffic control, manage public safety, and even command missile guidance. Paper maps have become obsolete. Instead, we have come to rely on the navigation screens we see before us every day and readily assume the computer knows best. In fact, around the world, there is example after example of drivers blindly following their navigation screens instead of their own two eyes and as a result turning down one-way streets or even off bridges. In Spain, when a GPS device suddenly told a man to turn right, he obliged and went off the road into La Serena, the largest reservoir in western Spain. Although his passenger survived, the driver drowned—all because he followed the directions on the screen before him.

A report from the U.S. Department of Homeland Security warned that America's critical infrastructure was "increasingly at risk from a growing dependency on GPS for positioning and navigation." The press release for a similar report from the U.K.'s Royal Academy of Engineering was even more stark in its assessment: "Society may already be dangerously over-reliant on satellite radio navigations systems like GPS . . . [S]ignal failure or interference could potentially affect safety systems and critical parts of the economy." As it turns out, just as cyber infrastructure is poorly protected, wide open, and wholly hackable, so too is our satellite and radio spectrum infrastructure.

The Global Positioning System is a brilliant technological accomplishment, but the actual satellite GPS signals we receive, although perfectly functional, are very weak, akin to viewing a car headlight from twelve thousand miles away. The signals cannot be boosted further because of the limited power supply on any one satellite, and, what's worse, they can easily be overpowered by broadcasting noise on the same frequency, thereby blocking other ground-based devices from receiving navigational information.

Previously, only military forces practiced in the art of "electronic warfare" had access to the technology and know-how to disrupt GPS signals. The strategic implications of doing so are obvious. If you can block your enemy's navigation systems, then you can interfere with the movement of his troops, ships, tanks, and navy. You can also severely harm an adversary's civilian critical national infrastructure. We had a taste of this in the United States in January 2007 in San Diego, California, when the entire city went on the "electronic fritz." Just about midday, air traffic controllers suddenly found that their systems were malfunctioning. At local hospitals, doctors' pagers ceased to work, and in the port of San Diego ship navigation faltered. For two full hours, cell phones in the city stopped working, and ATMs failed to dispense cash. About as close to a Bruce Willis movie as you can get in America's Finest City. What had caused this massive outage? For three days, the event remained a mystery until the navy finally came forward and admitted it had been conducting a training exercise and testing a new radio-jamming technology.

Sometimes the military jamming of GPS signals is not an accident. North Korea routinely lashes out at its neighbor to the south and blocks its GPS signals. Pyongyang uses three tractor-trailer-sized jammers that it can reposition in order to block satellite navigation to much of South Korea. The longest GPS attack carried out by the North occurred in early 2012 and ran for sixteen days, causing disruption to 1,106 aircraft and 254 ships. Owing to Moore's law, GPS signal technology is becoming smaller, cheaper, and more powerful. As a result, it's not just armed forces that can have access to navigational jammers; now every Tom, Dick, and Harry can get one, with notable results for your screens.

Though illegal in the United States, GPS jammers are widely available online on Web sites such as www.jammer-store.com. For a mere $50, any-body can buy a dashboard model that plugs into your car's cigarette lighter and create an electromagnetic bubble that will surround you as you drive. Their use is more popular than you might imagine. Increasingly, compa-nies are putting GPS on all vehicles in their commercial fleets. Doing so helps long-haul trucking companies, delivery firms, police departments, taxi outfits, armored car carriers, and cable providers track employees, manage operations, increase fuel efficiency, and quantify the productiv-ity of employees. To the workers driving these vehicles, the addition of GPS tracking feels as if Big Brother is always watching them. In response, employees began sabotaging the devices by cutting the wires or remov-ing them altogether. Of course, doing so got them in hot water with their employers. Now a $50 jammer does the same trick and leaves no evidence behind.

The problem with these portable jammers is that they can extend up to five hundred feet around the vehicles using them. That means that depend-ing on the power of the device, for every one truck driver who doesn't want to be seen napping by his boss, fifty to a hundred cars will also have their GPS signals blocked. But your car or phone's navigation is actually the least critical network disrupted by the jammer. As we saw in the San Diego incident, though not immediately obvious, cell-phone towers, power grids, air traffic control, and ATMs also depend on GPS-embedded systems to function properly. When local truck drivers go off the grid, they are tak-ing many other people and services with them, and hundreds of incidents of collateral damage have been reported annually. For example, in Lon-don, for ten minutes a day, traders were discovering that their trades were not going through because there was a problem with the time-stamping mechanism in the system. Baffled exchange personnel wondered whether they were under some kind of cyber attack by a foreign power. Nope, just a London truck driver who parked his truck next to the exchange when making his deliveries once a day for ten minutes.

The disruption in London is but one of many such global incidents. In New Jersey, the government had installed a new GPS-enabled landing system at Newark Liberty International Airport to allow aircraft to land in poor visibility. For reasons unknown, the system was shutting down twice a day, causing air traffic controllers to scramble to guide in arriving aircraft. Though it took several months, officials discovered the disruption was caused by a single truck driver on the New Jersey Turnpike. He was using his trusted portable GPS jammer to avoid paying tolls on the highway (and crippling the screens of air traffic control at the same time). Of course, there are more profound criminal uses of GPS jammers than avoiding tolls on the New Jersey Turnpike. Stung too many times by surprise police raids, both organized crime groups and even neighborhood thugs have learned that if you're going to steal a car, especially one with cargo valuable enough to warrant a tracking device, you better be prepared to smooth your getaway with a GPS jammer—and that's exactly what they are doing. Police in the United States, Germany, Russia, and England have all seen stolen vehicles they were tracking suddenly disappear off their radars when criminals activated their GPS jammers, providing a protective bubble of safety for them to make their escapes. In one case in the U.K., an organized crime group successfully used GPS jammers to steal more than forty large tractor trailers containing cargo worth in excess of $10 million.

Given the level of disruption small GPS jammers generate, imagine what could be achieved with a larger model. Costing just a few thousand dollars, commercial radio-frequency jammers are widely available for sale on the Internet. Deploying one or two of these devices around a major metropolitan area could cause widespread disruption and would make a worthy target of attack for any terrorist organization trying to grab the world's attention. The threat is serious enough that the U.S. government has issued a chilling warning on the topic: "A 'multiple agency approach must be urgently developed and executed' to counteract the 'alarming' rise in availability of GPS jammers . . . The threat to our national security could be 'devastating.'"

Of course, there is a more sinister threat to the global navigation system than preventing the signals from reaching your screens: changing them before they get there. Not only can GPS jammers block locational signals, but GPS spoofers can alter the positional data you receive. The diabolical plot envisioned in *Tomorrow Never Dies* back in 1997 has become a reality, and GPS spoofing devices are also widely available online, allowing those with the means and technical power to broadcast their own fake earth-based GPS signals. Because of the weak nature of GPS signals, spoofers fool navigational devices by overpowering the legitimate signal with a stronger counterfeit one. Once this has been accomplished, criminals, hackers, terrorists, and governments can take complete control of any GPS receiver and connect it to a low-cost simulator capable of re-creating any route desired on a Google Earth map. Emitting phony signals can send an oil tanker into a bridge or an army convoy into enemy territory. Given how reflexively obedient drivers have become to their GPS devices, what havoc might a mass spoofing attack against a large city's drivers bring?

To date, GPS spoofing attacks have occurred on numerous occasions around the world. Think of the impact on just one industry: global cargo. According to Cargo Security International, cargo theft costs business $25 billion annually, and 90 percent of global cargo crosses the world's seas. GPS is a critical component in ensuring the right goods get to the right place at the right time. Yet spoofed navigation systems are capable of putting a major chink in the armor of this arrangement. All passenger and cargo ships at sea (approximately 400,000 vessels worldwide) rely on the Automatic Identification System (AIS) to report their positions to other vessels and to port authorities, which can view all nearby craft in real time. In 2013, however, security research proved AIS lacked even modest security controls and that the system was vulnerable to spectacular spoofing attacks. An assault on these systems could make entire oil tankers and cruise ships vanish from view, crash into one another, or run aground. Because GPS and navigations are "invisible utilities," we tend to forget about them, but we do so at our own peril. While, as of this writing, the exact location of the missing Malaysia Airlines Flight MH370 still remains a mystery, one thing is clear. The navigational systems responsible for tracking the flight were wholly inadequate for the task. Location matters, and poor, missing, or inaccurate navigational information costs lives.

We saw the power of GPS spoofing in action in mid-2013 when an $80 million yacht was hijacked by spoofing GPS signals. The sixty-five-meter luxury super-yacht *White Rose of Drachs* was traveling just off the cost of Italy when suddenly it began to drift to the right. The vessel had been on a Mediterranean cruise from Monaco to Rhodes when hackers fired up their blue-box spoofing device. They aimed their briefcase-sized device at the ship's navigation systems and slowly, imperceptibly, began to emit a counterfeit location signal. At first, the strength of the fictitious beacon was deliberately weak. Gradually, its resonance increased until it first matched and then overpowered the legitimate GPS signals received by the *White Rose of Drachs*. The hackers now had complete control over the super-yacht and could direct it anywhere they pleased. On the bridge, no alarm bells went off, and the captain continued to believe, erroneously, that he was still in command of his ship.

The hackers' false signals were indistinguishable from the authentic, and their mission was complete. Though those on board could tell that the yacht had made a pronounced turn in direction, inside the ship's command room all the screens responsible for its navigation showed the vessel to be traveling in a straight line. Spoofing on the high seas was now a reality. Fortunately for the passengers and crew of the *White Rose of Drachs*, the ship's hijacking was carried out not by Somali pirates but by some graduate students from the University of Texas, Jahshan Bhatti and Ken Pesyna. The pair were working with Professor Todd Humphreys, who has for many years raised concerns about the deep insecurity of the Global Positioning System and our dependencies on it.

Just as criminals have mastered GPS jammers to facilitate their robberies and escape routes, so too will they undoubtedly employ spoofers to misdirect 18-wheelers to the incorrect delivery points and cargo ships to the wrong berths, where they will be met by criminal gangs dressed as employees happy to unload all goods and merchandise in their containers. A confused GPS unit equals a successful heist. If the idea seems far-fetched, recall Moore's law and the iPhone in your pocket. Smaller, faster, cheaper means all of this technology filters down to criminals, often long before it is in common use by the general public. By controlling the navigational screens of ships at sea, cargo trucks, passenger cars, and even aircraft, hackers can project an altered reality, one that is indistinguishable from the truth, allowing them unprecedented control of a world that is run by computer code and screens of all shapes and sizes.

Our undeniable faith in screens can be manipulated in new and novel ways as well, even by falsifying the locational data we see on the apps on our smart phones. In early 2014, students at the Technion–Israel Institute of Technology in Israel hacked the incredibly popular Waze GPS navigational app (purchased by Google in 2013 for a cool $1 billion). The app, which offers crowdsourced real-time traffic management, relies on users to report accidents, police checkpoints, and road hazards as a means of improving vehicle flow. Once the app is launched on your phone, it uses the device's GPS readings to report how slow or fast your car is going to the Waze network, providing moment-by-moment intelligence reports on the state of congestion in a city. Normally, the app works brilliantly and is a lifesaver in cities with heavy traffic (as supported by Google's purchase price). Your Waze screen, like all others, however, is up for grabs by hackers.

The students at Technion registered droves of fake Waze users in the system, using an automated scripting program that they wrote to impersonate thousands of smart phones (a sock puppet attack of sorts). Next those virtual smart-phone users connected to another application that reported falsified GPS coordinates to the Waze system, making all the users seem as if they were legitimately moving about the city. Last, the sock puppets intentionally submitted thousands of reports "claiming to be stuck in traffic at the false coordinates." The result: The Waze system did exactly what it was supposed to do. It rerouted thousands of legitimate users away from a fictitious traffic jam, thereby causing true gridlock in the city as the unsuspecting drivers all converged at the same time on the previously unobstructed routes. Done on a mass scale, such tactics could cause further panic and chaos in association with any other criminal or terrorist attack. Rerouting traffic is one way hackers can make their victims come to them, but some clever Chinese hackers tried a different and more delicious tactic.

When General Tso Attacks

In what investigators called a "'coordinated, covert and targeted' campaign of cyber espionage against major Western energy firms," Chinese hackers reportedly pilfered "gigabytes of highly sensitive internal documents, including proprietary information about oil and gas field operations, project financing, and bidding documents." The perpetrators used a variety of techniques, but on occasion enhanced security measures employed at some of the oil firms posed challenges for the Chinese. Their response was not merely to hit their targets harder but to cause what is known as a "watering hole attack." Named after a similar maneuver used by lions on the plains of the Serengeti for millennia, the strategy allows predators to merely lurk by a watering hole known to be used by local herbivores. When zebra, antelope, and gazelle arrive, the lions pounce, killing their thirsty prey. The online equivalent entails infecting a Web site frequently visited by a hacker's targets. Once the innocent stop by and unsuspectingly click on a link or download a file, the virtual predator has his prey. The only question is, which Web site to infect?

After monitoring the online activities of their intended target (an unnamed American oil firm), hackers uncovered a revealing pattern. Their targets loved ordering food from one particular eatery in close proximity to the energy giant's headquarters—a Chinese restaurant famous for its delicious General Tso's chicken. In response, hackers infected the online menu of the Chinese restaurant with malware, and when the workers viewed the fare on offer, "they inadvertently downloaded code that gave the attackers a foothold in the business's vast computer network." The fact that the Chinese government used a Chinese take-out menu to resurrect the power of one of their fiercest generals is simultaneously brilliant, hysterical, and deeply ironic. You'll be glad to know that when Wang Baodong, a spokesman for the Chinese embassy in Washington, was asked about the incident, he said that "allegations about Chinese hacking had been raised unfairly. 'China has very strict laws against hacking activities, and China is also a victim of such activity.'" I wonder what General Tso would have thought of Mr. Wang.

Screen Play: Hacking Critical Infrastructures for Fun and Mayhem

All data presented on screens are hackable, not just information on laptops, iPads, or even Chinese menus. From the Jumbotrons at the Lakers game to the bright lights and news tickers of Times Square, screens abound, and each and every one of them can be manipulated, including our television screens. In 2013, hackers were able to take control of the Montana Emergency Alert System and issue an alert on the CBS affiliate KRTV. Afternoon television programming was suddenly interrupted by the attention-grabbing three crackling beeps and long squelching tone of the nation's Emergency Alert System, meant to warn the public of impending disasters ranging from earthquakes to hurricanes. In this case, however, the warning in Montana notified the public, "Civilian authorities in your area have advised that the bodies of the dead are rising from their graves and attacking the living." The ominous announcer warned, "Do not attempt to approach or apprehend these bodies, as they are considered extremely dangerous." After dozens of terrified citizens placed calls to the local sheriff's office, the station admitted that the alert was not its own: rather, the station's feed had been hacked and somebody had wrestled control of the airwaves, hitting screens away from CBS.

Even everyday common road signs are fair game for hackers. In Russia, the hacker Igor Blinnikov was able to seize control of another screen, a twenty-by-thirty-foot electronic billboard on one of the main thoroughfares of Moscow, which he commandeered at the height of rush-hour traffic. From his home seven hundred miles away, Blinnikov compromised the server of the advertising agency that owned the massive sign and replaced its video file ads for vodka and high fashion with those of hard-core pornography. In response, "traffic jerked to a standstill as rubbernecking motorists ogled a sexual pornographic clip posted by hackers on big-screen video billboards" on the Garden Ring Road, which just happened to be next door to the Ministry of the Interior. Needless to say, authorities were not amused, and Blinnikov was given a six-year sentence.

Public signage screens are increasingly being taken over to display political messages as well, even racist ones. At the height of the 2012 tensions over the Trayvon Martin shooting in Florida, tempers flared across the nation. It was with that case as a backdrop that somebody chose to hack into the operating system of a digital road sign on Interstate 94 in Dearborn, Michigan, and change its message to read, "Trayvon is a Nigger." The sign stood for all passersby to see on the busy freeway for over an hour until workers were able to power down and reboot the device. Such incendiary messages could easily push an already tense situation over the edge. By manipulating what we see on the screens, televisions, and billboards all around us, hackers can cause bemusement, panic, or outrage.

Hacked road signs, emergency broadcasts, and GPS signals are of concern because they form part of our critical information infrastructures: "those core elements of a modern society whose destruction or incapacity would have a debilitating impact on national security, the economy, public health or community safety." The Department of Homeland Security counts among these sectors the energy, food, agriculture, health-care, oil, gas, water, transportation, emergency services, defense, financial services, and transportation industries. Yet the one thing all of these crucial service sectors have in common is their near-total reliance on computer technology and screens as core elements of their safe and secure operation. As we saw at the Iranian nuclear enrichment facility in Natanz, however, such systems can readily be compromised. This fact holds significance for nearly all citizens in both the developed and the developing world.

Though the threats against each and every critical infrastructure sector are far too numerous to list here, just a few examples from the transportation industry alone prove instructive. Screens manage vehicular traffic, railroads, naval shipping, and air traffic control, and at nearly every step of the way the system can be compromised. Consider air transportation: If there is a single false entry in a terrorism watch database about a passenger, a plane can be rerouted mid-flight for an emergency landing or receive an escort by two F-16 jet fighters. Even the security process to get onto a plane is heavily dependent on screens. Transportation security officials don't pat down every passenger or open every bag. Rather, they let technology do the heavy lifting for them: X-ray machines for carry-on luggage and a variety of metal detectors, millimeter-wave scanners, and backscatter radiation detectors for passengers. Inherent in these security procedures, however, is a layer of technology that intermediates and separates human security officials from the things and people they are investigating, an opportunity for hackers to ply their trade with potentially deadly consequences.

Though airport scanners look like highly complex and specialized machines, their core processing functions are connected to and carried out by ordinary PCs running software on top of a typical Windows installation, and like all Windows machines they are eminently hackable. Even in 2014, many of these devices, such as the commonly used Rapiscan 522B, use Windows variants such as Windows 98 or even Windows XP, operating systems for which thousands of security vulnerabilities have been documented and Microsoft itself has stopped issuing updates. In addition, the banks of scanners at airports are often networked to one another via either Ethernet cables or Wi-Fi, two protocols that are also routinely hacked. Shockingly, operator passwords on many airport security detectors are "stored in plain text, and there are multiple ways to log in to the system without any prior knowledge of user actual names." Even if a hacker were to enter a completely made-up account and password, after showing an error, the system on these machines would still log in an attacker, as the security researcher Billy Rios at Qualys discovered.

Given the number of zero days and exploits for the underlying software running these systems, were an airport X-ray machine infected with malware and had a rootkit placed on it, hackers could completely control the images security officials viewed on their screens. A Tumi bag containing a bomb or firearm can thus be made to appear on-screen as a Tumi bag with three suits and a pair of Bruno Maglis. Screens intermediate security officials from their task at hand and as such are subject to traditional man-in-the-middle attacks. In a typical airport security configuration, one official watches the bags as they go into the machine, where they are X-rayed by a second official, while yet a third individual supervises the removal of the bags as they came out of the device. With segmented responsibilities such as these, the first and third screeners could view the Tumi go in and out of the device, while the second screener was presented with a video image of a completely different bag. Because the person in the number two position rarely physically observes the object, he or she relies completely on its computer representation to determine whether or not the bag passes security screening.

By commandeering an airport video screen monitoring station, hackers could allow weapons to pass through without detection. And though TSA would rush to deny the possibility, Billy Rios and his team proved that devices such as the Rapiscan 522B already have an embedded supervisory capability that allows TSA managers to see and control dozens of the machines at airports around the country in real time, affecting what individual screeners observe on their monitors. Shockingly, using a common hacker tactic, Rios was able to get around the log-in screen at the supervisory console station and take control of banks of X-ray scanning devices.

Of course, why even bother with going after piddly airport X-ray machines if the goal is to cause a major air disaster? The global air traffic control system too depends on screens, ones that hackers have already successfully attacked on numerous occasions. According to the Department of Transportation's inspector general, "Hackers have hobbled air traffic control systems in Alaska, seized control of FAA network servers, stolen the personal information from 48,000 current and former FAA employees and installed malicious code on air traffic networks." The inspector general "warned that the FAA isn't well equipped to identify intrusions into its computer systems" and noted that the agency had "detection sensors at only 11 of its 734 facilities across the country." Moreover, a security audit of the FAA's air traffic control networks uncovered 763 high-risk technological vulnerabilities within the system.

In the United States, the Federal Aviation Administration is spending billions to upgrade the nation's air traffic control system. The new system, called the Next Generation Air Transportation System, or NextGen, "will be highly automated. It will rely on GPS instead of radar to locate planes" (yes, the same GPS vulnerable to widespread systemic jamming and spoofing attacks). The FAA's upgrade will allow for more planes, helicopters, and even drones into our overly crowded skies by using automatic dependent surveillance-broadcast (ADS-B) network, little bits of computer code that a plane will constantly emit over radio frequencies to announce its identity and position to the world. Unfortunately, these signals are both

unencrypted and unauthenticated. As a result, they can be spoofed, causing chaos on the screens of air traffic controllers. Were hackers to inject
one hundred extra phantom flights onto a controller's screen, panic would
ensue. If the ruse ran for just an hour, the effects would ripple throughout the world of civil aviation, crippling global air travel. Worse, air force
analysts published an article in the *International Journal of Critical Infrastructure Protection* that warned that systemic flaws in ADS-B "could have
disastrous consequences including confusion, aircraft groundings, and
even plane crashes if exploited by adversaries."

Hackers commandeering the screens of the world's air traffic controllers is indeed a frightening prospect, but even more mundane screen hacks
can have momentous consequences, such as those against our voting systems. Nowadays, even good old-fashioned ballot boxes are transforming
themselves into software programs and touch screens. While rigging elections is nothing new (Saddam Hussein and Kim Jong-un both achieved
100 percent voter approval using paper ballots), the transition to entirely
digital systems creates new opportunities to not just hack computers but
hack democracy as well. There are dozens of reports of electronic voting
systems being compromised around the world.

In Washington, D.C., officials wanted to make it easier for their citizens to vote, especially absentee voters on active military duty. In response,
the city spent hundreds of thousands of dollars on an electronic voting system. District officials were rightly concerned, nonetheless, about the possibility of online vote rigging. Thus, before they actually launched their
system, they put it live online and dared hackers to see if they could break
the integrity of the online voting mechanics. Within forty-eight hours,
researchers from the University of Michigan were able to take full control of the Board of Elections' server. Not only could they change any
votes that came in, but they were also able to see how every voter voted,
thereby breaching the secrecy of the ballot system upon which democracy
is based. Once the Michigan team had its way with the district's voting
technology, the final ballot tally wasn't even close: Bender, the antihero
robot of *Futurama* fame, was elected school board president by a landslide.
He wasn't actually running, but he was the successful write-in candidate,
garnering the most votes.

Interestingly, while roaming around on the computers they had compromised, the University of Michigan team encountered other hackers
from Iran, India, and China also trying to subvert the system. As a calling
card and final insult to the world of online voting, the Wolverine hackers
altered the district's software so that anytime voters clicked the submit
button for their ballots, their computers' speakers would be taken over,
and they would be treated to a resounding chorus of the University of
Michigan fight song. District officials did not realize the system had even
been compromised until two days after the breach, when an elderly citizen

called city hall to say she found the online process easier than making it to the polling station. Only when she informed district officials how much she enjoyed the song that played after she had voted did the Board of Elections realize it had a problem. The experience of the District of Columbia is not unique, nor is the integrity of electronic voting systems in America and around the world an esoteric question, but one that is central to democracy itself. When votes become electrons recorded in computers, there is an opportunity for malicious actors to exert influence.

The problem with voting and managing air traffic control on screens is that the systems running these critical infrastructures are wholly insecure. By adopting them into our everyday lives without reflecting on the plainly obvious consequences, we are growing ever more connected, dependent, and vulnerable to subterfuge and putting ourselves at grave risk for future catastrophes. Given the opportunity for nation-states to hack a nation's critical information infrastructures, it should come as no surprise that they are increasingly doing just that for the purposes of warfare and armed conflict.

Smoke Screens and the Fog of War

All warfare is based upon deception.
SUN TZU

Since the days of Sun Tzu, military forces have relied on the art of deception in order to obtain a tactical advantage over their enemies. In ancient Greece, it was the gift of a large wooden horse presented to the people of Troy that misled. During World War II, it was the fake radio transmissions and inflatable balloon tanks of Operation Fortitude that falsely signaled an Allied invasion on the beaches of Calais (vice Normandy) and allowed American and British troops to retake the European continent and defeat the Nazis. Given that today's soldiers experience the world through their computer screens, it is only logical that that information technology has become the latest battlefield in warfare. Screens tell battle commanders the locations of their aircraft, ships, tanks, and troops. Screens manage logistics and supplies, and screens provide up-to-the-moment intelligence on the plans, capabilities, and intentions of the enemy. Of course it should come as no surprise that they too are increasingly the targets of choice when attempting to deceive or defeat an enemy.

In modern military doctrine, there are many names for these types of activities, alternatively described as information operations, electronic warfare, computer network operations, information warfare, or psychological operations. Their common goal is to "influence, disrupt, corrupt, or usurp the decision making of adversaries." In years past, this might have been done by spreading false rumors and misinformation by word of mouth to one's adversaries or by dropping leaflets over civilian populations with propaganda messages. Today it's all about the screens. Screens and information technologies are perfectly suited for deception. The code is weak and readily corruptible, making the systems deeply vulnerable. These systems are almost all connected in one way or another to the global information grid, allowing them to be penetrated by enemies thousands of miles away. Finally, these technologies form part of any nation's critical information infrastructure, a dependency that makes a government and a people vulnerable when these systems are attacked or debilitated.

Some of these attempts at deception are simplistic. In the battle between the Syrian government and its rebel forces, a Reuters news site was hacked to disseminate a false news report suggesting the rebels had suffered tremendous defeat in Aleppo, which in fact they hadn't. Other digital smoke screens are much more sophisticated, such as the reported successful hacking of Syrian military radar capabilities by Israel Defense Forces preceding an attack against a nuclear site under construction in northern Syria. Dubbed Operation Orchard, the air strike successfully destroyed a secret military nuclear reactor being built with the assistance of the North Koreans. The raid required the Israelis to overfly Syria deep into the country's territory, nearly to the border with Iraq. In order to do so without a hot war breaking out and the Israeli jets being shot down, the Israelis hacked Syrian air defenses, effectively blinding the Assad government to the attack as it was happening. Though enemy jets were en route to their target deep inside Syria, all was calm and clear on the screens of the Syrian air force. The screens showed a different reality on the ground from that in the sky.

In the world of information operations, the players are many, and those perpetrating the attacks one day may find themselves victims the next. Such was the case at the height of the Israeli conflict with Hamas in the Gaza Strip in January 2009. Inside both Israel and Gaza, tensions were running high. As the Israelis began to mobilize troops in the south for a possible incursion into Gaza, hundreds of reservists began to receive their "Tzav Shmone," or emergency call for duty messages, via both voice mail and texts on their mobile phones. The reserves were being activated, and things were getting serious for all parties.

Many Israeli soldiers were ordered to report, however, not to the front along the southern border with Gaza but rather to the very north of the country, to an Israel Defense Forces recruitment center in Haifa. As it turned out, these Tzav Shmone were fictitious, likely perpetrated by Hamas. At a time when Israel needed its soldiers to report for duty near Gaza, they were being misdirected to the north, because they relied on their screens for instructions. Sun Tzu would have been proud. Both Israel and Hamas have mounted electronic psychological warfare against each other and Hamas claimed it was capable of sending seventy thousand text message an hour to Israeli mobile phones, proving technological tools developed by nation-states rapidly devolve into the hands of both non-state actors and terrorist organizations in time.

There is another way both governments and non-state actors are battling for supremacy over your screens—via sock puppetry. Remember those 140 million fictitious Facebook accounts? Not all of them are destined to be used as fake Likes for Shakira. As it turns out, military and intelligence officials around the world have flocked to social media in an effort to manipulate and deceive what is seen on our screens. It was widely reported that the U.S. government extensively uses sock puppets as part of its psychological operations (psyops) to counter "extremist ideology and propaganda." What that means is that the Americans monitor jihadist Web forums, and when "Abdul" says "death to the infidels," the Pentagon can have a virtual "Hassan" in its back pocket ready to respond with a verse from the Quran extolling peace, mercy, and understanding. Of course that's just the beginning of the capability. Fake personas scale too, and with thousands of sock puppets under one's control, influence and opportunity for deception grow exponentially.

In June 2011, it was revealed that the U.S. Central Command had awarded a $2.76 million contract to a California company to create fake online personas for the purpose of manipulating online conversations and spreading pro-American points of view in social media. Each fake online persona was contractually required to have a plausible personal history and that "up to 50 US-based controllers . . . be able to operate false identities from their workstations 'without fear of being discovered by sophisticated adversaries.'" The military's goal was to create an online persona management dashboard that would allow each human serviceman or servicewoman to control ten separate identities based around the world in order to "degrade the enemy narrative." Talk about an exponential projection of force! The sock puppets operated in Arabic, Farsi, Urdu, and Pashto and allowed U.S. service personnel working around the clock to manipulate online conversations as desired. The sock puppet contract was part of a much larger $200 million military coalition operation, ironically called Operation Earnest Voice (OEV). OEV was first developed in Iraq "as a psychological warfare weapon against the online presence of al-Qaida supporters and . . . jihadists across Pakistan, Afghanistan and the Middle East."

Once an exponential virtual deception engine has been constructed, those who run and operate it have tremendous power to quell dissent and "degrade the narrative" of their enemies. The only things preventing the use of such a tool for domestic oppression would be public policy and the law—both quite malleable and fungible themselves. According to Freedom House, an NGO founded in 1941 to advocate for democracy and human rights, at least twenty-two governments around the world manipulate social media for propaganda purposes, including Venezuela, Egypt, and Malaysia.

In Russia, an undercover investigation by the *St. Petersburg Times* revealed that numerous covert organizations exist that hire young tech savvy "Internet operators" to post pro-Kremlin articles and comments online and to smear opposition leaders. Each Internet operator is paid approximately $36 for a full eight-hour shift and is expected to write at least a hundred posts a day. Russia's president, Vladimir Putin, a former KGB lieutenant colonel, is well versed in the art of propaganda and reportedly uses an "invisible army of social media propagandists" to generate up to forty thousand comments a day on his behalf. Whether the international or domestic press in Russia is writing about gay rights or opposition candidates, armies of sock puppets are poised to strike back instantly and with a vengeance. In recognition of their outstanding service to the nation, particularly during the "liberation" of Crimea, Putin awarded many of these social media operatives "Orders of Service to the Fatherland."

Of course the Russian operation to shape what people see on their screens is paltry compared with the capabilities developed by the People's Republic of China. According to the *Beijing News* and state media reports, China employs approximately 2 million online propaganda workers to help shape online public opinion and manage domestic Internet surveillance. These commentators are paid to "blitz social media with state-approved news and ideas." In early 2013, China's propaganda chief, Lu Wei, whose official title is chairman of State Internet Information Office, directed his 2.06 million netizens to open up accounts on social media sites such as Weibo, a Twitter-like micro-blogging site, in order to spread "positive energy" and help guide online discussions of sensitive topics "in a positive direction." These workers also received training on how to frame online discussions and steer conversations away from political hot potatoes as well as to question the value of Western concepts of democracy.

Government sock puppetry is a powerful complement to censorship and Internet surveillance. Censorship ensures the greatest number of "undesirable" ideas never make it past a national firewall, and if they do, legions of sock puppets can be covertly unleashed to undermine any idea that does not sit well with those in power. In both instances, screens are heavily manipulated to ensure those in power stay there and that no threatening new ideas challenge their authority. Each and every day around the world, display wars are taking place as governments, multinational corporations, criminals, and terrorists battle to shape and control what is seen online. What ensues is a real but covert war on reality, one that is meant to blind us to the truth. Sadly, the situation is about to get much worse as new generations of even more powerful technologies come online, further separating us from ever experiencing a reality that has not been in one shape or form intermediated by somebody else.

Control, Alt, Deceit

One of the definitions of sanity is the ability to tell real from unreal.
Soon we'll need a new definition.
ALVIN TOFFLER

In 1865, Congress passed legislation allowing the director of the U.S. Mint to add the motto "In God we trust" to all gold and silver coins minted for circulation. The line, originally drawn from the fourth stanza of "The Star-Spangled Banner," has since become the official motto of the United States. While many Americans on a spiritual level have deep convictions about their trust in God, from a practical perspective something has shifted. They may go to temple on Friday nights or church on Sundays, but they look at screens every single day. It is as if we have transformed into an "in screen we trust" culture. If something is on a screen, whether it be a computer, iPad, industrial control system, street sign, GPS device, radar installation, or mobile phone, our first inclination is to trust what we see before us. However, we have shown time and time again that everything from our friends on Facebook to the numbers we dial on our mobile phones can be rigged to deceive us. The problem is that we are leading lives fully intermediated by screens and other technologies that, although they give the appearance of transparency, are in fact programmed, controlled, and operated by others. Worse, none of us have a freaking clue as to how any of it works.

Increasingly, we are living in a "black box" society, one in which magical boxes provide directions, report the news, execute stock trades, make phone calls, recommend restaurants, and put the world's knowledge at our fingertips. But how all this mystical technology operates is almost completely opaque to the average user. While most of us are pleased not to have to learn the intricacies of writing computer code in order to make a phone call, visit an ATM, vote, or apply the ABS brakes on our cars, those who possess this know-how are at a distinct advantage moving forward. They are poised to shape the world for the great unwashed masses who would rather leave such technical unpleasantries to others to sort out. In an exponentially changing world driven by Moore's law, Moore's outlaws very much have the upper hand.

As noted in the first chapter of this book, every day we are becoming more connected, dependent, and vulnerable. The overwhelming majority of our information systems can be penetrated in mere minutes, and there has been exponential growth in the number of viruses, Trojans, and zero days available to accomplish the task. The average time to discovery from the moment an intruder first breaks into a system until the hack is uncovered is measured not in minutes but in hundreds of days. We are being penetrated, digitally probed, spied upon, robbed, and virtually manipulated day in and day out, and most of us remain blissfully unaware of the threat. Welcome to the new normal, a world in which for every screen in your life governments, criminals, terrorists, and hacktivists have a plan of attack.

In the end, all the computer hacking, code manipulation, and screen shifting boil down to a fundamental issue of trust. Trust is at the core of all of these discussions, and currently in our world there is no such thing as trustworthy computing. The security, privacy, and reliability of technology are too easily disrupted, sabotaged, and undermined. The fact of the matter is we have no earthly idea what is going on inside our systems, the same ones we use every day personally, professionally, and to run the world. While we may still faithfully place our trust in God, placing our trust in screens is deeply misguided and will come back to bite us in ways we will regret.

The Heartbleed security bug that burst into prominence in early 2014 is emblematic of the challenges we face. In theory, cryptographic algorithms are meant to secretly encode and decode sensitive information passed between two parties. The most common encryption protocols on the Internet are the Secure Sockets Layer (SSL) and Transport Layer Security (TLS). In fact, a version of SSL, known as open SSL, is responsible for protecting more than two-thirds of all Internet traffic. Even if you don't know what cryptography or SSL is, chances are you use it every time you log in to your bank, check your e-mail, or buy something online. We've all been trained to look for the little green lock in our Web browser's address line and to search for HTTPS versus HTTP to ensure our connections to a given Web site are trusted and secure. Green means go, it's safe, all is okay—at least we thought so.

The core big reveal about the Heartbleed bug is that even though the little green locks on our browsers were showing us we were safe, in fact, we were not. The trust we had in closed SSL locks on our browser screens was in fact misplaced. Once again, "in screen we trust" deceives. Heartbleed is the largest and most widespread vulnerability in the history of the Internet to date. A programming flaw in open SSL meant that those secret cryptographic keys you thought you were privately sharing with your bank or social media company's server were in fact suddenly accessible by somebody else. Worse, the flaw was completely undetectable, even though it had been in existence since December 2011. That means that all the chat messages, e-mails, online purchases, Web site visits, and downloaded apps during the past several years were in fact fully accessible to somebody with the time, energy, and inclination to decipher them.

Open SSL is used by 66 percent of all Web sites on the Internet, which is why millions of Web sites around the globe had to inform their users that there was a big gaping hole that allowed hackers to circumvent the encryption between you and their sites. Instagram, Pinterest, Facebook, Tumblr, Google, Yahoo!, Etsy, GoDaddy, Foursquare, TurboTax, Flickr, Netflix, YouTube, USAA, and Dropbox are just some of the companies to have been affected by the problem. Moreover, 150 million apps downloaded on the Android mobile phone platform were also susceptible. Sadly, changing your password was not good enough to resolve the problem on the consumer's end. Each of these Web sites first needed to change its server software and update the version of open SSL it was using; otherwise any potential attackers would still be able to read your new password even after you changed it. Even a full month after the Heartbleed bug was announced, hundreds of thousands of sites remained vulnerable to the massive flaw in the cryptographic backbone running the majority of the Internet. Of course attackers lost no time in taking advantage of the Heartbleed opportunity, including the NSA, which had reportedly known about the vulnerability for years but kept it to itself in order to exploit the opportunities it provided. Criminals also took part in the gold rush created by Heartbleed, with attacks carried out against the Canada Revenue Agency (Canada's IRS equivalent) and dozens of e-commerce sites around the world.

Cryptographic keys and digital certificates are the means by which our online data and their underlying technologies are meant to be protected and secured. Yet Heartbleed was not the first time these systems themselves were successfully subverted. By and large the tools to make our technological world secure and trustworthy are simply not at hand. As a result, we do not have the means we require as a global society to make smart and reliable decisions in an increasingly confusing world. Human beings don't directly read the ones and zeros on our hard drives, nor do we think in binary code (at least not yet). We use a bevy of screens and other machines to interpret this information for us and, in doing so, sacrifice any real hope of understanding the innermost truth of anything. As long as others can intermediate our digital and virtual experiences, we remain deeply at risk for fraud, abuse, and attack—not a foundation on which to build any future civilization.

The biggest challenges we face in the "in screen we trust" world are not the problems of today, however, but the ones of tomorrow. Given the obvious implications of Moore's law, the number of screens in our lives today will pale in comparison to those to come. To paraphrase the rapper the Notorious B.I.G., "Mo' screens, mo' problems." We will have screens everywhere—on our wrists, in our glasses, on our contact lenses, and in our clothes, as so-called wearables become commonplace. In our homes, our dining room tables, picture frames, refrigerators, and washing machines will all be transformed into screens. As we travel about our daily affairs, we will have screens in our cars, on our trains, and on the headrest of every airplane seat. Menus at restaurants, mirrors in the ladies' room, and the walls behind the urinal in the men's room will all beacon us with visual information. Not only will all billboards become screens, but so too will the walls of homes, office buildings, and shops. Heads-up displays like those used by jet fighter pilots and augmented reality will become mainstream and will project layers and layers of virtual information into our line of sight, ever influencing our points of view. In fact, every possible flat surface will be transformed into an interactive screen, each serving as a filter for our reality, easily manipulated by those whom we allow to interpret the real world for us.

There are ghosts in the wires, screens, and data banks of our twenty-first-century world. As the digital and the virtual drown out the real, our lives will be intermediated by others, but at what cost? The global information grid that we are all increasingly connected to and dependent on is deeply vulnerable.

There is a gathering storm before us, and all the signs of disaster are there. The technological bedrock on which we are building the future of humanity is deeply unstable and like a house of cards can come crashing down at any moment. Despite this, we plod forward, adopting newer, brighter technologies, each promising to solve a new problem or deliver a particular convenience. The problem is not that technology is bad; in fact, science and technology hold the promise of profound benefit to humanity. The problem, as we have seen, is that those with technological know-how, be they criminals, terrorists, or rogue governments, can use their knowledge to exploit an exponentially growing portion of the general public to its detriment. Though today's technologies have been a boon for illicit actors, they will pale in comparison to the breadth and scope of technological change that will rapidly unfold before us in the coming years. Soon a plethora of exponential technologies now just in their infancy, such as robotics, artificial intelligence, 3-D manufacturing, and synthetic biology, will be upon us, and with them will come concomitantly profound, perhaps even life-altering, opportunities for harm.

Though criminals have taken advantage of the techno-tools available to them to date, the worst may be yet to come. Vulnerable and untrustworthy computing has prepped the battlefield for a future world replete with criminality and social insecurity. The gathering storm has gathered and the result may well be a destiny for which we are wholly unprepared. Welcome to the future of crime.

The
Future
of
Crime

CHAPTER 10

Crime, Inc.

**Organized crime in America takes in over forty billion dollars a year . . .
[and] spends very little for office supplies.**
WOODY ALLEN

nnovative Marketing was a small and promising start-up that created pioneering software products to address its clients' needs. The firm's young
founders incorporated their company in Belize because of its favorable tax
regimes, a smart move that they modeled on the business practices of well-
established tech giants, such as Apple, Google, and HP, each of which has
cleverly created subsidiaries in tax havens around the world. To further
reduce overhead costs, Innovative Marketing chose to establish its main
offices in Kiev, Ukraine, where highly competent technical graduates with
advanced degrees in computer science and mathematics were abundant and
employees could be hired for a fraction of the salaries offered in Silicon
Valley.

Like any good tech start-up, Innovative Marketing advertised its wares
across the Web using banner ads and paid to ensure its software appeared
high up in search engine query results. It attracted new customers by
turning to a well-honed and tested technique developed by Amazon.com
known as affiliate marketing: if a potential customer clicked on the affiliate
link, Innovative Marketing would pay the hosting Web site a small fee for
serving up the ad, and if any actual sales were generated, the affiliate would
receive a percentage referral fee. The system worked well for all parties: it
incentivized a commission-based workforce and drove software sales for
the young start-up.

The two entrepreneurs who founded Innovative Marketing, the India-born Shaileshkumar "Sam" Jain and the Swedish national Bjorn Sundin, had picked their software product lineup well. The pair decided to focus their creative energies on designing an entirely new class of antivirus and computer security software back in 2006, just as the world was growing increasingly concerned about cyber threats. Soon business was booming and sales of the company's products, such as Malware Destructor, System Defender, and Windows AntiSpyware, were growing year over year. Soon hundreds, then thousands, then millions of orders for its products flooded into the firm's Kiev offices.

Innovative Marketing, like so many successful start-ups, had more demand than it could supply and was struggling to keep up with its rapid expansion. Before long, the company occupied a full three stories of modern office space at 160 Severo-Syretskaya Street in the burgeoning industrial section of Kiev. Inside, dozens of highly talented computer geeks churned out code at a frenzied pace, as engineers laid out clusters of new Ethernet cables and added racks of computer servers trying to keep up with consumer demand.

In the lobby of Innovative Marketing's growing headquarters, workers hung a colorfully backlit five-foot-square glass logo that they suspended behind a bank of receptionists, busy answering phones and greeting employees at the start of their day. Beyond the ultramodern reception area, executives were abuzz establishing business processes and putting systems in place to provide the corporate structure required to grow the firm. Soon, department after department was added, including software development, quality assurance, finance, billing, marketing, human resources, translation and software localization, research and development, production, outsourcing, and technical support. Jain and Sundin, like any proud parents, were watching their baby grow.

Within short order, Innovative Marketing had become a massive success—a global multilingual company, operating around the clock, with more than six hundred employees and customers in sixty countries. Through its subsidiaries, it outsourced call center functions to India to handle technical support and customer service queries in English. German speakers had their questions answered by bilingual staff in Poland, and Francophone clients were routed over VoIP to Algeria. Sales of Innovative Marketing's software were all automated and distributed online. Customers could buy their products at the click of the mouse and product ID numbers were issued on e-mailed receipts, which offered money-back guarantees on goods sold. Innovative Marketing took customer service seriously and advised clients calling its 800 numbers that calls would be monitored for quality assurance. According to statistics kept by the call centers, over 95 percent of clients described themselves as "happy" with the service they had received.

Like all tech start-ups, Innovative Marketing was well represented on social media. Hundreds of its employees had established profiles on LinkedIn, including their positions and work histories. To bring in the talent required to grow the start-up, Innovative Marketing placed job ads on numerous career Web sites and used recruiters to help find project managers, UNIX administrators, search engine optimization specialists, researchers, support engineers, and business development associates. To manage its explosive growth, Innovative Marketing used a variety of techniques to address the human resources issues common in the start-up world. It offered prizes to the best salesmen and carefully selected its employees of the month.

To relieve stress caused by the frenetic pace of work, Innovative Marketing also rewarded its employees with staff outings to seaside resorts where employees would engage in team-building exercises, including footraces, wall climbing, rope exercises, and paintball competitions, in order to build morale and cooperation. By all accounts, Innovative Marketing was a great place to work and a wildly profitable business. From the customer's perspective, however, there was a slight problem.

The typical scenario went something like this. As users sat at their keyboards, pining away on Facebook, responding to an e-mail, or checking the latest quarterly report, suddenly a large red pop-up would appear on the center of their screens: WARNING: SERIOUS VIRUS DETECTED. Simultaneously, the computers' speakers would begin to wail, as a blaring siren sound let users know something was seriously wrong with their system. In an instant, the System Defender logo would appear on-screen next to a large magnifying glass that appeared to be scanning the files on the users' hard drive. One by one, long and complex system file names would fly by in rapid succession as a mounting tally of malware threats detected was displayed on a scoreboard on the bottom of the screen. In the end, System Defender might show twenty-three known viruses, seven worms, and eighteen pieces of spyware along with an ominous warning: YOUR COMPUTER IS AT IMMINENT RISK FOR A SYSTEM CRASH AND PERMANENT DATA LOSS. CLICK HERE TO REMOVE ALL THREATS.

As the siren continued to screech in the background on the computers' speakers, users generally chose the most obvious course of action, clicking on the "remove threats" button glaring before them. When they did, they were directed to a purchase page for Innovative Marketing's System Defender product, a $49 software program guaranteed to resolve all known computer issues. Those who foolishly opted to ignore the "remove threats" option and tried clicking anywhere else on the screen soon discovered that their computer had completely locked up, save for the obnoxious siren noise. The Escape key did not work, and users were permanently stuck on a red screen of death, unable to control their own computers. Savvy users thought rebooting might resolve the problem, but when they did, they were met with the blaring siren noise and the same implacable red alert screen. Paying the $49 fee was the only way to regain access to their own computers and data (a deluxe version with unlimited tech support was available for $79).

So what exactly was this pioneering software product Innovative Marketing had created? It was called crimeware, a whole new product category within the software industry—software that commits crime. Crimeware, sometimes called scareware, ransomware, or rogue antivirus, is nothing more than a malicious computer program that plays on a user's fear of virus infection. We've all been trained to be on the lookout for antivirus alerts and to run our security software when a problem is detected. Thus it seemed entirely logical that when System Defender's critical system pop-up message appeared on the screens of users around the world, the best and commonsense course of action was to click on the "remove all threats" button. There was only one hitch: the warning messages displayed were nothing more than an elaborate software hoax, a case of "in screen we trust" gone wrong.

Innovative Marketing's customers never actually had a virus; instead, their browsers and operating systems had been hijacked. The animated graphical image that gave the appearance that the user's computer was being scanned for viruses was just a visual ruse, no different from a Disney animation. No scan of the computer ever actually took place, and the "found" viruses and Trojans detected were virtual figments of the software's imagination, projected convincingly on the screen. Once users were tricked into paying for and downloading the System Defender product, the software had one primary mission: to remove a user's legitimate antivirus program, thereby allowing additional malware, back doors, and keystroke loggers to be installed on the affected hard drive. Worse, those credit card details provided to buy the bogus software were now up for sale to the highest bidder on the black market. Innovative Marketing, for all its call centers, gleaming offices, and employee retreats, was nothing more than an immensely successful front for modern organized crime.

Innovative Marketing was able to create the massive market for its felonious products by using its own teams and those of its affiliates to booby-trap legitimate Web sites with malware-infected ads sold by subsidiary front companies. When an unsuspecting user innocently visited an infected Web site or clicked the wrong link, a bit of malware code was downloaded to infect the machine, allowing the programmers at Innovative Marketing the access they needed to pull off their convincing red-screen scams. Eventually, after numerous customers complained to authorities in dozens of countries, the criminal enterprise was exposed, and the results of the investigation were shocking. Innovative Marketing kept in its offices copies of all the receipts it had issued to its crimeware customers around the world. In 2009 alone, it processed 4.5 million individual customer orders for an average sales price of $35. That works out to $180 million in revenue for Innovative Marketing back in 2009, handily beating the $106 million earned by Twitter two years later in 2011. In total, Innovative Marketing pulled in a jaw-dropping $500 million in global sales for the three-year period in which it sold its crimeware.

As it turned out, the founders of Innovative Marketing chose to locate their company in Ukraine not just because technical talent was cheap but because authorities asked few questions and law enforcement cooperation was easily purchased. There young workers, like twenty-year-old "Maxim," a former programmer for Innovative Marketing, admitted that frequent bonuses made it easy to ignore the ethical implications of the company. "When you're just twenty, you don't think a lot about ethics," he added. As for the company's founders, Messieurs Jain and Sundin, they are under indictment for their activities, wanted by both the FBI and Interpol. They were able to flee, however, to safe havens before they could be arrested, and their whereabouts remain unknown.

With their hundreds of millions of dollars stashed in secret bank accounts around the world, these Internet entrepreneurs achieved what most Silicon Valley entrepreneurs only dream of: a successful exit for their start-up. Though no longer in operation, Innovative Marketing was probably one of the most lucrative techno-centric criminal operations ever known. It is, however, by no means the only one. An estimated thirty-five million PCs worldwide continue to be infected with these rogue antivirus programs every month, putting $400 million a year in the hands of the remaining global cyber-crime syndicates. Welcome to the world of Crime, Inc.

The Cyber Sopranos

You been reading a lot of stuff about "Crime don't pay." Don't be a sucker!
That's for yaps and small-timers on shoestrings. Not for people like us.
JAMES CAGNEY, IN *ANGELS WITH DIRTY FACES*

Crime is big business and the United Nations estimates that transnational organized crime rakes in more than $2 trillion a year in profits. The money comes from the narcotics trade, intellectual property theft, human trafficking, counterfeit goods, child pornography, identity theft, wildlife smuggling, and of course cyber crime. In total, organized crime is believed to account for up to 15 to 20 percent of global GDP. Consider it the world's largest and most illicit of social networks, a constant cycling and recycling of people and contraband goods, spanning the globe, in operation twenty-four hours a day, seven days a week. Thanks to Hollywood, most of us have images of prototypical gangsters in mind when we think of organized crime, including mob bosses such as Tony Soprano, Vito Corleone, and

Tony Montana. Today's modern criminals, however, have mostly forgone the hierarchical structures of days gone by in favor of modern corporate organizations. Capos, dons, and lieutenants have been replaced with local, just-in-time, outsourced, ad hoc criminal networks that rapidly assemble and re-form to exploit any potential illicit opportunity.

Modern times call for modern crimes. As a result, the Tony Sopranos of the world have built and nurtured a much more powerful, far-reaching, increasingly profitable, and technologically competent criminal workforce. To this end, traditional crime groups—such as the Cosa Nostra (Italian Mafia), Japanese Yakuza, and Chinese Triads, as well as the Russian and Nigerian mobs—have all opened cyber-crime divisions to take advantage of the high-reward, low-risk profits available to them in a globally connected world. Cyber crime is borderless and offers great anonymity. Moreover, prosecutions are exceedingly rare, perhaps occurring in less than one one-thousandth of 1 percent of all cases.

The second major trend leading to the explosion in organized cyber crime has been the professionalization of hackers themselves. Their modus operandi has changed significantly since the good old days of the 1980s, when most hackers were tinkering with computer systems out of curiosity or to prove their own technical prowess. Hacking is no longer ruled by pimply teenagers wreaking havoc from Mom's basement; in fact, today more than 40 percent of organized cyber criminals are above the age of thirty-five. Long ago, individual hackers figured out there was money to be made in subverting technology, and criminal hackers such as Albert Gonzalez were born. They realized it was possible to earn a good living by illegally breaking into the computer systems of others. Over time, word got out, and soon hackers were unifying around the world in underground networks, collaborating, and competing for criminal profits.

Hacking became a fully monetized activity, and the shift from hacker hobbyists to for-profit criminal hacking gangs was complete. New transnational cyber-crime syndicates, such as the Russian Business Network, ShadowCrew, Superzonda, and of course Innovative Marketing, were established to meet the wide-ranging opportunities in next-generation crime. And business is booming. As if the threat from individual hackers' stealing credit cards and Mafia thugs' breaking kneecaps weren't bad enough, today traditional organized crime groups and highly talented hackers have united to combine forces, and the results for the general public and business are disastrous. While historically perhaps 80 percent of hackers were independent freelancers, today the opposite is true. According to a 2014 study by the Rand Corporation, a full 80 percent of hackers are now working with or as part of an organized crime group.

The Rand findings remind me of the great scene from the 1980s film *Ghostbusters* in which Bill Murray, Harold Ramis, and Dan Aykroyd have armed themselves with "proton pack" weapons to defeat the ghosts who have invaded New York City. At one point in the film, Ramis advises his two co-stars, "There's something very important I forgot to tell you . . . Don't cross the streams of your weapons . . . It would be bad." Murray asks, "How bad?" Ramis replies, "Try to imagine all life as you know it stopping instantaneously and every molecule in your body exploding at the speed of light." Murray's response is deadpan: "Okay, that's bad. Thanks for the important safety tip." Borrowing from Murray and Ramis, our online and off-line worlds are converging, proverbial criminal "streams" are crossing, and we are now entering the great age of digital crime. In this new realm of digital criminality, hackers and old-school gangsters have joined forces in a modern "Legion of Doom" focused on leveraging technology to the fullest extent possible to maximize their power and profits at the expense of you and me.

This criminal exploitation of technology is nothing new, per se. When most cops were on foot or horseback, Chicagoland gangsters began using automobiles for their getaways. When the average patrol officer was issued a six-shot revolver, George "Machine Gun" Kelly was using automatic weapons. Drug dealers were the first major demographic after physicians to use pagers and had access to mobile phones long before any police officer was using them. Technology makes crime more efficient and so criminals are perpetual early adopters of all things tech.

Outlaws have proven particularly adept at using and exploiting technologies created by others and co-opting them for their own purposes, always on the lookout for new opportunities. Just as smart phones with Internet connectivity were coming into fashion, organized crime groups in Mexico City began using them for research purposes. What were they researching? Whom to kidnap, of course. As wealthy executives landed at the Mexico City International Airport, there was a smorgasbord of potential kidnap victims, but criminals wondered which companies might pay the largest ransom (greatest ROI) to get their executives back? Such a difficult thing to know; that is, until the smart phone came around.

Organized crime teams deployed at the airport had stationed themselves in the arrivals area, next to baggage claim, where rows of smartly dressed chauffeurs waited for the business travelers who had reserved their services. Each chauffeur carried a large cardboard sign with the name and company of an expected passenger—Mr. Smith from Merck pharmaceuticals or Ms. Jackson from Goldman Sachs, for example. The criminal gangs at the airport used the information on the chauffeurs' signs to Google the executives on their smart phones and determine their corporate positions and net worth. Once they had found the biggest fish, kidnappers merely approached the chauffeur holding the most profitable sign and paid him to get lost, or else. A substitute criminal chauffeur kept the sign taken from the legitimate driver and held it calmly, awaiting his mark. The trap had been set and the deplaning executive walked right into the arms of the faux chauffeur, all because a cardboard "screen" had been hacked. Several executives were kidnapped and others killed using the smart-phone research technique.

Whatever the technical innovation, criminals are quick to adapt, either by mimicking legitimate Internet start-ups or by abusing their services. Borrowing a page from Uber, the ride-sharing phone app that connects crowdsourced drivers to passengers, a woman in the U.K. created her own SMS vehicle-on-demand service—for getaway cars. Sensing a market need by criminals without wheels, Nicole Gibson of Londonderry created a real-time "text a getaway driver" service to help robbers make a clean escape with the goods they had stolen from homes and businesses along the Irish border. In San Francisco, drug dealers in Dolores Park began using Square, a small white plastic device that connects to the iPhone and allows anybody to accept credit card payments, enabling hipsters who eschew cash to charge their ecstasy and pot. In New York, prostitutes tired of the cameras and overly inquisitive doormen at chic Manhattan hotels have turned to Airbnb to rent apartments for their trysts. The prostitutes pose as students or tourists, and the unsuspecting New Yorkers who rent their apartments have no idea their own beds are being used to entertain multiple clients and to host orgies. One escort service claimed it was saving "a fortune" by using Airbnb. "It's more discreet and much cheaper than The Waldorf," said a twenty-one-year-old sex worker. Whatever the technology or Internet service, criminals are there at the earliest stages, innovatively turning the newfangled tools to their advantage.

Crime, Inc.—the Org Chart

On the home page of Innovative Marketing's Web site, it, like many Internet businesses, had helpfully included both an "About Us" and "FAQ" section for visitors to their site. Those who clicked "About Us" learned that "Innovative Marketing has been working hard to develop several products that help the consumer adapt to the change technology brings." That's one way to put it. No doubt had it written, "Innovative Marketing works hard to rip off people around the world by tricking them into believing they have a virus and duping them into paying $49 to remove something that doesn't exist," fewer people would likely have purchased its product. While organized crime groups themselves are not forthcoming about their actual structure and business practices, a variety of undercover operations, law enforcement sources, and cyber-security intelligence firms have shed light on their business structure and organization, which are presented below.

Surprisingly, the org chart of Crime, Inc. would look remarkably familiar to anybody in the traditional corporate world. It's part Peter Drucker mixed with the latest cutting-edge business practices taught at Wharton or Harvard Business School. While there are elements of the digital underground that are not purely motivated by profit, such as hacktivists, Crime, Inc. is first and foremost about the money—shareholder value, if you will. These criminal enterprises go to great lengths to ensure their sustainability and as such are almost exclusively located in jurisdictional safe havens, places with weak governments, unstable political regimes, and police forces willing to look the other way, for a fee of course. Within these criminal syndicates, there are divisions of labor, supply chain management, department heads, outside consultants, and team deliverables. To understand the power and professionalism of Crime, Inc., we must first and foremost take a look at its org chart in order to deconstruct the modern criminal organization. Here are the most common roles and responsibilities based on undercover research:

CHIEF EXECUTIVE OFFICER

The CEO of any criminal enterprise is responsible for decision making and overseeing operations. He, like most traditional entrepreneurs, comes up with the "big idea" and provides the seed capital to see it through. He is often a "people person" and well connected to other elements of the criminal world and serves as the convener who assembles the right team of criminals to carry out any task. He is usually not deeply technical but hires others with the required coding and hacking skills to carry out his vision. The criminal CEO is not involved in any day-to-day dirty work or cyber attack that might be traced back to him. He sets goals and targets for his staff and oversees the distribution of criminal proceeds, especially at bonus time. The CEO is supported by a leadership team, including other C-suite executives.

CHIEF FINANCIAL OFFICER

The CFO keeps track of key crime syndicate metrics, including how much crimeware has been sold, how many accounts have been hacked, and what their balances are. He will use commercial business process tools, including financial reporting systems and databases to handle accounts payable (to crime contractors) and payroll for the criminal workforce. He also maintains a sophisticated network of clandestine financial contacts for purposes of money laundering, is responsible for managing front-company merchant accounts, and oversees global transactions in a variety of currencies, including online-payment service companies that eschew any "know your customer rules," such as Liberty Reserve.

CHIEF INFORMATION OFFICER

The CIO keeps the computer infrastructure of Crime, Inc.'s enterprise up and humming. He maintains so-called bulletproof untraceable computer servers and contracts with crooked Internet service provider hosting companies to ensure his crimeware remains beyond the reach of global law enforcement. The CIO helps maintain "customer" databases and botnet armies and is responsible for information security, including the management of "proxy networks" that preserve his employees' activities and ensure that they cannot be traced. The CIO also handles the encryption of corporate criminal data, ensuring it is unreadable and unusable by either the authorities or competitor criminal hacking organizations.

CHIEF MARKETING OFFICER

As many legitimate businesses have learned, having a great product is often not enough. Profits depend on a company's (or criminal enterprise's) ability to effectively promote its goods and services. As such, marketing executives help design effective advertising copy and provide it to criminal affiliate networks for distribution throughout the digital underground.

MIDDLE MANAGEMENT

These operational managers are often recruited through long-term friendship, proven in crime and blood over extended periods of time. They are responsible for managing the greater criminal workforce as well as command-and-control networks that carry out the organization's criminal technical operations.

WORKER BEES/INFANTRY

These are the ground forces in the war for crime, the equivalent of street corner dope dealers. They work with other elements of Crime, Inc. to help distribute malware via infected links, PDFs, and compromised Web sites. They will also break CAPTCHAs (those squiggly word designs that humans must type into a box to prove we're human) and help deploy credit card skimmers in retail stores and on ATM faceplates.

RESEARCH AND DEVELOPMENT

As with most enterprises, the way to stay ahead of your competition is through cutting-edge research and development (R&D), and crime syndicates are no different. The R&D department is constantly on the lookout for the latest exploits in desktop software, mobile apps, and networking systems—opportunities that can be monetized by the rest of Crime, Inc. In addition, the R&D teams can handle particularly difficult customized coding as required to go after particular targets or systems.

CODERS, ENGINEERS, AND DEVELOPERS

These are the technical brains of the criminal outfit, and they are the key ingredients to any online criminal enterprise. These techies must develop the computer code and software programs that will infect other systems.

They build Web sites and write the bulk of crimeware, ransomware, and scareware, including fake antivirus programs, to be distributed by the criminal network's operatives. These are the people who write the exploits and malware that infect and attack the world's information systems. Of course, before their code is released, it must first pass through quality assurance.

QUALITY ASSURANCE

The QA team is key to the success of Crime, Inc. It ensures the encryption shells in which the coders' malware is hidden are good enough to bypass current security systems, such as antivirus software and firewalls. The QA coders test all crimeware against known antivirus definitions to ensure their malware can avoid detection prior to its release. Tools such as avcheck.ru and Scan4You.net allow these teams to evaluate the possibility of detection by eighteen of the most popular antivirus programs. Importantly, these anti-detection models are updated daily and are fully automated. Criminal QA testers can even sign up for notifications to let them know when some of the prior malware authored by their coders has been identified as a threat by security firms. These alerts allow the coders to rapidly update and modify their malware so that it again becomes undetectable and business marches on.

AFFILIATES

Affiliated marketing, as noted previously, is incredibly popular and profitable in the online world. Commonly used by Amazon.com and others, it pays affiliates based on the number of customers they bring to a given retailer. Affiliate networks form the very backbone of the cyber-criminal enterprise and many of the very best are located in Russia. These so-called Partnerkas work day and night to drive as much traffic as possible to the Web sites of their criminal partners. These low-level criminals handle product placement, whether the product be fake antivirus software, child pornography, Rolex reproductions, or counterfeit Viagra. The affiliate's role is to introduce the criminal merchant to the unsuspecting consumer. Partnerkas spread their schemes via spam in e-mails, chat forums, blog comments, social media, and SMS messages. Crime, Inc. pays affiliates per click or per install each time an affiliate drives traffic to the criminal enterprise or when malware is downloaded to a victim's machine. Active criminal affiliates can easily earn $5,000 a day, with some clearing $300,000 a month. Hysterically, crime bosses advise affiliates on their underground Web sites that "the use of spam or other illicit methods of machine infection are strictly prohibited." That's right, Crime, Inc. too has adopted terms of service and end-user license agreements to protect itself and deflect any claims of criminal culpability against the C-suite executives.

TECHNICAL SUPPORT

Sometimes running criminal software campaigns can be hard. Just as we must frequently reboot our computers, ask for help from our corporate IT department, or visit the Best Buy Geek Squad, so too must criminals. Thus modern cyber-crime syndicates offer technical support for both their employees and their affiliates.

DIRECTOR OF HUMAN RESOURCES

Successfully running a global crime campaign worth hundreds of millions of dollars, such as Innovative Marketing, requires people, lots and lots of people. The HR team helps recruit the criminal foot soldiers and worker bees necessary to perform the day-to-day operations of the criminal enterprise. It sets up Web portals to handle "human capital management," including job applications, pay and benefits, and the online training required to carry out a successful malware infection campaign. The director of human resources will place ads in the digital underground to recruit the affiliates who very much know they are working as part of a criminal enterprise. HR will also help recruit another type of employee, so-called mules, who may or may not even know they are working for Crime, Inc. Ads for mules promise high earning potential, flexible hours, and the ability to work from home and are often placed on Craigslist or even on legitimate employment Web sites. The criminal HR staff handle inbound phone calls from prospective job applicants and are quick on their feet, ready to answer questions about job benefits and 401(k) plans (promised after the first successful year of employment).

MONEY MULES

Key to the growth of any illicit organization is the successful laundering of criminal proceeds. All the cash generated, whether through narcotics, scareware, or identity theft, must be properly transformed into ostensibly legitimate assets. To accomplish this goal, "money mules" are recruited via front companies to help move money anonymously from one account, bank, or country to another. Mules naively respond to ads for jobs with titles such as regional assistant, company representative, or accounts receivable. They are told they will be responsible for "payment processing" and are instructed to open two accounts in their own names—one for salary and one for the funds they will be processing, usually via Western Union. Mules, who generally receive between 3 and 10 percent of funds handled, must provide a photocopy of government ID, a completely logical legitimate business requirement that makes it easier for Crime, Inc. to track down any snitches at a later date.

Mules are the face of cyber crime and operate in true name, meaning they have a very short shelf life. Before long, police come calling, and it is only then that these housewives, students, and long-term unemployed, previously happy to look the other way and not ask too many questions, learn for certain that they have been involved in a criminal enterprise. By then, the money and their "bosses" operating under pseudonyms are long gone. According to one money-mule expert, the lack of available mules is the key bottleneck facing Crime, Inc. today. Breaking into systems is easy; where to cash the checks is the hard part. Experts estimate that the ratio of stolen account credentials to available mules could be as high as ten thousand to one. In other words, with sufficient mule and HR capacity, losses attributable to cyber crime could be ten thousand times worse.

The Lean (Criminal) Start-Up

The structure of Crime, Inc., like that of any modern techno-centric organization, is not fixed in time and space but rather constantly in flux. In his book *The Lean Startup: How Today's Entrepreneurs Use Continuous Innovation to Create Radically Successful Businesses*, Eric Ries outlines methods by which budding entrepreneurs can create new products "under conditions of extreme uncertainty." For criminals, uncertainty is where they excel, never knowing when the next police raid or rival gang drive-by shooting will take place. Outlaws are constantly adapting and innovating to overcome obstacles and meet the latest market demands. They build, measure, and learn by using data-driven Web analytics and keeping good metrics on their products and suppliers. But not all online criminal enterprises are directed from management down to worker bees; some are much more ad hoc and lean.

These criminal organizations are much more in line with the world Tim Ferriss describes in his *4-Hour Workweek*, which espouses streamlining business activities by eliminating overhead and automating systems. Heavy organizational structures and obvious leadership are shunned in deference to just-in-time products and services that can often self-assemble on demand. These underground online actors may be much more interested in work-life balance, or lifestyle design, allowing them to balance crime and play while maximizing the opportunities in both. They come together in swarms, groups of individuals in constant motion, contributing specific skill sets toward a common goal. Their assembly is both ephemeral and amorphous, making enforcement extremely difficult. Once the criminal task has been achieved, such as the takedown of a major data broker or retailer, the group can dissipate until reassembling with others for the next criminal engagement.

Actors in these online crime swarms sometimes form in hubs, based on criminal specialty. For example, an identity-theft ring might spontaneously form a hub using the skill sets of multiple swarms. One group of actors with deep technical skills might take responsibility for hacking into a corporate data system; the next group would serve as a data broker, distributing the stolen personal information to counterfeit-document experts, who would make driver's licenses, credit cards, checks, and passports with the information. The swarms of low-level thugs executing the actual financial frauds would forward any funds received to a mule network, which in turn would collaborate with a money-laundering network to ensure all criminal parties were paid for their services and received their cut of criminal proceeds.

In the worlds of both Crime, Inc. and swarm criminal networks, operational security is paramount. Work and communications are carried out remotely, obviating the need to ever meet in person. Work is compartmented and layered to ensure low-level participants don't know the true identities of other parties to the crime. Underground online hacking forums and communications channels serve as the main introduction, recruitment, and assembly points for the criminal conspiracies and enable coordination for the swarm as necessary to complete work on specific projects.

A Sophisticated Matrix of Crime

As the United States attorney in Manhattan, I have come to worry about few things as much as the gathering cyber threat.
PREET BHARARA, U.S. ATTORNEY FOR
THE SOUTHERN DISTRICT OF NEW YORK

Whether organized cyber-crime groups structure themselves along the lines of corporations, such as Innovative Marketing, or more nimble self-assembling swarms, one thing is clear: they are deeply sophisticated in their approach to business and their "customers." They have appropriated the latest legitimate corporate strategies and are well versed in supply chain management, global logistics, creative financing, just-in-time manufacturing, workforce incentivization, and consumer needs analysis. The result is the modern cyber-crime enterprise, a full-service, multiproduct, highly profitable global organization capable of taking down any individual, company, or government at will. As noted previously, there are at least fifty such online Crime, Inc. organizations currently in operation around the world.

I've seen this sophistication firsthand while working with Interpol and the Brazilian Federal Police on cases involving stolen credit cards throughout Latin America. In the favelas outside Rio de Janeiro, organized cyber-crime groups sold software programs on DVDs containing tens of thousands of compromised credit card numbers and user details. The crime start-ups sold their DVDs to other criminals, offering discounts when bought in bulk. They also included service-level agreements with their software, assuring that at least 80 percent of our stolen credit card numbers would work or "your money back!" The Brazilians even provided phone numbers for technical support for other criminals who were trying to figure out how to use the software but ran into technical difficulties. "Sir, have you tried rebooting your computer?"

Some Crime, Inc. organizations actually use customer relationship management (CRM) software to track customer requests and build brand loyalty among criminal clients, as was the case with the proprietors whose start-up created the banking Trojan Citadel. The malware, a variant of the infamous Zeus Trojan, allowed criminals to steal banking information, log user keystrokes, and install other forms of crimeware on a victim's machine. When the Citadel hackers sold their malware to fellow criminals, they wanted to ensure their customers were happy with the crimeware they had created. Borrowing a page from Marshall Field and Harry Gordon Selfridge, the Citadel gang pledged, "Our products will be improved according to the wishes of our customers," and they meant it. Their developers created a CRM user interface that allowed fellow criminals using the Citadel banking malware to file bug reports, propose and vote on new features for upcoming versions of the software, and even submit and track trouble tickets for the developers. Technical support was available via instant messenger on ICQ and Jabber, and trouble tickets were addressed in a timely manner. The Citadel "crime-trepreneurs" even built a social network to allow "like minded people" using their banking Trojans to come together to discuss "projects of mutual interest," such as robbing you and me.

Crime, Inc. can be strangely reasonable and rational, utilizing proven tactics to keep its competitive advantage and ensure the continuity of its operations. In the digital underground, this means keeping close track of the competition and potential business disruptions, in particular law enforcement. As seen previously, criminal hackers are not only monitoring the activities of relevant police agencies and officials but also gathering open-source intelligence to uncover any threats to their massive profits. One group of cyber thieves responsible for hacking JetBlue, 7-Eleven, JCPenney, and the Nasdaq stock exchange created a system of "trip wires" to provide an early-warning system to notify them if news of their exploits had become public. Specifically, they created a series of Google alerts with carefully selected keywords covering their targeted victims so that if any news stories were released about "Nasdaq hacked," they could pull up stakes and get out before tracked by police. Hackers have become the new Mafia and are contributing daily to the ever-increasing industrialization and professionalization of crime.

Honor Among Thieves: The Criminal Code of Ethics

**If you were going to be successful in the world of crime,
you needed a reputation for honesty.**
TERRY PRATCHETT, *FEET OF CLAY*

In order to maintain a well-ordered and functioning criminal underground economy, Crime, Inc. must observe certain rules of the road. As such, there is indeed honor among thieves, and some elements of Crime, Inc. actually publish "codes of conduct" to help reassure fellow criminal customers. These cyber black markets are well structured and self-policed, with buyers and sellers constantly reporting on and validating each other's reputations. Some digital criminal marketplaces actually have star-reputation systems so that fellow hackers can rate stolen credit cards, fake driver's licenses, and computer viruses with zero to five stars, just like on eBay or iTunes.

On the lower end of the online criminal marketplace, those levels that are easiest to access, violations of the code of conduct are not uncommon. These individuals are known as "rippers" and fail to deliver promised criminal goods or services as much as 30 percent of the time. Once identified, however, they are quickly reported, banned, and driven from the marketplace—just like a seller on eBay or Amazon who fails to deliver on his promise. To help alleviate these problems of trust, cyber criminals have actually established clearinghouses and escrow services, just like the ones you use when buying or selling a home. These honest, but criminal, brokers help verify that the illegal product or stolen data on offer are actually delivered—only then do they release the funds, after taking a 5 percent transaction fee for their services.

At the higher echelons of Crime, Inc., new entrants to the cyber underground are well vetted and must be vouched for by a trusted party, just like drug dealers working themselves up the food chain. Here, where the big boys play, violations of the code of conduct are rare and the consequences high. All parties know it is in their best interests to follow the rules. Just as retaliation is common with traditional organized crime, so too does it occur in the cyber underground. Though "whacking" the competition and dumping them with cement shoes in the East River is more a trademark move of old-school gangsters, their digital equivalents have their unpleasant methods as well. Digital drive-bys do occur, such as the two-day spree carried out by Max Ray Vision (a.k.a. Iceman), who infamously trained his keyboard guns on his competitors and wiped them out. From his apartment in San Francisco, Iceman was able to commandeer the information databases of his criminal competition, absorb their content, and use it to create his own massive site, CardersMarket, which grew to be six thousand members strong. Using the data stolen from his competition, Carders-

Market amassed more than two million pilfered credit cards and racked up $86 million in fraudulent charges. Superior technical skills count in the world of Crime, Inc., and hackers are forever studying and learning to improve their capabilities.

Crime U

Hackers are not born; they are trained, supported, and self-taught by an enormous amount of free educational material in the digital underground. Crime, Inc. is a learning organization, and there are online tutorials for everything from defeating firewalls to cloning credit cards. Criminals have access to their very own massive open online courses where they can learn how to launch phishing and spamming campaigns as well as how to use crimeware exploit kits. All of this training amounts to a sort of online criminal university (Crime U) that has accelerated the sophistication and skills of individual criminal hackers. Interestingly, student tutors, in the forms of fellow hackers, will often come together to help support newbies learn the art and craft of digital criminality. Numerous wikis are set up throughout the cyber underworld that provide detailed links, arranged by category, on how to hack every possible device, app, software, and operating system in existence.

Of course not all delinquent and illicit computer training takes place in the free world. Often thought of as "finishing schools" for criminals, prisons offer very little in the way of reform but much in the way of a graduate education in criminality. In fact, a study by Ohio University showed that "individuals with an incarceration history earn significantly higher annual illegal earnings than those who do not have such a history, bringing in on average an additional $11,000 per year of illicit income." Just as college improves the earning potential of those working in the lawful economy, so too does the graduate education received behind bars.

Thus it may be surprising that more and more prisons are offering computer and coding training to inmates. While such skills might be the key to a legitimate career post-incarceration, they can be useful for illegal purposes as well, even while still in jail. Such was the case with Nicholas Webber, who, while serving time in Her Majesty's Prison Isis in south London, used his computer skills during his IT training class to hack the prison's computer system. At the San Quentin maximum-security prison just outside Silicon Valley's backyard, corrections officials have even created a start-up incubator for those behind bars with entrepreneurial ambitions. With the support of the local technorati, inmates take part in "demo days" and pitch start-up ideas judged by Silicon Valley executives for their potential. While the intent of these programs is commendable, from a practical perspective the results may turn out differently from those expected.

Innovation from the Underworld

**A key ingredient in innovation is the ability to
challenge authority and break rules.**

VIVEK WADHWA

Criminals, forced to work outside the legitimate systems of power, have always been expert at innovating new solutions to difficult problems and thinking outside the box. Time and time again, they have shown deep inventiveness in their business practices and creative use of resources. In his short story "The Blue Cross," G. K. Chesterton summed it up nicely by declaring, "The criminal is the creative artist; the detective only the critic." The dark side of this creativity plays out daily in the world of Crime, Inc. The challenge for the rest of society is that technological innovation is proceeding at an exponential pace, and, importantly, Moore's law works for criminals too.

Technological innovation from the underworld is thriving, and the criminal hive mind is leaving antivirus companies, technology vendors, and law enforcement in the dust. No longer is hacking the province of a select few digital masters; rather, today it has become democratized with all necessary information readily available at Crime U. Modern criminals are innovating not just technologically but in their business models as well. Crime, Inc. has incorporated subscription models for malware services, gamification for staff members, and open-source software development for banking Trojans. To drive sales, Crime, Inc. offers fellow crooks stripped-down versions of illicit software tools or even provides them for free. If their felonious clients are happy with the product, they can pay more and upgrade to full versions—a strategy known as freemium pricing.

Organized cyber criminals have very much embraced Chris Anderson's "long tail" strategy and see the future of the crime business as stealing less from more. While criminals of yesteryear were always on the lookout for the single heist of a lifetime (think *Ocean's Eleven* or the *Pink Panther* diamond), today's cyber hoodlums have learned that they can reap massive profits by simply executing smaller operations over and over again against the masses. As we will see in the next chapter, much of this micro-thievery can be automated, leading to a steady stream of repeatable income, with lower risk of apprehension.

To motivate a diverse criminal workforce, the executives at Crime, Inc. have devised a number of encouragement schemes to keep business booming. For many hackers, cash isn't the only incentive; many enjoy the thrill of breaking the law, the challenge of cracking a sophisticated security system, or the bragging rights they gain when foiling such a system. Members of the cyber underground have established Web sites where fellow hackers can peer-review and rank their digital break-ins. RankMyHack

.com awards points to the best of the best and uses leaderboards to separate the wannabes from the hacker elite.

Cyber-crime bosses are well aware of these trends and have found a variety of ways to tap into employees' needs for recognition, challenge, and belonging by incorporating elements of gamification into their criminal activities. In Montenegro, the KlikVIP scareware gang threw a party for its most productive malware installers and offered a large briefcase full of euros to the affiliate who infected the greatest number of machines. In early 2014, in an effort to drive innovation and create new lines of nefarious business, an eastern European executive at Crime, Inc. offered a brand-new Ferrari for the hacker who invented the best new scam. The news of the prize was unveiled in a dark corner of the digital underground in a professionally produced video that featured several glamorous female "assistants" on the floor of the dealer's showroom. The boss's gamification strategy paid off and received widespread attention among his workers, with the Ferrari reserved for the chosen "employee of the month."

From Crowdsourcing to Crime Sourcing

Of all the business innovation techniques utilized by Crime, Inc., perhaps none has been as widely adopted as crowdsourcing. Crowdsourcing began as a legitimate tool to leverage the wisdom of crowds to solve complex business and scientific challenges. The concept of crowdsourcing first gained widespread attention in an article written in 2006 by Jeff Howe for *Wired*. Howe defined crowdsourcing as the act of "outsourcing a task to a large, undefined group of people through an open call." While hundreds of examples of crowdsourcing have been documented with great results, these very same techniques can be harnessed for criminal purposes as well.

YouTube is replete with great examples of apparent strangers suddenly breaking out into song, whether at Heathrow Airport or Times Square. But these flash mobs can rapidly devolve into "flash robs," wherein less charitably inclined strangers gather not for art's sake but for crime's. Though flash robs are mostly a tool of low-end thugs, they are highly successful. In Washington, D.C., thirty young adults, all coordinating on social media and via SMS, simultaneously rushed into the G-Star Raw store and ran out with $20,000 worth of clothing, easily overpowering shopkeepers. If any of the participants involved are arrested, they are unlikely to be able to "rat" on their co-conspirators, whom they met for the first time at the scene of the crime. Similar incidents have taken place in Chicago, Philadelphia, and Los Angeles.

Some crowdsourcing techniques are meant to give potential lawbreakers a leg up on the police. In the United States, mobile apps such as DUI Dodger, Buzzed, and Checkpoint Wingman allow those who have had too much to drink to crowdsource the location of DUI checkpoints, view them on an interactive map on the iPhone or Android device, and receive alerts when checkpoints are moved or newly established. When the 2011 London riots against government spending cuts turned violent, protesters created an app called Sukey, which allowed them to photograph police and upload their geo-tagged images to a crowdsourced interactive map. When other protest participants launched Sukey on their mobiles, they knew which areas contained riot police and were shown interactive compasses advising them how to avoid the cops (green pointed to safe areas, red to police danger zones).

Hacktivists too have taken good advantage of crowdsourcing techniques. At the height of its dispute with Sony and News Corp, LulzSec brazenly established a crime-request telephone hotline asking whom the hacktivists should target next. The group established a phone number in Ohio and recorded a greeting message with a French accent advising callers, "We are not available right now as we are busy raping the Internet," and asking callers to leave their hacking requests after the beep. This new modus operandi in crime sourcing allowed the public to vote, *American Idol*–style, on who shall be the next victim of a crime. The group later released a statement noting that it had successfully launched DDoS attacks against eight sites suggested by callers. Crime sourcing can be defined as taking the whole or part of a criminal act and outsourcing it to a crowd of either witting or unwitting individuals. By aggressively adopting crowdsourcing techniques, Crime, Inc. is able to build largely anonymous distributed criminal networks that can self-organize and assemble with amazing rapidity. To put these capabilities into perspective, in 2013 Crime, Inc. bosses in Russia and Ukraine were able to unleash a hundred money mules on a hospital in Washington State that they had hacked. As a result, more than $1 million was stolen from the hospital's payroll system and laundered through ninety-six separate accounts in just a few days. As noted previously, many of these mules might have unknowingly been co-opted by organized crime, believing they were "working from home" as "regional accounts receivable representatives."

Technology makes it easier than ever for Crime, Inc. to crowdsource its work to unwitting co-conspirators who have no idea they are taking part in an illicit plot. For example, criminals need a constant stream of new e-mail accounts by which to send their spam and phishing attacks, but CAPTCHAs can slow them down. As a work-around, criminals created a software system that automatically took the CAPTCHA image they were shown on Yahoo! or Hotmail and provided it to random strangers to solve for them. But why would any stranger do this? Simple. They were properly incentivized, with pornography. To crowdsource its problem, Crime, Inc. just created dozens of free porn sites and told visitors they would have to solve a CAPTCHA to prove they were over eighteen to gain access. The riddle the horny public was solving, however, was actually the CAPTCHA the criminals needed to create their spam e-mail accounts, cut, pasted, and switched in real time. A win-win situation, free high-quality porn in exchange for unwitting crowdsourced participation in a phishing scam.

Though the CAPTCHA scheme was clever, it pales in comparison with a criminal casting call posted in an online ad. In Seattle, Washington, a bank robber had carefully plotted out the day and time an armored truck was scheduled to deliver a large haul of cash to the local Bank of America. On the Tuesday in question, at precisely 11:00 a.m., the robber, wearing a yellow safety vest, goggles, a blue shirt, a tool belt, a hard hat, and a respiratory mask, walked up to the armored car guard as he was carrying several large bags of cash into the bank and squirted him in the face with pepper spray. The guard was disabled and dropped the bags of money, which the crook shoved into a large duffel bag he was carrying before making his escape with what Monroe police called "a great amount of money." When the guard regained his composure, he put out a help call on his radio and described the bank robber to a T. Soon half a dozen police cars were en route with lights and sirens to the scene of the crime on the lookout for the construction worker who had just pulled off the heist.

The first police car on the scene noticed the construction worker, and the cops drew their guns on him, ordering him to put his hands up and drop to his knees. Then another police car spotted the construction worker culprit and then another and another. In fact, there were dozens of construction workers at the scene matching the description provided by the armored car guard. What authorities did not realize is that the actual bank robber had carefully crowdsourced his escape well in advance. A few days prior to the robbery, the true bandit placed an ad on Craigslist in the help wanted section purportedly looking for construction workers to participate in a road maintenance crew. The pay was great at nearly $30 per hour and interested parties were told to appear on Tuesday at 11:00 a.m. at the intersection where the Bank of America was located. Oh, they were also told to bring their own equipment—in particular a yellow safety vest, goggles, a blue work shirt, a tool belt, a hard hat, and a respiratory mask. Dozens looking for work showed up at the appointed place and time, having no idea they were unwittingly suckered into a crowdsourced bank robbery. In the world of "in screen we trust," the public is easy to deceive. Only when all the construction workers were rounded up and detained did police realize what had transpired; of course by then, the actual bank robber was long gone.

Not only is Crime, Inc. rapidly adopting witting and unwitting forms of crime sourcing, but it is also using another white-hot trend in the start-up community: crowdfunding. Crowdfunding is a process by which money is collected from a crowd of backers who agree to support either a new start-up company or a nonprofit project, usually described in great depth on a Web site. The most popular of these sites are Kickstarter and Indiegogo, and tens of thousands of projects have successfully been funded, raising in excess of

$1 billion from the crowd. Criminals are of course happy to hack anybody raking in that much money and have already successfully compromised the Kickstarter Web site. That said, criminal hackers have much bigger and more nefarious crowdfunding plans in mind, such as hacking the iPhone in your pocket. When Apple released its iPhone 5s mobile phone, it included a feature known as Touch ID, a fingerprint-recognition scanner touted as a "convenient and highly secure way to access your phone." Though Apple probably spent years and millions of dollars developing its patented biometric technology, by introducing the feature, Apple was in effect throwing down a gauntlet challenging hackers to defeat its "highly secure system."

Around the world, security professionals and hackers alike wondered, who would be the first to crack the uncrackable, and how long would it take? The answer was the Chaos Computer Club in Germany, and it took a day. Using elements of both crowdfunding and gamification, hackers set up a Web site called IsTouchIDHackedYet.com, offered a $20,000 bounty, which was contributed by fellow hackers, and used a leaderboard to show progress toward the $20K goal. In the end, the prize went to a hacker known as Starbug of the Chaos Computer Club who cleverly figured out how to subvert Apple's multimillion-dollar investment. Starbug took a high-resolution twenty-four-hundred-DPI photograph of the fingerprint oils left behind on the Touch ID screen by the device's legitimate owner. He then imported the picture to Photoshop, cleaned it up, inverted it, and printed it on a transparency film using a thick toner setting. Finally, white wood glue was smeared onto the pattern and, when dry, could be held over the Touch ID sensor to unlock the phone. Mission accomplished.

As if crowdfunding hackers weren't serious enough, recently yet another crowdsourced enterprise surfaced in the digital underground: the Assassination Market. Regrettably, the service is not some sort of deeply disturbing joke. Rather, it is the work of a dedicated anarchist who goes by the pseudonym Kuwabatake Sanjuro. As of late 2014, eight U.S. government officials have been selected via crowdsourced voting for assassination, with the former Federal Reserve chairman Ben Bernanke receiving the greatest number of votes. Donations have been made via encrypted and untraceable online currencies, and Sanjuro has crowdfunded $75,000 for the murder of the former Fed chairman to be paid to any hit man who comes forward upon completing the act.

Though the $75,000 raised is profoundly alarming, it is not even close to the most successful criminal crowdfunding exercise ever to take place, one in which neither the victims nor the crowd funded the activity willingly. In what was perhaps the most masterful single heist ever carried out by Crime, Inc., thieves around the world crowdsourced a robbery in twenty-seven separate countries, carried out simultaneously. The massive larceny occurred in early 2013 when coders, engineers, and the R&D team at Crime, Inc. in eastern Europe broke into the network of two credit card processors in India and one in the United Arab Emirates. Crime, Inc. stole prepaid MasterCard and Visa debit card numbers and then hacked the processors' internal computer systems to remove any and all account withdrawal limits on the cards they had pilfered. As a result, the master hacker-criminals had hundreds of debit cards, each capable of withdrawing unlimited funds from the global ATM network.

Crime, Inc. then sent encrypted messages via the digital underground to crime associates in more than two dozen countries. Those receiving the stolen data used their own criminal professional-grade credit card printers to print the debit cards and encode the card numbers on the magnetic strips on the reverse. What happened next is perhaps one of the greatest feats in crime sourcing, or even crowdsourcing, history. The cards were distributed to hundreds of teams of worker-bee criminals around the world. When Crime, Inc. gave the signal, the race was on, and the infantry of outlaws went on a synchronized withdrawal spree, hitting as many ATMs as humanly possible. In the ten-hour time span that Crime, Inc.'s crowdsourced operation ran, thieves carried out thirty-six thousand ATM transactions in twenty-seven countries and walked away with over $45 million in cash. Because Crime, Inc. had already hijacked the banks' computers and had the debit card numbers they had assigned, they could watch exactly how much was being taken out and, importantly, how much each criminal worker bee had to kick back prior to taking his "service fee." Though a small handful of low-level thugs were caught by police, the Crime, Inc. masterminds behind the plot remain unidentified and at large, probably organizing their next massive crowdsourced caper. Ten hours, thirty-six thousand transactions, twenty-seven countries: an amazing logistical feat that few corporations or governments could actually execute. Welcome to the world of network-distributed criminality.

Crime, Inc. is a business and a highly profitable one. Unencumbered by moral considerations, it is free to profit without limit and use the very latest business practices to do so. Crime, Inc. uses freemium pricing, gamification, crowdsourcing, crowdfunding, reputation engines, just-in-time manufacturing, online training, swarms for distributed project management in pursuit of the long tail of crime victims around the world. Global criminal syndicates such as Innovative Marketing in Kiev have earned upward of half a billion dollars (tax-free of course) in just three years. These outlaws, Moore's outlaws, are fully networked and capable of leveraging and subverting any technology at will. They do so with near impunity, and their actions imperil a world that is both increasingly connected and profoundly dependent on technology to function. The result is an ever more powerful criminal underworld, one that is growing exponentially in its capabilities. This thriving criminal superorganism lives, breathes, and is controlled from within the deepest, darkest recesses of the Internet—the Dark Web, the inner sanctum of the digital underground and the nerve center of Crime, Inc.

CHAPTER 11

Inside the Digital Underground

Our representation of the standard criminal might be based on the properties of those less intelligent ones who were caught.

NASSIM NICHOLAS TALEB, *THE BLACK SWAN*

Dread Pirate Roberts (DPR) was the most wanted man in the digital underground. From within the darkest reaches of cyberspace, the mysterious outlaw ran a vast empire of covert criminality. He was the subject of a global manhunt, actively pursued by special agents from the FBI, Drug Enforcement Agency (DEA), ATF, Homeland Security, the Royal Canadian Mounted Police, Scotland Yard, and Interpol. Precious little was known about Dread Pirate Roberts, save for the fact that his alias was taken from a character in the cult classic film *The Princess Bride*. DPR was the mastermind behind Silk Road, a massive online criminal marketplace painstakingly hidden from public view where any and all manner of illicit goods were for sale in a secret Web: "If you can smoke it, inject it, or snort it, there's a good chance Silk Road has it."

Named after the ancient Asian trading route, Silk Road was a place where buyer and seller could anonymously come together to exchange goods and services in a dizzyingly large emporium of contraband. Known as the "eBay of drugs and vice," Silk Road offered every possible illicit product imaginable, neatly organized by category such as drugs or weapons, each with accompanying photographs and descriptions. Other items on offer included stolen bank accounts, counterfeit currency, AK-47s, armor-piercing ammunition, stolen credit cards, computer viruses, keystroke loggers, compromised Facebook accounts, tutorials on hacking ATMs, child pornography, and even hit men for hire. Under the category of forgeries, there were more than two hundred listings for fake driver's licenses, passports, Social Security cards, utility bills, credit card statements, diplomas, and other identity documents.

At its core, however, Silk Road was all about the drugs, with more than thirteen thousand postings for controlled substances for sale listed. The "narcocopia" of merchandise included heroin, Oxycontin, powder and crack cocaine, morphine, LSD, ecstasy, Molly, marijuana, crystal meth, 'shrooms, syringes, precursors, steroids, stimulants, and a panoply of prescription pills, from Adderall to Xanax. Narcotics were sold both in individual quantities and in bulk, including multi-kilo offers for heroin, cocaine, and methamphetamines. Clicking on any particular link brought up a picture of the product in question as well as descriptive ad copy such as "Nod's Black Tar HEROIN—ships sweetness express to the veins—or to the lungs if you prefer to smoke it and chase the dragon."

For nearly three years, Dread Pirate Roberts operated the largest online criminal marketplace in the world, successfully attracting over 950,000 users to create accounts on Silk Road. But how could such a flagrant violation of the law operate for such an extended period of time without any productive police intervention? Simple, they had no idea how to stop it. Silk Road was not a standard Web site, readily accessible by typing www-something in a browser's address bar. Rather, it operated in the digital underground, hidden away behind layers of secrecy enabled by specialized encryption and obfuscation software known as The Onion Router, or Tor for short (more on this later). Using the Tor software, all parties buying and selling illicit goods could remain anonymous, only identifying themselves via a chosen made-up screen name. To further protect users and their illegal activities, the only form of payment accepted on Silk Road was Bitcoin, a new type of electronic currency that allowed parties to exchange funds online with strong privacy protection.

Among those in the know, the frequent references to Silk Road as an eBay for drugs were incredibly apt. In keeping with the latest Crime, Inc. techniques, DPR instituted a robust online reputation system that allowed users to evaluate and trust one another prior to transacting. Yes, you can rate your drug dealer. Thus Basehead888 could see that DealioInThe312 had conducted over forty-six hundred cocaine sales and earned a 97 percent customer approval rating from his adoring cracked-out fans. Specific comments left by customers touted "how fast shipping was" or "how solid the stealth packaging was—no drug sniffa [sic] dog gonna find this!"

As time went on, the popularity and infamy of Silk Road grew, and before long nearly 600,000 private messages were being exchanged monthly between buyers and sellers. Eventually, the volume of traffic and transactions became more than DPR alone could handle. In response, the crime boss hired a small staff of system administrators who were paid $1,000 to

$2,000 a month to help with the day-to-day operation of the site, including monitoring user activity for problems, performing customer service, and acting as a mediator when buyers and sellers were in dispute. Of course, the founder of the world's largest underground illicit drug marketplace made significantly more than his employees, something his low-level system admins quickly realized. To right the perceived injustice of a low salary, one of the Silk Road employees began to embezzle from the company. As anybody who has ever watched *Scarface*, *The Sopranos*, or *The Godfather* could tell you, stealing from Mr. Big is never a good idea.

When Dread Pirate Roberts realized he was being ripped off, it was a betrayal he could not tolerate. In response, he reached out to one of the many professional assassins on his site and negotiated an $80,000 fee to have the employee whacked (50 percent due up front per the standard assassin code of conduct). DPR was so outraged by the lack of respect shown by his employee he provided specific instructions to the hit man to torture his soon-to-be-former system administrator prior to his death. DPR forwarded the assassin his employee's address in Utah and agreed to pay the balance of the kill fee after photographic proof of the homicide was received. A few days later, the Silk Road CEO received the verification he had been seeking in the form of a JPEG photograph. Ever the man of his word, DPR wired the remaining $40,000 balance to the killer and even sent a thank-you note for the hit, lamenting in an encrypted e-mail, "I'm pissed I had to kill him . . . but what is done is done . . . I just can't believe he was so stupid . . . I just wish more people had some integrity." Yes, the founder of Silk Road, the world's largest illicit marketplace, the man who had just ordered a hit on his own employee, was disturbed by the lack of integrity in this world.

But this wasn't the only time Dread Pirate Roberts had ordered a hit on somebody who had crossed him. His exploits were an open secret in the digital underground, and even the U.S. Senate held hearings demanding police action. Of course, the FBI and others were already on the case and had completed more than a hundred undercover buys on the site. Before long, they were also on the trail of Silk Road's godfather, the Internet entrepreneur-cum-murderous drug lord who started it all. The global manhunt for Dread Pirate Roberts eventually led the FBI's Silk Road task force to the Glen Park branch of the San Francisco Public Library.

There, on a cool, sunny day in the fall of 2013, a man in his late twenties with brown wavy hair settled down with his laptop in the quiet of the science fiction section and began typing away as the surrounding patrons read their books and leafed through magazines. Suddenly the silence was broken when a young woman charged toward the young man screaming, "I'm so sick of you!" In an instant, she was upon him and grabbed the laptop right off the table. As he struggled to regain control of the computer, the fellow patrons at his table, rather than helping him, threw him up against the wall, allowing the strange woman to abscond with his most prized possession.

This was no random theft. Many of those masquerading bibliophiles had been patiently waiting for both this particular twentysomething and his laptop. As soon as the young man booted up and entered all the necessary passwords to decrypt the computer's hard drive, the assailants pounced. The confrontation, however, was over in an instant, once the would-be thieves reached underneath their shirts and one by one revealed their hidden gold FBI badges. Stunned librarians, jaws agape, looked on as the young man with wavy brown hair was placed under arrest and taken to the Glenn Dyer Jail in Oakland for booking. Dread Pirate Roberts was no more.

Though DPR had worked hard to protect his identity by using Tor and Bitcoin to cover his tracks, he made a series of rookie operational mistakes that eventually led the Feds to his frequent log-ons at the San Francisco Public Library. According to a federal indictment, Dread Pirate Roberts was in fact Ross William Ulbricht, a twenty-nine-year-old from Texas who had moved to San Francisco a few years earlier.

The U.S. Attorney for the Southern District of New York charged Ulbricht, a.k.a. DPR, with a variety of offenses, including "conspiracy to commit narcotics trafficking, computer hacking, money laundering and running a criminal enterprise." Oh yeah, Ulbricht was also charged with attempted murder and the "Use of Interstate Commerce Facilities in the Commission of Murder-for-Hire." Turns out the professional hit man DPR thought he had hired was actually an undercover federal agent. Prosecutors have accused Ulbricht of ordering a total of five additional assassinations. When Ulbricht reportedly paid the sums demanded by the would-be assassin, the FBI knew he was dead serious and intervened to save those targeted. The Feds obtained the cooperation of all those who were to be "whacked" and took staged photographs of the alleged victims covered in fake blood and wearing the ashen face makeup of a dead body that they forwarded to DPR as the proof of killings he demanded.

Who was this criminal mastermind behind Silk Road? Not at all whom you would expect. Ross Ulbricht was the kind of kid any parent would be proud of, an Eagle Scout from Austin, Texas, who had earned a master's degree in science and engineering. In grad school, Ulbricht eventually lost interest in science in favor of a new passion for libertarianism. He wrote on his LinkedIn profile that he now wished to "use economic theory to abolish the widespread and systemic use of force by institutions and government against mankind." To that end, Dread Pirate Roberts was born, and the Internet's Silk Road became the canvas on which he could test and perfect the limits of his free-market ideals. The result, much like the fictional Walter White of TV's *Breaking Bad* fame, was the real-world story of a scientist who turns his passion for drugs and crypto anarchism into the world's largest online purveyor of contraband ever known. In the process, our antihero also made money, lots and lots of money.

Like eBay, Silk Road charged a commission for every transaction, ranging from 8 to 15 percent depending on the size of the sale. Amazingly, according to the charges filed against Ulbricht, Silk Road processed more than $1.2 billion in transactions between February 2011 and July 2013 alone, netting its twenty-nine-year-old founder a cool $80 million in fees. Not bad for a two-year-old start-up. At the height of its operation, according to a study published in the journal *Addiction*, nearly 20 percent of the drug users in the United States had purchased narcotics on Silk Road.

Ulbricht has pleaded not guilty to all charges, and his friends and family have all resoundingly declared him "such a nice guy," even launching a crowdfunding campaign to help pay for his legal expenses (Bitcoin gladly accepted). The federal government, however, paints a much more troubling picture of Ulbricht in its indictment of him, that of a drug kingpin, cold-blooded killer, and mad criminal mastermind who completely reinvented the business model of Crime, Inc. Eagle Scout or villain, one thing is clear. Ulbricht, a.k.a. Dread Pirate Roberts, has now added yet another alias to his long list of names—that of inmate ULW981, confined to a cell twenty hours a day and facing life imprisonment. In the meantime, like a multiheaded Hydra, Silk Road, which was only ever briefly shut down, has roared back to life, under new management, flourishing and spreading across the vast expanses of the Dark Web that is the digital underground.

Passport to the Dark Web

In order for Dread Pirate Roberts's criminal buyers and sellers to transact in his Silk Road marketplace, they first had to figure out how to get there. Just like the real world, you can't simply knock on the door of any house on the block and expect to score a kilo of meth. The same holds true for the digital underground. You don't get there by merely typing an address in your Firefox browser and hoping to be magically transported to the inner sanctum of Crime, Inc. Rather, you need a passport and a Sherpa to guide you. That journey begins with Tor—The Onion Router, a software tool that provides the closest thing to actual anonymity on the Internet.

Tor works by routing your Web connections through a worldwide array of five thousand computer servers in order to hide the source and destination of your connection. Without Tor, your online activities are easy to trace, and every time you visit sites like CNN or ESPN.com, you reveal your location and home network. Bad guys don't like this; it makes them easy to catch. So instead, they obfuscate and route their traffic through services like Tor. This way, the cops can't see that gangsters are selling AK-47s online using the Comcast server in Chicago (one mere subpoena away from identifying the Comcast customer assigned the Internet protocol address in question). Instead, any experienced hacker, let's say in Moscow, will route his Internet traffic first through London, Cape Town, Tokyo, Austin, and Milan before popping out to attack a target in Manhattan. Doing so makes the proverbial "call" nearly impossible to trace.

While the Tor software client can be used to anonymously visit any common Web site such as Google, its true power lies in enabling connections to Tor's hidden services—Web sites specifically configured to only receive inbound connections through the Tor network. Without the Tor software client, you simply cannot access the vast content available hidden away within The Onion Router network. With Tor's hidden services, not only can the site visitor maintain privacy, but so too can any underground Web site visited. Rather than using a standard Web address such as Facebook.com, all of Tor's hidden services have their own domain names, which end in an "onion" suffix. This system of dual anonymity allows both buyer and seller on Silk Road to transact by visiting a unique hidden domain (in the case of Silk Road, silkroadvb5piz3r.onion) without ever revealing their true identities to each other.

Though most people have never seen or used it, the Tor software is freely available for download via the Tor Web site, www.torproject.org. It can be installed in just a few minutes, and running the program stealthily transports users well off the beaten path of the mainstream global information grid. Oddly, Tor was originally created and funded as a project of the U.S. Naval Research Laboratory in 2004, with backing from the Electronic Frontier Foundation and the State Department as a means of helping overseas political dissidents and democracy activists safely organize and communicate with one another. There are any number of completely legitimate uses for Tor, and those behind the Great Firewalls of China, Iran, and elsewhere routinely depend on it to access everything from Facebook to the *New York Times*. Tor is also increasingly being used by journalists to securely communicate with sources and whistle-blowers, such as those within the WikiLeaks community.

While Tor might have been created for good, given its powerful ability to facilitate clandestine communication, it should come as no surprise that criminals have adopted the tool in droves, enabling the creation of services such as Silk Road. While precise numbers are hard to come by, a 2013 study of forty thousand hidden Tor sites found that nearly 50 percent were involved in illicit activities such as selling stolen credit cards, hacked accounts, weapons, drugs, and child pornography. Some security and law enforcement experts privately estimate that as much as 85 percent of Tor's hidden services may be unlawful, with the rate of criminal adoption far outpacing that of privacy activists.

As of early 2014, the Tor software has been downloaded nearly 150 million times and is used by two million people daily. Assuming the more conservative figure of 50 percent illicit use, that means every day 300,000 criminals are getting up and going to work on the digital underground using Tor's hidden services. According to Metcalfe's law, the value of a tele-communications network is proportional to the square of the number of users connected to the system; as such, the threat from a fully networked and anonymous criminal workforce is profound.

Crime, Inc. may not be the only dark force using Tor to access hidden Web services. A number of reports have noted that al-Qaeda and its affiliates too leverage the secrecy and anonymity afforded by Tor's encryption protocols to communicate, recruit new members, raise funds, spread propaganda, and even plan operations. After the former NSA contractor Edward Snowden leaked details of his agency's vast communications interception capabilities, evidence emerged suggesting that numerous terrorist groups reevaluated their communications strategies and in numerous missives stressed the ongoing importance of online operational security to their members.

Organizations such as al-Qaeda in the Arabian Peninsula and Ansar al-Mujahideen have even produced training materials and YouTube videos encouraging their members to use Tor for all online activities.

Given Snowden's revelations, as well as the widespread assaults on privacy previously noted, it is absolutely logical that ordinary citizens would turn to a powerful tool like Tor to maintain their online dignity, freedom, and human rights. That said, Tor's hidden services have been thoroughly usurped by Crime, Inc., and the innovation they have and continue to unleash in the digital underground is mind-boggling for its size, scope, and scale.

A Journey into the Abyss

The Internet provides a delivery system for pathological states of mind.
PHILLIP ADAMS, AUSTRALIAN BROADCASTER AND AUTHOR

You thought you knew the Internet, but you don't. You while away day after day watching videos on YouTube, posting status updates on Facebook, and shopping on Amazon, believing you are in a boundless online Garden of Eden, but you are not. From that very first time you ever ventured online, you've only ever really been visiting the surface Web. You've been trapped in a walled garden, one carefully manipulated and manicured just for you while those in the know have entered the Matrix, the other online world. This is the Internet most of us will never see. It goes by many names—the Deep Web, the Dark Net, the Secret Web, the digital underground, and the Invisible Internet, to name but a few. This is the shadow Internet, and Google definitely won't take you there.

The Deep Web technically refers to those online information resources that search engines such as Google, Yahoo!, and Bing cannot index, because they are password protected, are behind paywalls, or require special software to access. Because the sophisticated Google Web crawler that searches for all Internet content has itself no ability to type, it can't enter passwords, complete CAPTCHAs, or register for private sites and thus never catalogs vast swaths of the world's information. Much of the Deep Web's unindexed material lies in academic databases such as LexisNexis or in topical data sets such as those held by the Patent Office or the Census Bureau. But beyond the mundane, there also lies much more salacious material.

Shockingly, the Deep Web is a massive five hundred times larger than the surface Web you use and search every day. While the Deep Web contains seventy-five hundred terabytes of information, the Googleable universe contains a paltry nineteen terabytes. According to a study published in *Nature*, Google captures no more than 16 percent of the surface Web and misses all of the Deep Web. As a result, when you search Google, you are only seeing 0.03 percent (one in three thousand pages) of the information that actually exists and would be available online if you knew how to get it. In other words, a Google search misses 99 percent of the World Wide Web's data. Searching the Web today is akin to only fishing across the top two feet of the world's vast oceans. Though you may catch something in your net, you are missing the monumental bounty available just below those two feet, down to the very depths of the seas. For the intrepid, there lies a digital equivalent of the Mariana Trench, a veritable undiscovered treasure trove of data just waiting to be explored.

But like a Russian matryoshka doll, nested within the Deep Web is another hidden world, a smaller but significant community where malicious actors unite in common purpose for ill. Welcome to the Dark Net, also called the Dark Web, a vast digital underground within the Deep Web where hackers, gangsters, terrorists, and pedophiles come to ply their trades. The Dark Net holds some of the greatest secrets the Internet has to offer, and like the back alleys and black market bazaars of any big city it is where criminals connect to conduct their illegal activities. The Dark Net uses encryption and peer-to-peer Internet relay channels specifically designed to hide the IP addresses of its users, thereby providing an anonymous, untraceable, and secure platform for Crime, Inc. to communicate and transact without fear of government or corporate interference.

Though Tor is the largest and most popular gateway to the Dark Net, it has competitors, including Freenet and I2P (the Invisible Internet Project). Moreover, Silk Road is just one of dozens of online criminal super marketplaces; others include Black Market Reloaded, OpenMarket, Sheep Marketplace, Agora, BlackBank, Atlantis, and the Pirate Market, and new channels come and go daily. Importantly, criminals, like any good entrepreneurs, learn from their past mistakes, and thus in the wake of the Silk

Road takedown a new generation of Dark Net bazaars has risen in its place, most notably DarkMarket. Whereas Silk Road had centralized control in the hands of DPR and the computer servers he administered, DarkMarket is a completely decentralized online black market with no single owner. For the FBI to successfully take down DarkMarket, it could not just arrest its leader, as none exists. Rather, police would be forced to go after every contraband buyer and seller one by one, a near impossibility, making the Dark Net a veritable Elysium for Crime, Inc.

To help newbie criminals and hackers navigate the Dark Web, illicit marketplaces have established helpful hidden wikis—think *Crimeopedias*—neatly organized by category with links to other .onion sites. Categories include hacks, phreaks, anarchy, warez, viruses, markets, drugs, and erotica, each with links and descriptions of what will be found there. Even with the help of wikis, the Dark Net can be challenging to navigate, and finding the exact drug, gun, or assassin you've been looking for can prove difficult. To that end, in mid-2014 one highly innovative hacker created the Dark Web's first distributed search engine, known as Grams. Modeled after Google, Grams can only be accessed via the Tor anonymizing browser and using an .onion address. With Grams, those searching for contraband can now enter their keywords and search simultaneously across eight separate dark markets for goods and services. The search engine returns a vendor's name and allows for comparison shopping. Like Google, it has an "I'm Feeling Lucky" button, which, when clicked, might take users to a site for "high-quality crystal meth." As the prototypical Google for crime, Grams even accepts advertising, and various cartels can compete for customers by buying keyword search terms. That's right, your search for "Afghani Brown Heroin" in Grams will return all available results on the Dark Web, but those members of Crime, Inc. who pay a fee will have their search returns listed higher up just like sponsors in Google. The fact that Grams offers an AdWords program for the digital underground demonstrates both the keen technical business acumen and the sophistication of Crime, Inc.

Though ad-supported search has come to the digital underground, not all illicit traders are enthralled by being readily locatable by the criminal masses and, by extension, law enforcement. Thus in the more discriminating preserves of the Dark Web, just as in the real world, online criminals need introductions and must be vouched for before they can transact. Here "the distribution of goods and services is organized through thousands of illicit chat rooms and invitation-only forums." To access the most exclusive of these illicit domains, one need be forearmed with the secret alphanumeric address, one that is not cataloged or listed anywhere else online but must be passed from person to person. Certain criminal forums like the Russian carder site Maza, a massive online stolen credit card exchange, prevent aspiring entrants to their clandestine worlds unless they are unanimously approved by a vote of the organization's senior membership and then only after an eight-day waiting period. Once you are accepted into the elite realm of the criminal digerati, however, the world is your oyster.

Browsing the cornucopia of taboo and illicit wares available in the digital underground can feel much like a slow descent through Dante's nine circles of hell, with each step leading further and further into a frightening and deeply disturbing abyss. What follows is but a cursory sampling of the goods and services available from within the darkest recesses of the Internet, listed from the banal to the horrific.

PIRATED CONTENT

Numerous illicit "torrents," or peer-to-peer file-sharing sites, exist such as the Pirate Bay, which ranked among the top one hundred most visited Web sites on the Internet. Another such site, New Zealand's Megaupload.com at its peak was visited by fifty million "customers" a day and accounted for a full 4 percent of global Internet traffic. International law enforcement authorities maintain the site was run by a six-foot-seven, 350-pound expatriate German national hacker who used the alias Kim Dotcom. According to authorities, Megaupload's main products were fifty petabytes (fifty-two million gigabytes) of stolen movies, songs, video games, books, and software. Business was good at Megaupload, and the firm generated an estimated $25 million a year from online ads and an additional $150 million in fees paid by users who wanted to steal their downloadable content more quickly. These profits allowed Kim Dotcom to enjoy a wildly opulent lifestyle down under in a $24 million mansion, with sixty acres of manicured lawns and its own tennis courts and golf course. Some of Kim Dotcom's other possessions included a helicopter, a mega-yacht, fifteen Mercedes, a Rolls-Royce Phantom Drophead Coupé (MSRP from $474,600), and a Swedish horsehair Hästens bed, custom-made at a cost of $103,000. Piracy is indeed a profitable business.

DRUGS

As we saw with Silk Road, illicit and prescription drugs of every type are available in the digital underground in quantities ranging from individual use to bulk dealer-to-dealer sales. Silk Road, however, is far from the only narcotics marketplace on the Dark Web, and hundreds of such sites exist. Not only do they sell the standards like marijuana, heroin, ecstasy, and cocaine, but they also offer much rarer fare such as Scopolamine, the so-called Devil's Breath powder used as an offensive zombification drug, which, when blown into a victim's face, leaves him or her coherent but with no free will. Once ingested, within minutes the odorless and tasteless dust allows burglars, robbers, and rapists complete control over their victims and, worse, completely wipes the victims' memories of the details of the incidents.

COUNTERFEIT CURRENCY

Counterfeit currency is widely available in the digital underground, and costs vary by quality, amount purchased, and currency type, including dollars, euros, pounds, and yen. Tor hidden sites such as Guttemberg Print, Cheap Euros, and WHMX Counterfeit offer high-quality notes for twenty-four cents on the dollar ($600 real dollars to buy $2,500 counterfeit ones). Vendors promise all notes will pass pen and UV light tests meant to uncover counterfeit bills.

STOLEN LUXURY GOODS/ELECTRONICS

Dark Web sites such as Tor Electronics, CardedStore, and Buttery Boot-legging all offer factory-new brand-name electronics and luxury goods at a Deep Web discount. Their advertisements offer "expensive items from major stores at a fraction of the price!" Of course these items have all been stolen, diverted from the factory, or mysteriously fallen off their delivery trucks.

CARDS/ACCOUNTS

Perhaps no item is more abundant in the digital underground than stolen credit cards, widely available in so-called carder forums, where individuals buy and sell credit and debit cards from virtually every bank and country in the world. All the financial data stolen via malware, hacking, and credit card skimmers eventually end up for sale on the Dark Web via "dumps." Dumps refer to the data contained on the magnetic strip of a credit card and include details such as the cardholder's name, card number, expiration date, and CVV (card verification value). Once stolen, this information is used by criminals to make online purchases or even to encode the data onto new plastic counterfeit cards, which they use to go shopping for "high-dollar merchandise that can be resold quickly for cash." Given the vast amounts of credit card theft, prices per stolen card on the digital underground have been dropping (from about $3 in 2010 to $1 in 2013). Stolen credit card dumps demonstrate market elasticity, and after a major break-in like the 2013 Target store hacking, prices for stolen cards drop further because of an oversupply beyond market demand. The cards are sold on Dark Net sites such as Mazafaka, Tortuga, CarderPlanet, Shadow-Crew, Approven.su, and my personal favorite, IAACA (the International Association for the Advancement of Criminal Activity). Nearly all these sites require registration, and members are often vetted to keep out the cops. Carders promise "high validity rates" and offer guarantees that 95 percent of their stolen credit cards will work or "your money back!" For the carders engaged in this business, profits are staggering, with over $11 billion lost to global payment card fraud annually. The United States is the largest victim of these thefts, accounting for 47 percent of all fraudulent card activity globally.

IDENTITY THEFT

The illicit use of personally identifiable information (PII) is widespread in the digital underground. The information leaks from insecure data brokers, social media sites, and the unsound handling of your medical, financial, educational, tax, and online shopping transactions. These stolen identities are often referred to as "fullz" by hackers and contain names, addresses, Social Security numbers, dates of birth, workplaces, bank account numbers, bank routing numbers, state driver's license numbers, mothers' maiden names, e-mail addresses, and additional online account names and passwords. Nearly 20 percent of American and EU citizens have been the victims of identity theft, and the profits facilitated through the Dark Web PII sales are monumental. Medical identity theft (false claims with stolen IDs) costs the U.S. health-care system $5.6 billion annually, while tax refund identity theft (somebody submitting a fraudulent tax return on your behalf and pocketing the refund) will cost the IRS $21 billion over the next five years—all because we're leaking massive amounts of data from deeply insecure systems that can easily be traded at tremendous profit on the Dark Net.

DOCUMENTS

Any number of documents can easily be purchased online, including passports, driver's licenses, citizenship papers, fake IDs, college diplomas, transcripts, immigration documents, and even diplomatic ID cards. Dark Net firms such as Onion Identity Services are glad to sell passports and ID cards for Bitcoin. These documents in turn are used by criminals and terrorists to facilitate their free movement across international borders, to establish new identities, and to launder money. U.S. driver's licenses (any state) of extremely high quality often ship from China or Russia and cost approximately $200, while passports from the United States or the U.K. will sell for a few thousand dollars.

WEAPONS, AMMUNITION, AND EXPLOSIVES

Just about any weapon you would like to purchase is available online via the Dark Web on sites such as the Armory, Black Market Reloaded, and LiberaTor. Handguns such as Glocks, Berettas, and 9 mm machine pistols with silencers are readily traded here. So too are assault rifles, including AK-47s and Bushmaster M4s (used by Special Forces in Afghanistan and capable of firing 700–950 rounds per minute in full-auto mode)—no waiting period and no background checks required. Also in stock are C-4 explosives from a provider who cheerfully notes "we ship worldwide." Of course shipping is a bit of a problem: one can't just FedEx an Uzi without raising some alarm bells. Thus gun purveyors on the Dark Net have adapted well to this challenge and send their products via shielded packaging, disguised to look like other products. The firearms are broken down into smaller component parts and forwarded via special shipments. Weapons merchants can even arrange for "dead drops" where they hide the fully assembled firearms buried in a park, in a rubbish bin, or in an alley. After payment, buyers receive GPS coordinates and descriptions of where the items are hidden. Even military-grade weaponry is available online these days. One Dark Web user going by the name Bohica offers mostly small arms for sale but noted, "If you need artillery, MANPADS [man-portable air-defense systems], ordnance, APCs, Helos we do have resources and can make certain introductions for a fee. Please send your next message through PGP encryption, our public key is on our profile page."

HIT MEN

As we saw with Silk Road, assassinations are just a click away on the Dark Net. Service providers such as Killer for Hire, Quick Kill, Contract Killer, and C'thulhu all advertise "permanent solutions to common problems." The mercenaries offering these services proudly note their training with militaries such as the French Foreign Legion or their sniper capabilities honed in Iraq or Afghanistan. Each service has different rules and regulations. One has a strict "no minors policy" and refuses to murder children under eighteen, while another demurs when it comes to political assassinations. No worries, though, there are just as many services that are dedicated to killing government officials, such as the crowdsourced Assassination Market mentioned previously. Prices range from a low of $20,000 to more than $100,000 to kill a police officer. The sites request you provide a recent photograph of the target, as well as home and work addresses, daily routines, and frequent hangouts. Bitcoin gladly accepted, and photographic proof of murder is included standard.

CHILD SEXUAL ABUSE IMAGES

Disturbingly, the Dark Net is a sanctuary for merchants of child pornography, widely offered on underground Tor sites such as Hard Candy, Jailbait, Lolita City, PedoEmpire, and Love Zone. The Family Album offers "exclusive private child sex material by the members" and is a veritable YouTube of self-produced and videotaped child sexual abuse. Kindergarten Porn provides links sorted by a child's age and gender. E-books are also for sale with titles such as *Producing Kiddie Porn for Dummies* and *How to Practice Child Love.* Many sites allow pedophiles to connect and communicate with one another in order to share fantasies or images. They also exchange detailed tactics on how to effectively target, seduce, and engage in sexual acts with children. The volume of these activities is startling. Just one Dark Web site alone had more than twenty-seven thousand registered pedophile members in its forums. Moreover, the National Center for Missing and Exploited Children examines nearly twenty million child sexual abuse images annually, a 4,000 percent increase since 2007. Law enforcement sources report that 19 percent of pedophiles have sexual abuse images of children younger than three years old, 39 percent younger than six years old, and 83 percent younger than twelve years old. Across the Dark Web, experienced pedophiles teach one another how to evade law enforcement authorities and discuss encryption and anonymity techniques to avoid detection online.

HUMAN TRAFFICKING

The Dark Web also facilitates the trafficking in human beings, and numerous Web sites specialize in the sale and trade of both adults and children. The U.S. Department of Justice estimates that trafficking generates upward of $32 billion a year in cash for Crime, Inc. While traditional channels of trafficking remain, online technologies provide traffickers the "unprecedented ability to exploit a greater number of victims and advertise their services across geographic boundaries." In the United States alone, nearly 200,000 children are trafficked for sex, and a pimp can generate $150,000 to $200,000 a year per child. Nearly 70 percent of the survivors of child trafficking said that they were advertised online at some point during their trafficking situation and forced to have sex up to twenty times a day. These activities are transacted not only via the Dark Web but also relatively openly on social media and in surface Web classified ads. Web sites such as BackPage.com promise "escorts" and "massages," and advertising by pimps generates nearly $45 million a year in revenue for online classified firms. Not only are children for sale online, but so too are immigrants and other at-risk populations.

HUMAN ORGAN TRAFFICKING

Around the world, there is a vibrant and gruesome black market for body parts. Kidneys can fetch up to $200,000, hearts $120,000, livers $150,000, skin a mere $10 per square inch, a shoulder $500, and a pair of eyeballs $1,500. Where do these bits of humanity come from? Both the living and the dead. Grave robbing is alive and well in the twenty-first century, and many mortuaries around the world are selling parts of those entrusted to their care, correctly assuming the family will never know. Worse, the living poor are at particular risk, actively targeted online and off to sell their organs to wealthy patients in desperate need. In the United States alone, there are more than 100,000 people on the list to receive kidneys, which works out to a ten-year wait. Most will die in just half that time. As a result, wealthy patients head overseas in search of "donors," driving a flourishing illicit trade in human organs. Crime, Inc. has a full division of dedicated organ brokers who act as middlemen across continents to connect buyers and sellers, whom they guide to broker-friendly physicians, surgeons, and medical facilities. The World Health Organization estimates an illicitly obtained organ is sold every hour within these underground networks. The organs may come from executed prisoners in China, from women in India, forced by their husbands to sell body parts in order to contribute to the family's income, or from newly arrived Syrian refugees sheltering in Lebanon desperate for cash. Though a kidney may sell for $200,000, those donating the organ are paid far less, perhaps only $2,500 to $10,000, leaving a massive profit for criminal organ brokers. Sadly, those selling body parts receive little or no post-op care, and many die from the operations. Increasingly, these activities are taking place online, in chat rooms and via the Dark Web. The impoverished in places such as India, Bulgaria, and Serbia are posting desperate pleas, such as the one from a homemaker who listed her blood type and phone number and wrote, "I will sell my kidney, my liver, or do anything necessary to survive." In China, organ brokers are particularly targeting young people in Internet forums with slogans such as "Donate a kidney, buy the new iPad." At least one seventeen-year-old boy, now in ill health, is known to have taken the deal, after his mother discovered a new iPad in the impoverished family's home and contacted police.

LIVE CHILD RAPE
Like Dante's *Inferno*, the Dark Web has its own ninth circle of hell, and here is where you will find the most abominable acts of violence against the youngest and most vulnerable members of our society. In a deeply disturbing report issued by Europol—the European Union's police agency—law enforcement officials noted the growing number of underground Web sites now providing live-streaming videos of the abuse and rape of children. Crime, Inc. and pedophile networks are actually organizing pay-per-view child rape on demand. Organized criminal networks in Asia in particular provide pedophiles with the ability to connect live via Tor to video feeds with built-in instant messaging functions where users around the world can direct the rapists who have abducted the children to carry out specific acts of abuse. Yes, nauseatingly, they do take requests. In one incident investigated by police, those connecting via Tor could order a group of men to rape an eight-year-old girl, directing their actions in real time, all for about $100. Because these activities take place on the Dark Web and because the video images are streamed instead of downloaded, evidence to prove the crimes is not recorded anywhere, save for in the permanent and traumatic memories of the victims who survive the depraved brutality of their fellow man.

All of these illicit goods and services offered for sale in the digital underground drive tremendous profits to Crime, Inc., a trend that is accelerating thanks to new forms of illicit finance that greatly facilitate its clandestine business operations.

Dark Coins

Bitcoin's got its issues. But it is not competing with perfection.

DAN KAMINSKY, SECURITY RESEARCHER

Technology is enabling new forms of money, and the growing digital economy holds great promise to provide new financial tools, especially to the world's poor and unbanked. These emerging virtual currencies are often anonymous and none have received quite as much press as Bitcoin, a decentralized peer-to-peer digital form of money. Bitcoins were invented in 2009 by a mysterious person (or group of people) using the alias Satoshi Nakamoto, and the coins are created or "mined" by solving increasingly difficult mathematical equations, requiring extensive computing power. The system is designed to ensure no more than twenty-one million Bitcoins are ever generated, thereby preventing a central authority from flooding the market with new Bitcoins. Most people purchase Bitcoins on third-party exchanges with traditional currencies, such as dollars or euros, or with credit cards. The exchange rates against the dollar for Bitcoin fluctuate wildly and have ranged from fifty cents per coin around the time of its introduction to over $1,240 in November 2013.

People can send Bitcoins to each other using computers or mobile apps, where coins are stored in "digital wallets." Bitcoins can be directly exchanged between users anywhere in the world using unique alphanumeric identifiers, akin to e-mail addresses, and there are no transaction fees. Anytime a purchase takes place, it is recorded in a public ledger known as the "blockchain," which ensures no duplicate transactions are permitted. Bitcoin is the world's largest crypto currency, so-called because it uses "cryptography to regulate the creation and transfer of money, rather than relying on central authorities." Bitcoin acceptance is growing rapidly, and it is possible to use Bitcoins to buy cupcakes in San Francisco, cocktails in Manhattan, and a Subway sandwich in Allentown. They can also be used to purchase a new Tesla Model S, to pay your DIRECTV bill, to sign up with OkCupid, or even to book a ticket on Richard Branson's upcoming Virgin Galactic space flight.

Because Bitcoin can be spent online without the need for a bank account and no ID is required to buy and sell the crypto currency, it provides a convenient system for anonymous, or more precisely pseudonymous, transactions, where a user's true name is hidden. Though Bitcoin, like all forms of money, can be used for both legal and illegal purposes, its encryption techniques and relative anonymity make it strongly attractive to criminals. Because funds are not stored in a central location, accounts cannot readily be seized or frozen by police, and tracing the transactions recorded in the blockchain is significantly more complex than serving a subpoena on a local bank operating within traditionally regulated financial networks. As a result, nearly all of the Dark Web's illicit commerce is facilitated through alternative currency systems. People do not send paper checks or use credit cards in their own names to buy meth and child sexual abuse images. Rather, they turn to anonymous digital and virtual forms of money such as Bitcoin.

In the days of Al Capone's Prohibition-era racketeering, the Feds' mantra became "Follow the money," and it was ultimately tax evasion charges, not murder convictions, that brought down the world's biggest crime boss of the 1930s. Though "follow the money" has been the core credo in law enforcement ever since, cops may soon have to find a new motto. There are now more than seventy virtual crypto-currency competitors to Bitcoin, such as Ripple, Litecoin, and Dogecoin, and it is estimated nearly $10 billion in virtual currencies were transacted in 2013 alone. Given the vast sums at play, it should come as no surprise that criminals are not only transacting Bitcoin but also targeting the crypto currency for theft. Hackers have been able to steal millions and millions of dollars in virtual money from one another, with the largest attack to date directed against Mt. Gox, a Tokyo-based Bitcoin exchange that had $470 million pilfered from its digital coffers in early 2014. This is undoubtedly the future of bank robbery, and, no, FDIC insurance will not cover you for your Bitcoin losses.

Beyond crypto currencies, there are numerous other forms of electronic payment favored by Crime, Inc., including Liberty Reserve, E-gold, and WebMoney. Just one of these companies, Liberty Reserve, is accused of laundering more than $6 billion over several years, according to federal prosecutors. Known as the "PayPal for criminals," with no personal account details required, Liberty Reserve facilitated a broad range of Crime, Inc.'s activities across the Dark Web, including "credit card fraud, identity theft, investment fraud, computer hacking, child pornography, and narcotics trafficking." It is also thought to have played a central role in the previously noted $45 million crowdsourced ATM heist that took place over a ten-hour time frame in 2013. Though Liberty Reserve, like Silk Road, was ultimately taken down by the FBI and its founder arrested, many competitors have sprung up in its place, and these new marketplaces generally have decentralized peer-to-peer structures and favor next-generation iterations of crypto currencies. They promise not just pseudonymity as recorded publicly in the Bitcoin blockchain but completely untraceable anonymity. One such new currency, Darkcoin, can be viewed as the ultrasecret shadowy cousin of Bitcoin, created specifically to obfuscate users' purchases by combining any single transaction with those of other users so that payments cannot be tied to any particular individual. The popularity of Darkcoin is increasing rapidly, and its value has skyrocketed from seventy-five cents a coin to almost $7 shortly after its introduction.

Another tool, Darkwallet, created by an organization referring to itself as unSYSTEM, aims to take Bitcoin back to its libertarian roots by enabling "hyper-anonymized" transactions. Operating under the motto "Let there be dark," Darkwallet "aims to be the anarchist's Bitcoin app of choice," and its creators explicitly describe it as "money laundering software." By combining and encrypting user payments, Darkwallet "enables practically untraceable flows of money" across the digital underground. Armed with these new financial tools, criminals are primed and ready to go shopping, and there is much to buy.

Crime as a Service

With an untraceable illicit monetary system in place, crime is no longer something that you just commit; it is something you can buy. Crime as a Service (CaaS) is the new business model and allows all or part of an offense to be carried out by others, while the crime-trepreneur who organized and invested in the scheme is ensured the profit. Just as large corporations are increasingly using Software as a Service to carry out their enterprise operations beyond their core competencies, so too are criminals.

One of the most oft-purchased services is that of IT infrastructure—the technological guts and pipes required to run any successful modern enterprise. But Crime, Inc. has special technological infrastructure needs, specifically for what has become an exceeding rare commodity these days: privacy and anonymity. Criminals have flocked to the Dark Web because it allows them the best chance to evade both the surveillance business models popularized by Facebook and Google and the state-level capabilities disclosed by Edward Snowden. Because both their livelihood and their lives depend on assuring this anonymity, members of Crime, Inc. dedicate significant resources to preserving their privacy prior to attacking their targets or selling their contraband.

Practically, this means that illicit actors in the digital underground make extensive use of virtual private networks (VPNs) and proxy servers that hide their Internet protocol addresses and conceal their locations. They also rely heavily on so-called bulletproof hosting services, companies that provide Web hosting in jurisdictions such as Russia or Ukraine and welcome all illicit content, make no attempts to know their customers' true identities, accept anonymous payments in Liberty Reserve and Bitcoin, and routinely ignore subpoena requests from law enforcement. One such CaaS company, Freedom Hosting, was the largest Web host on the Tor network and was accused by the FBI of being the most prolific facilitator of child sexual abuse images in the world, supporting more than 95 percent of the world's child pornography. Hundreds of crime-trepreneur purveyors of child sexual abuse images paid Freedom Hosting to anonymously host their underground Web sites, with each of these individual sites having thousands of registered users.

In addition, just as companies have rapidly adopted cloud computing to store their files on services such as Google Drive and Amazon, so too has Crime, Inc. In an interesting turn of events, not only are hackers targeting the data you've stored in the cloud, but they are increasingly benefiting from the ease of its use to store their own less sensitive files online. The cloud is particularly well suited to the computing needs of the members of Crime, Inc. who use stolen credit cards, fake identities, and front companies to rent space with legitimate companies in order to host malware on their servers. By using reputable firms to host their crimeware, hackers are much less likely to have their traffic blocked or detected by third parties. The trend is accelerating, and a 2013 study suggested that 16 percent of the world's malware distribution channels were hosted in the Amazon Cloud while another 14 percent emanated from GoDaddy's servers.

Moreover, the cloud puts tremendous computing power at the disposal of legitimate users and hackers alike. As a result, we've entered the age of weaponized computing, where literally anybody with a few dollars to spare can have access to previously unimaginable levels of computing power to use for good or ill. For example, the hackers who broke into the Sony PlayStation Network used the vast computing power of Amazon's cloud-computing services to break several of Sony's encryption keys, providing access to hundreds of thousands of user accounts and credit card details. This "cloud cracking" significantly reduces the time it takes to break even the strongest passwords and in the process leaves us all less secure. Today, using the distributed computing power of the cloud and tools such as CloudCracker, you can try 300 million variations of your potential password in about twenty minutes at a cost of about $17. This means that anyone could rent Amazon's cloud-computing services to crack the average encryption key protecting most Wi-Fi networks in just under six minutes, all for the paltry sum of $1.68 in rental time (sure to drop in the future thanks to Moore's law).

Just as legitimate companies can hire computer coders to help them build Web sites and write software, so too can Crime, Inc. A firm such as CrimeEnforcers (a play on the term "law enforcers") describes itself as a "private organisation for your special developing requests . . . [i]f you need special hardwares [sic] . . . [or] software that can not be done or even discuss [sic] in your Country . . . We are offering absolutely anonymous & offshore developing [sic] for your projects. We dont [sic] care what you want to do with hardwares and softwares you requested to be done by us." No questions asked in the world of criminal software development. Other Crime, Inc. firms can be hired to break into any system of your choosing and may have powerful capabilities to do so. For example, China's Hidden Lynx organization comprises up to a hundred professional cyber thieves known to have penetrated systems belonging to Google, Adobe, Lockheed Martin, and others. Frighteningly, the membership of Hidden Lynx includes military and intelligence officers working for the Chinese government during the day to carry out offensive cyber operations on behalf of the state. Off duty, however, many of these officials supplement their income considerably by moonlighting as cyber fraudsters and hackers for hire, distinguishing themselves for their advanced skill sets far beyond those of the average hacker. Welcome to the world of cyber mercenaries, now available as one of the many CaaS offerings in the digital underground.

In addition to hacker-for-hire services, Crime, Inc. subcontracts out for a wide variety of administrative services such as banking, translation, travel, and call center operations.

For instance, companies such as CallService.biz fill a niche in the digital underground by providing on-demand English-, French-, and German-speaking stand-ins to help crooks contravene bank security measures required to initiate wire transfers, unblock hacked accounts, or change address contact information with the banks. Staffed 24/7, the multilingual crime call center will play any duplicitous role you would like, including providing job and educational references, for a mere $10 per call. Just about any professional service a crime-trepreneur might need can be found in the digital underground. Increasingly, however, these services are being bundled, packaged, and sold in the form of criminal software, widely available in the depths of the Dark Web.

Crimeazon.com

The economy of the digital underground is a complex one. Not only do criminals sell directly to consumers (drugs, fake driver's licenses, pirated content, and so on), but they also sell in bulk directly to one another. While much of Crime as a Service is about maintaining the support infrastructure and anonymity required to keep the crime factory humming, the underground economy has been bolstered as members of Crime, Inc. began to offer prepackaged tools for phishing, spam, fraud, DDoS, and data theft.

Top-notch criminal coders have recognized that the offensive hacking tools they have created for themselves can bring additional profits when sold to their criminal brethren—short on time or expertise—to launch their own attacks. As a result, less skilled criminals can simply buy the tools they need on demand to identify system vulnerabilities, commit identity theft, compromise servers, and steal data—crime at the click of a mouse.

The Dark Web has thus become a virtual "Crimeazon.com"—the world's largest online marketplace where criminals go to shop. There they will find a Turkish bazaar of forbidden fruits, all neatly arranged for purchase. Like other purveyors of e-commerce, Crime, Inc. has created Dark Net–product storefronts complete with online shopping carts, checkout management systems, coupon codes, payment processing, technical support, live customer service chats, and escrow services. Vendors offer one-stop shopping, and you can leave your American Express card at home; they gladly accept Bitcoin.

As an example, the malware responsible for the massive invasion of Target's point-of-sale system in late 2013 was perpetrated by a crimeware tool kit known as BlackPOS. Some of the most popular criminal software tool kits for sale in the digital underground include the following:

Zeus Builder: Ranging in price from $5,000 to $7,000, the program has many functions ranging from surreptitiously capturing a user's keystrokes to the theft of digital encryption certificates required for online banking. Over the years, Microsoft has estimated the Zeus Trojan has infected more than thirteen million computers worldwide and been used to steal more than $100 million.

Bugat: Priced at a mere $1,000, Bugat specializes in spoofing bank account and wire transfer requests. In 2010, Bugat was used in a phishing e-mail sent to tens of millions of LinkedIn users with an "update your account" message. When they did, the Bugat Trojan installed malware in their Web browsers in under four seconds, lying stealthily in wait to steal their financial details the next time they logged on to their bank accounts.

SpyEye: For just $500, SpyEye offered all the features of Zeus and more. Its introduction in late 2009 set off a crimeware pricing war, and its market share grew rapidly. In a fascinating turn of online gang warfare, the inventors of SpyEye actually included an anti-virus module to detect the presence of the rival Zeus Trojan on the infected machines of users in the general public. Once it found it, SpyEye would happily remove the competitor Zeus threat and repair the point of entry to ensure SpyEye remained the only malware operating on the targeted machine. Like its rival Zeus, SpyEye is believed to have generated hundreds of millions of dollars in proceeds for its architects.

The software tool kits sold on Crimeazon.com are continually developed, and Crime, Inc. sells updates to their "latest versions" to ensure the most current computer exploits are included in its programs. Of course, there is also Crimeazon Prime, a program that offers fellow thugs the opportunity to "subscribe and save" on their purchases. Once such example is the Blackshades tool kit available on an ongoing rental basis, providing users with unlimited free updates and technical support. The tool, perhaps one of the world's most popular and notorious malware exploit kits, combines remarkable technical agility with a highly evolved business model that could have come straight out of a Harvard Business School case study.

Crime-trepreneurs who bought the Blackshades tool kit could select the manner by which the malware would attack a machine in question, such as embedding the Trojan in a document, concealing it on a Web site, or placing it on a USB drive that would deliver its deadly payload when inserted innocently into a target's computer. Because Blackshades was an advanced remote-access Trojan (RAT), it gave its developers complete and total control of an infected machine's functions. As a result, Blackshades could capture keystrokes, steal passwords, launch denial-of-service attacks, hijack Facebook accounts, and install additional malware on the affected system. Worse, it was the tool of choice for would-be stalkers because it allowed its masters to remotely turn on any computer's microphone and camera to capture any audio and video in its field of view, without giving any notice such as a little green recording light. So good was the Blackshades RAT that Bashar al-Assad's Syrian regime used it to spy on democracy activists within the country. Though point-and-click crime and espionage tool kits are widely available for purchase at Crimeazon.com, each and every computer attack begins with the initial system penetration and malware infection, vulnerabilities that are widely available for sale in the digital underground.

The Malware-Industrial Complex

Nuclear scientists lost their innocence when we used the atom bomb for the very first time. So we could argue computer scientists lost their innocence in 2009 when we started using malware as an offensive attack weapon.

MIKKO HYPPONEN

In order for criminals, spies, militaries, and terrorists to carry out their offensive cyber attacks, they must first figure out how to exploit the information system they wish to target. As we saw with the Stuxnet attack against the Iranian nuclear enrichment site at Natanz, such operations can take years of planning and cost millions of dollars. Fortunately for those without the time and budget to devise their own cyber weapons, there is a vast shadowy black market where spies, soldiers, thieves, and hacktivists can shop for so-called zero-day exploits. As mentioned previously, these zero-day bugs have not yet been discovered by software and antivirus companies and thus handily defeat common security and firewall measures without sounding an alarm.

In the old days, hackers used to hold on to these exploits for their personal use or attempt to sell them to software giants such as Microsoft, Yahoo!, and Google via company-established "bug bounty" programs. The rewards, however, were paltry—a mere $500 for uncovering major security holes. In frustration, hackers realized there were much better options available, including selling their security flaw vulnerabilities on the open market to criminals and governments. This realization has led to the establishment of a highly complex network of buyers, sellers, and brokers of cyber exploits in what has come to be known as the malware-industrial complex.

Before Crime, Inc. can sell its fully packaged cyber-crime tool kits such as SpyEye and Zeus, it must amass a series of malware vulnerabilities and package them into crimeware for use by the general criminal public. It does this by funding exploit buying sprees and has the budgets to do so. One criminal hacker known as Paunch reportedly contracted with a third-party exploit dealer and provided him with a $100,000 budget to gather vulnerabilities for use in his malicious Blackhole exploit kit. Not to be outdone, another hacker, using the alias J. P. Morgan, posted a message in the Darkcode crime forum advertising that he had a budget of $450,000 to spend on zero-day exploits for use in his proprietary crimeware tool kit. Dark Net chat rooms are replete with malware shopping requirements, and posts such as "Do you have any code execution exploit for Windows 7? . . . If yes, payment is not an issue" are commonplace.

The trade in cyber arms is not restricted to criminals; government security services are also frequent buyers of these tools, turning to third-party brokers to obtain their technical weaponry. One such middleman by the name of the Grugq has established himself as an exploit broker of choice, capable of negotiating significant deals between those who uncover security flaws and those who are looking to buy them for operational exploitation. In 2012, the Grugq sold an exploit for the iOS mobile phone operating system to a U.S. government contractor for a cool $250,000 (minus his standard 15 percent commission).

A number of professional firms have emerged whose sole business model is the trafficking in computer malware exploits to governments. Companies such as Vupen in France, Netragard in Massachusetts, Endgame of Georgia, Exodus Intelligence in Texas, and ReVuln in Malta are all heavily involved in selling offensive exploits to customers around the world. While some zero-day trafficking firms vet their clients, others will sell to anybody, from Crime, Inc. to notorious dictators, no questions asked. The result, as pointed out by the noted security researcher Tom Kellermann, is that now anybody can download a cyber Kalashnikov or cyber grenade from a myriad of sites.

Many zero-day exploits enable particularly stealthy and sophisticated attacks against specific targets, giving rise to what security researchers have termed the advanced persistent threat, or APT. APTs use extensive targeting research combined with a high degree of covertness to maintain command and control of a marked system for months or years at a time, and their use is growing. Hide, watch, and wait is the modus operandi for these cyber attacks and good hackers always erase the system logs so you never know they were even there. Whether it has been developed by the U.S. government, China, or Crime, Inc., the likelihood of a consumer-grade antivirus product detecting one of these advanced persistent threats is effectively nil.

Stuxnet is perhaps the most infamous of APTs, but it has cousins such as Flame and Duqu, along with many others yet to be discovered. Worse, now that Stuxnet, a tool developed to attack industrial control systems and take power grids off-line, is out in the wild and available for download, it has been extensively studied by Crime, Inc., which is rapidly emulating its techniques and computer code to build vastly more sophisticated attacks. The deep challenge society faces from the growth of the malware-industrial complex is that once these offensive tools are used, they have a tendency to leak into the open. The result has been the proliferation of open-source cyber weapons now widely available on the digital underground for anybody to redesign and arm as he or she sees fit. How long will it be before somebody picks up one of these digital Molotov cocktails and lobs it back at us with the intent of attacking our own critical infrastructure systems? Sadly, preparations may already be under way.

Net of the Living Dead: When Botnet Zombies Attack

A zombie apocalypse isn't the most jovial situation.
DANAI GURIRA (MICHONNE) IN *THE WALKING DEAD*

One of the most powerful tools in a hacker's arsenal is a botnet, a robot network of infected computers under the remote control of the hacker. These so-called zombie-infected machines are taken over and enslaved to unite in botnets, which can be used for a variety of criminal services such as spreading malware, perpetrating DDoS attacks, disseminating spam, or hosting illicit content. Computers and even mobile phones can be drafted into a botnet army upon infection by malware, particularly that served up by the prefab crimeware tool kits such as Blackshades and SpyEye, widely available for sale in the digital underground.

Unfortunately, the malicious payload for victims of these tool kits is twofold: not only will they steal your credit card details, banking log-ons, and identity, but they also leave behind a persistent back door in your system that gives Crime, Inc. perpetual access to your machine for it to do as it pleases.

As you sit there writing a Word document or reading CNN online, the botnet master may be simultaneously and surreptitiously using your machine for any number of criminal services. Ever wonder why your computer runs so slowly? You may be unwittingly participating in an ongoing cyber attack against others, and you have no idea it's happening. Thank *you* for your service.

Hackers have crowdsourced and off-loaded their attack to you and your computers, involuntarily embroiling you and them in their international criminal conspiracy. Crime, Inc. can even draft your computer into a peer-to-peer child pornography network, hiding sexual abuse images on your hard drive. After all, why should it risk keeping them on its own networks? As a result of your network's insecurity and your inability to protect your own digital devices, you too are now participating in the cyber-crime economy. Just as Facebook is monetizing you and your online life, so too is Crime, Inc.

Some of the most notorious botnets include Mariposa, Conficker, and Koobface, with new entrants such as Gameover Zeus rapidly gaining market share. According to the FBI, Gameover Zeus alone controlled more than one million computers worldwide and resulted in $100 million in financial losses. As of mid-2014, the largest botnet known to be in existence was called ZeroAccess, which on any given day had nearly two million zombie computers under its complete control. With larger and larger botnets come increasing offensive power, as these millions of computers can be trained on any target of interest selected for a distributed-denial-of-service attack. DDoS offensives work by flooding a computer system or Web site with tens of thousands of sham requests for information, thereby crashing the targeted site, leaving it off-line, unable to send e-mail, serve up Web pages, process orders, or clear bank transactions.

Like all of Crime, Inc.'s tools and services, zombie botnets can be purchased or rented online, bringing this offensive capability into the mainstream on the cheap. In the Russian digital underground, powerful DDoS botnets can be purchased for $700 or rented for just $2 an hour, long enough to take down the typical Web site or call center. On average, nearly three thousand such attacks are launched around the world daily. Moreover, the threat is growing in sophistication as both Crime, Inc. and state actors such as Iran and China increasingly turn to the massive distributed computing power of the cloud to carry out DDoS attacks. One zombie network, known as Storm.bot 2.0, for sale on the digital underground in mid-2014 for a mere $3,000, has usurped fifteen cloud servers around the world and is capable of generating an unfathomable three hundred gigabytes per second of attack traffic, advertised as being more than sufficient to "knock small countries off-line." The result of these botnet zombies has been the weaponization of cyberspace by Crime, Inc.

The toll of victims affected by this type of botnet cyber extortion is growing, and even high-profile companies such as Evernote and MeetUp .com have been attacked. The panoply of malware tool kits and the millions of botnet zombies around the world are providing Crime, Inc. with powerful tools of domination that can be used as offensive weapons, cash-making machines, or both. Consequently, we've entered the Industrial Age of Crime, with malicious computer code churned out in assembly-line fashion, specifically developed and scripted to run on autopilot, toiling away day and night committing offenses while hackers earn healthy profits in their sleep.

Committing Crime Automagically

Though Crime, Inc. engages in constant business process improvement, it is not committing new crimes from scratch each and every time. In the age of Moore's law, these tasks have been readily automated and can run in the background at scale without the need for significant human intervention. Crime automation allows transnational organized crime groups to gain the same efficiencies and cost savings that multinational corporations obtained by leveraging technology to carry out their core business functions. That is why today it's possible for hackers to rob not just one person at a time but 100 million or more, as we saw with the Sony PlayStation and Target data breaches.

Exploit tool kits like Blackhole and SpyEye commit crime "automagically" by minimizing the need for human labor, thereby dramatically reducing costs to Crime, Inc. They also allow hackers to pursue the "long tail" of opportunity, committing millions of thefts in small amounts so that victims don't report them and law enforcement has no way to track them. While particular high-value targets (companies, nations, celebrities, high-net-worth individuals, or objects of affection or scorn) are specifically and individually targeted, the way the majority of the public is hacked is by automated scripted computer malware—one large digital fishing net that scoops up anything and everything online with a vulnerability that can be exploited. Given these obvious advantages, as of 2011 an estimated 61 percent of all online attacks were launched by fully automated crime tool kits, returning phenomenal profits for the Dark Web overlords who expertly orchestrated them. Modern crime has become reduced and distilled to a software program that anybody can run at tremendous profit.

Not only can botnets and other tools be used over and over to attack and offend, but they are even enabling the commission of much more sophisticated crimes such as extortion, blackmail, and shakedown rackets. In an updated version of the $500 million Ukrainian Innovative Marketing solutions "virus detected" scam, Crime, Inc. has unleashed a new torrent of malware that can hold your computer hostage until a ransom is paid to regain access to your own files. Known as ransomware, these attack tools are included in a variety of Dark Net tool kits, such as Gameover Zeus. There are several varieties of this scam, including one that purports to come from law enforcement. Around the world, users who become infected with the Reveton Trojan suddenly have their computers lock up and their full screens covered with a notice, allegedly from the FBI. The message, bearing an official-looking large, full-color FBI logo, states that the user's computer has been locked for reasons such as "violation of the federal copyright law against illegally downloaded material" or because "you have been viewing or distributing prohibited pornographic content."

To unlock their computers, users are informed that they must pay a fine ranging from $200 to $400, only accepted using a prepaid voucher from Green Dot's MoneyPak, which victims are instructed they can buy at their local Walmart or CVS. To further intimidate victims and drive home the fact that this is a serious police matter, Crime, Inc. prominently displays the alleged violator's IP address on their screen as well as snippets of video footage previously captured from the victim's Webcam. The scam has successfully targeted tens of thousands of victims around the world, with the attack localized by country, language, and police agency. Thus users in the U.K. see a notice from Scotland Yard, other Europeans get a warning from Europol, and victims in the United Arab Emirates see the threat, translated into Arabic, purportedly from the Abu Dhabi Police HQ.

Another, even more pernicious type of automated extortion has emerged in the form of CryptoLocker, a Trojan that actually encrypts all the files on a victim's computer so that they can no longer be read or accessed. Alarmingly, the malware presents a ticking-bomb-type count-down clock advising users that they only have forty-eight hours to pay $300 or all of their files will be permanently destroyed. Akin to threat-ening "if you ever want to see your files alive again," these ransomware programs gladly accept payment in Bitcoin. The message to these victims was no idle threat. Whereas previous ransomware might trick users by temporarily hiding their files, CryptoLocker actually uses strong 256-bit Advanced Encryption Standard cryptography to lock user files so that they become irrecoverable. Nearly 250,000 individuals and businesses around the world have suffered at the hands of CryptoLocker, earning an esti-mated $30 million for its developer.

Automated ransomware tools have even migrated to mobile phones, affecting Android handset users in certain countries. Not only have indi-viduals been harmed by the CryptoLocker scourge, so too have companies, nonprofits, and even government agencies, the most infamous of which was the Swansea Police Department in Massachusetts, which became infected when an employee opened a malicious e-mail attachment. Rather than losing its irreplaceable police case files to CryptoLocker, the agency was forced to open a Bitcoin account and pay a $750 ransom to get its files back. The police lieutenant Gregory Ryan told the press he had no idea what a Bitcoin was or how the malware functioned until his department was struck in the attack.

As we have seen throughout this chapter, a journey into the abyss can be a dark and scary place. Yet within this world, Crime, Inc. has evolved highly sophisticated methods of operation to sell everything from methamphetamine to child sexual abuse live streamed online. It has rapidly adopted tools of anonymity such as Tor to establish Dark Net shopping malls, and criminal consulting services such as hacking and murder for hire are all available at the click of a mouse. Untraceable and anonymous digital currencies, such as Liberty Reserve and Bitcoin, are breathing new life into the underground economy and allowing for the rapid exchange of goods and services. With these additional revenues, Crime, Inc. is becoming more disciplined and organized, significantly increasing the sophistication of its operations. Business models are being automated wherever possible to maximize profits and botnets can threaten legitimate global commerce, easily trained on any target of Crime, Inc.'s choosing. Fundamentally, it is done. The computing and Internet crime machine has been built. With these systems in place, the depth and global reach of Crime, Inc.'s power mean that crime now scales, and it scales exponentially. Yet for as bad as this threat is today, it is about to become much worse, as we hand Crime, Inc. billions of more targets for them to attack as we enter the age of ubiquitous computing and the Internet of Things.

CHAPTER 12

When All Things Are Hackable

We're still in the first minutes of the first day of the Internet revolution.
SCOTT COOK, INTUIT

E ven in the age of the Internet, buying a car can be an expensive, frustrating, and laborious process. It's even worse if you are unemployed or have limited resources. Fortunately, Texas Auto Center in Austin caters to just these customers, promising a car for everyone, "no matter if you have good credit, bad credit, a bankruptcy, repossession, or no credit at all." Of course when times are rough, people do get behind in their loan payments, and repossession rates at some dealerships run as high as 45 percent. Repossessing cars is never fun, either for those who are about to lose their primary means of transportation or for the dealers who have to send out a fleet of tow trucks in search of the car. These vehicles are often purposefully hidden by those who know they are facing repossession. When the repo man and his tow truck eventually come calling, tempers flare, and many repo men have been punched, kicked, spit upon, bitten, stabbed, and even shot to death trying to recover the dealer's property. Surely there had to be a better approach, and Texas Auto Center thought it had found just the solution.

The dealership purchased a new technological tool from the Cleveland-based Pay Technologies that promised a far superior alternative to the confrontational repossessions of yesteryear. Pay Technologies' product was known as the WebTeckPlus, a system that allowed car dealers to install "a small black box, about the size of a deck of cards, cleverly concealed underneath a vehicle's dashboard." The devices were controlled remotely via a central Web site that relayed signals over a wireless network to the cars' black boxes. When activated, the signal allowed the dealership to "disable a car's ignition system or trigger the horn to begin honking," a nice, if not too subtle, way to remind owners their payment was overdue. Texas Auto Center began slowly installing the boxes in its entire fleet, and before long more than eleven hundred cars had the system in place. In charge of administering the new high-tech repo management system was Omar Ramos-Lopez, a young credit collector at the dealership with an affinity for technology.

All seemed to work well with the new system until February 2010, when suddenly a few of Texas Auto Center's customers' cars just stopped running and would not restart. They had no idea why. A check of company records indicated that the clients were all current with their payments. Throughout the day, the number of complaints began to increase, and by the fifth day more than a hundred owners had flooded the dealership with their irate grievances. What was going on?

Customers throughout Texas suddenly had their cars bricked, completely un-drivable and unable to start. Randomly, in the middle of the night horns began honking out of control around the city of Austin, and police were called with numerous noise complaints. When the cops arrived, they discovered the horns could not be shut off until physically disconnected from their car battery cables. Worse, these hundred customers found themselves without transportation, forced to miss work and desperately needed paychecks.

Though the incident was initially dismissed as a "systemic mechanical failure," something much more nefarious was at play. An intruder illegally accessed Texas Auto Center's Web-based remote vehicle immobilization system and one by one began turning off their customers' cars throughout the city. Attempts by the dealership to turn the cars back on were stymied because the hacker had also altered the records in its database, changing vehicle identification numbers and replacing the names of legitimate customers with those of celebrities, such as the long-dead rapper Tupac Shakur and the pop star Jennifer Lopez.

Clearly something was amiss, and eventually suspicions fell upon twenty-year-old Omar Ramos-Lopez, who had been fired from the dealership in the days prior to the widespread vehicular paralysis for "not meeting company standards." Law enforcement officials alleged Ramos-Lopez used his knowledge of his former employer's system and the password of a former co-worker to exact revenge for his firing by disabling cars en masse throughout Austin. The police investigation showed that the former collection agent logged in to Pay Technologies' servers in Ohio from the AT&T broadband network leading to his home. Ramos-Lopez was arrested and charged with felony breach of a computer system.

As for Texas Auto Center, it is far from unique in its decision to install remote repo-man technology in its vehicles; today there are more than two million cars with the technology. But as we shall see later, there are tens of millions of vehicles around the world that can be controlled one way or another online, with thousands more being added to the global information grid every day. With such black boxes installed in more and more automobiles, it is becoming increasingly clear that there may be more back doors in your car than you ever realized.

Where the Wireless Things Are

Throughout the short history of modern computing, we have come to think of computers as big boxes of one size or another. In the 1950s, a single computer occupied an entire building. By the 1970s, a mainframe computer had been reduced to the size of a refrigerator. The 1980s brought the personal desktop computer and the 1990s the introduction of the laptop. At the turn of the millennium, mobile phone usage exploded, and by 2007 Steve Jobs had given the world his iPhone, a small but powerful handheld computer. As always, Moore's law marches on, but in the very near future our concept of what constitutes a computer will be blown away as the boxes that have always caged the processor disappear and we enter the reign of ubiquitous computing.

Unlike the stationary desktops of yesteryear, the post-PC era promises a world in which computer processing will take place anywhere, everywhere, and in all things. We are already well into this transition. Laptop sales supplanted desktop sales back in 2005, and in 2015 the number of tablets, such as the iPad, sold worldwide will outstrip sales of desktops and laptops combined. In 2014, we saw more cell phones in use than people on the planet. Of course the smart phones and tablets in our homes have company, joined by gaming consoles, DVRs, cable boxes, and smart TVs, all networked and connected online. But a stroll down the aisles of a local retailer such as Best Buy, Lowe's, or Home Depot reveals yet another trend already under way. At these stores and elsewhere online, a whole new array of digital devices are vying for a position on our home networks—things such as Internet-enabled thermostats, lightbulbs, music speakers, baby monitors, and security systems. Together they represent the first steps in a rapidly emerging new paradigm of computing known as the Internet of Things (IoT), and when it takes off, it may very well change the world we live in forever.

The Pew Research Center defines the Internet of Things as "a global, immersive, invisible, ambient networked computing environment built through the continued proliferation of smart sensors, cameras, software, databases, and massive data centers in a world-spanning information fabric." The term was first coined in 1999 by the MIT researcher Kevin Ashton, who, when working on a project for Procter & Gamble, realized that "if all the objects in daily life were equipped with identifiers and wireless connectivity, these objects could communicate with each other and be managed by computers. . . . 'If we had computers that knew everything there was to know about things—using data they gathered without any help from us—we would be able to track and count everything, and greatly reduce waste, loss and cost.'" Ashton's concept was both simple and powerful and had a major impact on manufacturers and retailers like Walmart, dramatically improving their supply chain management and cutting costs for consumers. Back in 1999, however, the technology did not exist to make the IoT a reality outside very controlled environments, such as factory warehouses. Today that has changed, and a confluence of developments has come together to enable major leaps forward in the world of ubiquitous computing, allowing for the first time the widespread "embedding of miniature computers in objects and connecting them to the Internet using wireless technology." Indeed, according to the Semiconductor Industry Association, as of 2004, human beings were producing more transistors than grains of rice—and at a cheaper cost.

Thanks to advances in circuitry, software, and miniaturization, it is possible to build an Internet of Things whose devices fall broadly into one of two categories: sensors and microcontrollers. Microcontrollers are tiny programmable computer processors measuring just millimeters across. They are low-powered, ultracheap computer chips, some as small as the head of a pin, that can be built and embedded in an infinite number of devices, some for mere pennies. These miniature computing devices only need milliwatts of electricity and thus can run for years on a minuscule battery or small solar cell. As a result, it is now possible to make "a Web server that fits on (or in) a fingertip for $1."

These microchips will receive data from a near-infinite range of sensors, minute devices capable of monitoring anything that can possibly be measured and recorded, including temperature, power, location, hydroflow, radiation, atmospheric pressure, acceleration, rotation, magnetic force, altitude, sound, and video. This global array of sensors will allow us to perceive, analyze, and interact with the world around us as never before humanly possible. Once gathered, these data will not sit idle but rather be processed by a bevy of new IoT microcontrollers such as those mentioned above—miniature switches, actuators, valves, servos, turbines, and engines—all capable of autonomously interacting with the physical world around them. Thus, for example, when a sensor detects excessive temperature or pressure in a gas pipeline, its microcontroller receiving the information will be preprogrammed to react by shutting down or rerouting the flow of natural gas, thereby averting a catastrophic explosion.

Expansive growth in high-speed wireless data networks will allow these sensors to speak to the world using a variety of communications protocols and technologies such as Wi-Fi, broadband, GSM, CDMA, Bluetooth, radio-frequency identification (RFID), near-field communication (NFC), ZigBee, Z-Wave, and power lines. They will communicate not only with the broader Internet but with each other, generating unfathomable amounts of machine-to-machine (M2M) data, which will be stored and processed at greater speed and lower cost thanks to cloud computing and its near-unlimited data storage capabilities. The result will be an always-on "global, immersive, invisible, ambient networked computing environment," a mere prelude to the tidal wave of change coming next.

However, one thing had to be fixed first—the basic communications protocol that routes nearly all traffic on the Internet. The backbone of today's Internet runs on what is known as Internet Protocol Version 4 (IPv4). The communications architecture has been around since 1981 and provides for about 4.3 billion separate network addresses, each one representing a connected device. Back when IPv4 was introduced in the late 1970s, nobody could have imagined that 4.3 billion addresses would be insufficient to meet the demands of the very few major universities and corporations that were online back then. Yet today the unthinkable has happened: we're out of Internet addresses. Just as New York City needed to create new area codes when it ran out of 212 phone numbers to serve its residents, so too has the Internet.

The Internet's answer to this problem is IPv6, which will supplant IPv4 and profoundly increase the size of addressable space available online. The new protocol resolves this problem by increasing the length of the "phone number" from 32 bits to 128 bits. Mathematically, IPv4 can only support about 2^{32} or 4.3 billion connections. IPv6, on the other hand, can handle 2^{128} or 340,282,366,920,938,463,463,374,607,431,768,211,456 connections. The implications of a number this large are mind-boggling. There are only 10^{19} grains of sand on all the beaches of the world. That means IPv6 would allow each grain of sand to have a trillion IP addresses. In fact, there are so many possible addresses with IPv6 that every single atom on our planet could receive a unique address and we would "still have enough addresses left to do another 100+ earths." It is in the wake of these changes that the Internet of Things will be born.

To help put these gigantic numbers in perspective, we can think of today's Internet metaphorically as about the size of a golf ball. Tomorrow's will be the size of the sun. That means that within the coming years, not only will every computer, phone, and tablet be online, but so too will every car, house, dog, bridge, tunnel, cup, clock, watch, pacemaker, cow, streetlight, bridge, tunnel, pipeline, toy, and soda can. Though in 2013 there were only thirteen billion online devices, Cisco Systems has estimated that by 2020 there will be fifty billion things connected to the Internet, with much more room for exponential growth thereafter. As all of these devices come online and begin sharing data with one another, they will bring with them massive improvements in logistics, employee efficiency, supply chain operations, energy consumption, customer service, and personal productivity.

As noted previously, Metcalfe's law dictates that the value of a network increases exponentially with the number of nodes or computers attached. As IPv6 adds 340 undecillion (340 trillion trillion trillion) new potential nodes to the global information grid, the concomitant explosion in economic value will be incalculable. The McKinsey Global Institute predicts that the innovation enabled across multiple sectors by the Internet of Things is expected to drive as much as an additional $6.2 trillion in value to the global economy by 2025. The IoT may very well be where the next Google, Facebook, or Apple is found, and the number of sensors, consumer devices, and industrial control systems online has already surpassed the number of mobile phones. Early entrants to the IoT such as Fitbit, Jawbone, Oculus Rift, Withings, Estimote, and Sonos have generated significant buzz and market valuation. Indeed, one such firm, the smart-thermostat company Nest Labs, was acquired in 2014 for an astounding $3.2 billion just 854 days after the launch of its first product. And while there is undoubtedly big money to be made in the IoT, its social implications may even outstrip its economic impact.

Imagining the Internet of Things

The Internet of Things is a way of saying that more of the world will become part of the network . . . We are assimilating more and more of the world into the computer.
GORDON BELL, MICROSOFT RESEARCHER

The promise of the Internet of Things sounds rosy. Because chips and sensors will be embedded in everyday objects, we will have much better information and convenience in our lives. So, for example, because your alarm clock is connected to the Internet, it will be able to access and read your calendar. It will know where and when your first appointment of the day is and be able to cross-reference that information against the latest traffic conditions. Light traffic, you get to sleep an extra ten minutes; heavy traffic, and you might find yourself waking up earlier than you had hoped. When your alarm does go off, it will gently raise the lights in the house, perhaps turn up the heat or draw your bath. The electronic pet door will automatically open to let Fido into the backyard for his morning visit, and, most important, the coffeemaker will begin brewing your first cup of coffee right on time. You won't have to ask your kids if they've brushed their teeth; the chip in their toothbrush will send a message to your smart phone letting you know the task is done. As you walk out the door, you won't have to worry about finding your keys; the beacon sensor on the key chain makes them locatable within a two-inch radius in your home. It will be as if the Jetsons era has finally arrived.

While the hype-o-meter on the Internet of Things has been blinking red for some time, everything described above is already technically feasible today. To be certain, there will be obstacles, in particular in relation to a lack of common technical standards, but a wide variety of companies, consortia, and government agencies are hard at work to make the IoT a reality. The result will be our transition from connectivity to hyperconnectivity, and like all things Moore's law related, it will be here sooner than we realize. Ubiquitous computing will affect every area of human endeavor, including transportation, energy, finance, government, agriculture, education, public safety, travel, and commerce.

The IoT means that all physical objects in the future will be assigned an IP address and be transformed into information technologies. As a result, your lamp, cat, or ficus will be part of an IT network. Things that were previously silent will now have a voice, and every object will be able to tell its own story and history. The refrigerator will know exactly when it was manufactured, the names of the people who built it, what factory it came from, and the day it left the assembly line, arrived at the retailer, and joined your home network. It will keep track of every time its door has been opened and which one of your kids forgot to close it. When the refrigerator's motor begins to fail, it can signal for help, and when it finally dies, it will tell us how to disassemble its parts and best recycle them. Buildings will know every person who has ever worked there, homes every person who has ever lived in them, and streetlights every car that has ever driven by.

All of these objects will communicate with each other and have access to the massive processing and storage power of the cloud, further enhanced by additional mobile and social networks. We will be living in a world where everything is programmable and interactive. Objects will become "smart" and be able to describe their own location, proximity, velocity, temperature, flow, acceleration, ambient sound, vision, force, load, torque, pressure, and interactions. The first generations of smart phones, smart meters, smart watches, and smart cards are already here today, but in the future all objects may become smart, in fact much smarter than they are today. As these devices become networked, they will develop their own limited form of sentience, resulting in a world in which people, data, and things come together. As a consequence of the power of embedded computing, we will see "billions of smart, connected 'things'" joining a global neural network in the cloud that "will encompass every aspect of our lives."

While the "old" Internet allowed desktops, laptops, and servers to share information, the "new" Internet will make it possible to remotely control any object on earth. As Joi Ito, director of the MIT Media Lab, explains, there is a "phenomenon of convergence, where bits from the digital realm are fusing with atoms here in the physical world." Every object will have an identity and a life in both the physical and the virtual worlds, and when this happens, the difference between online and off-line, previously a meaningful distinction, goes away. To this point, the CEO of Cisco, John Chambers, recently predicted that the Internet of Things would have an impact five to ten times as large as the Internet itself.

In this world, the unknowable suddenly becomes knowable. For example, groceries will be tracked from field to table, and restaurants will keep tabs on every plate, what's on it, who ate from it, and how quickly the waiters are moving it from kitchen to customer. As a result, when the next *E. coli* outbreak occurs, we won't have to close five hundred eateries and wonder if it was the chicken or beef that caused the problem. We will know exactly which restaurant, supplier, and diner to contact to quickly resolve the problem. The IoT and its billions of sensors will create an ambient intelligence network that thinks, senses, and feels and contributes profoundly to the knowable universe.

Not only does the unknowable become knowable, but the impossible suddenly becomes possible. Things that used to make sense suddenly won't, such as smoke detectors. Why do most smoke detectors do nothing more than make loud beeps if your life is in mortal danger because of fire? In the future, they will flash your bedroom lights to wake you, turn on your home stereo, play an MP3 audio file that loudly warns, "Fire, fire, fire." They will also contact the fire department, call your neighbors (in case you are unconscious and in need of help), and automatically shut off flow to the gas appliances in the house. You're not the only one who might have your life saved by the Internet of Things—so too might your plants. Cheap moisture sensors placed in the soil of house plants have been using home Wi-Fi networks to send out tweets since 2009 screaming, "URGENT! Water me!"

Not futuristic enough? What about an interspecies Internet—one that links elephants, dolphins, and great apes for "the purposes of enrichment, research, and preservation"? Though it may sound crazy, it's already here. In Australia, for example, there are over 300 sharks on Twitter (no, they did not sign up themselves). Researchers fitted 338 sharks, including many great whites, with acoustic tags that send an electronic signal to shore-based receivers when the animals come within half a mile of the beach. For a country that has suffered more fatal shark attacks than any other, this IoT development is saving human lives, and the sharks have attracted nearly forty thousand beach-going Twitter followers as a result.

The by-product of the Internet of Things will be a living, breathing, global information grid, and technology will come alive in ways we've never seen before, save for in science fiction movies. While this future may seem far-fetched, M2M communications have already supplanted all human-originated online activities, with more than 61.5 percent of worldwide Internet traffic being generated by things as of late 2013. As we venture down the path toward ubiquitous computing, the results and implications of the phenomenon are likely to be mind-blowing. Just as the introduction of electricity was astonishing in its day, it eventually faded into the background, becoming an imperceptible, omnipresent medium in constant interaction with the physical world. Before we let this happen, and for all the promise of the Internet of Things, we must ask critically important questions about this brave new world. While its multifold benefits seem manifest, an Internet of everything also poses tremendous risk. For just as electricity can shock and kill, so too can billions of connected things networked online.

Connecting Everything—Insecurely

**Connecting products to the Web will be the
twenty-first-century electrification.**
MATT WEBB, CEO, BERG CLOUD

In order for things to go online and communicate with one another, they must first be enabled with the technological equivalent of speech. As we saw with Texas Auto Center and technologies such as the WebTeckPlus black box, cars today can indeed "talk," squawking data about their location, condition, and state. Central to achieving the vision outlined by the proponents of the Internet of Things is giving everyday objects the capacity to speak to us and to each other.

In order to make this happen, the IoT relies on a series of competing communications technologies and protocols. At a distance, cellular and mobile data transmission standards such as LTE, 4G, GSM, and CDMA will connect devices to the mobile phone network. Many larger things will be able to communicate via fixed wired lines such as Ethernet and optical fiber, but for price and convenience perhaps the largest number of connections will take place via wireless networks. The result will be billions of embedded chips in things using standards such as Wi-Fi, Bluetooth, ZigBee, Z-Wave, near-field communication, and radio-frequency identification in order to communicate. As the price of these tools drops, new consumer products such as Apple's iBeacon and Tile's location tags may soon become an omnipresent feature in our daily lives, allowing us to track objects with centimeter precision.

The first of these IoT enabling technologies, RFID, was patented in 1983 and is a wireless low-energy device that can be embedded into any object to make it "smart," or able to interact with RFID readers. RFID tags are printed electronic circuits no thicker than a piece of paper, often come in sticker format, many the size of a dime, and can be produced for under a penny. They are capable of performing real-time, constant data exchange and can be read by scanners, some as far as up to one hundred meters away. Even if you are unfamiliar with RFID technology, chances are you have already encountered it in your life, whether it's the security ID card you use to swipe your way into your office, your "wave and pay" credit card, the key to your hotel room, your subway pass, or the little box you use to pay for highway tolls, such as E-ZPass. Though the convenience of RFID, considered by many the gateway to the Internet of Things, sounds great, there's one problem: it's eminently hackable.

There have been dozens of exploits against RFID technology, whose electronics can be readily hacked, spoofed, and jammed, and there is an active "RFID underground" continually working on improving its offensive techniques. The overwhelming majority of today's RFID tags have no effective security, encryption, or privacy protocols in place. These shortcomings have allowed the security hacker Francis Brown to build his own RFID readers for under $400 that can scan, copy, clone, and steal data from your smart cards. As a result, while you're standing in line at the grocery store, sitting in a crowded subway car, riding the elevator up to your office, or waiting for your morning latte at Starbucks, Brown can conduct a "brush pass" attack. As he stands there smiling and perhaps even chatting with you, the concealed portable RFID reader in his backpack can query the office key card you have in your wallet, pocket, or purse and abscond with all the details encoded in it. So what?

Here's why it matters. Brown can then plug his RFID reader into his computer at home and use it to clone RFID cards all day long. That means he can get into your office, hotel room, or home anytime he likes. Every Fortune 500 company in America uses RFID in its employees' badges to control access to its office buildings, and Brown has a 100 percent success rate in cloning the cards. The implications of this for everything from industrial espionage to common burglary to employee safety are enormous. Relying on insecure RFID identity cards as the primary system we use for security and identity in the workplace means the current system is completely broken. Worse, these cards cannot simply be updated like your home computer by downloading new software. Each of them would need to be replaced—an expensive proposition for a corporation with 100,000 employees.

Even if you don't use an RFID card for work, there's a good chance you either have it or will soon have it embedded in the credit card sitting in your wallet. Hackers have been able to break into these as well, using cheap RFID readers available on eBay for just $50, tools that allow an attacker to wirelessly capture a target's credit card number, expiration date, and security code. Seconds later, using a $300 card-magnetizing tool, the data is then encoded on a new card, allowing fraudulent purchases in a process that can be completed in just a matter of minutes. Welcome to pocket picking 2.0, where the thieves don't even need to stick their hands in your pocket anymore.

The techniques to hack RFID are easy to emulate, and there are hundreds of instruction sites and videos online telling hackers exactly how. Troubling given that billions of things coming online will be using RFID as their primary language to speak and interact with the world. RFID chips can also be infected with viruses, and just like GPS signals RFID can be jammed, preventing you from getting into your office and allowing thieves to shoplift expensive goods electronically tagged by retailers. Another popular IoT communications technology is RFID's younger brother, known as near-field communication (NFC) and currently built into 20 percent of mobile phones, particularly Android models, as well as the latest iPhone 6 devices. There are many uses for NFC, but one of the most common is for mobile payment services such as Google Wallet.

Just swipe your phone past an NFC reader to pay for a product, and the funds will be deducted from your phone's virtual wallet or charged to your credit card. But like RFID, NFC has been compromised on many occasions, with hacker apps such as NFCProxy capable of copying NFC credit card data in real time and replaying them later when the bad guy uses them to buy goods and services of his own choosing. Google Wallet has also been hacked repeatedly, including reading your PIN without authorization and accessing funds stored on your phone. And now that the iPhone is enabling mobile payments with Apple Pay, it's likely that criminals will turn their attention to getting around Apple's security system too.

In another instance, a hacker successfully targeted the NFC chip in a nearby mobile phone to take command of the device and make phone calls, text, and access files—all without the knowledge of the device's rightful owner. NFC apps on mobile phones are already being used to pay for local transit systems, and crooks in San Francisco and New Jersey have hacked the NFC turnstiles by using an app called UltraReset, which automatically replenishes any fares deducted by the train operators, equating to free subway rides for life.

Another IoT wireless communications technology that has surged in its usage and popularity is Bluetooth, but like RFID and NFC it too is easily subverted. There are dozens of easy-to-use free apps and programs such as Blue Scanner, Blue Bugger, BT Browser, and Blue Sniff that make it simple for any malicious individual to connect to a Bluetooth-enabled device and take control of it. These tools provide unauthorized access, known as Bluesnarfing, via the Bluetooth port to any data stored on smart phones, desktops, and laptops. They can also intercept data you type on your wireless keyboards, read your text messages, snap photographs without your knowledge, and even eavesdrop on your Bluetooth headset as you're seated in the airport awaiting your flight.

The gold rush of the Internet of Things is upon us, and there will be no turning back. While connecting everything to a global Internet of Things may indeed have tremendous value, connecting everything insecurely does not. Before we add billions of hackable things and communicate with hackable data transmission protocols, important questions must be asked about the concomitant risks with regard to the exponential implications for the future of security, crime, terrorism, warfare, and privacy.

Obliterating Privacy

Items of interest will be located, identified, monitored, and remotely controlled through technologies such as radio-frequency identification, sensor networks, tiny embedded servers, and energy harvesters— all connected to the next-generation Internet using abundant, low-cost, and high-power computing.

DAVID PETRAEUS, CIA DIRECTOR (RETIRED)

In the same way our every move online can be tracked, recorded, sold, and monetized today, so too will that be possible in the near future in the physical world. Real space will become just like cyberspace, and as all the objects around us join the Internet of Things, any meaningful distinction between the online and the off-line worlds will disappear.

With the widespread adoption of more networked devices, what people do in their homes, cars, workplaces, schools, and communities will be subjected to increased monitoring and analysis by the corporations making these devices. Of course these data will be resold to advertisers, data brokers, and government alike, providing a heretofore-unprecedented view into our daily lives. Unfortunately, just like our social, mobile, locational, and financial information, our IoT data will leak, providing further profound capabilities to stalkers and other miscreants interested in persistently tracking us. While it would certainly be possible to establish regulations and build privacy protocols to protect consumers from such activities, if past is prologue, the greater likelihood is that every IoT-enabled device, whether an iron, vacuum, refrigerator, thermostat, or lightbulb, will come with terms of service that grant manufacturers access to all your data. More troublingly, while it may be theoretically possible to log off in cyberspace, in your well-connected smart home there will be no "opt-out" provision. As a result, more of what happens behind closed doors will be open to scrutiny by parties you would never invite into your home, and in this world pulling down the shades won't keep out the twenty-first century's Peeping Toms.

We may find ourselves interacting with thousands of little objects around us on a daily basis, each collecting seemingly innocuous bits of data 24/7, information these things will report to the cloud, where it will be processed, correlated, and reviewed. Your smart watch will reveal your lack of exercise to your health insurance company, your car will tell your auto insurer of your frequent speeding, and your garbage can will tell your local municipality that you are not following local recycling regulations. This is the "Internet of stool pigeons," and though it may sound far-fetched, it's already happening. Auto insurance companies such as Progressive are offering discounted personalized rates based on your driving habits. "The better you drive, the more you can save," according to its advertising. All drivers need to do to receive the lower pricing is agree to the installation of Progressive's Snapshot black-box technology in their cars and to having their braking, acceleration, and mileage persistently tracked. As we move forward, however, it's not unreasonable to believe that drivers who do not consent to such devices in their cars will face such outrageously high premiums that they will de facto become obligatory.

The IoT will also provide vast new options for advertisers to reach out and touch you on every one of your new smart connected devices. That means every time you go to your refrigerator to get ice, you will be presented with ads for products based on the food your refrigerator knows you're most likely to buy. Screens too will be ubiquitous, and marketers are already planning for the bounty of advertising opportunities. In late 2013, Google sent a letter to the Securities and Exchange Commission noting, "we and other companies could [soon] be serving ads and other content on refrigerators, car dashboards, thermostats, glasses, and watches, to name just a few possibilities." Knowing that Google already reads your Gmail, records your every Web search, and tracks your physical location on your Android mobile phone, what new powerful insights into your personal life will the company develop when its entertainment system is in your car, its thermostat regulates the temperature in your home, and its smart watch monitors your physical activity?

Not only will RFID and other IoT communications technologies track inanimate objects, but they will be used for tracking living things as well. Many pet lovers are already familiar with companies such as PetLink, HomeAgain, and AKC Reunite, which provide implantable RFID chips to veterinarians so that lost dogs and cats can be identified and returned to their homes if they run away. What may be less well known, however, is that increasingly human beings too are forcibly monitored via RFID wristband systems, such as those becoming commonplace in jails and prisons from Los Angeles to Washington, D.C. In some countries, such as the U.K., government officials are even considering implanting RFID chips directly under the skin of prisoners, just as is common practice with dogs. While many might not object to convicted criminals being subjected to such RFID tracking, they may feel differently when similar techniques are applied to their own children.

School officials across the United States have begun embedding RFID chips in student identity cards, which pupils are required to wear on their persons at all times. In Contra Costa County, California, preschoolers are now required to wear basketball-style jerseys with electronic tracking devices built in that allow teachers and administrators to know exactly where each student is. According to school district officials, the RFID system saves "3,000 labor hours a year in tracking and processing students." Of course, when people are forced to join the Internet of Things, a wide variety of other privacy and public policy issues arise. For example, the same RFID system that enables constant student monitoring will be able to identify those students who move around "too much" and therefore may be deemed hyperactive, disruptive, and better suited for "alternative schools." Students who do not wish to be tracked are being told "tough luck," and in 2013 the sophomore Andrea Hernandez in San Antonio, Texas, was suspended when she refused to wear her RFID device on campus.

Meanwhile, the ability to track employees, how much time they take for lunch, length of their bathroom breaks, and the number of widgets they produce will become easy. Moreover, even things such as words typed per minute, eye movements, total calls answered, respiration, time away from desk, and attention to detail will be recorded. The result will be a modern workplace that is simultaneously more productive and more prison-like. Though it won't only be your employer who will access data from the IoT for reasons of efficiency and control, so too will the government. Already police agencies query local utilities to uncover customers with unusually high electrical bills, thought to correlate to indoor marijuana farms. Based on nothing more than electrical bills, search warrants have been issued and suspected growers arrested. In the future, law enforcement may be able to completely bypass the subpoena and just remotely query your smart meter to see if your energy usage is "out of profile" for homes in your neighborhood.

At the scene of a suspected crime, cops will be able to interrogate the refrigerator and ask the equivalent of "Hey, buddy, did you see anything?" Child social workers will know that there hasn't been any milk or diapers in the home and the only thing stored in the fridge has been beer for the past week. The IoT also opens up the world for "perfect enforcement." When sensors are everywhere and all data are tracked and recorded, it becomes more likely that you will receive a moving violation for going twenty-six miles per hour in a twenty-five-mile-per-hour zone and get a parking ticket for being seventeen seconds over on your meter. As today's red-light cameras have already shown, when everything is connected, nothing can be hidden, particularly when infractions translate into revenue for government agencies and their business partners.

The former CIA director David Petraeus has noted that the Internet of Things will be "transformational for clandestine tradecraft." While the old model of corporate and government espionage might have involved hiding a bug under the table in the conference room to listen in on your conversation, tomorrow the very same information might be obtained by intercepting in real time the data sent from your Wi-Fi lightbulb to the lighting app on your smart phone. Thus the devices you thought were working for you may in fact be on somebody else's payroll, particularly that of Crime, Inc.

Hacking Hardware

A much rarer breed of hacker targets the physical elements that make up a computer system, including the microchips, electronics, controllers, memory, circuits, components, transistors, and sensors—core elements of the Internet of Things. These hackers attack a device's firmware, the set of computer instructions present on every electronic device we encounter, including TVs, stereo receivers, mobile phones, game consoles, digital cameras, hard drives, printers, automobiles, avionics, heating and air-conditioning systems, network routers, alarm systems, CCTVs, SCADA industrial control systems, USB drives, traffic lights, parking meters, gas station pumps, digital watches, sensors, smart home management systems, robotics, and programmable logic controllers (such as those used by the Iranians in Natanz). The overwhelming number of "smart" objects are dead dumb and have no capacity whatsoever to have their firmware upgraded.

Indeed, the small embedded computers that comprise the IoT, and most of our everyday electronic devices, have very limited processing power and memory. As a result of these limitations, they must be built to exceedingly tight specifications that barely accommodate the functions their designers need to make the devices work, leaving precious little room for anything as "trivial" as security, often a far afterthought in the manufacturing process. Most firmware lacks a common automatic mechanism for updating itself to fix any functionality or security issues detected after the device has been shipped, meaning that a preponderance of devices already online for five to ten years are sitting ducks. For some more expensive items, such as smart phones, the device's firmware is meant to be upgradable so that improvements and security patches can be downloaded. Yet for the majority of other electronic devices, manufacturers rarely change a device's firmware over its useful lifetime, as doing so might require the integrated circuits on the item to be physically replaced—a profoundly expensive economic nonstarter. But even if your phone has the latest firmware, there are still dangers to consider.

While many iPhone or Android users may understand that downloading the wrong app or computer file might give their phones a virus, few if any understand that their choice of mobile phone charger can do the very same thing. Hackers have already successfully built a hardware virus directly into a compromised USB charger capable of targeting Apple devices. The mere plugging in of your phone to one of the rogue power cords is all you need to get infected. By modifying the firmware and electronics of the innocent little plug we use to charge our phones, attackers were able to bypass the iPhone's security safeguards and infect the phone. No pop-up alert was provided, and the stealthily running malware was not visible anywhere on the list of running programs. In the background, however, the rogue charger installed a back door on the device that allowed hackers to make phone calls, read texts, steal banking information, capture account passwords, and track the movements of phone users. The phenomenon is known as juice jacking, and the malicious charger was built for under $50—something to consider the next time you're plugging in your battery-starved smart phone at a public charging station kiosk at an airport, hotel, or local shopping mall (the very places hackers would place these devices to infect the greatest number of victims).

Illicitly modified chargers aren't the only hardware surprises we need to be on the lookout for today. Just about anything with a microcontroller or sensor can arrive in your home with "enhanced features" that nobody would ever want. In 2013 in Russia, customs officials noticed that a series of consumer goods manufactured in China, including electronic teakettles and clothing irons, arrived with modifications that Russian authorities were none too pleased about. The devices contained hidden miniature Wi-Fi cards capable of spreading malware to any open Internet network within two hundred meters and were able to "phone home," relaying secret messages back to China. Not only could these irons and teakettles surreptitiously join your Wi-Fi network (something nobody would ever expect from an ordinary everyday iron), but they could use your own network to spread viruses to the other computers in your home and disseminate spam messages to your neighbors and the rest of the world. While we'd like to believe that spying irons and hacked iPhone chargers are some bizarre oddities, the fact of the matter is they are the harbingers of much more widespread and serious threats posed by the rapid assimilation of billions of networked objects to the worldwide information grid.

More Connections, More Vulnerabilities

For all the untold benefits of the Internet of Things, its potential downsides are colossal. Adding 50 billion new objects to the global information grid by 2020 means that each of these devices, for good or ill, will be able to potentially interact with the other 50 billion connected objects on earth. The result will be 2.5 sextillion potential networked object-to-object interactions—a network so vast and complex it can scarcely be understood or modeled. The IoT will be a global network of unintended consequences and black swan events, ones that will do things nobody ever designed or purposely planned. While there may be serendipitous benefits of such a network, there is also every chance many of its developments will be undesirable, negatively affecting global security, personal privacy, and human rights. Moreover, if you think the number of error messages and application crashes we face today are a problem, just wait until the Web is embedded in everything from your car to your sneakers to your microwave. Having to reboot your refrigerator, your thermostat, and your garage door in order to get them to run won't be much fun either.

If ever there were a technology that embodied the butterfly effect, it is surely the Internet of Things. In this world, it is impossible to know the consequences of connecting your home's networked blender to the same information grid as an ambulance in Tokyo, a bridge in Sydney, or a Detroit auto manufacturer's production line, and yet it will all be connected in one way or another.

While some of the world's smartest research and technology firms are rushing forward to build the Internet of Things (and claim their share of its multitrillion-dollar economic bounty), their colleagues back in the IT security department are frantically working to combat yesterday's zero-day attack or the malware vulnerability crisis du jour. There is little time to speculate and prepare for what is coming next. The vast levels of cyber crime we currently face make it abundantly clear we cannot even adequately protect the standard desktops and laptops we presently have online, let alone the hundreds of millions of mobile phones and tablets we are adding annually. In what vision of the future, then, is it conceivable that we will have any clue how to protect the next fifty billion things to go online? Given our inability to secure today's global information matrix, how might we ever protect a world in which every physical object, from pets to pacemakers to self-driving cars, is connected to the Net and hackable from anywhere on the planet? The obvious reality is that we cannot.

The Internet of Things will become nothing more than the Internet of Things to be hacked, a cornucopia of malicious opportunity for those with the means and motivation to exploit our common technological insecurity. The IoT and its underlying insecure protocols will open a Pandora's box of security vulnerabilities on an unprecedented scale, potentially creating systemic malfunctions whose reach will be simultaneously unpredictable, extraordinary, and terrifying.

Houston, we have a problem, particularly with our threat surface area—that is to say, the sum of the different points or attack vectors through which an enemy can strike. The challenge with the Internet of Things is that our technological threat surface area is growing exponentially and simply stated we have no idea how to defend it effectively. The logic is clear: the more doors and windows you have, the more places a burglar can enter your home—particularly one connected to the Internet.

CHAPTER 13

Home Hacked Home

We estimate that only one percent of things that could have an IP address
today have one, so we like to say that ninety-nine percent of the world
is still asleep. It's up to our imaginations to figure out what will happen
when the ninety-nine percent wakes up.

PADMASREE WARRIOR, CHIEF TECHNOLOGY OFFICER, CISCO

Blake Robbins, a student in Pennsylvania's Lower Merion School District, couldn't imagine why he'd been summoned to the principal's office. When the assistant principal accused the sixteen-year-old student of "inappropriate behavior," Robbins responded that he had no idea what the school official was talking about. The assistant principal then clarified: she knew the student was dealing drugs and threatened him with suspension. The sixteen-year-old vigorously denied the allegations until suddenly the administrator turned around her laptop and showed Robbins several pictures of himself *in his own bedroom* holding small oblong-shaped pills in his hand that he then proceeded to ingest. Shocked, the boy asked where the photographs had come from, a point the school official did not feel she needed to address.

Robbins went home and told his parents about the incident, who then confronted the school district. As it turned out, Robbins was not using or dealing drugs, but had merely been eating red-colored Mike and Ike candy, a fact known to his parents. But how the hell did school officials get a photograph of a sixteen-year-old boy in his own bedroom eating candy? Through an elaborate spying program purportedly meant to protect school property.

Officials in the affluent school district had provided twenty-three hundred high school students MacBook laptops to support their studies. What they failed to disclose to either students or parents, however, was that the laptops had secret software installed that gave administrators remote access to all student activities on the devices, including student chat logs and records of the Web sites they visited. It also allowed officials to remotely commandeer the laptop's camera to photograph and record students anytime the devices were open—all allegedly to help track lost or stolen laptops. The remote spying system was set to silently snap photographs automatically every fifteen minutes whenever any student's laptop was up and running, though officials could set the time-frame intervals to every sixty seconds for students suspected of "inappropriate behavior."

The photographs were uploaded to the school district's Web server, where they were individually reviewed by district officials. District officials captured over fifty-six thousand images, including photographs of naked children in their bedrooms, bathrooms, and any other venue they traveled with their laptops. Administrators covertly snapped more than four hundred images of Robbins alone once he came under suspicion of bad behavior, though police were never notified and no search warrants were ever obtained for the school district's invasive activities. Once news of the school district's egregious behavior became public, numerous lawsuits were filed, including by Robbins's parents, and an FBI criminal investigation against the district ensued. As Robbins told *Good Morning America*, "They might as well be sitting in my room watching me without my knowing." Unfortunately, at the time the sophomore had received his "free" laptop, he was too young to have yet studied the prophetic warning about Greeks bearing gifts, a topic he would likely cover two years later when assigned Virgil's *Aeneid* in his senior English class.

Candid Camera

You can't assume any place you go is private because the means of surveillance are becoming so affordable and so invisible.
HOWARD RHEINGOLD

When a public school district, an organ of the state, has the ability to spy on us in our homes at will and without warrant, it is clear the age of pervasive universal surveillance has upon us. From London to New York and Chicago to Beijing, massive video surveillance, or CCTV, networks have been installed to protect us from threats, real and imagined. In one city alone, Chongqing in southwest China, officials have installed 500,000 cameras to deal with religious and political unrest, in addition to other "organized crimes." Though once upon a time government had a monopoly on such security systems, today we are as likely to encounter cameras in grocery stores, gas stations, car dealerships, hospitals, schools, office buildings, bridges, tunnels, bars, taxicabs, buses, trains, doctors' offices, and dry cleaners. We also have them on our laptops, mobile phones, game consoles, televisions, tablets, nanny cams, and home security systems, and the more ubiquitous they become, the less we're aware they're even there. Given the near-zero cost of these cheap video sensors, their presence in our lives is about to vastly expand as the Internet itself develops its own sense of vision.

The capabilities and quality of today's cameras are improving to unimaginable levels, long since ditching the grainy black-and-white images of yesteryear. The Defense Department has already deployed a 1.8-gigapixel camera that can be attached to a drone and spot targets "as small as six inches at an altitude of 20,000 feet" (technology that will undoubtedly be available commercially in the near future). Moreover, today's cameras don't just watch and record; they can see and understand, by linking their sensors to cloud-computing algorithms and big-data analytics. As a result, cameras can perform facial recognition, read your license plate, and even determine that a package (potential bomb) has been left alone in one place too long. This analysis can be done in real time as well as retrospectively, making it possible to unlock millions of hours of long-ago-recorded video footage to search for "a woman with a red hat."

Unfortunately, the tools that are meant to protect us can provide a false sense of security, as the hundreds of millions of cameras around the world going online prove themselves vulnerable to attack by hackers with ill intent. As previously discussed, your mobile phone's camera can easily be turned on remotely without your knowledge, using widely available tools such as Mobile Spy (of which sixty thousand copies have been sold).

One young woman who learned this lesson the hard way was Cassidy Wolf, Miss Teen USA, whose open laptop in her bedroom was commandeered by a hacker, capturing nude photographs and video of her as she walked around in her own bedroom after coming out of the shower and while getting dressed for school. Her tormentor watched her daily for several months until one day he sent her a "sextortion" e-mail demanding she perform a series of sex acts on camera for him "or I upload these pics and a lot more (I have a LOT more and those are better quality) on all your accounts for everybody to see and your dream of being a model will be transformed into a pornstar [sic]." Upon receiving the e-mailed threat, Cassidy slammed her laptop shut and burst into tears, eventually deciding to go to the police. Three months later, an FBI investigation revealed one of her high school classmates, Jared Abrahams, was her extorter. Abrahams carried out his attack using Blackshades, a Crime, Inc. tool kit readily for sale on Crimeazon .com, malware he used to target eight other women in Southern California.

Meanwhile, modern baby cams, which allow parents to monitor their kids not only in the next room but over the Internet, are just another point of presence on the Net waiting to be cracked. Hackers and pedophiles routinely compromise these devices, the majority of which require no password or use a standard well-known one provided by the manufacturer, leading to a lurid trade in nanny cam images in the digital underground, including those of young mothers breast-feeding their infants. The cameras allow for full pan, tilt, and zoom and include two-way audio with a built-in speaker and microphone so parents can both listen and speak to their children. Convenient for mom, dad, and hacker alike, as Heather Schreck of Cincinnati found out when she was awoken from a deep sleep one midnight.

"All of a sudden, I heard what sounded like a man's voice, but I was asleep so I wasn't sure." Confused, Heather checked the baby camera in the bedroom of her ten-month-old daughter, Emma, using her cell phone. Strangely, the camera was moving, but it wasn't the mom who was doing it. Suddenly Heather could now hear a man's voice across her house screaming, "Wake up baby, wake up!" Heather and her husband, Adam, ran into Emma's room and upon entering saw the baby camera turn from their now crying daughter and directly focus on Adam. The male voice coming through the device observing the parents unleashed a tirade of obscenities on the half-awake couple before Adam had the presence of mind to yank the camera cord from the wall. The baby monitor's manufacturer, Foscam, later admitted the device had a "firmware vulnerability" that had allowed the creep into the cradle of the Schrecks' sleeping infant. These incidents are far from isolated, as another family in Houston discovered when they awoke to a man's voice screaming the name of their two-year-old daughter, Allyson, cursing at her and wailing, "Wake up . . . you little slut." The virtual intruder knew the girl's name because it was written in pink on the wall. It is both ironic and troubling that the devices families purchase to protect themselves can actually be used as weapons to target them and invite trouble into their homes.

In addition to baby cams, home and office security camera systems are just as vulnerable, and researchers have found widespread flaws in more than twenty major brands, most of which are sold with remote Internet access enabled by default and weak security features. Nearly 70 percent of users never change the default user names, such as "user" and "admin," nor do they reset the manufacturer's preset password, such as "1111" or "1234." As a consequence, tens of millions of Internet-connected cameras are wide open to interception by parties unknown, and hackers are delighted to share their voyeuristic discoveries. Without the consent or knowledge of those surveilled, thousands of these live feeds are available online for all to see: a Laundromat in Los Angeles, a man in Newark watching football on the couch, patrons at a bar in Virginia, a living room in Hong Kong, or an office in Moscow—take your pick. Given the opportunities, it wasn't long before Crime, Inc. began exploring how to best use IoT-enabled cameras to its advantage.

Why not hack the bank's cameras before the robbery to learn employee patterns of conduct, when cash deliveries arrive, and the times the guard is on his break? Of course knowing that most bank robberies only yield a paltry sum, with high risk, there are bigger fish to fry. That's exactly what a team of criminals from Crime, Inc. did in March 2013 when they carried out an *Ocean's Eleven*–style attack against the Crown Casino in Melbourne, Australia. Hackers took over the casino's own security system and used the resort's own security cameras to spy on the house, including its VIP gambling rooms. The prime suspect, described only as a foreigner, was known as a "whale," a high roller who regularly bet large amounts of money. Except this time he had an edge. Because he and his accomplices had hacked the live video feeds, they were able to see all the cards held by the dealer and fellow players at his poker table. When the criminal whale sauntered up to play with other high rollers, his hidden hacker compatriots were feeding him betting instructions into a concealed wireless earpiece. Confident of his bets, the hacker was able to win more than U.S.$33 million in just eight hands of cards. Rather than tempt fate, he left a wealthy man and flew back home to his own country before authorities ever realized what had happened. As the exponential march toward the Internet of Things drives on, more people will discover that the trusted things they expected to protect them, whether security cameras or air bags, can be commandeered by others and used against them in surprising and even deadly ways.

From Carjacking to Car Hacking

Most people would rather have malicious software running
on their laptop than inside their car's braking system.
PROFESSOR CHRISTOF PAAR, EMBEDDED SECURITY RESEARCHER

Cars used to run on gasoline. Today they run on code. Sure, you still need gas or electricity for power, but without functioning computer code any modern car is dead in its tracks. Though your dad's 1957 Chevy might have been a purely mechanical device, these days automobiles are little more than computers on wheels. A car rolling off the assembly line in 2015 has between seventy and a hundred onboard computers, known as electronic control units. Together they manage the auto's engine, cruise control, ABS brakes, climate, transmission, entertainment, windshield wipers, power seats, locks, navigation, fuel efficiency, and air-bag deployment, to name but a few. Though automakers do a good job of making it all look relatively seamless, today's cars are remarkably complex systems containing nearly 100 million lines of computer code (versus the relatively paltry 1.7 million lines of code running the avionics on the U.S. Air Force's F-22

Raptor frontline jet fighter). All these embedded electronics account for 50 percent of a new vehicle's cost on average (nearly 80 percent for hybrids). Together these microchips form the controller area network (CAN), the onboard computer network that is the lifeblood of any recent automobile and is responsible for improved safety, reduced emissions, and better mileage for our cars than at any time in history.

These embedded technologies are not just communicating internally with one another via the CAN but also increasingly sharing this information online with the outside world via a variety of radio and cellular networks built into the car itself. Doing so provides tremendous conveniences for drivers: BMW's TeleServices network enables a vehicle's internal sensors to continuously self-diagnose and report malfunctions to the local dealer. When a problem is spotted, owners receive a call telling them their car is sick and to come in for an appointment. GM's OnStar will call for an ambulance automatically if its air-bag and motion sensors detect that a car has been in an accident.

While event-data-recording black boxes in cars can help crash investigators and reduce insurance premiums, they can also "narc" on your every move, generating hundreds of megabytes of data per second. These devices continuously track a horde of vehicular data, including your location, seat belt use, speed, and turn-signal operation. As Jim Farley, global vice president of marketing and sales for Ford Motors, admitted in early 2014, "[We know] everyone who breaks the law, we know when you're doing it. We have GPS in your car, so we know what you're doing." It's not just Ford of course. GM's OnStar caused outrage when it unilaterally updated its terms of service to grant itself the lifelong right to monitor all its vehicles, including location and odometer reading, and to share this information with third parties even after the service had been canceled by the car's owner. Oh, and that convenient built-in car microphone that allows you to ask OnStar for directions and listens in after a crash, it can also be turned on remotely without your knowledge and secretly listen in on your private conversations, as the FBI has been doing since at least 2003 in its mob-related investigations.

There may be even greater concerns for the future of your car than privacy alone. The growing complexity of the modern automobile is leading to massive recalls because of system failures and tragic loss of life. In just the first six months of 2014, GM was forced to recall twenty-nine million cars, with millions more recalled from Nissan, Hyundai, Ford, Honda, and BMW. When the deeply complex electronics in a car control all of its major functions, system failures can have unintended consequences, such as the spate of problems at Toyota late in the first decade of the twenty-first century that resulted in the deaths of as many as thirty-seven drivers. A jury found that many of the crashes could have been caused by software deficiencies in Toyota's electronic throttle control system, which caused the accelerator pedal to remain depressed and vehicle brakes to fail. Toyota was accused of covering up the defects and in 2014 agreed to a record $1.2 billion fine by the U.S. Department of Justice for putting profits ahead of safety. Of course accidental safety issues with vehicle electronics are just part of the problem. When cars become computers, they, like all other systems, can make attractive targets for malicious hackers.

The days of thieves using clothes hangers to break into cars are quickly becoming history. No need to stick a gun in somebody's face to steal a car either; carjacking has joined the modern age, replaced by car hacking. In the United States, all cars manufactured since 1996 have been required to have standardized electronic onboard diagnostics ports, which provide direct physical access to a vehicle's central computer systems, and a cluster of new IoT communications protocols such as RFID, Bluetooth, and mobile telephony provide such access at a distance. Newer vehicles even come with USB ports, and as always more connections mean more vulnerabilities. According to the London Metropolitan Police, nearly half the eighty-nine thousand vehicles stolen in London in 2013 were hacked, with criminals' using a variety of electronic devices to open and start the cars. The gadgets crooks use in the attacks can be purchased on Crimeazon .com, mostly from suppliers in Bulgaria. The operation takes less than ten seconds to pull off, and of course there are videos on CrimeU explaining the whole process.

Using mobile-phone-sized gadgets originally designed for locksmiths to help individuals who had lost their electronic car keys, thieves can merely program a new blank electronic key to replace the original. This spoofing technique fools the car into thinking the owner's original key fob is present and can be accomplished by either wirelessly intercepting the radio signal you use when opening or locking the car or by targeting the car's onboard computer directly.

Using nothing more than a laptop and an SMS text message with the correct encoded instructions, thieves can unlock your doors, start the car, and drive off. Your musical tastes could put you at risk too, as several security researchers proved in 2011 when they added malicious computer code to an MP3 music file and burned a list of songs to a CD. When played through the car's audio system, the infected song file warped the vehicle's firmware, allowing hackers entry to all the car's main control systems. In a situation like this, car theft might be the best of all possible outcomes, for once a vehicle's onboard computer systems have been compromised, the possibilities are near limitless.

For just under $30, hackers can build a hardware device, such as the CAN Hacking Tool, which, when plugged into your car's onboard computer network, allows them to remotely seize control of your vehicle's lights, locks, and steering and brake systems. Because nearly every single element of your car is managed by a computer system, devices such as these mean it is now possible to reach out and touch any car on the road from halfway around the world by subverting mobile phone receivers embedded in the car itself. More closely, remote hacks are also possible via Bluetooth and Wi-Fi. Dozens of demonstrations by hackers and security researchers have proven it is entirely possible for criminals fifteen hundred miles away to seize control of your car when you are driving sixty-five miles per hour down the highway. What they do with your hacked vehicle is only limited by their imaginations. Change your odometer to zero and your speedometer to 160, even when your car is standing still? Easy. Honk your horn, blast the radio, tighten your seat belt, and turn on the windshield wipers? Simple. Turn off the engine or jerk the steering wheel sharply to the left so that you lose control of the car at high speed? Yes. Suddenly deploy your air bag so that you careen out of control with your kids in the backseat? Entirely possible. If a computer controls your car, it can be controlled by an attacker.

The challenge with these vulnerabilities is that they needn't target any single car but instead could affect all cars or vehicles of a particular make, model, and year simultaneously. In the Texas Auto Center case, a rogue employee was able to remotely shut down a hundred cars. But companies such as OnStar have their technology installed in millions of vehicles, including the ability to remotely block an engine's ignition from starting and actually disable a moving vehicle in case of theft. Couldn't a rogue employee at OnStar thus turn off a hundred thousand or a million cars? Though GM would surely deny it, once the back door has been built into the car, protecting it from abuse becomes a profound challenge and creates the opportunity for widespread infrastructure attacks by both hackers and nation-states alike.

As ambient sensor networks proliferate and vehicle technology improves, human beings will be turning more control of their driving responsibilities over to machines. Renault Nissan's CEO, Carlos Ghosn, has publicly stated his company will have a fully autonomous self-driving vehicle available for the mass market by 2020, and Volvo's plan is to have such cars by 2017. The biggest proponent of such technologies has been Google, whose own self-driving test cars have logged over 700,000 miles without a single crash or accident. The point is an important one because, as it turns out, human beings are terrible drivers and more than thirty-three thousand Americans are killed in car accidents annually. A fully automated, well-functioning autonomous vehicle network could avoid thousands of needless deaths and save billions in associated economic costs. As the price of these technologies plunges, you can expect UPS drivers and taxis to be replaced by autonomous and cheaper non-union alternatives.

But modern cars, whether driven by people, artificial intelligence, big data, or sensor networks, are still just computers on wheels, powered by insecure data systems, communicating via entirely hackable transmission protocols. As such, things might not turn out quite as rosy as proponents of autonomous vehicles suggest. When the majority of vehicles join the IoT, it won't be long before some rogue attacker seizes control of a car and turns it into a multi-ton weapon of metal, glass, and explosive fuel. In the same way both Crime, Inc. and crazed exes are targeting computers and mobile phones, it's only logical that they will go after cars in the future too, bringing scenes like those in Stephen King's 1983 horror thriller about a possessed car named Christine many steps closer to reality. Law enforcement officials clearly see the threat, and in July 2014 the FBI warned in an internal report that driverless cars could be used as "lethal weapons, with terrorists potentially packing explosives into a self-driving car aimed at a specific destination." Autonomous vehicles could also potentially be turned off en masse, bringing traffic to a complete standstill in a city or country.

To be certain, some of these vehicular attacks require a high degree of computer savvy to pull off, but as we have seen with other exploits, soon there will be point-and-click crimeware options for car hacking as well. Automakers are starting to take notice, particularly as "most hackable car" lists come out. Just as vehicles were rated for their crash safety in the past, now security researchers are ranking which cars are most hackable (the answer is Jeep Cherokee, Cadillac Escalade, Infiniti Q50, and Toyota Prius). In a nod to these growing concerns, Tesla, creator of some of the most technologically advanced vehicles on the road today, hired a high-profile security guru away from Apple to make the point. But what new threats will be enabled by these technologies in the future when Crime, Inc. remotely seizes control of your self-driving car, locks the doors, and speeds you off toward an abandoned warehouse on the wrong side of town? Though you might futilely attempt escape, the last thing witnesses reported seeing was you screaming in horror and banging your fists against the vehicle's interior windows, impotent to respond to the next generation of kidnapping. Of course, assuming you make it home alive in your potentially hacker-possessed car, you may find more troubles waiting, as while you were out, your house joined the Internet of Things as well.

Home Hacked Home

Since the days of the Jetsons, we've been promised a space-age home filled with robotic contraptions and whimsical electronics meant to guarantee the good life, all at the touch of a button. While we don't yet have our flying cars, the Hanna-Barbera cartoon from the early 1960s was prophetic in predicting flat-panel TVs, video chats, and automatic sliding doors. In theory, the modern networked home sounds great. Security systems and video cameras will protect us from burglars and call police if a window is smashed. Digital thermostats will interface with weather reports for your home's specific GPS coordinates and intelligently adjust heating and air-conditioning to ensure maximal efficiency, comfort, and cost savings. Smart sensors in the basement will detect the water on the ground after a pipe bursts and automatically turn off the flow to the affected area. Your smart phone will lock your front door's dead bolt over the Internet so you won't ever have to worry again about whether or not you remembered to do so while en route to the airport. Smart refrigerators will warn us when our milk is about to spoil, and the mere act of dropping an empty Cheerios box in the garbage pail will automatically use your stored credit card details to order more cereal without your lifting a finger. But do you really want your garbage pail to have your credit card number?

The home automation market in the United States is "expected to reach $16.4 billion by 2019," and all the major technology firms are vying for a piece of the pie. Elements of your home might have already joined the IoT, with an increasing number of utilities installing smart meters to measure and regulate water, electricity, and gas usage. But perhaps some of the largest opportunities are in the consumer products space where Google, Apple, Samsung, and Microsoft, to name but a few, are fighting it out to become the central hub and operating system for your house, allowing you to remotely monitor and manipulate your humble abode using your home automation gateway while on the go.

Apple's recently unveiled HomeKit, included with iOS 8, brings the Cupertino giant's design flair to home automation, enabling users to lock their doors, dim their lights, and play their stereos by merely tapping their iPhones or by voicing a request to the company's AI voice agent, Siri. By your merely speaking the words "going to bed," HomeKit will know to automatically carry out a series of actions, such as drawing the curtains, lowering the temperature, and turning off the lights, although given the experience some have had with Siri's voice recognition, hilarity may ensue when the TV suddenly blares, your car starts, and your front door unlocks. Eventually, though, the kinks will be ironed out and centralized digital hubs in our homes managed by our smart phones will become reality in the very near future. So what could possibly go wrong?

Well, for one, you've never had to upgrade the firmware in your washing machine, reinstall the OS for your house, and reboot your home in order to get the front door to work. While connecting lightbulbs, toasters, washing machines, DVRs, game consoles, refrigerators, set-top boxes, ovens, dishwashers, televisions, door locks, security systems, baby cameras, thermostats, toilets, lamps, and bathtubs to the Internet of Things may offer Jetsonian convenience, joining all of these objects to the IoT will of course bring its own set of privacy and security risks. Many such systems use no authentication or encryption when communicating between an appliance, your mobile device, and the home system. As a result, they can easily be spoofed, hacked, intercepted, and subverted. A July 2014 study by HP found that 70 percent of the devices connected to the Internet of Things were vulnerable to attack, with each object on average containing twenty-five unique security flaws.

Once your house goes fully online, there's no reason to think hackers won't consider it a viable target, and all evidence suggests they are already actively working on it. Each attacker will have his or her own motivation—the neighbor kid you told to get off your lawn, the ex-boyfriend with a jealous streak, the Peeping Tom who saw you once at the grocery store, or the foreign government intent on exploiting cyber-espionage capabilities. For Crime, Inc., however, it's mostly about the money, and it will exploit weakness in your home's IoT devices to gain access to valuable data stored on your network or for the purposes of everyday burglary. Oh, and remember CryptoLocker—the ransomware that seizes control and locks laptops and mobiles by encrypting them? Well, you can expect the digital underground to sell crimeware tool kits that lock you in or out of your home, forcing you to pay a Bitcoin ransom to get your house up and running again.

Your kids may face similar threats while "playing house." Major toy makers, such as Disney and Mattel, are already studying the IoT, and there are a slew of Wi-Fi-enabled dolls, stuffed animals, and miniature robots coming your way in this "Internet of Toys." But toys too can be subverted, and at least one, the Karotz plastic interactive bunny, which can be controlled by a smart-phone app and includes a camera, microphone, and RFID chip, has been hacked, allowing an attacker to conduct video surveillance on your kid.

Other technologies, including the 135-year-old lightbulb, are getting their IoT makeovers, and systems such as the Philips Hue LED lighting system allow consumers to turn lights on and off from their smart phones. They allow hackers to turn off your lights as well, exploiting a known security flaw in the Philips system, troubling given the obvious link between lighting and physical security. Additional systems, such as the LIFX energy-efficient smart lightbulb, actually leak your home's Wi-Fi router password once plugged into a lamp, exposing it to any hacker who merely queried the "master bulb" on your home network. Lamps and lightbulbs can also have back doors built into them, similar to the Chinese irons discovered in Russia. In early 2014, hackers created an eavesdropping lamp capable of live tweeting your private conversations. The device, known as Conversnitch, costs less than $100 and resembles a standard lightbulb; the only difference is this one has a hidden microphone that listens in on all nearby chats. To prove a point, the creators of the device recorded a video showing themselves easily placing Conversnitch devices in libraries, offices, McDonald's restaurants, and a bank branch—all without any interference or notice from company personnel—foreshadowing a powerful new IoT tool for industrial espionage.

As smart devices proliferate, they will be controlled by centralized home automation gateways, the majority of which have already been compromised, allowing hackers to take over all the devices on your local network. It'll be like Nightmare on Connected Home Street. Though we might sleep better at night believing we're safe and sound in our IoT-enabled homes, home invasion 2.0 is way easier than you might expect, as proved by the *Forbes* reporter Kashmir Hill. While working on a story about the IoT, Hill merely Googled the term "smart homes" and was able to quickly uncover eight families using the popular Insteon home automation system, which controls appliances such as "lights, hot tubs, fans, televisions and garage doors."

Because Insteon did not require a user name or password and allowed its products to become crawlable by search engines, the *Forbes* reporter was able to find them without difficulty—yes, people can now Google your smart refrigerator and communicate with it from afar. Hill then contacted the innocent parties in question, introduced herself, and said, "I can see all of the devices in your home and I think I can control them." She asked if she had their permission to give it a try, and freaked-out homeowners thousands of miles away reluctantly agreed as the reporter easily commandeered their appliances. The Insteon Hub was not alone, and a 2013 study found hackers were able to easily break into 80 percent of all the smart home hubs commonly available, including the VeraLite Controller, which is compatible with over 750 smart home products.

The number of vulnerabilities in home automation systems is so great that the Department of Homeland Security's Computer Emergency Readiness Team was forced in 2014 to issue a public alert to 500,000 users of Belkin's popular WeMo smart home device, identifying five separate vulnerabilities in the product. The warning noted a "remote unauthenticated attacker may be able to sign malicious firmware, relay malicious connections, or access device system files to potentially gain complete access to the device." DHS added, "We are currently unaware of any practical solution to this problem." A report into the incident noted that "once an attacker has established a connection to a WeMo device within a victim's network[,] the device can be used as a foothold to attack other devices such as laptops, mobile phones, and attached network file storage." The last admonition is an important one. Hackers are not going to try to break into the most secure device on your network, such as a locked-down encrypted laptop using a software firewall. Instead, they will always go for the weakest link, the trusted WeMo Internet-enabled coffeepot on your home network, the one with inadequate or absent security protocols. Once they've compromised the coffeepot, they've broken the virtual Maginot Line perimeter of your network: from there it's just a hop, skip, and jump to infect and attack the more secure and profitable devices in your home.

One of the most common online objects in many homes and businesses is their security alarm systems, and more than thirty-six million Americans rely on them to keep themselves and their families safe. But whether it's the simple door sensors or the keypads, they too are readily hackable, just as we've seen in all those Hollywood *Mission Impossible* movies. The majority of the alarm systems, including those from companies like ADT and Vivint, used legacy wireless communications protocols from the 1990s that fail to encrypt or authenticate their transmission signals. As a result, their cameras that are meant to protect can be turned against their owners to spy on their activity and their alarms suppressed so that they fail to go off when an intruder enters the home.

It's not just older alarm systems that are vulnerable; newer IoT radio communications protocols such as Z-Wave can also be hacked, which is troubling, given that there are 160 manufacturers using the protocols, in use at thousands of companies, such as Las Vegas's Wynn hotel, which has deployed sixty-five thousand Z-Wave devices throughout its guest rooms. Hilton Hotels too announced that it would allow guests to use their smart phones as keys to unlock their rooms at four thousand hotels worldwide by the end of 2014. As more and more home front door locks and dead bolts go online, they may be opening their doors to home invasion 2.0. Crooks will now be able to hack open your front door from their smart phones and kill your panic alarms, ensuring that in these homes nobody will hear your scream.

Not only can centralized home automation hubs be attacked, but so too can individual "smart" devices such as televisions. Indeed, numerous reports indicate that as you sit there watching your smart TV, it may be watching you right back. The majority of mid- and high-range televisions today are IoT compatible and come preloaded with apps such as Netflix, Skype, Facebook, and Hulu, not to mention embedded cameras, microphones, and USB ports. Worldwide nearly ninety million smart TVs were sold in 2013, and soon legacy "dumb" TV sets will be hard to find, a potentially troubling trend for those who value privacy and security. Many brands have been found to contain security vulnerabilities, such as Samsung Smart TVs, which allowed hackers to remotely turn on the built-in camera meant for Skype calls and surreptitiously snap photographs and watch viewers in their living rooms and bedrooms.

The hackers were also able to steal the log-in credentials and account details stored on the Samsung TV's smart apps to take control of users' Facebook and other social media accounts. For those unsuspecting consumers who had also used the TV's USB port to attach an external hard drive so as to be able to stream music and video directly to their televisions, there was another nasty surprise. Hackers were able to view, download, and erase those files via the TV, bad news for those who store any financial details or personal documents on their external hard drives. These additional in-home connections open users up to a Mat Honan–style attack where precious photographs and other data stored locally can be remotely erased by those with ill intentions.

Crime, Inc., just like Silicon Valley, is experimenting with the best ways to monetize the Internet of Things and, in doing so, has updated its tried-and-true tactics for the era of ambient computing. In early 2014, hackers commandeered over 100,000 everyday "smart" objects—including home routers, burglar alarms, Webcams, multimedia boxes, and refrigerators—and united them to create the first-ever home appliance botnet. The attackers used the devices to send more than "750,000 malicious spam and phishing emails," each one meant to turn a profit for Crime, Inc. Refrigerator spam (the nonedible kind) is troubling enough, but it is important to recall that smart devices are full-fledged computers and once compromised can do everything a hacked desktop can, from hosting child pornography to flooding targeted Web sites with vast volumes of useless data. Linking one million of today's computers into a botnet army is bad enough, but adding the next fifty billion smart objects to the Net, each with poor or no security, will open up phenomenal opportunities for offensive computing attacks.

Botnets will grow in size from millions of compromised machines to potentially billions, bringing with them new forms of WMD—weapons of mass disruption. Using these cyber arms, Crime, Inc. will have powerful new tools in its arsenal to extort businesses and individuals alike, keeping them off-line until a "tribute" is paid in Bitcoin. The computing power embedded in smart objects scattered throughout your home and office can be profitable to criminals in other ways as well. In early 2014, researchers discovered tens of thousands of Internet-enabled DVRs that had been hacked with the Linux.Darlloz worm to use their processing power to mine for crypto currencies such as MinCoins and Dogecoins. In doing so, hackers can keep your appliances running at full speed, generating virtual currencies for them while sticking you with the electric bill for spinning your devices 24/7. In theory, the new smart meter in your home might catch the excessive electricity use, but of course it too can be hacked.

What the Outlet Knows

Smart meters will be at the core of the global IoT, and their two-way communications abilities will record and track details of electricity usage in homes and businesses in order to increase the overall efficiency and reliability of an outdated and overburdened electrical grid. As of mid-2013, smart meters had been installed in over forty-six million homes in the United States, and the U.K. anticipates their deployment throughout all of Britain by 2020. Smart-meter information, much of which is transmitted in an unencrypted format, can actually reveal details such as the brand and age of your appliances and when you are using them in which rooms of your home. Extrapolating such data reveals how much time you spend cooking and when you turn on the TV in the bedroom. But the deep granularity smart meters can provide on your activities extends far beyond simply knowing you used the microwave at 7:26 p.m. on Thursday.

Researchers in Germany revealed that smart meters could also tell what television programs people were watching at what times, because of the specific electricity required in order to display the scenes of each show on your screen. By measuring these in the aggregate, the researchers were able to create individual profiles for all television programs, and it turns out episode 71 of *Star Trek* has a different power signature from episode 17 of *Modern Family*. Of course, there are potentially billions to be made selling all of these data to third parties. Indeed, in May 2014, WPP, the world's largest advertising agency, announced it was teaming up with the London-based data analytics company Onzo to study ways to collect smart-meter data in order to finally "open the door of the home" to advertisers.

The threats from smart meters extend well beyond their deep privacy implications, and criminals have attacked insecure smart utility devices for a variety of purposes, in particular financial fraud. In Puerto Rico, for example, Crime, Inc. employed large teams of techno-thugs to take advantage of the widespread deployment of smart meters on the island. Using software widely available in the digital underground and a simple laptop, criminal hackers began making "service calls" to both businesses and the general public. For fees ranging from $300 to $1,000 for residential customers and $3,000 for commercial clients, Crime, Inc. successfully reprogrammed the smart meters in order to save its "clients" up to 75 percent off their monthly electricity bills. According to an investigation into the incident by the FBI, the Puerto Rican electrical and power authority affected lost nearly $400 million in revenues annually as a result. Like all computers, smart meters are also vulnerable to malware attacks, and security researchers at IOActive devised a worm capable of rapidly spreading from one infected smart meter in a home to another, eventually infecting a whole neighborhood and plunging it into darkness.

Working hand in hand with your smart utility meter will be your smart thermostat on the wall in your home, and one company above all others is revolutionizing the field: Nest Labs. Founded by two former Apple executives, Nest has completely reimagined the clunky old thermostat, which hadn't changed much since the 1950s. Leveraging the profound design expertise garnered at Apple, the Nest founders created a beautiful Wi-Fi-enabled thermostat replete with cutting-edge sensors, including temperature, motion detection, humidity, and light. Nest employs adaptive artificial intelligence algorithms designed to learn what temperatures make you happy and when. Nest also has an auto-away mode that determines when there hasn't been any motion or light near the device, correctly deducing when you are on vacation or not at home. Nest's thermostats have been extremely popular with the public, and a hundred thousand units a month are flying off the shelves along with other Nest products, such as its talking multi-sensor IoT Wi-Fi-enabled smoke alarm. The widespread enthusiasm hasn't gone unnoticed by other tech giants, and in 2014, just a few years after its founding, Nest was purchased by Google for $3.2 billion. Good news for Nest's founders and the hundred or so employees, but why would an Internet advertising firm buy an IoT device maker?

Google clearly sees the opportunities in the Internet of Things, and Nest is a powerful hardware product to anchor its ambitions in the battle for what it is calling the "conscious home." But Nest thermostats and smoke detectors with all their embedded sensors are prodigious producers of data, and just as the Android mobile phones brought new advertising and data sales opportunities, so will Nest Labs products. Google, however, is far from done with its acquisitions, and in June 2014 it announced it was purchasing Dropcam, a large video camera security start-up, for $555 million. Dropcam makes high-definition Wi-Fi and Bluetooth security cameras that stream live video to mobile apps and send alerts based on predetermined activities sensed by the devices. With the purchase of Dropcam, Google now owns not only your Web searches, e-mail, mobile phone, maps, and location but also your movements inside your own home through live-streaming video feeds. As a result, your thermostat, smoke detector, and security system all come with lengthy terms of service. Could the privacy implications be any more obvious?

Of course an insecure and accessible smart meter is a great way to tell when you are away from home for extended periods of time. Rather than search your Facebook postings, tomorrow's burglars will just be able to tap into your video feeds, query your refrigerator to see when the last time its door was opened, or simply ask the smart thermostat if it is in extended vacation mode. Google's Nest thermostat has already been successfully hacked allowing just that, giving hackers potential remote access to the device, including monitoring whether an owner is home via the embedded motion detector or even cranking up the heat full blast. Nest's other main product, the Nest Protect smoke and carbon monoxide alarm, has also had difficulties, and 440,000 of the devices had to be recalled because of a software glitch that could delay the alarm from going off in case of an actual fire. Dropcam cameras too have had their security vulnerabilities, which hackers can exploit to watch videos remotely, turn on the camera's microphone, and inject fake video into the device's online live video stream, in case thieves want to cover their tracks in an *Ocean's Eleven* scheme. Needless to say, Crime, Inc. too is eager to learn what your outlets know: you may find that with each new Wi-Fi lightbulb and door lock you buy, you are unwittingly providing hackers all they need to find new ways to haunt your house from afar.

Business Attacks and Building Hacks

Businesses too are jumping on the IoT bandwagon to further drive cost savings, and though the majority of corporations do have chief information security officers, the technological battleground that is the office is proving extremely difficult to navigate. Unbeknownst to most, since 2002, nearly all photocopiers have come with internal hard drives that store every document copied or scanned. Because many of these devices are leased or eventually sold, the data they contain is wide open for pilfering, as a CBS News investigative report demonstrated. A visit to a warehouse in New Jersey found six thousand used copiers for sale, all loaded with penetrating government and corporate secrets. Researchers and reporters purchased just four used copiers to see what they might recover, and the results were scandalous. Using simple, widely available data recovery tools, investigators found "tens of thousands of documents," including "95 pages of paystubs with names, addresses and social security numbers"; $40,000 in copied checks; "300 pages of individual medical records" from the Affinity Health Plan, including everything from drug prescriptions to cancer diagnoses; "detailed domestic violence complaints and a list of wanted sex offenders" from the Buffalo Police Department's sex crimes unit and a "list of targets in a major drug raid" from its narcotics squad.

Needless to say, in the networked world of the IoT, physical access to copiers is not even required, as many of these documents can just be pulled from office photocopiers remotely. Hackers have been able to gain access to networked photocopiers (the vast majority of which are online in any modern office) and watch what was being copied in real time. Moreover, office printers, such as the HP LaserJet Pro, have been remotely hacked to gain unauthorized access to your Wi-Fi network and its administrator passwords, which the device stores in plain, unencrypted text. An embedded firmware attack discovered in 2011 demonstrated that millions of HP printers could remotely receive update instructions from hackers that sent the devices into overdrive, resulting in their catching fire. By exploiting a vulnerability in the machine's fuser element, the hackers were able to get the printer to overheat, burn the paper passing through the device, and ultimately burst into flames. Thanks to the IoT, it is now possible to commit arson from thousands of miles away, and I wouldn't count on your IoT-enabled smoke detector to save you, because any hacker motivated enough to burn your office or home will likely turn off any safety systems to detect the fire as well.

Other common office equipment can be hacked, including the video-conferencing equipment commonly found in most offices and boardrooms, where some of the most closely guarded corporate secrets are discussed. Just as the cameras in your home can give a hacker a bird's-eye view into your activities, so too can the digital eyes in the workplace. Videoconferencing systems, such as those from Polycom and Cisco, widely used in most offices today, have proven easily vulnerable to attacks. To prove the point, one hacker wrote a script to detect as many insecure videoconference systems as he could find and in short order had uncovered more than "five thousand in conference rooms at law firms, pharmaceutical companies, oil refineries, and medical centers." Among the live video feeds he was able to tap into were a meeting between a jailhouse lawyer and a prison inmate, "an operating room at a university medical center, a venture capital pitch where a company's confidential financials were being projected onto a screen," and even the Goldman Sachs boardroom. The experiment proves that in the office too, when everything is connected, everyone is vulnerable. Because many of the Polycom and other videoconferencing systems are sold, installed, and maintained without any serious security protocols in place and with auto-answer enabled by default, hackers can just remotely dial in and boot up the cameras and speakerphones to spy on you and your company.

Meanwhile, around the world construction crews are busy creating new "smart" buildings—skyscrapers, warehouses, and factories—and retrofitting those already in existence. Putting a building online offers profound potential savings for property owners, who can leverage complex automation systems to save on water, electricity, and gas costs in buildings that sense our presence and learn to turn themselves off appropriately as people come and go. Modern heating, ventilation, and air-conditioning (HVAC) systems are all going online and are being integrated with a variety of alarms, sensors, security card readers, cameras, and even physical objects, such as vending machines, water pipes, parking gates, and elevators, all centrally controlled via building management operating systems. Evidence of these "improvements" is everywhere, and in many large office buildings, such as those in Manhattan, elevators do not even have individual numbered buttons for you to select your desired floor anymore. Instead, the data encoded in your RFID badge or the controls at a central security station predetermine to what floors the elevator will take you.

Just like your home automation hub, commercial building management systems can be hacked, and when they are, the results can be surprising. In April 2012, students at MIT hacked the twenty-one-story Green Building, home to the university's Department of Earth, Atmospheric, and Planetary Sciences, and used its compromised electrical systems to create a giant, playable, and multicolor *Tetris* game. A wireless gaming console connected to the building allowed "players" to move, rotate, and drop blocks that corresponded to the lights in various offices. From across the street and throughout Cambridge, the windows in the building's offices lit up and moved as if the hackers were playing the famous Russian puzzle video game. While some building hacks can be fun, others can be much more costly.

Systems that had historically operated as stand-alone entities are now being unified, and the far-reaching interconnections of the IoT can prove extremely difficult to foresee, map, and protect. To deal with these challenges, many organizations are turning to centralized management for their building systems, choosing to hire, for example, an outside security contractor to remotely monitor all the security feeds for a particular company across numerous sites. When everything is connected, other services, including HVAC, can also be centrally managed, and one such company to do so is the retailer Target, which had outsourced its heating and cooling responsibilities to a vendor known as Fazio Mechanical Services of Pennsylvania. From their headquarters, Fazio technicians directly interfaced with Target's supplier and contract management system, a pathway to the mother lode Crime, Inc. found too tempting to resist.

When an employee of Fazio Mechanical inadvertently opened up a phishing e-mail with a malware-infected attachment (a variant of the Zeus banking Trojan produced by Crime, Inc.), he infected himself and the rest of his company. But because Fazio was tied into Target's network, the Trojan also made it possible for hackers to peer into the network of their ultimate quarry: the giant retailer, Target Corporation. The result was the attack against Target mentioned earlier and the massive breach of personal information and payment card details for 110 million American consumers. Once hackers had compromised Fazio Mechanical and stolen an employee's log-in credentials there, they were able to use them to fish around Target's network until they struck gold.

There they found information on Target's suppliers' portal and data from Target Facilities Management. Ultimately, hackers learned that these systems were not segmented off from other major IT systems used by the retailer, including, shockingly, its payment and financial systems. Armed with all the details they needed, the hackers burrowed like rats through a multitude of interconnected networks until they arrived at the company's internal server responsible for controlling the tens of thousands of individual point-of-sale terminals where customers swipe their credit cards at the register. Once there, attackers installed malware known as Trojan .POSRAM, which copied all the card swipes taking place throughout Target stores nationwide and secretly exfiltrated the data to Russia, a breathtaking fraud that continued until the story was broken by the security researcher Brian Krebs. No doubt the Target attack is the highest-profile penetration of an HVAC system to date, but it is not the only one.

We might like to believe that the government could do a better job in protecting its buildings from remote attacks, but evidence does not seem to suggest that is the case, even at those facilities one might expect to be among the most secure. In 2011, researchers were able to successfully hack the Federal Bureau of Prisons' industrial control system network and remotely take over the facilities. Hackers could unlock individual cell doors or an entire cell block at will, even though the computer screens at the central guards' stations indicated they were still locked. The prison's communications network could also be turned off so that individual guards could not call for help in case of emergency. Worse, it was possible to electronically "destroy the doors" by overloading the electrical system controlling them, thereby leaving them permanently open for the entire prison. Using these techniques, Crime, Inc. could potentially free compatriots to put other prisoners at risk by opening their cell doors for retaliation attacks. These threats are not just theoretical.

In mid-2013, an unknown computer "glitch" at the Turner Guilford Knight Correctional Center in Miami, Florida, caused all doors in the maximum-security wing to suddenly and simultaneously open, setting prisoners free, causing a riot, and allowing gang members to seek revenge on rivals. According to surveillance video at the prison, one inmate in particular seemed prepared for the incident that stunned guards and other prisoners alike. At the moment the doors unexpectedly opened, the prisoner calmly walked down the passageway to the cell of a long-standing enemy and "shanked" him with a homemade prison knife before returning to his own lockup pod. The cause of the "glitch" was still under investigation in late 2014, and the incident suggests that not every building in our society need be connected to the Internet.

Our growing threat surface area brought about by the Internet of Things creates opportunities not just for Crime, Inc. but for nation-states as well, as the U.S. Chamber of Commerce has discovered. As the leading business group lobbying on behalf of America's corporate interests, the chamber often took stances on international affairs and foreign trade issues, positions that were frequently critical of China, in support of its three million business members. While the chamber had successfully blocked cyber attacks against its main network emanating from the People's Republic in the past, its luck ran out in late 2011 when it discovered a recently installed Internet-enabled thermostat at one of its offices on Capitol Hill had inadvertently created a back door to its internal corporate network. Chamber officials made the discovery when they found the energy-saving device had been secretly communicating with an Internet address in China.

While attackers might have been clever in their use of the thermostat as a means of breaking into the chamber's primary network, they were less skilled at routing their own print jobs. Their carelessness caused a printer used by chamber executives to spontaneously start printing pages of information with Chinese characters on them, something officials at the FBI viewed as a useful clue of something amiss. Once inside the chamber's network, attackers searched for financial and budgetary information, compromised e-mail systems, and focused on employees working on trade policy matters in Asia. Make no mistake, there are deep geopolitical implications to the Internet of Things, and those nations capable of leveraging these technologies to their fullest will have access to unparalleled intelligence and strategic advantage. As the Chinese premier, Wen Jiabao, noted in a speech in August 2009 in the city of Wuxi, "Internet + Internet of Things = Wisdom of the Earth."

The Smart City Operating System

Those skilled in war are able to subdue the enemy's army without battle.
They capture his cities without assaulting them and overthrow
the state without protracted operations.
SUN TZU

In 1964, Marshall McLuhan presciently predicted that by "means of electric media . . . all previous technologies . . . including cities . . . [would] be translated into information systems." It might have taken fifty years, but his forecast was spot-on. The Internet of Things has the full potential to transform cities into living, breathing ecosystems of ambient intelligence and connected sensors, vastly improving the quality of life for their inhabitants. In the utopian vision of smart cities, trash cans with embedded sensors will notify the rubbish collectors when they are full, immediately dispatching the closest GPS-equipped garbage truck to whisk them away. The growing numbers of "municipal sensor networks" can measure the pollution produced by individual buildings, the air quality on a particular block, or the number of pedestrians walking on a given street, creating the first-ever "Fitbit for the city." Better sensors in our streetlights will mean municipalities will be able to provide just the right level of lighting, appropriately adjusted for time of day, season, and weather conditions, reducing energy costs by up to 30 percent. That is of course if things go well.

The less sanguine perspective on a citywide operating system would be a municipal network of IoT-enabled devices, always on and subject to attack from hackers anywhere in the world. Using a wireless traffic-detection system commonly deployed in cities throughout the world, an Argentinean hacker, Cesar Cerrudo, was able to control traffic lights in Manhattan by hacking the underlying sensors embedded in the roadways, a technique that enabled him to reroute traffic and cause traffic jams at will. Hacking buildings and a city's operating system can compromise physical safety as well as allow attackers to gain control of elevators, air ducts, door locks, lighting, bridges, tunnels, water treatment facilities, and other vital systems. If smart meters can be hacked, so too can smart grids, and the ability of a hacktivist collective, organized crime group, or rogue nation to shut off power to the masses now becomes a reality. In July 2014, a security researcher was able to seize control of the power supply to Ettlingen, a town of forty thousand people in southern Germany. A hacker using the same exploit could have switched off all municipal utilities, including power, water, and gas.

Creating the Internet of Things holds the possibility for immense improvements in both our quality of life and the global economy, particularly as objects become "smart" and learn to automatically interact with one another for our benefit. Putting aside major privacy concerns for the moment, with billions of cars, coffee machines, buildings, mobile phones, elevators, dishwashers, and toys talking to each other and taking commands from the Internet at large, we have provided attackers innumerable points of contact to reach into our lives and affect them for the worse.

We can't even protect the relatively few things we have online today, yet day in and day out we bring new smart objects into our homes and into our lives without ever bothering to stop and ask what the potential risks and pitfalls might be. As a result, much like practitioners of the ancient martial art of judo, attackers can now use the weight and strength of our own overgrown connections to defeat us. In effect, we've wired the world but failed to secure it—a decision we may well come to regret, especially as we begin connecting the human body itself to the Internet.

CHAPTER 14

Hacking You

The Internet of Things, sometimes referred to as the Internet of Objects, will change everything—including ourselves.
DAVE EVANS, FORMER CHIEF FUTURIST AT CISCO

Steve Austin, astronaut, a man barely alive. Gentlemen, we can rebuild him, we have the technology. We have the capability to make the world's first bionic man. Steve Austin will be that man. Better than he was before. Better. Stronger. Faster." Those were the lines spoken in the opening credits of the hit 1970s television show *The Six Million Dollar Man*. Like a lot of boys growing up at the time, I was amazed by the tremendous superhuman strength the starring bionic superhero possessed and longed for the ability to run that fast, jump that high, and see that far. As cool as the bionic man was, adults assured me it was all just made up, pure fantasy from the wildest annals of science fiction. But as I later learned, science fiction can rapidly become science fact.

I first met Bertolt Meyer in mid-2012 when he was filming a television documentary for Channel 4 in the U.K. titled *How to Build a Bionic Man*. Meyer, a thirty-three-year-old social psychologist from the University of Zurich, was exploring both the possibilities and the ethical implications of the latest bionic technologies. His interests in the topic were motivated not just by scientific curiosity but also by personal destiny: Meyer was born without his lower left arm. As a child, he was fitted with a variety of primitive prosthetics, all of which made him feel different and self-conscious, not to mention had extremely limited functionality. Meyer aspired to have the same anatomical functionality others had and wanted to move beyond the non-opening hands and metal hooks with which he had been fitted as a child. In 2009, that dream became a reality when he was outfitted with one of the most advanced prosthetic devices in existence: the Touch Bionics i-limb hand.

Meyer had in fact become a real-world bionic man. He was thrilled with his new physical abilities, including for the first time the capacity to clap his hands and to hold a fork and carry a heavy shopping bag with his left hand. The new bionic hand included an advanced aluminum chassis for "improved durability, increased grip strength and a more anatomically correct" design than any of his previous prostheses. Meyer controlled the device by sending myoelectric pulses from the human flesh of his arm just above the prosthetic limb to electrode sensors attached to his skin, enabling the hand to open, close, rotate, and pick up objects. It was a true breakthrough for Meyer personally and led to a profound interest in the world of bionics, the subject of his documentary, in which I was interviewed.

As the bionic man cum filmmaker and I discussed the ethical implications of these technologies, the conversation turned to the matter of digital security, a topic Meyer had yet to consider. As it turned out, myoelectric impulses from Meyer's own body were not the only way his bionic hand could be operated. His bionic hand was also Bluetooth enabled and could be controlled, adjusted, and reprogrammed by a mobile app that he had downloaded from the manufacturer to his iPhone. I discussed with Meyer the inherent well-known insecurities of the Bluetooth protocol and how many times it had been subverted by hackers in the past. Suddenly, in an instant, Meyer understood the implications of his vulnerability and turned ashen gray, jaw agape at the vulnerabilities that nobody had previously disclosed to him.

I asked Meyer if I could see his mobile phone and the app he used to control his bionic appendage. He dutifully complied, passing the device to me. As I examined the Bluetooth app, I saw it offered a variety of grip positions and reprogramming options. If I pushed one button, his hand would open, another it would close; individual fingers could be positioned and his thumb and wrist manipulated. I was now in control of the bionic man and his body. Through my mere possession of his iPhone, Meyer's body would now do my bidding. Of course, I needn't have physical access to his phone, given its use of the insecure Bluetooth protocol. I could just hack it and remotely take control. Though Meyer hadn't realized it, his hand had joined the Internet of Things, and when it did, its usage no longer belonged solely to its rightful owner. After Meyer overcame the initial shock, we continued our discussions and eventually became friends. We also learned an important lesson together. Now, for the first time ever in the history of humanity, the human body itself was subject to cyber attacks.

"We Are All Cyborgs Now"

You know, anyone who wears glasses, in one sense or another, is a cyborg.
EVGENY MOROZOV

The term "cyborg"—short for "cybernetic organism"—conjures up images of a scary world populated with humanoid aggressors, such as the Cylons in *Battlestar Galactica*, the Borg in *Star Trek*, or the Cybermen in *Doctor Who*. Though the term is relatively new, the act of augmenting the limitations of the human body dates back millennia, with ancient peoples using wood, copper, and iron to replace missing or deformed limbs. Since then, prosthetics have come a long way, not merely replacing partial bodily functionality lost because of injury or disease, but actually improving the capabilities of well-functioning biological equivalents. These advances were highlighted in the case of the gold-medal-winning South African sprinter Oscar Pistorius, the double below-the-knee amputee, who was persistently targeted by other athletes who complained his "blade runner" artificial limbs gave him an unfair advantage.

Today, technology is artificially augmenting not only our limbs and senses but our minds as well. Over 90 percent of smart-phone owners report keeping their mobile phones within three feet of them, all hours of the day, a number that will surely increase in the future. These devices amount to not just external brains but also phantom limbs to which we are persistently attached, deeply anxious when they are far away or accidentally left behind. We use our mobiles as external sources of memory (they can remember the thousand phone numbers we cannot) and as additional means of communication, sharing our thoughts across the planet by SMS, status updates, tweets, and e-mails. We will also increasingly wear and eventually embed intelligent devices in our own bodies, and when that happens, we too will join the Internet of Things. These wearable computers, implantable medical devices, bionics, and exoskeletons will be interacting with the world around us, providing new physical and mental capabilities, as well as continuous health monitoring and feedback. Just as the number of microchips in our cars has increased over time, united in a single controller area network, so too will all the devices on us and in us, to form their own body area network in the future. With these changes will come the profound security and privacy issues affecting the broader Internet of Things at large, except this time we ourselves will be nodes on the Internet.

Whether our cyborg future more closely resembles the horror portrayed in Mary Shelley's *Frankenstein* or the heroic possibilities of Tony Stark in *Iron Man* remains to be seen. One thing is clear, however: Crime,

Inc. has shown time and time again its willingness and ability to leverage any emerging technology to its advantage, and hacking you and your body may be an opportunity too good to pass up.

More Than Meets the Eye:
The World of Wearable Computing

Perhaps one of the earliest wearable-computing devices to gain wide acceptance was the hearing aid, which has transformed from a deck-of-cards-sized transistor unit visibly worn over the chest with attaching shoulder straps to a fully contained digital-microprocessor-enabled unit, small enough to fit inconspicuously inside a user's ear canal. Not surprisingly, today's modern hearing aids use Bluetooth technology and are capable of streaming multiple audio sources and amplifying them for the hearing impaired. Using mobile phone apps, users can control and adjust hearing aid settings on their phones to choose whether to listen to ambient sound, a telephone conversation, or the music on an iPod, all at the press of a button. But now even the humble hearing aid, just like the Bluetooth headsets worn by the general public, can be hacked using a variety of widely available underground Bluetooth programs previously mentioned. As a consequence, not only is it feasible to remotely intercept what another person is hearing in real time, but it would also be possible to play sounds or noises directly into the ears of the hearing impaired. Whether it be heavy metal music or threatening voices that the person wearing the device alone could hear, these noises are sure to cause annoyance and consternation on the part of those affected.

Hearing aids have now been joined by a panoply of additional choices when it comes to the sensors, trackers, and computers available to be worn on our bodies today. Many of these developments have been driven by the "quantified self" movement, which employs a variety of methodologies for collecting data about an individual's life using technological tools. Every day, millions of quantified-self adherents record every aspect of their lives, thoughts, and experiences via self-tracking tools in search of a better life through "life logging." They track and measure their sleep, weight, calories burned, biofeedback, heart rate, brain waves, EKG rhythms, happiness, number of steps in a day, all in an effort to improve mental and physical performance, easily gathered through the introduction of wearable-computing devices known as wearables.

By providing measurable feedback collected by small computers worn on the body, the devices let dieters know with precision how many steps they have taken and how active they have been. The information can be displayed on beautifully designed data dashboards on a computer, clearly delineate fitness trends, and even offer elements of gamification with lead-erboards and badges when predetermined goals have been met. Armed with this information, dieters can make behavior changes such as eating less or moving more in order to meet their weight loss goals. The devices may also play important roles in disease prevention and improving general health.

Over 100 million wearables were sold globally in 2014, expected to grow to 485 million units by 2018. Wearable devices fall into several broad categories such as bracelet activity trackers, including the Fitbit Flex, Jaw-bone's UP, Nike FuelBand; smart watches (the Pebble, Samsung's Galaxy Gear, the Apple Watch); or even eyewear such as Google Glass. Though wearables have been mostly a niche item until now, they are poised to go mainstream in the near future.

Most wearable devices sync with a user's mobile phone, via Blue-tooth or Wi-Fi connectivity, and when they do, your personal health information joins the Internet of Things as well, easily hackable just like other IoT objects. Moreover, many wearables are tightly integrated with social networks so that, for example, your Fitbit tracker can auto-matically post the number of daily steps you have taken directly to your Facebook page. Doing so, however, raises a variety of privacy concerns, in particular who owns your data, how are they being secured, and how can they be shared with third parties? Surprisingly, however, 52 percent of fitness apps had no available privacy policies. And as we've learned, information that seems harmless now can come back and bite you later. Poor sleep patterns documented automatically by your wearable can be directly relevant in a court case about a traffic accident. Will your health insurance company require you to don an activity tracker to get its best rates, just as car insurance companies are doing with their black boxes in your car?

One of the latest trends in wearable computing is the incorporation of video cameras into the devices, whether it be the popular GoPro HD Wi-Fi-enabled camera used in extreme action photography or something more subtle such as the camera embedded in Google Glass. While the idea of most people walking about our streets wearing Internet-enabled video camera eyeglasses may seem preposterous at the moment, keep in mind that the same thing was said about the personal computer and the mobile phone. Google has already partnered with the eyewear giant Luxottica to build Glass into Oakleys and Ray-Bans, and Deloitte has predicted mil-lions of pairs of smart glasses will be sold in 2015. Devices like Google Glass will offer myriad technical conveniences, all in one highly porta-ble device, such as the ability to take pictures, send photographs, record video, make phone calls, search the Internet, send SMS messages, and read e-mail. These capabilities are at the pinnacle of what is possible in today's wearable-computing marketplace and will be empowered through a variety of Wi-Fi, Bluetooth, and GPS connections, also conveniently tethered to the mobile data plans on our smart phones. As noted in earlier chapters, with observations by both Mr. Burns of *The Simpsons* and Mr. Chertoff of Homeland Security, with all of Google Glass's power and connectivity come a host of privacy and public policy issues. But there are important security threats to be considered as well.

The fear of filming has led to Google Glass's being banned in a number of public venues, including sporting events, concerts, gym locker rooms, bars, restaurants, strip clubs, casinos, hospitals, and U.K. movie theaters. Cited reasons for the prohibitions against the device include everything from card counting to film piracy and industrial espionage. But there is another concern. Google Glass can be hacked to secretly take photographs and record video, silently streaming the data to Crime, Inc. anywhere in the world, all without the knowledge of the device's owner. Just as we saw with the malware used to subvert your mobile phone or laptop, IoT eyeglasses can be switched on without any visible indication they are recording.

In fact, hackers had already cracked Google Glass's security before the device even went on sale to the general public. The security holes in Google Glass mean that the device can be "rooted," subverted to transmit everything you see and hear in real time, including your account details and password as you type them for your online bank account. The GPS features of Glass mean that Crime, Inc. will also be able to tell your precise location, such as when you are at an ATM typing in your PIN number. While your grandma never needed antivirus programs for her eyeglasses, you may. A variety of malware and spyware tools have already been created for Google Glass, and as a result, now for the first time in human history, our eyeballs can be hacked too.

Given the pace of technological progress, wearing a "bulky" computer in our eyeglasses will soon become too cumbersome to bear for the next generation, and so surely the next iteration of these devices will be the Internet on a contact lens. While Google has yet to publicly confirm a contact lens version of Google Glass, it did surprise the world in mid-2014 when it announced that it was working on an IoT "smart contact lenses" project with the pharmaceutical firm Novartis. The companies' lens will offer an array of microchip sensors and antennas that will for the first time make it possible to continuously monitor a diabetic's blood sugar levels without the need for the painful needle pricks required by today's glucose-testing systems. The device is in early stages of testing with the FDA. Not to be outdone, Samsung is developing its own full-fledged Internet contact lens, which will display all the Web data currently available with Google's eyeglasses but in contact lens format, using mounted light-emitting diodes and a mix of graphene and silver nanowires. Yet for as advanced as wearable computing promises to become, there is still a further frontier in the quest to fully integrate man and machine—implanting computers inside the body itself.

You're Breaking My Heart:
The Dangers of Implantable Computers

The first time an electronic medical device was successfully implanted into the human body was 1958. The historic operation was performed by two Swedish surgeons on Arne Larsson, an engineer who went on to live another forty-three years—a lifetime of memories and experiences he would never have lived to see had it not been for the hockey-puck-sized computer installed in his abdominal cavity driving his heart to beat normally. Now, nearly sixty years on, the world of medicine has made phenomenal strides in advancing the breadth and capabilities of implantable medical devices (IMDs). These devices have seen multifold increases in portability, battery life, and efficacy and today remotely transmit critical information to a patient's doctor over the Internet. The first Wi-Fi pacemaker in the United States was implanted in the chest of Carol Kasyjanski of Roslyn, New York, in 2009, and when the surgery was complete, her beating heart became the first to join the Internet of Things.

In addition to pacemakers, there are a variety of other IMDs in common use around the world today, including implantable defibrillators, diabetic pumps, cochlear implants, and neuro-stimulators. While each device has its own therapeutic purpose within the body, IMDs communicate with the outside world via familiar radio-frequency protocols such as Bluetooth, Wi-Fi, NFC, and RFID. Millions of Americans have been equipped with IMDs, and approximately 300,000 patients receive wireless implantable medical devices annually. The devices have become pervasive in modern medicine given their shrinking sizes, their growing capabilities, and the manifest clinical benefits they provide. Wireless medical devices, such as the implantable cardioverter-defibrillators (ICDs), allow physicians to remotely monitor patients' heartbeats and EKGs in real time, greatly reducing the need for expensive office visits. Should a problem be detected by the ICD, doctors can immediately contact their patients and notify them to come in for treatment. The vast lifesaving potential of these advances cannot be overstated, but as we increasingly integrate information technology with our own biology, more and more people join the cyborg nation—with significant implications for their safety, privacy, and security.

Malfunctioning medical devices are one of the leading causes of serious injury and death in the United States, and the number of device recalls doubled between 2004 and 2014. Nearly 25 percent of these recalls were because of computer-related failures, of which 94 percent "presented a medium to high risk of severe health consequences." Even in hospitals, a wide variety of therapeutic devices such as MRI, X-ray, and anesthesia machines, IV pumps, CT scanners, and ventilators have been found to be riddled with computer viruses and remotely exploitable by hackers with ease. Indeed, in 2013, the Department of Homeland Security issued an alert advising medical facilities that more than three hundred devices from forty different vendors had vulnerabilities that could readily be exploited by those with ill intentions. As it turns out, just as your Windows computer or iPhone can crash, so too can a medical device on which your life depends. There's one important difference with IMDs, however. Unlike with your smart phone, you can't simply download new firmware over the air for your pacemaker; instead, surgeons have to cut open your chest or abdomen and get physical access to the device for a full firmware update or replacement.

Of perhaps even greater concern is the fact that the more we implant tiny computers inside ourselves to monitor and improve our health, the more we create opportunities for others to hack into our bodies and subvert these machines for nefarious purposes. Many medical devices are sold without any security mechanisms in place. Instead, manufacturers of IMDs, like other objects connected to the IoT, tend to rely on security through obscurity—after all, why would anybody want to hack a pacemaker? The flawed logic neglects the fact that there are indeed a tiny minority of cruel and odious people in the world who would be happy for the chance to prove their own technical prowess at the expense of others. Such was the case when hackers in 2008 altered the national Epilepsy Foundation's Web site to include hundreds of rapidly flashing animated images, causing violent seizures among epileptics innocently visiting the site for medical advice.

A team of researchers from the Universities of Massachusetts and Washington also showed the threat to medical devices was quite real when they successfully compromised the wireless security of Medtronic's combination heart defibrillator and pacemaker. After gaining unauthorized access to the device, they not only were able to read confidential patient information but, much more troubling, were fully capable of delivering jolts of electricity to a normally functioning heart—an act that would prove fatal to a hapless innocent. For hackers, IMDs represent an irresistible new yardstick by which to measure their talents, and the topic is among the most popular at the annual Black Hat hacker conference in Las Vegas. One well-known hacker, Barnaby Jack, had particularly good success in subverting a range of IoT devices from ATMs to pacemakers. In 2012, Jack uncovered serious software flaws within the IMDs produced by several manufacturers that allowed him to commandeer the devices. From fifty feet away, using nothing more than his laptop, the hacker was able to remotely order an implanted defibrillator to deliver 830 volts of electricity directly to a person's heart—a shock so powerful it would surely kill anyone with an implanted pacemaker.

Fearing the profound risk of such an attack, the cardiologist to the former vice president Dick Cheney physically altered the VP's ICD to remove its wireless capabilities lest terrorists attempt to send a lethal shock to the almost commander in chief's already ailing heart. In a case of art imitating life, a fictionalized but entirely viable version of just such an attack was memorably portrayed on the Emmy Award–winning Showtime drama *Homeland*, in which the terrorist bad guy Abu Nazir directs the assassination of the veep over the Internet by fatally compromising his implanted cardiac defibrillator. But pacemakers are not the only wireless IMDs that hackers have cracked. Hundreds of thousands of people in the United States also rely on diabetic pumps, a device meant to dispense insulin in carefully controlled amounts to those needing help regulating their blood sugar levels. Once again, the talented Mr. Jack proved his technical expertise and readily defeated the weak security protecting some of the most common diabetic pumps on the market. Using a specialized radio antenna that he devised, Jack was able to locate and compromise any insulin pumps within a three-hundred-foot radius, causing the forty-five-day supply of insulin held within the device to be released instantaneously and all at once, a remote cyber attack almost certain to result in death without immediate treatment.

Though no known criminal attacks against IMDs have been uncovered to date, we can fully expect that Crime, Inc. will turn its attention to these devices. Indeed Europol, the European police agency, predicted that online murder via IMDs could become reality by the end of 2014. Some of these incidents may just be potentially run-of-the-mill cyber attacks, for just as a scripted botnet attack can commandeer your computer and mobile phone (and even refrigerator as we saw in the last chapter), so too may it ensnare your pacemaker. Hacked medical devices might just look like any other available IP address on the Internet of Things, and once your implanted defibrillator or diabetic pump has been infected, the spam it has been ensnared into sending might well drain the very limited and precious battery life desperately needed to regulate your heartbeat and insulin dosages, requiring surgical intervention for replacement.

Needless to say, even more sinister plots will become possible when the numbers of connected medical devices going online grow exponentially. In effect, there will be new ways to commit murder from afar by hacking insecure medical devices, ushering in the unwanted era of medical cyber crime. Though there may be no muzzle flash with a laptop, in today's modern world it is a device that can kill all the same. Crime, Inc. will also look for ways to monetize attacks against IMDs. Just as ransomware such as CryptoLocker can destroy your computer's hard drive or your mobile phone, rendering it unusable, it would not be unreasonable to expect similar extortion attempts against medical devices. "You have sixty minutes to transfer $10,000 in Bitcoin to this account or we will deliver an 830-volt shock to your heart." Tick, tock, tick, tock, might go the refrain. Worse, consider the ramifications if hackers were to infiltrate the industrial control systems at the factory where implantable defibrillators were made and insert a zero-day exploit into the devices. The minute software changes might go unnoticed for months or years, until hundreds of thousands of devices had been implanted in patients around the world. Only then might Crime, Inc. strike with the first-ever critical infrastructure assault to use information technology to attack our own human biology, demanding millions in ransom to avert a global crisis. There would be no way to operate on the thousands and thousands of individuals who had a literal ticking time bomb in their chests in any reasonable time, leaving no alternative but to accede to the demands.

Given the pace of Moore's law, we can surely expect implantable medical devices to shrink even further in size, delivering amazing clinical benefits to patients. For instance, biomedical engineers at Stanford University have even created a wireless, battery-less robotic device so small that it can swim through the bloodstream, performing diagnostics and even microsurgery. Welcome to the world of *Star Trek* medicine, but even these new miracle treatments may face threats from hackers, who might be able to falsify results so that drugs are released into the bloodstream when they should not be or micro-robots attack healthy tissue instead of a tumor in the case of cancer. Should ingestible and injectable computers become compromised, how would anybody know? What detectable evidence, if any, might be left behind?

When someone with an IMD passes away, the medical examiner tasked with determining the cause of death will face multiple questions: Was this an accidental death caused by an IMD malfunction? Was the device specifically targeted for criminal purposes? Or was this a suicide wherein the patient himself subverted his own IMD to end pain and suffering, hoping that his family would receive life insurance funds for his apparent natural death? As modern medicine evolves and the proliferation of IMDs increases, one vital question must be answered: When a technologically enhanced body shows up at the morgue, who will be capable of performing the autopsy? Physicians and forensic pathologists have absolutely no training in computer forensics. How, then, will they possibly be able to determine the cause of death? They won't, and the threat we face from medical device insecurity means that in the future it may be even more possible to get away with murder.

When Steve Austin and Jaime Sommers Get a Virus

A smartphone links patients' bodies and doctors' computers,
which in turn are connected to the Internet, which in turn is connected to
any smartphone anywhere. The new devices could put the management
of an individual's internal organs in the hands of every hacker, online
scammer, and digital vandal on Earth.

CHARLES C. MANN

Shortly after television audiences in the 1970s fell in love with Steve Austin, the rebuilt astronaut star of *The Six Million Dollar Man,* he received a female companion in the form of Jaime Sommers, the world's first bionic woman. While both faced and defeated a variety of villains, none ever attempted to thwart the superheroes by compromising the electronics of their bionics. Why not? Perhaps because computer viruses and wireless technologies were not in the zeitgeist at the time, but as we saw with Bertolt Meyer, when your hand, leg, or arm is wirelessly controlled and online, it, like all other objects in the Internet of Things, can be targeted by hackers. Though relatively uncommon today, bionic prosthetics will grow tremendously in the coming years, particularly spurred on by the unfortunate needs of thousands of young soldiers returning from Iraq and Afghanistan who have been gravely injured in war.

In response, the Pentagon and DARPA, the Defense Advanced Research Projects Agency, have launched the Revolutionizing Prosthetics program, a $100 million investment with over three hundred scientists to completely transform the world of bionics. One such triumph has been the inventor Dean Kamen's Luke Arm/DEKA prosthesis, whose name was inspired by Luke Skywalker's robotic arm in *Star Wars.* The device is controlled by electrical signals from electrodes connected to the wearer's muscles and is so precise its fingers can pick up a quarter lying flat on a table. Other efforts are under way as well, including MIT's Human Bionic Project, which catalogs a "repository of every FDA-approved replacement part for amputees, to make it easier for them to find the best ways to rebuild their bodies." Even implantable bionic organs, such as the bionic pancreas, have been created to help diabetics regulate their glucose levels, conveniently driven by an app on their smart phones that wirelessly connects to the bionic organ.

Another field of bionics that is developing rapidly is the commercialization of exoskeletons or wearable robots, such as the Ekso Bionics system, which, when worn externally, can allow those paralyzed because of stroke, spinal cord injury, or disease to actually walk again. The exoskeletal suit supports those who cannot walk and actually moves their limbs for them, smoothly allowing them to stand and ambulate. Ekso's designs can also be used by those who are not physically impaired, providing tremendous support and strength by lightening the workload on functioning muscles, allowing soldiers, for example, to carry hundreds of pounds of weight over extended distances without tiring. Graduate students at NYU's Interactive Telecommunications Program have even developed "an open-source API that allows you to move someone else's arm remotely using a keyboard, a joystick or even an iPhone," in order to help those with paralysis or limited control of their own limbs function normally. The result is nonautonomous body control, allowing others to control your own arm or leg via the Net.

Of course the future of bionics won't merely be limited to restoring human capabilities lost to illness or injury. The much larger market opportunity will be focused on enhancing human capabilities, giving us powers we never had before and augmenting others we already have. Who wouldn't want the superhuman powers envisioned by *Iron Man*'s Tony Stark? But our growing ability to transform the human body by enhancing our own biology through information technology brings with it a host of risks and ethical questions that will have to be addressed in the future. Sure, your robo-legs may get a virus and your bionic hand can be hacked, but what happens when robotic exoskeletons are available to the common man and crooks begin using the superhuman strength to rob others? Imagine the future of street gang warfare when both Crips and Bloods have access to those tools and begin battling it out on the streets of your city or when Crime, Inc. sends an exoskeleton-clad enforcer knocking on your door to collect on that gambling debt you owe. Though those scenarios may seem fantastical, there is a long history of military technology eventually being adopted by the general public, whether it be firearms, night-vision goggles, GPS navigation, or the Internet itself. In the future, it is clear that there will be numerous ways for hackers to take advantage of present and forthcoming developments in both wearable and implantable computing. But there are other ways in which our biology is being used for identification and security purposes, and here is where the next battlefield for control of our bodies and ourselves will take place.

Identity Crisis: Hacking Biometrics

We tend to think of our face, eyes, voice, fingers, heartbeat, legs, and palms as unique elements of our own biology and anatomy that, without question, belong to us and us alone. If only it were true. Whether we realize it or not, we are sharing growing volumes of information about our physical and behavioral traits with others. These biometric identifiers are distinctive physical characteristics, the most common of which is the standard fingerprint, which police have used for more than 125 years to identify criminals.

For over a century, biometric fingerprint analysis could only be performed manually by specially trained human technicians. Times are changing, however, and rapid advances in data-processing power and sensor technology mean that computers too can now perform biometric identification. As a result, biometric systems are proliferating and becoming much more common in our everyday lives. Biometrics will fundamentally shift the way we are identified in the future. Unlike traditional forms of identification where you needed to carry something with you such as your driver's license or passport or remember something, such as your password or PIN, biometrics are something you always have on you and never have to worry about forgetting. Biometrics are *you*.

Biometric identification systems use computer sensors to measure things such as the ridges on your fingerprints, the distance between features on your face, or the tone and quality of your voice. All of this information is translated into ones and zeros that can be compared, sorted, and reidentified so that your particular set of fingerprints can be matched against a database of hundreds of millions of others in mere seconds. Given their falling costs and growing capabilities, biometrics is expected to grow to a $23 billion global market by 2019, with more than 500 million biometric sensors potentially joining the Internet of Things by 2018. Biometrics will be massive, they will be everywhere, and the movement has already begun.

Today gym goers at 24 Hour Fitness locations are encouraged to use their fingerprints for identification at the chain's clubs. Patients at New York University's medical center needn't carry their insurance cards anymore, because the hospital has enrolled more than 125,000 individuals in its PatientSecure system, which uses a specialized biometric scanner to measure the unique vein patterns in the palm of the hand as the primary means of identifying patients. But if hospitals can't keep malware out of their MRI machines, why would they be any more successful in protecting your biometric information? And is it really a good idea for the staff at your local gym (whose expertise is unlikely to be in biometric security) to have access to your fingerprints?

The biometric scanners we see in Hollywood spy thrillers such as *Mission Impossible* all look so high-tech and capable—eye scanners, fingerprint readers, and facial-recognition systems that perfectly distinguish friend from foe. With press like this, it is easy to understand why people think biometric authentication systems are impossible to defeat. As it turns out, biometrics are not as safe or foolproof as is often thought, and a 2010 report from the National Research Council concluded that such systems are "inherently fallible." Not only can certain biological markers be copied, but the databases where all of the biometric information is stored (the digital representations of your eyes, face, and fingers), like all other information systems, can be compromised. Risks aside, both government and the private sector are racing to wring any possible security advantage or economic benefit from the collection of your biometric details, data that can be gathered without your permission or knowledge.

The largest government biometric identity database in the world is run by the Indian government. The project, known as Aadhaar (meaning "Foundation") is an ambitious attempt to fingerprint, photograph, and capture iris scans on the nation's 1.2 billion citizens. Over 500 million Indian nationals have already received their Aadhaar identification numbers and had their biometric details conveniently stored in a national database. Not to be outdone, the U.S. government has dedicated significant resources across Homeland Security, the Department of Defense, and the Department of Justice, each of which has established vast biometric programs in the post 9/11 world.

While a national government biometrics database sounds as if it might be a useful tool in catching criminals and terrorists, it is not without its own privacy and security risks, as the government of Israel discovered in 2011. Authorities in the Middle Eastern country announced that its entire national biometrics database had been stolen, including the names, dates of birth, social security numbers, family members, adoption details, immigration dates, and medical records of nine million Israelis. The information was stolen by a contractor and sold to Crime, Inc., eventually ending up posted in full online in the digital underground. The obvious opportunities for a wide variety of fraud, identity theft, and security issues are manifest.

By 2016, Gartner estimates that 30 percent of companies will be using biometric identification on their employees. Biometric sensors will be built into most high-end mobile phones by the end of 2015, and by 2018 it is estimated that 3.4 billion smart-phone users will unlock their phones with their fingers, faces, eyes, and voices. Biometrics are the future of identity, security, and authentication. They will replace the common password, which, as we have seen throughout this book, is easily hacked, can be stolen by the millions, and has long outlived its useful life span.

Biometric security will offer many advantages; while you may forget your password or driver's license, you'll always have your fingerprints on you. Though biometrics will solve some problems, they will create others. Today if you are one of the tens of millions of victims affected by identity theft, it is possible to get a new credit card or even Social Security number. If your Facebook or bank account is hacked, you can reset your password. But when your fingerprints are stolen, there is no reset. They are permanent identification markers and, once snagged by hackers, are out of your control forever. When your gym, mobile phone company, and doctor all have your biometric details and those systems become hacked—as they undoubtedly will—remediation of the problem will prove much more difficult, if not impossible. If the future of identity is all about biometrics, then the future of identity theft will involve stealing and compromising biometrics, and thieves and scammers are already hard at work circumventing these systems.

Fingers Crossed (and Hacked)

If someone hacks your password, you can change it—as many times as you want. You can't change your fingerprints. You have only ten of them. And you leave them on everything you touch.
SENATOR AL FRANKEN

Apple's decision to launch its flagship iPhone 5s in late 2013 with fingerprint authentication was a seminal moment in biometric identification. Known as Touch ID, the embedded fingerprint sensor could be used to unlock the phone as well as make online purchases. Beginning with iOS 8, Apple has made the technology available to third-party vendors, allowing you to use your finger in lieu of a log-in to many other services and apps. The potential convenience of using a swipe of your finger to instantly authenticate yourself and securely access myriad other online services is appealing. Similarly, Samsung launched a fingerprint scanner with its top-of-the-line Galaxy S5 phone, and just like the iPhone it was hacked. Samsung's biometric scanner allowed mobile phone users to use their fingerprints to authenticate services such as a PayPal account stored on the device—thus a hacked fingerprint could open a biometric wallet to an unauthorized money transfer to the accounts of Crime, Inc.

Fingerprint sensors have dropped significantly in cost over the past ten years, and some low-end models can be purchased for about $10. The decline in prices means more manufacturers are embedding these security technologies in a wide variety of devices, including laptop computers. Samsung, Dell, Lenovo, Sony, Acer, and ASUS have all embedded fingerprint readers into their laptops and encouraged consumers to use fingerprint biometrics to lock their Windows machines and even encrypt their hard drives. Great in theory, but the implementation was poor, and hackers were able to see the digital representations of the fingerprints in plain, unencrypted text, making them easy to crack. Crime U has dozens of online videos showing how to hack fingerprint scanners, and Crime, Inc. figured out long ago how to hack fingers—by hacking them off. Gangs in Malaysia, for example, have defeated the fingerprint-recognition ignition systems in Mercedes S-Class vehicles by cutting off the fingers of the luxury cars' owners with machetes. Though stunts like this have been commonplace on TV shows like *24*, the days of cutting off an adversary's finger to get into a secret building or log in to a computer may soon no longer be necessary. Tsutomu Matsumoto, a security researcher at Yokohama National University, has devised a method allowing him to "take a photograph of a latent fingerprint (on a wineglass, for example)" and recreate it in molded gelatin. The technique is good enough to fool biometric scanners 80 percent of the time. Other hackers have used everyday child's Play-Doh to create a fingerprint mold good enough to fool 90 percent of fingerprint readers. As biometric access controls become more prevalent, so too will the reasons to defeat them.

Though government and business are trying to persuade the public of the superior safety and security offered by biometrics, many remain unconvinced, citing a wide array of privacy and vulnerability concerns. In Germany in 2008, a public debate erupted over the issue when the country's chief cop and interior minister, Wolfgang Schäuble, began strongly advocating for the greater use of fingerprint biometrics. In response, our friends at the Chaos Computer Club were able to lift the minister's fingerprint off a water glass that he had left behind after delivering a public speech at a local university. The hackers successfully copied the print and reproduced it in molded plastic—four thousand times. The replica prints were distributed as a special insert in the club's hacker magazine along with an article encouraging readers to use the print to impersonate the minister, opening the door to planting his fingerprints at crime scenes.

Proponents of biometric security argue it is inherently more secure because nobody can steal your fingerprints (incorrect, as noted above) and because fingerprints are an immutable physical attribute that can't be altered by criminals. Turns out, that's not true either, as the twenty-seven-year-old Chinese national Lin Ring proved in 2009. Lin paid doctors in China $14,600 to change her fingerprints so that she could bypass the biometric sensors used in Japan's airports by immigration authorities. Lin had been deported previously and was eager to return to Tokyo, something that would not have been possible if she provided her actual fingerprints upon arrival at Narita International Airport. In order to sneak back in, she paid Chinese surgeons to swap the fingerprints from her right and left hands, having her finger pads regrafted onto the opposite hands. The ploy worked, and she was successfully admitted. It was only weeks later, when she attempted to marry a fifty-five-year-old Japanese man, that authorities noticed the odd scarring on her fingertips. Japanese police report that doctors in China have created a thriving business in biometric surgery and that Lin was the ninth person they had arrested that year for surgically medicated biometric fraud.

Needless to say, such draconian measures would not be necessary if hackers could merely intercept the fingerprint data from the IoT-enabled biometric scanner as it was sent to the computer server for processing—something the security researcher Matt Lewis has already demonstrated at the Black Hat hacker conference in Europe. Lewis created the first-ever Biologger, the equivalent of a malware keystroke logger, which, rather than capturing all the keystrokes somebody innocently typed on her computer, could effectively steal all the fingerprint scans processed on an infected scanner. Lewis demonstrated that his biologging device allowed him to analyze and reuse the data he had captured to undermine biometric systems, granting access to supposedly "secure buildings." While it is tempting to believe that biometric authentication is inherently more impenetrable than legacy password systems, the assumption only holds true if the new systems are actually implemented in a more secure fashion. Otherwise, it's just old wine in a new bottle.

Your Password? It's Written All Over Your Face

In the sci-fi film *Minority Report*, Tom Cruise plays a Washington, D.C., police officer in the year 2054. In one scene, as Cruise's character, John Anderton, strolls through the local shopping mall, his face is recognized by the interactive billboards that greet the detective by name, serving up ads based on his prior purchase history. It seems 2054 has arrived sooner than expected. For just as fingerprints can uniquely identify an individual, so too can face prints, biometric scans of the features on your face, such as the distances between your eyes, nose, ears, and lips. Not only can these biometric characteristics reveal your personal identity, but they also allow others to profile you by gender, age, race, and ethnicity. All these data are like manna from heaven for marketers eager to re-create Mr. Anderton's targeted-advertising experience.

Back in today's world, billboards in Japan are already peering back at passersby, comparing their facial features in real time with an NEC database of more than ten thousand pre-identified patterns to accurately sort individuals into various consumer profile categories and change the ad messages displayed in real time based on demographic assessments. Beyond advertising, there are numerous other uses for facial-recognition technology. FaceFirst, a California biometrics firm, allows retailers to scan the faces of all customers in their stores to identify known shoplifters. If one is detected, the software immediately sends e-mails and text messages to all store personnel with a photograph of the suspected thief so employees can take "appropriate action." A similar system employed by Hilton Hotels uses facial recognition to scan the faces of all guests, allowing employees to greet visitors by name, especially VIP Gold card members.

It's not just advertisers who are getting access to facial-recognition data, so too are other consumers. Many of us might have noticed a security camera when walking into our neighborhood bar and innocently presumed the device was there in case the place got robbed. But when an old-school camera is linked to the IoT and big-data analytics, a new powerful smart sensor is born. In 2012, a company in Austin, Texas, partnered with local pubs and nightspots to take those "dumb" video feeds and do real-time facial analytics on all customers in their bars. The result is an app called SceneTap, which enables those looking for a good time in Austin to pull up live statistics on each establishment and see which nightspots are full, gender mix, and how old or young the crowd is. For example, the app's dashboard may show that the Main Street Bar & Grill is 47 percent full; 68 percent women with a mean age of twenty-nine and 32 percent men, average twenty-six. The very useful app takes the guesswork out of barhopping and allows drunk frat boys to successfully avoid "sausage fests," selecting only those bars with the greatest number of young women in them. Might future in-app purchases allow users to obtain further demographic data such as a bar patron's height, weight, and ethnicity? They might, as there are no laws or regulations in the United States protecting Americans from invasive biometric technologies or governing the limitations of their use.

Facial-recognition technologies have improved their match rates significantly and can now approach 98 percent accuracy, a 20 percent improvement from 2004 to 2014. All the major Internet companies, including Apple and Google, have made substantial investments in facial biometrics, but none known to be quite as large as that of Facebook, which acquired the Israeli biometrics start-up Face.com in 2012 for nearly $100 million. Facebook has long been performing facial recognition on every photograph you have ever uploaded (something you agreed to in its ninety-three-hundred-word ToS). The acquisition of Face.com allowed Facebook to greatly improve its "Tag Suggestions" feature, identifying all the people in the pictures you post using biometric algorithms and by encouraging you to tag your friends, thereby confirming their biometric identities for Zuckerberg. Facebook's automatic facial-recognition technologies have not been without controversy given the obvious privacy implications, and regulators across the EU have banned the feature. In the meantime, back in the United States, there is no legislation whatsoever that prohibits running facial-recognition software against Facebook's inventory of product, which as noted previously is you. More than a quarter of a trillion photographs have been uploaded since Facebook was founded, which means Facebook—not India's Aadhaar program—is the single largest repository of biometric data on earth, vastly exceeding that held by any government in the world.

You can expect increasing pressure from Wall Street to monetize biometric data, and there will be no shortage of potential customers, including the government. In his series of revelations, the NSA leaker Edward Snowden alleged that his agency had directly tapped into the servers of nine of the largest Internet companies, including Facebook, potentially giving the intelligence community access to the company's biometric gold mine. In a separate disclosure, Snowden divulged that the NSA was already sucking down millions of additional photographs posted online daily and was capable of processing at least fifty-five thousand images a day of "facial recognition quality." What might police and government security agencies do with this biometric data? In democratic societies, the hope is that it might be used to capture violent criminals and terrorists. But once such a biometric dragnet is built, its intended uses are controlled by those in power. In a classified world with little oversight, abuses are bound to occur, and in the hands of tyrants and dictators the tools become the foundation of a Stasi-like Orwellian dystopia.

Police forces in the U.K. have become among the first to implement a widespread automated facial-recognition program using NEC's Neo-Face technology to match faces from any crime scene photograph or video against a database of images. When facial recognition is combined with the highest density of CCTV cameras of any country in the world, the constantly recording body-worn cameras on police officers, and the *CSI*-style smart-phone apps capable of performing both facial and fingerprint recognition in the field, it seems as if the days of *Minority Report* criminal tracking have already arrived. So how far has facial-recognition technology progressed? Far enough to match your face to your Facebook profile within sixty seconds as you walk down the street and to your Social Security number sixty seconds later. The program that makes this possible is called PittPatt and began as a research project at Carnegie Mellon University in the wake of 9/11 with millions of dollars in funding provided by DARPA.

As more police forces use CCTV to monitor groups of individuals walking on downtown streets, in a football stadium, or at the airport, software such as PittPatt can run in the background in real time to identify every face as it passes, neatly placing a cartoonlike bubble above each individual's head with a Web link to more information. A simple click on the data bubble can show a person's Facebook profile, Social Security number, credit history, and past photographs of him posted online, whether it's a picture of the family trip to Disneyland, him holding a martini at the office Christmas party, or his dating profile on Match.com. While some might support law enforcement having access to these advanced image-recognition technologies for the purposes of public safety, they might feel different if such potent surveillance capabilities were in the hands of the private sector. Too late.

In mid-2011, Google acquired PittPatt, opening the door for the search giant to implement the formidable facial-recognition technology across the suite of its products, including YouTube, Picasa, Google+, and Android. Perhaps the most obvious candidate that could benefit from embedded facial-recognition technology is Google Glass. Using the tool, it would be possible to immediately identify that hot girl or guy at the party, and you'd never have to worry about forgetting whatshisname from the accounting department ever again. Concerned about a potential backlash, Google has banned the use of facial-recognition Google Glass apps for the time being, but with the acquisition of PittPatt the technical capability fully exists. Of course hackers have jailbroken their Glass devices and created a number of facial-recognition apps, including the popular NameTag app.

NameTag allows users to scan the faces of those before them and compare them with millions of publicly available online records, returning the person's name and social media profiles, including those on Facebook, Twitter, and Instagram, and other relevant identifying details. Such facial-recognition apps are not unique to Google Glass and can just as readily be used with the camera on your smart phone. Just like in *Minority Report*, we are all now living in the age of facial recognition. As a consequence, nobody is just a face in the crowd anymore. Indeed, your face is now an open book, one that can be read at a glance by anybody else—including the government.

The FBI's billion-dollar Next Generation Identification (NGI) system has as many as 52 million searchable facial images, including 4.3 million images of noncriminal background check applicants. NGI also contains 100 million individual fingerprint records, as well as millions of palm prints, DNA samples, and iris scans. Not only can the system scan mug shots for a match, but NGI can also track suspects by picking out their faces in a crowd from standard security cameras or compare them with photographs publicly uploaded on the Internet. Of course no biometric technology is foolproof, and the issue of false positives, being identified with a criminal based on a biometric match, when in fact no such match existed, can have profound consequences for the innocent, as we've seen previously with the proliferation of terrorism watch and no-fly lists.

Though facial recognition may sound like an anticrime and security panacea, it is not without its problems. Just as fingerprint sensors can be hacked, so too can face-printing systems increasingly be used to unlock your phone or computer or to gain access to your office. All it takes to defeat some systems, such as those on Lenovo laptops or smart-phone password apps such as FastAccess Anywhere, is to hold up a photograph of the person you wish to impersonate. The same technique has also worked with iris scanning, allowing hackers to reverse engineer the biometric information stored in a secure database and use it to print a photographic iris good enough to fool most commercial eye scanners.

Another challenge with facial-recognition algorithms is that even the very best systems are "only" approaching 98–99 percent accuracy. Though the error rate seems small, those errors can add up. Consider a facial-recognition system linked to a terrorist watch list installed at Chicago's O'Hare International Airport. With 50 million passengers transiting annually, a 1 percent false-positive rate means 500,000 travelers a year (more than 1,300 a day) could be erroneously detained or arrested because of computer error. The problem will almost certainly be compounded by human errors in data entry, as we have seen with the existing security watch lists, leading cameras to correctly match an individual's face to a mistyped name in a wanted-persons database.

The consequences of incorrect biometric identification could even prove fatal. The U.S. Department of Defense has begun implementing biometric targeting and recognition capabilities in its drone fleet. The defense contractor Progeny Systems Corporation, working in conjunction with the army, has already developed a drone-mounted "Long Range, Non-cooperative, Biometric Tagging, Tracking, and Location" system that allows an unmanned aerial vehicle (UAV) to positively identify a human target using biometrics prior to detonating ordnance on the target's head. In such a case, a false-positive biometric identification would be disastrous. The future of warfare is autonomous, with drones hunting, identifying, and killing an enemy based on calculations made by software, not on decisions made by human beings.

Given the growing ubiquity of cameras and facial-recognition software, we can fully expect criminals to adopt these tools and use them to their full advantage. Pedophiles could use biometrics to identify that kid at the park playground they like. For terrorists, such as the Mumbai attackers from Lashkar-e-Taiba, a facial-recognition app on their mobile phones would have allowed them to identify the bank president K. R. Ramamoorthy without having to play guessing games with their terrorist command center back in Pakistan as to his identity.

Even Crime, Inc. has begun to explore facial-recognition technologies, according to the commissioner of the Australian Federal Police (AFP). At the graduation ceremony of hundreds of new police recruits from the AFP in 2011, officials spotted a man who stood out from the crowd of families watching their loved ones receive their police badges. The individual had a professional camera with a telephoto lens and appeared to be snapping face pics of all the graduates. Upon his detention and questioning, officials learned the shutterbug was a member of an organized crime outlaw motorcycle gang. He was apparently working on behalf of Crime, Inc. to build a photographic facial-recognition database so that fellow gangsters would be able to identify any officer who attempted an undercover investigation against their organization in the future. Biometric tools will have profound implications not just for undercover cops but for witness relocation programs as well. Anybody who has had a prior life that he wishes to conceal for personal or professional reasons may find it impossible moving forward, and it is not just your physical attributes that may betray you—so might your imperceptible behaviors.

On Your Best Behavior

A lot of jobs today are being automated; what happens when you extend
that concept to very important areas of society like law enforcement?
What happens if you start controlling the behavior of criminals or people in
general with software-running machines? Those questions, they look like
they're sci-fi but they're not.

JOSÉ PADILHA, BRAZILIAN FILM DIRECTOR

When most people think of biometrics, they commonly focus on the measurement of anatomical traits such as our fingers, faces, hands, or eyes. But there is another category of biometrics known as behavioral biometrics, or behaviometrics, which measures the ways in which we and our bodies individually perform or behave, traits that can be just as revealing as human fingerprints. Our keyboard typing rhythms, voices, gait and walking patterns, brain waves, and heartbeats can be quantified in ways that provide unique signatures that singly identify us. Just as anatomical biometrics are being used more frequently for security, identification, and access control, so too will the growing field of behaviometrics; in fact, it is already happening.

Voice biometrics are already being used by companies and call centers around the world to uniquely voice print customers. That recording you hear when on hold telling you that "your call may be recorded for quality assurance purposes" fails to disclose the fact that one of the ways companies are measuring call satisfaction is by the tone, tenor, and vocabulary you use during your call. Moreover, in an effort to fight fraud, companies are building vast recorded-voice databases of consumers, generating unique voice prints that can be used in future calls to ensure the person on the line matches the original biometric vocal print taken. If the voices don't match, callers are asked further verification questions in a process that is completely non-transparent to the general public.

DARPA is working to develop "active authentication" techniques focused on a user's cognitive processes, personal habits, and patterns we all have for doing things, which in combination can uniquely identify us. One such field of behaviometrics is known as keystroke dynamics—measuring the variances with which we each type individual characters on a keyboard. Minute differences between when each key is hit, in what sequence, with what force, and even how we cut and paste can serve as our online fingerprints to the world. Companies such as the online education platform Coursera use keystroke recognition to ensure the same student "attends" each virtual class before issuing a certificate of completion.

Watchful Software's TypeWATCH product is designed to run on networks in the background to constantly monitor a user's typing rhythms to uncover and block attempts at unauthorized access. Other firms such as the Sweden-based Behaviometrics AB have created tools that note how each mobile phone or tablet user holds his phone, at what angle, the way he types on the virtual keyboard, and even how he swipes and pinches the screen, revealing minute millisecond pauses between various actions. Any variation from an established "cognitive footprint" will set off alarm bells at a bank and block access to the account, one of the reasons Denmark's largest bank, Danske Bank, adopted the technology. Banks believe biometric tools such as this may be able to cut fraud rates by as much as 20 percent, and thus you can expect the ToS at your online retailer or financial institution to be amended in the near future, requiring your consent to such detailed monitoring in order to use your bank's iPhone app.

New forms of behaviometrics are emerging all the time. The Nymi wristband uses a voltmeter to read the beating of your heart and uses its unique electro-cardiac rhythm to unlock your computer, smart phone, car, and house. Scientists at the U.K.'s National Physical Laboratory have developed the walking-gait-recognition system that can be used in conjunction with CCTV monitors to uniquely identify an individual based on the way she walks. There is, however, an even easier way to identify you by the way you walk—using the accelerometer in the smart phone that you carry with you twenty-four hours a day, thereby sharing this information with your mobile phone company, handset manufacturer, and app developers.

If these technologies seem intrusive now, the reality is that they may well be even more so in the future. Already Motorola has partnered with the firm MC10 to "extend human capabilities through virtually invisible wearable electronic RFID tattoos" that can be used for password authentication. Proteus Digital Health has created a pill that you can swallow and is powered by the acid in your stomach to create a unique eighteen-bit signal in your body to turn your entire person into an authentication token. Though many of these biometric security products offer great promise, hackers and Crime, Inc. will not just give up their self-enriching efforts and go home defeated. However, rather than merely hacking your computer, Crime, Inc. will be hacking the Internet of You.

Security vulnerabilities aside, biometric and behaviometric technologies bring with them a host of public policy and privacy issues with which society has only just begun to struggle. What does it mean to have any individual typing on a keyboard anywhere in the world be remotely identified based solely on the way he or she bangs away at the keys? Great for locating the world's most wanted hacker, but bad news for the leader of the opposition movement during the Arab Spring. The challenge with biometric surveillance, whether conducted by advertisers at the local mall or by the state's security apparatus, is that it affects our behavior. When we know we're being watched, we behave differently, more likely to conform and thus more easily be controlled. Whether at the hands of an out-of-control government or a monopolistic megacorporation, self-censoring behavioral modifications brought about by omnipresent surveillance can rapidly lead to a dystopian future for all. It is not just our physical selves that can be subjected to such persistent observation but our virtual selves as well.

Augmenting Reality

As the Internet of Things advances, the very notion of a clear dividing line between reality and virtual reality becomes blurred, sometimes in creative ways.

GEOFF MULGAN, U.K. NATIONAL ENDOWMENT FOR SCIENCE, TECHNOLOGY, AND THE ARTS

In the movies, Tony Stark amazes us with the capabilities of his all-powerful Iron Man suit, which, among its many features, benefits greatly from a plethora of real-time augmented-reality information streaming before his eyes from the suit's head-mounted display. The technology in the film is based solidly on reality. Augmented reality (AR) provides a live direct view of a physical, real-world environment through a computer screen, such as the one on your mobile phone or embedded in Google Glass, and overlays additional digital information such as images, sound, video, or GPS data on the real-world environment. Some of the earliest AR applications were those used in heads-up displays for jet fighter pilots, allowing them to see critical system information on their cockpit window screens without looking down at their instruments during a dogfight. Today that technology has come into civilian life, with car manufacturers such as Mercedes-Benz and Range Rover projecting vehicle speed and turn-by-turn directions directly onto a car's windshield. Unlike virtual reality, which can supplant the real for the virtual or even create an entirely fictional world, augmented reality enhances one's perception of reality by laying useful data on top of the things we see in the real world.

AR can be used with any screen that has embedded sensors and cameras, whether it be your mobile phone, tablet, eyeglasses, or even contact lenses. It is expected that 2.5 billion AR apps will be downloaded and installed on our devices annually by 2017. The benefits of AR will be astounding, and major companies are already showing us the possibilities. In a Google ad, a user wearing Glass is about to descend into a subway in Manhattan, only to receive a pop-up alert with the MTA's 6 train logo that train service has been suspended—data that are projected into his field of view on his Glass screen. Tools like this will allow travelers around the world to dump their bulky Fodor's travel guides and use an AR app to show them around the city.

As you walk down the street, these apps can overlay data so that you can see the Yelp reviews of restaurants when you pass and *Wikipedia* entries on statues and historical buildings in your field of view. Of course, AR will bombard us with advertisements as we walk about town, with our Google Glass recognizing all the physical objects around us and placing ads on top of them. Ikea even incorporated AR into its 2013 catalog, allowing users to snap photographs of couches or any other pieces of furniture with their smart phones and then place them in their own homes (with the correct dimensions) to see how they might look before actually making purchases. AR will be the way we interact with the world around us and the IoT in particular, allowing us to query physical objects to better understand their history, intended use, and context. It will connect the online and off-line worlds and will change every aspect of life and work.

AR will also bring with it a host of security and privacy questions that need to be addressed. A future malicious app might overlay an incorrect speed limit on a highway road sign or place a fake sign where none actually exists on our car-mounted AR windshield display. Worse, it could show a traffic lane as being clear, when in fact it is not, causing an accident when you change lanes into another car. As noted previously, the more we disconnect from reality and accept the virtual in place of the real, the more we open ourselves up to manipulation via "in screen we trust"–type attacks.

In addition, just as Crime, Inc. has created crimeware, such as Blackshades, to automate criminality, we can expect it to release any number of AR crimeware apps in the future. For example, using an iPhone or Google Glass, hackers might be able to visually interrogate all the IoT devices in your office or home and see information displayed on their screens about which devices had known vulnerabilities or perhaps even see your poorly secured password, making hacking the IoT even easier than it is today. Reality-altering technologies such as AR will open the door further to even more immersive virtual environments, such as virtual reality systems, which also can be subverted and abused in powerful ways.

The Rise of *Homo virtualis*

Reality is merely an illusion, albeit a very persistent one.
ALBERT EINSTEIN

Increasingly, as we live our lives through avatars—in video games, online worlds, and social networking sites—our online personas are standing in for us in social situations, commercial transactions, and even sexual encounters. They are there representing us online 24/7, compressing time and space, to interact on our behalf with the rest of the world even as we sleep. The renowned game designer Jane McGonigal has noted that "the average young person racks-up 10,000 hours of gaming by the age of 21," the vast majority of which is in the persona of an avatar or game character. As they do, we witness the rise of *Homo virtualis*, perhaps the next evolution of *Homo sapiens*, a species that is pulled away from the constraints of our natural physical world in favor of the immediacy and perceived unlimited potential of the virtual.

Virtual reality (VR) uses computers to create simulated environments, worlds real and imagined, in which we can insert a representative physical presence of ourselves and our senses. Even the sense of touch can be re-created as haptic or tactile feedback technologies apply "force, vibration or motions" to the user. As Mark Zuckerberg commented upon Facebook's $2 billion acquisition of Oculus Rift, a highly responsive virtual reality head-mounted display, in early 2014, "Strategically we want to start building the next major computing platform that will come after mobile." Tools like the Oculus Rift headset can transport us in an instant to immersively experience a beautiful Tuscan villa, a courtside seat at an NBA game, or an imagined but realistic battle with Klingons and Romulans.

One of the earliest virtual worlds was Second Life, which was launched by Philip Rosedale of Linden Lab in 2003 and allowed users to represent themselves in the form of highly customized avatars. In Second Life, it was possible to make friends, shop, learn, and even attend a U2 rock concert performed by the actual avatars of the band's members. Another common form of virtual worlds are known as MMORPGs (massive multiplayer online role-playing games). MMORPGs are video games that "allow thousands of players to simultaneously enter the virtual world and interact with one another. Players can run their own cities and countries, stand up armies" to engage in battle, and go on a "variety of quests with their own avatars." The largest MMORPG is Blizzard Entertainment's *World of Warcraft*, which has drawn up to twelve million subscribers, each paying monthly fees to inhabit a virtual world. Yet for as intricate and multilayered as these virtual spaces are today, Rosedale points to a near future wherein hardware and software advances, such as the High Fidelity platform, will deliver us the next-generation virtual world—one potentially as large and as complex as the real world is today.

In order to understand virtual worlds, one needs to comprehend the mind-set and psychology of those who *inhabit* virtual spaces. Many genuinely view their "second lives" as "first lives," and 20 percent of MMORPG players regard the game world as their "real" place of residence. To them, earth is nothing more than "meatspace," a secondary home in which the meat of their physical bodies can eat and sleep, while most of their interpersonal, commercial, and sexual relations take place online. While the overwhelming majority of VR users do not feel this way, the feelings may become commonplace as we spend more time in highly immersive and pleasurable virtual environments.

But there is a downside to this technophoria, as evidenced by a South Korean couple who spent so much time at a local cybercafe obsessively caring for their virtual daughter in the online world known as Prius that they failed to return home for days to feed their actual three-month-old, resulting in the real-world infant's death. While this case is extreme, dozens of such incidents have been reported over the years, and even more may be yet to come.

The line between man and machine, online and off-line, is becoming increasingly blurred. Anybody who has ever played a hyperrealistic first-person shooter video game such as *Doom* or *Call of Duty* will know that the virtual experience definitely leads to physiological changes, including a quickened heart rate and sweaty palms in the heat of battle. Because avatars are virtual representations of ourselves and because people are spending thousands of hours in the personas of their avatars, our real-world psyches are becoming increasingly enmeshed with our virtual representations. In effect, what happens to our avatars leaves a mark on us, and within virtual worlds nearly any crime that can take place in our physical space can be replicated. Virtual worlds have their own currencies, such as Linden dollars or *World of Warcraft* gold, which like Bitcoin can be converted into "real money," and have become a favorite target of Crime, Inc., which launches 3.4 million malware attacks daily in pursuit of online gaming accounts.

As strange as it may sound, crimes by and against avatars are becoming more common, and in virtual worlds you can be subjected to everything from cyber bullying to identity theft, with police in Japan having arrested a man for a series of avatar muggings. Even "sexual assaults" have been reported in virtual worlds, as was the case in 2007 in a matter investigated by the Belgian Federal Police. The incident involved a woman whose avatar was infected with malware by a man she met in Second Life. The computer virus allowed the aggressor to take control of the female avatar and violently and graphically sexually assault it. Ultimately, the case was investigated as an incident of "unauthorized access to a computer system," and while some may find it easy to dismiss a case of "virtual rape" out of hand, doing so in the future will prove more difficult given the ever-improving immersiveness of virtual space and the very likely real trauma such incidents may cause moving forward. These incidents might be further exacerbated by the growing number of corporeal haptic feedback devices that are increasingly being connected to online worlds, allowing partners to use the science of *teledildonics* to remotely stimulate each other over the Net. Like any other IoT-enabled object, these will be subject to hacking with unpredictable consequences.

The rise of VR may have not only criminal implications but terrorism and national security ones as well. A 2008 report by the U.S. director of national intelligence suggests that terrorists may well be using virtual spaces for covert communications, to spread propaganda, train members, launder virtual currency, and even recruit new followers. According to an eighty-two-page document leaked by Edward Snowden and published on the *New York Times* Web site, both the NSA and the U.K.'s GCHQ have been spying on gamers in virtual worlds, including *World of Warcraft*, Second Life, and various games hosted by Microsoft's Xbox platform. The spies have created undercover avatars "to snoop and to try to recruit informers, while also collecting data" and performing mass interception of communications between players, including the forty-eight million individuals using the Xbox Live console network. Concerns about terrorist organizations' using gaming platforms for fund-raising and recruitment are not without foundation. Hezbollah has produced its own first-person shooter video game titled *Special Force 2*, which is used as a radicalization medium for young jihadis. In the game, players earn points by launching Katyusha rockets at Israeli towns, and they win by successfully becoming "suicide martyrs."

As virtual reality continues to improve exponentially, the distinctions between our virtual and our physical selves will continue to erode as well. The result will be a world in which it will be increasingly difficult to tell where the physical you ends and the virtual you begins. This is the Internet of You, and it is entirely hackable. Throughout this chapter, we have seen numerous examples of how the technology around us is becoming the technology on us and in us. Wearables, embeddables, ingestibles, and implantables mean that to one extent or another we have all joined the cyborg nation—opening up our physical bodies to cyber attacks for the first time. Adding to these challenges is the fact that our anatomy and physiology can now be measured at a distance, with or without our knowledge, via biometrics and behaviometrics that can profile and uniquely identify us. As a result, digital bread crumbs have come to physical space, while we, our bodies and ourselves, are integrating with cyberspace as never before. But as we shall see, the converse is also true. Computers and other stationary techno-objects will soon leave the virtual world behind and join us in moving about real space. Machines are finally coming to life. After a long era of hibernation, they are ready to descend upon our physical world, and when they do, they will bring with them a tidal wave of threats for which we are wholly unprepared.

CHAPTER 15

Rise of the Machines:
When Cyber Crime Goes 3-D

It is only when they go wrong that machines remind you how powerful they are.
CLIVE JAMES

R ezwan Ferdaus was raised in Ashland, Massachusetts, an upscale town in the suburbs of Boston. His parents had emigrated from Bangladesh in search of a better life in America and had high hopes for their son, whom they had raised to respect Allah and their Muslim faith. After graduating from high school, Ferdaus earned a bachelor's degree in physics from Northeastern University in 2008. Unable to find significant work in his field, he moved back in with his parents. Like many his age, he spent a lot of time online. He began to frequent radical Islamist Web sites and watched numerous al-Qaeda videos calling on young Muslims to rise up in jihad against the great Satan—America.

As time went on, the twenty-five-year-old grew increasingly disillusioned with the United States and decided it was time for action. He told a man at his local mosque he wanted to join al-Qaeda and was eventually introduced to several "brothers" who could help him on his quest. In 2010, Ferdaus began to plan his own violent attack against the infidels he saw all around him in America. While the thought was not particularly original for a terrorist, his plot to use killer robots was. Ferdaus purchased three unmanned aerial vehicles (UAVs) that he intended to load with C-4 explosives and fly into the U.S. Capitol and the Pentagon.

The UAVs were actually remote-controlled aircraft, perfect model replicas of the navy's F-4 Phantom fighters, built to precise one-tenth scale, and available on the Internet from drone hobbyist Web sites. The planes were capable of carrying up to a forty-pound payload and could travel at speeds of 160 miles per hour powered by the onboard jet engines. They could be directed remotely by an operator on the ground using a handheld radio transmitter or, as Ferdaus envisioned, flown autonomously along a predetermined flight path using onboard GPS sensors that would crash each UAV precisely into its intended target. The plan had other advantages as well: the robotic aircraft could take off and land almost anywhere, and the small low-flying planes would be nearly impossible to detect on radar. Ferdaus shared his plans with his al-Qaeda affiliates, who offered enthusiastic support and funding for his efforts.

Using a fake name and cover story, Ferdaus ordered three of the model airplanes costing $3,000 each from different online sources. He paid for them with a PayPal account he had created under an alias and had the drones shipped to a storage facility in nearby Framingham that he had rented, paying cash. There Ferdaus began covertly assembling the devices before moving on to the second phase of his project, the acquisition of the explosives. For this goal, his new friends in al-Qaeda proved extremely helpful. They provided him with twenty-five pounds of C-4 explosives, numerous hand grenades, and six fully automatic AK-47 assault rifles, which he hid in his ten-by-ten storage locker.

Ferdaus traveled to Washington, D.C., to carefully scope out his targets, taking photographs and plotting attack points on a map. He decided to launch his drones from East Potomac Park, conveniently situated nearly equidistant from his two targets. First to be hit would be the Pentagon, with two drones approaching from opposite sides of the building, both aimed at the fourth floor. No need for Ferdaus to be on board this flight, though. He had constructed a high-torque robotic servo actuator for his self-guided UAVs, a device that would simultaneously pull the pins on the sixteen hand grenades that he would place on board each remotely controlled aircraft. The robotic assistant on the UAV would be programmed to act moments prior to impact and mechanically pull the pins for maximum effect.

Ferdaus's plan called for a ground assault in addition to his drone attack that would use two teams of three people armed with AK-47s to shoot innocents as they frantically fled the explosions rocking their building. The next phase of the plot called for another robotic remotely controlled precision-guided miniature jet aircraft laden with C-4 explosives to fly into and destroy the dome of the U.S. Capitol, blowing it to smithereens.

Ferdaus returned to Boston and wrote up an incredibly detailed plan of his mission, which included aircraft specifications, software protocols, hardware configurations, maps, pictures, diagrams, payload limitations, and budget requirements. He provided the document on a USB thumb drive to his al-Qaeda handlers, who expressed great admiration for his proposal. They asked him how he had learned so much about robotics and drones, and he replied, "UAV technology is quite simple. Sure you need to have a certain aptitude, but I've been doing this type of stuff since I was a little kid." All agreed the plot would move forward, and Ferdaus returned to Framingham to check on his cache of weapons and explosives. As he unlocked the storage unit, he was rushed by a bevy of special agents from the Federal Bureau of Investigation, who prevented the first-ever terrorist drone attack on U.S. soil.

As it turned out, the fellow Muslim whom Ferdaus had approached at his local mosque for an introduction to al-Qaeda was an upstanding citizen who contacted police upon hearing the request. The "brothers" he introduced to Ferdaus were in fact undercover FBI agents. In July 2012, Ferdaus pleaded guilty to charges of attempted destruction of a federal building with explosives and material support to a foreign terrorist organization and was sentenced to seventeen years in prison. Though we have seen the military employ drones to great effect around the world, criminals and terrorists are perfectly capable of building and using these devices as well. Multiple inbound UAVs launched from East Potomac Park and traveling at 160 miles per hour under the radar would strike their targets in mere minutes, leaving no time for evacuation or response. As the use of drones and other robotic technologies becomes more commonplace, we can expect them to be leveraged by all members of society, for both good and ill. While 9/11 1.0 was about human beings' seizing aircraft and flying them into occupied buildings for terrorist effect, 9/11 2.0 makes it possible to disintermediate the humans and use robots in their stead.

We, Robot

In the future, I'm sure there will be a lot more robots in every aspect of life. If you told people in 1985 that in 25 years they would have computers in their kitchen, it would have made no sense to them.
RODNEY BROOKS

Throughout the history of film and television, we've seen robots presented in a variety of lights. Some were lovable and helpful such as WALL-E, Johnny Number 5 of *Short Circuit*, and C-3PO and R2-D2 from *Star Wars*. Other robots were dangerous and out to destroy mankind, such as Gort from *The Day the Earth Stood Still* and the T-800s from *The Terminator*. Thanks to advances in Moore's law, robots are leaving the silver screen and joining reality. Exponential progress in silicon chips, digital sensors, cloud computing, and high-bandwidth communications means that robots, just like computers and mobile phones before them, will soon become omnipresent in our lives.

Robots are increasingly outfitted with advanced features such as high-definition cameras, touch sensors, and laser range finders, all united and run by computer brains. Robots move through their actuators, electrical motors connected to gears that power and drive their wheels, legs, and arms, just as muscles move human beings. Vast improvements in robotics have been driven in no small part by the smart-phone revolution, because robots depend on many of the same computer chips, batteries, and sensors as does the increasingly powerful mobile phone in your pocket.

Until now, robots have largely been used in manufacturing to handle repetitive tasks that are "dangerous, dirty, or dull"—such as those on an automotive assembly line. Robots today are becoming more sophisticated, endowed with advanced dexterity, senses, and intelligence, allowing them to handle significantly more complex tasks. They can walk, talk, dance, read our facial expressions, and respond to our verbal commands. There are robots that care for the elderly, detonate bombs, drive cars, work on the International Space Station (ISS), and kill terrorists around the world. In the years to come, they will increasingly fight fires, deliver our packages, respond to crimes, perform surgery, assist in disaster recovery, and provide companionship. The number of robotics start-up companies is exploding, and some estimate industrial robots alone could be a $37 billion market by 2018.

Robots are computers, automated systems that can reach beyond the purely two-dimensional digital plane of their ancestors in order to touch, influence, and interact with the corporeal world that surrounds them. Most can be remotely controlled over the Internet and via smart-phone apps, leading to legions of robots joining the Internet of Things. The implications are momentous. As Joi Ito, director of MIT's Media Lab, has observed, we are living in a period of convergence, a time "where bits from the digital realm are fusing with atoms here in the physical world."

Robots are entering our three-dimensional space—space that they will share with us. Like all objects connected to the IoT, robots are subject to hacking, though the consequences may be much more far-reaching. Throughout its short history, cyber crime has always been hidden behind computer screens—a two-dimensional problem that might affect your wallet or your bank account. No more. As a result of advances in robotics, cyber crime will finally escape its virtual confines and explode onto our physical space. And we are wholly unprepared for what is coming next.

The Military-Industrial (Robotic) Complex

For decades, industrial robots have toiled side by side with human workers in warehouses and on factory floors, but modern industrial robots are marvels of engineering, capable of lifting hundreds of pounds and moving objects repeatedly to within 0.006 inch's accuracy, a feat no human being could match. Initially, these machines were expensive, often costing hundreds of thousands of dollars and requiring months of highly customized computer programming before they could perform their assigned tasks. Despite the costs, no industry has benefited more from robotics than automobile manufacturers, which accounted for 40 percent of worldwide robotic sales in 2013. Robots make vehicle production faster, safer, cheaper, and more efficient, and all major manufacturers from Ford to BMW use them to automate production. In just one Hyundai factory in Alabama, five hundred robots work tirelessly welding, painting, bolting, and transporting auto parts in order to crank out more than a thousand cars a day. Not to be outdone, Amazon announced in 2014 that it employs ten thousand Kiva Systems robots that it uses to navigate its massive warehouses to fetch individual items and bring them to human employees who package them before turning them over to more robots for shipping. These robots work three shifts a day, 365 days a year, and never take a coffee break.

Industrial robots are growing exponentially cheaper, more efficient, and more user-friendly, and perhaps no other robot exemplifies this trend as much as Baxter, the cute low-cost industrial bot from Rethink Robotics. At $22,000, it is a tenth of the price of its predecessors. More impressive is the fact that it works right out of the box and can be up and running in just an hour, as opposed to the eighteen months it took to integrate the previous generations of industrial robots into a factory operation. Baxter can learn to do simple tasks, such as "pick and place" objects on an assembly line, in just five minutes. It has an adorable face on its head-mounted display screen and two highly dexterous arms, which can move in any direction required to get a task done. Baxter requires no special programming and learns by using its computer vision to watch an employee perform a task, which the bot can repeat ad infinitum. As costs drop even further, these robots will be competitively priced compared with cheap overseas labor, and many hope a rise in domestic robotics use may lead to a renaissance in American manufacturing.

Today robots are showing up everywhere from restaurants to hospitals. In more than 150 medical centers, Aethon's TUG robots can be summoned by a smart-phone app to autonomously travel throughout the corridors to deliver medicines, patient meals, and laundry, replacing work previously done by orderlies. Other medical bots, such as Intuitive Surgical's da Vinci robot, allow surgeons to operate on patients using robotic arms. Using a viewfinder screen and joystick controls, physicians can see a 3-D view inside a patient and manipulate small surgical instruments to carry out procedures ranging from hysterectomies to heart valve repairs. Absent the need to place large human hands inside the patient's body, robotic surgery can be performed in a minimally invasive fashion, with 80 percent fewer complications and significant reductions in recovery times. Over 500,000 such operations are performed annually worldwide. Using similar technology, a surgeon can remotely operate on a patient over the Internet via tele-surgery, with the first such operation having taken place back in 2001, a case in which a surgeon in New York performed a cholecystectomy across the Atlantic on a women in Strasbourg, France.

Though the gains in industrial and medical robotics have been impressive, the growth of military robotics has been astounding. In 2003, the Pentagon had fewer than 50 UAVs in its arsenal. Today the United States has the greatest number of military bots of any country, "deploying some 11,000 UAVs and 12,000 ground robots around the world." These machines are well armed and lethal and have killed thousands. In 2011, it was estimated that one in fifty troops in Afghanistan was a robot, and by 2023 there may be ten robots per human soldier in the U.S. military.

Unmanned ground vehicles (UGVs), such as iRobot's PackBot, routinely help with the detection and disposal of improvised explosive devices (IEDs). Foster-Miller's TALON is a "man-portable robot which operates on small treads" like a miniature tank. It can be outfitted with machine guns, .50-caliber rifles, grenade launchers, and antitank rockets, all while being remotely controlled via joystick. Boston Dynamics' Sand Flea weighs only eleven pounds but can jump up to thirty feet high, landing on the roof of a building or precisely leaping through an open window, capturing all it sees with its HD camera. The company has also created BigDog, a four-legged robot that can carry up to four hundred pounds of gear and weapons, easily walking over rugged terrain and obediently following its soldier master. Other UGVs like RiSE, a six-legged robo-cockroach, can climb walls, the Cheetah can run at nearly thirty miles per hour (faster than Usain Bolt), BEAR can lift and carry an injured soldier off the battlefield, and a briefcase-sized bot from iRobot can use facial recognition to identify a man in a crowd and follow him.

In the skies, pilotless aircraft, or UAVs, can perform imagery collection and communications interception and launch missiles against their targets. Remote pilots sitting halfway around the world can kill enemies (and sometimes innocents) with the click of a mouse. According to Peter Singer, a noted expert on military robotics, at least fifty-five other countries have military robotics programs. UAVs have become central to the military arsenal, and it is expected that "global spending on drones, military and civilian, could cumulatively reach $89 billion" by 2023. There are large drones, small drones, helicopter drones, handheld drones, and insect drones. Drones such as the MQ-9 Reaper cost about $12 million, one-tenth the price of an F-22 jet, with most of the same capabilities. Military officials note that drones such as the Reaper and the Predator are designed to carry out the full "kill chain" against their high-value targets—"find, fix, track, target, execute and assess."

The leviathan of the drone fleet is the Global Hawk. With a wingspan of 130 feet and weighing thirty-two thousand pounds, it can stay airborne for nearly two days at an altitude of sixty thousand feet. The sensors on the UAV fleet are equally impressive and include tools such as the ARGUS-IS, the world's highest-resolution camera, capable of taking 1.8-gigapixel photographs. The ARGUS comes equipped with a "persistent stare" capability equivalent to a hundred Predator drones, allowing it to track all ground movements across an entire medium-sized city. The images are of such high quality that the drones can generate one million terabytes of data daily, the equivalent of five thousand hours of HD footage, which record every single movement on the ground (car, bus, person, dog) and can be played back DVR-style at will.

Importantly, UAVs have long ago left the theater of war and can now be found flying domestic missions over the continental United States surveilling drug traffickers, organized criminals, and illegal border crossers. Traditional military contractors such as Northrop Grumman, Boeing, and Lockheed Martin were early entrants into the world of robotics, followed by smaller specialized firms such as Boston Dynamics and iRobot (yes, the same people who make your Roomba vacuum make the IED-disposal PackBot). But now another deeply disruptive player has entered the world of robotics: Google.

The search giant is on a robo-buying binge and purchased or acquired eight separate robotics companies in a six-month period through 2014, including companies that specialize in humanoid walking robots, robotic arms, robotics software, and computer vision. Its largest and most surprising robotics acquisition, however, was the military robotics company Boston Dynamics, the same folks who make BigDog, Cheetah, Sand Flea, RiSE, and PETMAN (a biped humanoid robot that might well be the soldier of the future). Google also bested Facebook's offer to buy Titan Aerospace, a maker of jet-sized solar-powered drones that can remain aloft for three years without landing. Why are two Web giants battling for air superiority? They claim the drones can be used to provide Internet access to parts of the world that are not yet online. Yet when one of the world's largest data and artificial intelligence companies enters the robotics realm and becomes capable of launching its own drone armies, important questions must be asked about its intentions and capabilities.

A Robot in Every Home and Office

Your living room is the final frontier for robots.
CYNTHIA BREAZEAL, MIT MEDIA LAB

In a seminal article in *Scientific American*, Bill Gates compared industrial robots to mainframe computers and predicted that miniaturization, common technical standards, and better sensors would bring a robot into every home in the coming years. There are signs that he's right. We already have domestic robots that clean our floors, water our plants, clean our BBQs, and feed our pets. iRobot has sold over ten million of its Roomba vacuums since its launch, and they are commonly available at the local Walmart. Children are enjoying growing numbers of robotics toys such as Lego's Mindstorms, WowWee's Robosapien X, and Sphero's Robotic Ball. Even the homemaker extraordinaire Martha Stewart has purchased a DJI Phantom quadcopter drone with HD camera, which she enjoys flying around her expansive 153-acre New York estate. The market for consumer and office robots is skyrocketing, growing seven times faster than demand for industrial robots.

Until now, most home robots have been created to carry out a single task such as vacuuming. But in the future, we will have multifunctional bots capable of much more, such as clearing the table after meals, loading the dishwasher, ironing our shirts, and picking up toys after the kids, all easily controlled from the familiar screen of our smart phones. While such dream home assistants have yet to materialize and may be years in coming, progress is being made. An Indiegogo campaign led by MIT's Dr. Cynthia Breazeal successfully crowdfunded a helpful and intelligent social robot named Jibo that can identify individual members of the household, snap family photographs, read e-mails, tell bedtime stories to the kids, and change its facial expressions to show emotions. Willow Garage's PR2 can already fold clothes, grab a beer from the fridge, clean up after the dog, bake cookies, and cook a complete breakfast. From Japan to Europe and the United States, there are unprecedented amounts of research-and-development dollars flowing into robotics.

Admittedly, some of these new developments sound like something out of a Philip K. Dick novel. For instance, nanny bots have already been launched in South Korea and Japan. They can play games and carry out limited conversations with speech recognition. Many use the robot's eyes to transmit live video of your children to your computer or smart phone. NEC's PaPeRo robot nanny also allows you to speak with your children directly or via text messages, which the robot can read to your child, and SoftBank's Pepper proclaims that "it can read your child's emotions and facial expressions and respond appropriately." Though robo-nannies may prove helpful to sleep-deprived, overworked parents everywhere, another area of personal robotics that is expanding even more rapidly is that of elder-care bots. Given demographic trends and aging populations in developed countries around the world, there is a dearth of caretakers to provide the emotional and physical support required for the elderly. Nowhere is this challenge as great as in Japan, where nearly 25 percent of the population is over sixty-five. To help alleviate the problem, Prime Minister Shinzō Abe's government allocated ¥2.39 billion in 2013 to assist the national development of elder-care robots. One such example is Paro, an adorably cute white baby harp seal robot meant to keep the elderly company. Paro "can recognize individual voices, track motion and remember behaviors that elicit positive responses from patients." When petted, it responds by cooing and cuddling up to any person touching it. Thousands of Paro units have been sold globally, and they have proven particularly useful with advanced dementia patients in reducing levels of violence and improving mood. Recognizing the market need for elder-care robots, iRobot (maker of vacuums and killer bots) has opened a new division specifically to serve seniors.

One of the fastest-growing types of elder-care bots are telepresence robots—machines that allow people to "move virtually through a distant building by remotely controlling a wheeled robot equipped with a camera, microphone, loudspeaker and screen displaying live video" of the person's face controlling the bot over the Internet. Robots such as the MantaroBot and the EU's GiraffPlus allow children to "beam in" from thousands of miles away and remotely drive a wheeled bot with an iPad-type face in order to interact with aging parents. Relatives can check on their elderly loved ones, eat meals with them via Skype-like video conversations, and even ensure that they have awoken and not fallen in their own homes. It's not just worried adults who are using telepresence bots to check in on their parents; increasingly, they are becoming mainstays in hospitals as well. iRobot's RP-VITA (Remote Presence Virtual + Independent Telemedicine Assistant) is allowing doctors, particularly specialists, to appear at their patients' bedsides and diagnose them without having to be physically in the same room. With the push of a button on an iPad, a doctor across town or around the world can direct the robot to the patient's bedside, zoom in on his pupils, and even have a nurse place a stethoscope on his chest to remotely hear his heartbeat. Whether robots have better bedside manner is yet to be determined.

Businesses too are starting to realize the value of having telepresence robots in the office, allowing employees to abstract their physical presence through remotely controlled devices. Companies such as Suitable Technologies and Double Robotics have models that cost around $3,000 and allow employees to work from home while their robotic alter egos wander the hallways at the office, walk up to colleagues at their desks, or catch up on all the latest gossip in the lunchroom. Even the famed NSA leaker Edward Snowden used a telepresence bot to give a presentation to an audience of thousands at TED 2014 in Vancouver, all without the bother of leaving the safety of his undisclosed location in Russia.

Humans Need Not Apply

As time moves on, we will see robots emerge for every possible job and purpose. Already Starwood hotels have introduced robotic butlers, "on call day and night." They can find their ways to any guest's room and deliver that toothbrush you forgot or the room service you ordered, freeing up staff to work on other tasks. Momentum Machines' burger bot can crank out 360 perfectly cooked-to-order hamburgers per hour, each with the precise toppings (lettuce, ketchup, onions) requested by the customer.

A 2013 study by Oxford University on the future of work conducted a detailed analysis of over seven hundred occupations and concluded that 47 percent of U.S. employees are at high risk of losing their jobs to robotic automation as soon as 2023. Those working in the transportation field (taxi drivers, bus drivers, long-haul truck drivers, FedEx drivers, pizza delivery drivers) face particular risk, with up to a 90 percent certainty that their jobs will be replaced by autonomous vehicles. But it's not just low-level positions that are at risk. News outlets such as the Associated Press and the *Los Angeles Times* are using bots and algorithms to automatically write thousands of articles on topics as diverse as homicides, earthquakes, and the latest business earnings. Biopsies can be "analyzed more efficiently by image-processing software than lab techs," and QuickBooks can handle the majority of tasks performed by an accountant. Many believe that it is the growth of automation and robotics that has led to the deep wage stagnation we have seen since 2004. Bill Gates was prescient in his predictions regarding the future of robotics and the presence of a robot in every home and office. But whether your job is flipping burgers, driving a truck, or writing breaking news, anybody who has read or seen John Steinbeck's *Grapes of Wrath* knows that industrial transitions are brutal for those left behind.

Now even foreign outsourcing may be replaced by robo-sourcing, eliminating more jobs for human beings both domestically and overseas. As machines become smarter and more capable, the human race may enjoy an incredible renaissance in which all our daily chores are carried out by bots, leaving us to a life of leisure, with unlimited free time to sing, dance, and paint while sunning our atrophying muscles on a beach somewhere. Alternatively, society might descend into chaos as the mass unemployed and unemployable revolt against the few human czars controlling the world's robots. The scenario could tip in either direction depending on the public policy, legal, economic, and ethical decisions we make today.

Robot Rights, Law, Ethics, and Privacy

A man without ethics is a wild beast loosed upon this world.
ALBERT CAMUS

While nobody would argue your Roomba should be covered under the UN's Universal Declaration of Human Rights, as robots grow more intelligent and potentially sentient in the distant future, such questions will undoubtedly be raised. In the meantime, the robots in our world bring with them a wide array of public policy, legal, and ethical issues beyond their impact on the workforce. If a robotic surgeon accidentally punctures an artery, leading to a patient's death, can the surviving family sue the robot or its manufacturer for malpractice? When a self-driving car gets in an accident, who will be at fault? Can the non-driving passenger be sued? The car company? The firm that wrote the driving and navigation software? When it's clear an autonomous vehicle is about to become involved in an unavoidable collision, should its crash-optimization algorithm cause it to hit the telephone pole (killing the passenger), the motorcyclist to the left, the Chevy on the right, or the pedestrian straight ahead? Though our ability to build and field robots is racing ahead exponentially, ethically we remain infants.

While ubiquitous robots are on the horizon, there is a paucity of robo-ethicists, policy experts, and legislators capable of keeping up with the complex questions these scientific developments will pose for humanity. In particular, we will see new and previously unimaginable assaults upon our privacy. Just like social media sites, apps, and mobile phones before them, robots will come with terms of service that will detail conditions that protect robo-manufacturers and affect your privacy. Though your robot vacuum, elder-care bot, or play toy may sit in the corner looking innocuous and cute, ready to serve at a moment's notice, it is armed with an array of cameras, microphones, and sensors capable of seeing and recording everything you do in the privacy of your own home.

Hobbyist drones equipped with HD cameras are already posing privacy threats previously not encountered. In mid-2014, a young woman in Seattle living on the twenty-sixth floor of an apartment building was surprised to see a quadcopter (a small helicopter with four rotors) hovering just outside her window filming her as she changed in her bedroom—a robotic Peeping Tom for the twenty-first century. In another Seattle incident, a man decided to hover his camera-equipped personal drone over a neighbor's backyard. When a woman heard the noise, which she thought was a garden weed whacker, she opened the curtains of her second-story bedroom to investigate, only to see a drone hovering outside her window just a few feet away. She sent her husband to investigate, and he found a neighbor flying the drone, but when the robo-invading pilot was asked to stop filming immediately, he refused, claiming it was legal for him to do so. He may be right.

While walking on somebody else's lawn is trespassing, flying over it with a helicopter (large or small) is not as the result of a 1946 Supreme Court decision that declared, "Air is a public highway." Of course the Seattle cops called to the scene of these incidents were confused, and they aren't the only ones. According to a 2012 government report on private drones flying over the United States, the GAO concluded, "Currently, no federal agency has specific statutory responsibility to regulate privacy matters relating to Unmanned Aircraft Systems for the entire federal government. Given the ability of these devices to house high-powered cameras, infrared sensors, facial recognition technology, and license plate readers, some argue that drones present a substantial privacy risk." Ya think?

Raising issues about who owns the air rights above property and who can be filmed where is just the very beginning of a deeply complex set of legal, ethical, and public policy matters that will undoubtedly arise with much greater frequency as the number of robots in use throughout our society proliferates. Perhaps the earliest attempts to deal with these foundational questions came in 1942 from Isaac Asimov when he published his short story "Runaround," in which he coined the term "robotics" and presented his famed Three Laws of Robotics:

1. A robot may not injure a human being, or, through inaction, allow a human being to come to harm.
2. A robot must obey the orders given it by human beings except where such orders would conflict with the First Law.
3. A robot must protect its own existence as long as such protection does not conflict with the First or Second Law.

While Asimov gives us an excellent starting point by which to consider these issues, we certainly could not program a machine at this point in time to concretely understand the concept of breakfast, let alone a construct as abstract as "harm." Robots would likely require a much more flexible and adaptive code of ethics, one that we have not even come close to constructing thus far. Yet the drive toward widespread industrial, military, medical, and personal robotics continues, and accidents are bound to happen.

Danger, Will Robinson

"Danger, Will Robinson!" was the phrase oft repeated by the protective robot guarding a young space adventurer to warn the boy of impending threats in the 1960s television show *Lost in Space*. If only all robots took such precautions in their interactions with all human beings. As people interact more and more with robots, there are unforeseen consequences, not the least of which are serious injuries or even death at the hands of machines, even those meant to help. In 2013, the FDA launched an investigation into numerous incidents of harm caused by Intuitive Surgical's da Vinci medical robot, incidents that the company allegedly failed to report to the government as required by law. In one instance, a man suffered a perforated colon during his prostate surgery; in another, the robot grabbed a patient's abdominal tissue during a colorectal surgery, refusing to release it despite the human surgeon's efforts to open the jaws of the machine's hand. It wasn't until the da Vinci was fully rebooted that it finally let go. In another case, a woman was struck in the face by a surgical robot during her hysterectomy.

The overwhelming majority of injuries in human-robot interaction occur as a result not of surgical bots but of industrial ones. Though no comprehensive statistics on robo-accidents exist globally, there are numerous reports of such accidents. In 2007, for example, a worker in Stockholm who thought he had turned off the power to a robot approached the machine to repair it. Unfortunately, the power was still on, and the robot suddenly came to life, firmly grabbed the man by the head, lifted him off the ground, and broke four of his ribs before he was able to struggle free. In a collision between man and machine, it is the machine who is likely to win, and many cases have resulted in death. One of the first cases of robotic homicide occurred in 1981 when a thirty-seven-year-old employee of Kawasaki Heavy Industries named Kenji Urada was working to repair a robot that he had not turned off completely. Unable to sense him, the robot's powerful hydraulic arm accidentally knocked the man into a nearby grinding machine, where he was crushed to death. Back in the United States, a car factory employee was killed in 2001 when he entered a robot's unlocked cage to clean it. The machine, thinking the worker was an auto part, grabbed the man by the neck, pinning him until he was asphyxiated. According to the Occupational Safety and Health Administration, in the United States alone at least thirty-three deaths have occurred—a number that is likely to go up as robots leave their cages and begin walking among us. Apparently, not all robots have yet to hear of Mr. Asimov and his three laws.

Robotic accidents become much worse once it is decided that it is a good idea to provide fully automatic weapons to robots, as members of the South African National Defence Force discovered in 2009 during a live-fire training exercise. A computer-controlled Oerlikon MK5 twin-barreled anti-aircraft gun suffered an apparent software glitch, causing the device to fire in full-auto mode, at the rate of 550 rounds per minute while spinning around wildly in 360-degree circles like an out-of-control garden hose. When it was all over, nine soldiers, including several female officers, were dead, and another fourteen were gravely injured, leaving behind a blood-splattered scene reminiscent of a *Terminator* movie. The incident goes to show that when a robot suffers a computer "blue screen of death," it can actually lead to death and have far-reaching impact in our common 3-D physical space. It's not just industrial- or ground-based bots that can fail, so too can flying ones.

According to a *Washington Post* report, over four hundred military UAVs have accidentally fallen from the sky, domestically and overseas, "slamming into homes, farms, runways, highways and in one case an Air Force C-130 Hercules cargo plane while mid-flight." While nobody has died in any of the reported incidents, it is only by a miracle that it is so.

In 2009, a drone pilot lost control of an armed Reaper UAV with a sixty-six-foot wingspan, flying uncontrollably across Afghanistan. The renegade flying robot was only stopped when U.S. jet fighters intervened and shot it down before it entered the airspace of Tajikistan.

Closer to home, nearly fifty drones have crashed in the United States, including a 375-pound army drone that smashed into the ground next to a Pennsylvania elementary school, "just a few minutes after students went home for the day." Robotic accidents are the exception, occurring relatively infrequently, and active measures are being taken to arm robots with collision detection and avoidance systems to prevent many of the industrial-type accidents. Nevertheless, given the expected tremendous growth in home bots, work bots, factory bots, doc bots, and war bots, the potential for harm is far from trivial—a risk that will grow significantly when robots join the IoT and can be hacked from afar by malicious actors.

Hacking Robots

In the future, when Microsoft leaves a security-flaw in their code it won't mean that somebody hacks your computer. It will mean that somebody takes control of your servant robot and it stands in your bedroom doorway sharpening a knife and watching you sleep.
DANIEL H. WILSON, ROBOTICIST AND AUTHOR

There are dozens of robotic operating systems, mostly proprietary, running everything from military weapons systems to SCADA industrial control systems. But just as desktop computers and smart phones coalesced around a few leading operating systems, the same is happening in robotics using ROS—the Robot Operating System. Doing so will have a tremendous positive impact on the future of robotics as programmers will not have to reinvent the wheel every time they want to encode a particular function in a robot. ROS is free and open source, providing modules for robotics simulation, movement, vision, navigation, perception, facial recognition, and so forth. It is exactly these types of open-source community efforts and shared experience building, barely conceivable just a few years ago, that allow companies like Rethink Robotics to offer Baxter for $22,000 instead of $200,000.

ROS, originally developed at Willow Garage in 2007, is now maintained by the Open Source Robotics Foundation and runs on everything from small toys to large industrial robots. As noted numerous times throughout this book, there has never been a computer that could not be hacked, a dictum that applies to robots as well, with important implications for our common security. The hackers' task will be unwittingly assisted by a standardized Robot Operating System, which will give them a unified target to attack. Standardization of a universal ROS paves the way for large-scale cyber attacks, just as we saw with PCs. Importantly, there is a momentous difference between hacking robots and hacking other computing systems and other objects on the IoT: robots will be perennially moving about our physical space, walking, driving, running, flying, and swimming all around us. Robots, connected to the Internet, can be hacked and redirected in any number of dangerous and sinister ways—a fact that has not escaped the attention of criminals and terrorists alike. When robots are commandeered, not only can hackers use the machine's sensors to spy, but they can use the device's robotic actuators, arms, legs, and wheels to follow, hit, kick, push, shoot, stab, drag, and kill.

In essence, robots are nothing more than moving computers, computers that will liberate cyber crime from behind today's two-dimensional screens and launch it into our everyday physical world. Researchers at the University of Washington examined three home bots, including the Erector Spykee and WowWee's Robosapien and Rovio, and uncovered significant security flaws in each, including a lack of passwords and poorly implemented or absent encryption. As a result, third parties could take over the devices from afar, move them, and remotely capture audio and video. Researchers described security in these devices as "merely an afterthought." But as robots become more prevalent in society, moving about our world, they will join the billions of other objects connected to the IoT. As we saw previously, tens of thousands of videoconferencing systems used at law firms, pharmaceutical companies, and medical centers are deeply insecure and have been successfully hacked, even inside the Goldman Sachs boardroom. Why would telepresence bots—moving videoconference devices—be any different? These robots could follow you around, listening in, or sit there silently during meetings observing everything, excellent tools for industrial espionage. When your factory closes and the lights go out, hackers halfway around the world could commandeer the bots to case the joint. Though you might have a security guard to keep criminals out, a robotic one may already be in the building.

Hacking robots raises a number of important questions. How private is that robotic bedside consultation your doctor is providing over the Internet? Worse, those industrial bots cooking up hamburgers and slicing tomatoes will be armed with sharp knives—how do we teach them to use caution when around humans? Though most industrial robots have safety systems, as we have seen, accidents happen, including deadly ones. But robotic safety routines are encoded in computer programming, programming hackers can interfere with and disable. The next generations of powerful household robots may well be misused in ways their designers never envisioned. Just as smart-phone users jailbreak their iPhones today to remove annoying software restrictions, so too will they with their robots, opening the door to a variety of "robots gone wild" scenarios.

Consider an "in screen we trust" attack in which an employee dutifully turns off the robot before cleaning it, but a hacker has interfered to keep it on. Though the screen shows the robot and its massive industrial arms are powered off, the unsuspecting worker approaching the device finds himself grabbed by the neck, picked up, and asphyxiated, a great way to deal with that co-worker you never liked in sector 3B. To the world, it might just look like another accident. If these scenarios seem far-fetched, evidence of hacking some of the most secure robots in the world—military and police bots—already exists.

Game of Drones

You need to put drones under control.
You need to lay out certain rules of engagement in order to prevent or minimize collateral casualties. It is extremely important.
VLADIMIR PUTIN

In late 2009, as the war raged on in the Middle East, U.S. Predator drones flew nearly constantly above the skies of Iraq. Their missions varied from intelligence collection to "kinetic operations against high value targets" such as launching Hellfire missiles against insurgents. The drone pilots remotely carrying out these operations seven thousand miles away in the Nevada desert intently watched live video feeds of their targets as they navigated their UAVs in pursuit of their quarry. As it turns out, they weren't the only ones watching. Shia militants had figured out a way to hack the American flying robotic fleet and capture its live video feeds. Using a $26 piece of Russian hacker software known as SkyGrabber, commonly sold in the digital underground to steal satellite television signals, the insurgents were able to intercept the video footage emanating from the classified Predator drones. Thus as the Americans were watching the insurgents, the insurgents were watching back, providing them with a tactical advantage and vital intelligence on coalition targets. If the militants saw their house coming into close video focus, they knew it was definitely time to rapidly consider alternative housing options.

This was certainly not the only time a drone was successfully hacked; it's even happened over the continental United States. The Department of Homeland Security uses a fleet of these UAVs to protect the border and in 2012 found out they were not nearly as secure as it had presumed. Students at the University of Texas at Austin had discovered a way to hack the drones and tried to inform DHS, which refused to believe them, saying its UAVs were "unhackable." After months of back-and-forth, officials were finally persuaded to participate in a demonstration by the students, at which point the UT wunderkinder seized the flying robot and began flying it sharply off course, leaving DHS officials with their jaws agape. The students carried out their attack by successfully spoofing the drone's GPS and changing its coordinates, all using hardware and software they had built at school for under $1,000. Their professor Todd Humphreys (the same man responsible for hacking the GPS on the super-yacht *White Rose of Drachs* mentioned earlier) astutely noted after the DHS incident, "In five or ten years we've got 30,000 drones inhabiting the national airspace . . . Each one of these could be a potential missile to be used against us."

Others have taken notice, including the Iranians, who successfully used the same technique to jam the communication links of an American RQ-170 Sentinel drone overflying their country, forcing it into auto-pilot mode. The drone followed its programming and returned to base in Afghanistan, or so it thought. In reality, the Iranians had successfully spoofed the UAV's GPS signals, flying the robotic soldier right into the hands of the Islamic Revolutionary Guard Corps. The capture of the drone and its classified technology was a significant intelligence coup for the Iranians and provided yet further evidence that the day of robo-hacking has arrived. It's not just the drones themselves that can be hacked; so can their command-and-control systems. In 2011, a potent computer virus struck the U.S. drone fleet, infecting the cockpits of America's Predator and Repeater UAVs, logging every keystroke of drone pilots as they flew missions over Afghanistan. The source of the breach remained unknown as of late 2014, and the incident was still under investigation.

In 2013, the serial hacker Samy Kamkar devised an attack (and posted it online for others to exploit) that allowed him to fly his own aerial drone that would seek out other flying robots in the sky, hack them, and turn them into a physical botnet army of UAVs under his control. The software, dubbed SkyJack, compromises the smart-phone wireless connections controlling drones, such as the wildly popular Parrot AR model, commonly sold at Costco, and allows hackers to commandeer a victim drone's flight control and camera systems. Over 500,000 Parrot UAVs have been sold, and Kamkar's technique should prove useful to hijack other drones, such as those that will undoubtedly be delivering goods around cities in the coming years—misdirecting packages and pizzas in real time. The future of robotic crime looks promising indeed to Crime, Inc., and it is beginning to dedicate significant resources to the effort.

Robots Behaving Badly

In 1982, on the streets of swanky Beverly Hills, California, police took a rather unusual perp into custody—a DC-2 robot that was illegally distributing advertising flyers in the city's business district without a permit. When officers approached the four-foot rogue bot on wheels, they discovered a machine with an old CRT monitor and keyboard for a chest and a head shaped like an astronaut's helmet. Police demanded the robot's mysterious operator identify himself, but instead they were met with a barrage of insults spouting from the robot's onboard speaker. Unamused, cops tried to disassemble the bot and take it into custody, and as they did, the robot began loudly screaming to a crowd that had gathered, "Help me! They're trying to take me apart." Eventually, the robot was "arrested" and transported to police headquarters by tow truck. A few hours later, Gene Beley, owner of the $30,000 robot and founder of the Android Amusement Corporation, appeared before cops with his two teenage sons dragged by their ears in tow. The boys had taken the professional robot for a "joyride" without their father's permission. Though police considered citing Beley for the incident, they instead released the robot on his own recognizance. When Beley was interviewed by the Associated Press after arriving at his house, he noted he was glad to have the DC-2 back home, adding, "We sort of felt like a member of the family was in jail." Though perhaps the first, the DC-2 will certainly not be the last robot arrested.

In time, robots will be used to assist in bank robberies, street holdups, and even kidnappings. Hackers have already created the R2B2, the Robotic Reconfigurable Button Basher, a machine capable of trying repeated passwords on locked, lost, or stolen iPhones and Android devices at the rate of one attempt per second. The hacking bot was built for under $50 from several servomotors, a plastic stylus, and a Webcam that "watches the phone's screen to detect if it successfully defeated the phone's password" (even criminals will use robots for jobs that are repetitive or dull). Robots can also be a criminal's best friend, as police in Taiwan discovered in mid-2014 when they attempted to arrest a known armed drug dealer who had tightly protected his home with a series of surveillance robots streaming video, meant to give early warning of police presence.

As we saw in the opening of this chapter, terrorists are also using robots as weapons, and they aren't limited to consumer-grade UAVs with small payloads. In both Iraq and Afghanistan, terrorists have turned to VBIEDs (vehicle-borne improvised explosive devices), commonly known as car bombs, to destroy multiple buildings and rock entire neighborhoods, with some vehicles' containing up to seven thousand pounds of explosives. VBIEDs are powerful weapons and have destroyed numerous targets around the world, including Khobar Towers in Saudi Arabia, the U.S. Marine Corps barracks in Beirut, and the Murrah Federal Building in Oklahoma City.

Now terrorists are turning to robotic weapons to supplant their previous VBIED capabilities. In a video discovered online, kaffiyeh-clad engineers from Ansar al-Islam can be seen bragging about their technical abilities while huddled over, soldering computer circuit boards. In the next scene of the four-minute clip, a pickup truck is seen driving in the middle of the desert with a tripod-mounted automatic machine gun in its bed. As the camera zooms in, it is clear there is no driver in the cab, which is being operated via crude robotic controls on the steering wheel and floor pedals. Moments later numerous rounds are fired from the machine gun, as a remotely controlled robotic actuator pulls the weapon's trigger.

Using such systems, jihadists no longer need to martyr themselves. While they may miss out on their promised seventy-two virgins, they would remain capable of coming back to fight another day. The potential for criminal abuse of self-driving vehicles has not escaped some in law enforcement, and the FBI issued an internal report citing fears about their forthcoming use as lethal weapons. Officials predicted that robotic conveyances could be used as VBIEDs preprogrammed to autonomously drive across town to detonate at their intended targets. Those fears we've always had about killer robots, depicted in films such as *Westworld, Blade Runner, RoboCop, The Terminator,* and *I, Robot,* may unfortunately be at the early stages of already materializing.

Attack of the Drones

Drones are scary. You can't reason with a drone.

MATT GROENING

When Jeff Bezos, the CEO of Amazon.com, announced in late 2013 that the world's "everything store" would soon be using octocopter drones to deliver packages to its customers, the world sat up and noticed. Sure, others had beaten Bezos to the punch, such as the entrepreneurs who launched the TacoCopter and the Burrito Bomber, not to mention the Vegas hotel that delivers chilled champagne poolside to its guests via drone, but Bezos's announcement was different. Amazon has perfected its logistics, and getting drones to go the last mile for its customers would undoubtedly be a game changer in business. In the fall of 2014, Google successfully began delivering goods on a pilot basis via a small five-foot-wide single-wing aircraft. Dubbed Project Wing, Google's drone can fly within a ten-mile radius of its warehouses, delivering everything from candy to dog food. The UAV also has rotors and can hover a hundred feet over a customer's home and lower products to the ground via a cable winch before flying back to the company's offices. Undoubtedly, there are plenty of kinks, technical and regulatory, to be worked out with these services, but in one form or another it's a done deal: like it or not, the era of commercial and civilian drones is upon us.

Though most often associated with the military and warfare, drones can also be a force for good. Drones are being used to catch poachers in Africa and help farmers maintain their crops in America. They surveyed the damage at the Fukushima nuclear disaster site and helped after the earthquake in Haiti. Today, UAVs are chasing storms to provide early warnings of hurricanes, putting out wildfires, and transporting medicine to remote villages. Real estate agents are using them to photograph properties, and parents such as Paul Wallich of Vermont are flying quadcopters over their kids as they walk to the local school bus stop to make sure they get there safely. The Royal Canadian Mounted Police have even used their quadcopter robo-Mountie to record the first-ever case of a life saved with a UAV when they flew it over a remote area of Saskatchewan to locate a missing injured man who became lost and disoriented after his car crashed and went off the road in freezing temperatures.

The day of the drone has arrived, and Web sites such as DIY Drones have established massive communities dedicated to building personal UAVs. For consumers, businesses, and government, UAVs have become easily affordable, costing only a few hundred dollars for basic models, and come loaded with high-powered sensors such as HD cameras whose video feeds users can view on their mobile phones. Though drones are becoming increasingly popular and can be used for good, they bring with them a host of concerns beyond the privacy matters mentioned previously. Soon our skies will grow crowded with these devices. We will fondly recall the days when we could look up and see the heavens, absent legions of quadcopters pulling banners for Pepsi, Viagra, and Coppertone in what is becoming the growing field of "drone-vertising." The problem will become much worse when the world of big-data analytics converges with robotics. Then, rather than showing you online banner ads based on your search history, cookies, and Facebook Likes, drones displaying carefully targeted ads will show up outside the window of your home or follow you down the street carrying actual banner ads. Also, more flying robots mean more accidents. If trained military pilots can have four hundred of their UAVs fall out of the sky, what will happen when the drunken kids at the frat party start playing with them?

Of course if Martha Stewart can figure out how to use a drone to surveil her property and photograph it, so too can Crime, Inc. Not only will camera-equipped UAVs be used for the obvious things like industrial espionage and casing joints for burglary, but they may also be used to help jealous husbands and wives stalk their exes, including in cases of domestic violence. Hackers have also figured out how to use drones for the purposes of communications interception, both listening in on your phone calls and tracking your every move online, with devices such as the WASP—the Wireless Aerial Surveillance Platform.

Unveiled in Las Vegas in 2011, the WASP is a small remote-controlled airplane with a six-foot wingspan. It has eleven antennas and is equipped with a variety of communications tools and sensors, including an HD camera. The WASP was designed to fly over your neighborhood and intercept the Wi-Fi signals of all those around, even those on encrypted networks. The UAV has a small onboard Linux computer that runs a variety of hacking tools, including a custom-built 340-million-word dictionary, which the drone can use to generate passwords to get brute-force access to your network in real time. The WASP also carries a rogue cell-phone tower that it can use to "impersonate" GSM mobile phone carriers. The fake cell tower tricks your mobile into connecting to the WASP and allows hackers to record all phone calls and text messages that pass through the device. Not long ago, signals intelligence capabilities such as these would have cost tens of millions of dollars and were only available to the world's most advanced militaries. The WASP was built for $6,000.

With basic drones equipped with HD cameras costing so little, they are beginning to show up in a variety of unexpected places, including at protests and riots. In Warsaw, Poland, demonstrators from the Occupy movement launched a quadcopter to document the activities of aggressive riot-clad police as they attempted to control a crowd of thousands with tear gas. The so-called Occu-copter flying a hundred feet off the ground provided protesters with stunningly clear images of police officers as they moved in column formation to try to encircle the demonstration, a powerful and previously unimaginable countersurveillance tool now in the hands of the common man. Needless to say, cops won't be the only ones to struggle with how to respond appropriately to drones flying overhead.

Crime, Inc. has taken to flying robots as the tool of choice to smuggle weapons, cell phones, and narcotics into correctional facilities around the world. At the Provisional Detention Center São José dos Campos in São Paulo, Brazil, correctional officers observed a quadcopter drone fly over the prison walls and drop a small package in the recreational courtyard of the facility and discovered 250 grams of cocaine in the package. Outside Moscow, it was a remote-controlled helicopter that flew 700 grams into the Tula prison. In Greece, it was a box of mobile phones, and similar prison intrusion incidents have been reported from Canada to Australia and the United States. Crime, Inc. is slowly building up its robotic air force.

Importantly, the rise of the criminal UAV is completely incompatible with our current security paradigms. Prisons use tall, sharp, often electric fences to isolate criminals for reasons of public safety, a system that worked relatively well for hundreds of years. But our security and defense mechanisms were meant to protect us from offending human criminals, not robotic ones. It may be time to rethink that. Drones can circumvent not only prison fences but any fence, including those protecting your backyard, office building, or even national borders, as the narco-syndicates of Latin America are demonstrating. In Mexico, for example, Crime, Inc. has hired assembly-line workers from local aircraft factories who moonlight for the cartels designing UAVs. In the Santa Fe district of Mexico City, just near the Bombardier factory, a covert narco-drone factory was uncovered, according to Mexico's Public Security Secretariat. Building on American, European, and Israeli designs, these ultralightweight autonomous aircraft are much larger than the average quadcopter, weighing a hundred pounds, and have foldable wings so they can easily be transported by and hidden in trucks on either side of the border. They fly low and are undetectable by radar. Each drone can carry a hundred kilos of cocaine per trip, coke that costs $1,700 a kilo in Colombia, $8,000 in Mexico, and $30,000 in the United States, netting traffickers more than $2 million a flight. Since 2012, the DEA has documented at least 150 such narco-drone crossings carrying multiple tons of cocaine. With profits like this, cartels from Cali to Sinaloa are reinvesting their proceeds into advanced research and development, spending millions to ensure a much more prominent role for their criminal robotic workforce moving ahead.

Apart from drugs, there are much more troubling items criminal techies can attach to drones, including firearms. YouTube is already replete with hobbyists showing self-built remote-controlled flying robots that perform elaborate tasks such as tracking and shooting people with water pistols and paint balls—a pastime ideal for criminal or terrorist adaptation. Other videos show hobbyists flying drones with stun guns on board, shooting electrical prongs at their prey and Tasing them down to the ground with eighty thousand volts of electricity. But the trend of course does not stop there; real guns have been used too. The earliest video of an actual firearm, a .45-caliber handgun, mounted on a remote-control helicopter and firing appeared way back in 2008. Since then, numerous other videos showing smart-phone-controlled UAVs with guns have appeared online, including an HD version of an octocopter carrying a .45 Colt handgun fired repeatedly using a remotely controlled robotic trigger finger. With so-called automatic "follow me" technologies, they can even autonomously track a particular individual as he runs down the street. Using a smart phone to shoot an actual gun mounted on a flying robot that costs a few hundred dollars means that first-person shooter games have entered three-dimensional space and become a reality. How long will it be before a criminal or mentally unstable individual uses such a device to kill others?

For as dangerous and frightening as such a scenario might be, other, much more pernicious payloads can be loaded on board UAVs as well, including explosives and even weapons of mass destruction, such as a biological, chemical, or radiological weapon. For under $20, bomb-drop systems for remote-control aircraft are available online, similar to the bomber doors on military aircraft that open up when directed via remote control or when reaching a particular GPS navigation point. Might drones become the next suicide bombers? Al-Qaeda, Lashkar-e-Taiba, and numerous other terrorist organizations already have active drone development programs. Several YouTube videos have shown farmers, tired of working in the sweltering heat, who have converted their remote-controlled helicopters into crop dusters. If a terrorist were to use the same concept to spread a lethal agent over a crowd instead of pesticides over rice fields, the potential for harm could be enormous.

Drones, as the military has shown us, can also be used in highly targeted ways against particular individuals, whether for personal revenge, criminal attack, or terrorism. We're already starting to see high-profile people come under assault in both strange and dangerous ways. In late 2013, Chancellor Angela Merkel of Germany found herself under drone attack during a campaign rally in Dresden when a quadcopter UAV charged toward her onstage, only to crash at her feet. The attack was carried out by Germany's Pirate Party, which said that it wanted to make sure the Chancellor knew "what it's like to be subjugated to drone observation." Her security protective detail surely got the message. Though nobody was injured as a result of the stunt, the story could have had a much unhappier ending had the device been armed or carrying explosives.

Drones can also cause damage when launched at other conveyances, shocking automobile drivers and causing crashes. There are already numerous reports of hobbyists intentionally flying drones directly into the flight path of jet aircraft around the world, causing pilots to take drastic evasive actions to avoid collision, including planes from American Airlines, US Airways, Alitalia, and Virgin Blue. Had any of these flying robots been sucked into the airliner's engines, they could easily have caused a crash similar to the one that downed the US Airways jetliner into New York's Hudson River. As time marches on and flying, swimming, rolling, and walking robots enter our lives, we will have to figure out ways to safely and peacefully coexist with them, but the future of robotics itself may bring with it even greater risks to be managed.

The Future of Robotics and Autonomous Machines

Robots will become faster, smarter, and smaller. Already great advances are being made in micro-robotics. The devices, some small enough to fit on your fingertip, are flown remotely and can be equipped with an HD camera and microphone, taking surveillance privacy concerns to a whole new level. Dragonfly drones were already reportedly used to spy on antiwar protesters in Washington, D.C., back in 2007, and the air force unveiled robo-bumblebees that could not be detected in hostile environments while flying into buildings to "photograph, record and even attack terrorists."

Another forthcoming breakthrough in robotics will be that of "swarm" capabilities—getting multi-robot systems to act in unison via collective behavior that mimics nature in the same way ants cooperate and birds flock. Using advanced distributed computing power to self-organize and solve problems, swarms of bots could coordinate their efforts to achieve incredible things whether in disaster relief, search and rescue, oil spills, or manufacturing. Much progress is being made in swarm intelligence, and in mid-2014 researchers at Harvard University created the largest robot swarm ever, using 1,024 tiny robots the size of a penny that could find one another and collaborate to assemble themselves into various shapes and designs, including stars and letters, like a mechanical flash mob. But there may be a gathering swarm on the horizon when it comes to self-organized cooperating bots: they can also be a force for ill. One drone with a hand-gun chasing you down the street is bad enough, but a swarm of thirty of them is terrifying and likely unsurvivable. Moreover, once swarming robots come into common use, any hack they suffer or virus they get could be disastrous, as it would affect all the bots on the network, just like the scenes portrayed in the television show *Star Trek* wherein the crew of the USS *Enterprise* uses a computer virus to successfully destroy the Borg collective of cybernetic organisms, except we might be the Borg that is destroyed. When the military starts widely using armed UAVs operating in swarm formation to attack the enemy and they get a virus (as the U.S. drone command already has), how easy might it be to turn the weaponized flying robots against their masters or innocent civilian populations?

Moving ahead, not only will we have ever-smaller bots swarming about us, but we will have robots that are increasingly autonomous—intelligent machines capable of performing tasks and making decisions in the real world on their own without explicit human control. An autonomous robot, like the Roomba vacuum, makes decisions based on its programming, but on its own in real time, using "bump and go" algorithms to move and avoid obstacles, enabling it to analyze and adapt to unfamiliar environments.

It is, however, with military robotics that particularly difficult questions about autonomy arise: How much is too much? A ground-based robotic army medic that can autonomously rescue a wounded solder from the battlefield and provide lifesaving first aid sounds like a great idea. But a UAV that can find its target and autonomously make the firing decision to kill may give many pause. Yet that is exactly where we are heading. As robotics, artifi-cial intelligence, and computer processing speeds improve exponentially, at some point human beings just aren't going to be fast enough to keep up, particularly in the realm of warfare. Once your enemy goes to fully autono-mous warfare, you will be compelled to do the same or face destruction.

Though they may seem like the stuff of apocalyptic dystopian films like *The Terminator*, autonomous killing machines are already here. The BAE Systems Taranis is a fully autonomous aircraft that can "fly deep into enemy territory to collect intelligence, drop bombs and 'defend itself against manned and other unmanned enemy aircraft.'" In the demilita-rized zone across the Korea Peninsula, South Korea has deployed Samsung SGR-1 border control sniper robots that can detect intruders with heat and motion sensors and automatically fire on targets up to a kilometer away with their embedded 5.5 mm machine guns and 4 mm grenade launchers. Though the border bots currently require human permission to attack as a matter of policy, technically they can go fully autonomous at the flip of a switch. Lethally autonomous killer robots will take many forms: walking, swimming, flying, and driving machines that can chase their prey, or just lie in wait. Yet despite our growing technical capability to outsource kill decisions to machines, doing so is fraught with a panoply of legal, ethical, moral, technical, and security implications.

Industrial accidents with robots are bad, but as we saw with the horrific South African National Defence Force computer glitch, accidents involving robots with automatic weapons can be catastrophic. As robots proliferate, we will suffer the consequences as Moore's law clashes with Murphy's law. Poor programming, inaccurate data, and software errors will undoubtedly lead to tragedy when robots themselves can decide to kill. Moreover, armed robots connected to the IoT will be hackable, as will their safety features and protocols, adding yet another notable peril to be considered. So too is the matter of repressive governments using killer robots to quell dissidents or narco-cartels to kill cops and rival drug gangs. It may seem like hyperbole to suggest that Crime, Inc. will one day have autonomous killing robots, but of course it will, just as it has adopted any number of other prior military technologies, including night-vision goggles, the Internet, and UAVs. Experts in both human rights and technology are concerned about delegating kill decisions to machines. The topic has also been raised by the United Nations, Human Rights Watch, and new organizations such as the International Committee for Robot Arms Control and the Campaign to Stop Killer Robots. The science fiction author Daniel Suarez and the roboticist Noel Sharkey have both even given impassioned presentations on the topic at TED calling for a global prohibition on robots' autonomously killing or injuring a human being, a rather sensible idea indeed, as Asimov first proposed decades ago.

That robots will enter our lives for everything from elder care to meal preparation and surgery is a given. They may well be a tremendous force for good, but as we have seen throughout this chapter, robots are also being used by street thugs, Peeping Toms, narco-cartels, and terrorists, a trend that will surely accelerate as their functionality improves and their prices drop, particularly in response to incredible new and complementary technologies such as 3-D printing.

Printing Crime: When Gutenberg Meets Gotti

Restrictions are difficult to enforce in a world where anybody can make anything.

HOD LIPSON

3-D printing, or, as it is sometimes called, additive manufacturing, promises to bring the *Star Trek* replicator to life. At the push of a button, a magical machine can make physical objects before your very eyes using a wide array of materials, including plastic, metal, wood, concrete, ceramics, and even chocolate. Just as you can send a photograph to your 2-D ink-jet printer, so too can you download or create a design on your laptop and send it to a 3-D printer, which, using a variety of techniques, can build objects in three dimensions, layer by layer, with incredible precision. These digital manufacturing techniques are making it easier and cheaper to build not only robots but a whole host of products ranging from airplane parts to fully functioning SLR cameras and lenses.

Goldman Sachs has noted that when compared with traditional manufacturing, 3-D printing will drive greater customization and reduce costs for complex designs, and others have forecast a 500 percent growth in the 3-D printer market to $16 billion by 2018. Today, inventors such as Scott Summit, founder of Bespoke Innovations, are using 3-D printers to create next-generation customized prosthetics that are not only perfectly fitted but beautifully designed. Digital fabrication can be used to print entire homes, concrete, electrical wiring, plumbing, and all. NASA has even purchased a 3-D printer for the International Space Station from the Silicon Valley start-up Made in Space to ensure it never has to worry about a missing part on board endangering the lives of astronauts, as was the case with *Apollo 13*. Bio-fabricating printers have taken things to the next level with machines that can even print human tissues and organs, such as capillaries, kidneys, ears, and hearts, potentially doing away with organ transplant lists and saving lives.

Prices on home 3-D printers—machines that used to cost tens of thousands of dollars—are dropping precipitously, and models such as the popular Cube 3 made by 3D Systems can be purchased at Staples for $999. Amazon has created its own 3-D printing store, and Web sites such as Thingiverse have become the go-to locations for users to freely share and customize their digital design files to make everything from jewelry to iPhone cases, and MakerBot offers kits to even build your own 3-D printer. Autodesk's free 123D software and apps can turn any 3-D digital model into a real-world object, and its open-source Spark operating system may do for 3-D printers what Android did for smart phones. These developments may shift manufacturing away from mass production toward mass customization, where people can print the exact shoe, table, or toy just as they like it. Chris Anderson, the former *Wired* editor in chief, has deftly documented this so-called DIY maker movement in his book *Makers*, in which he cites open-source design and digital manufacturing as the foundations of the new industrial revolution.

Another remarkable aspect of 3-D printers is that the devices are moving toward total self-replication. Today most 3-D printers can print more than 50 percent of the parts required to make another 3-D printer—a percentage that is increasing rapidly. The printers allow physical objects to be transmitted over the Internet and printed on demand. 3-D printers, just like robotics and the IoT, are ushering in the age when the analog and the digital are merging and becoming indistinguishable from each other. Bits and bytes are becoming atoms, and 3-D scanners such as Microsoft's Kinect can turn physical objects back into ones and zeros. The result may well be massive disruptions in manufacturing, retail, and even geopolitics. Local fabrication and assembly could have deeply positive impacts for the environment. When you can print the things you need at home, why run down to the local store? And if American firms can print more of what they need here, does it make sense to have tons of cheap plastic crap shipped across the oceans from China? Regardless of how these transformations play out, there is one group of individuals who have already wholeheartedly embraced the maker movement: Crime, Inc.

Just as robotics has brought new cyber risks into our three-dimensional world, so too will digital manufacturing. The first area criminals will pursue in the world of 3-D printing is intellectual property theft. Previously, it was only digital intellectual property that could be perfectly pirated and duplicated—music, video, games, and software programs. That's about to change. Though crooks have been making fake Gucci handbags and knockoff Cartier watches for some time, they were relatively easy to spot because of shoddy design and cheap manufacturing. But in the future these objects will easily be subjected to ultrahigh-resolution 3-D scanning and printing, making copies visually every bit as good as the original. The Gartner group has already predicted that 3-D printing will result in over $100 billion in intellectual property loss globally, per year, by 2018.

Digital manufacturing will also be a boon to burglars and stalkers who can now just take a high-resolution photograph of the home or office keys that you casually left on your desk and use a service like KeyMe to have duplicate keys printed via the 3-D printing marketplace Shapeways. There are apps too, such as Keys Duplicated, which will do the same thing, providing the keys to your castle to more people than you might like. If this bothers you, you're not alone. In 2012, cops uncovered computer-aided design files online that allow criminals to digitally manufacture police handcuff keys, including ultra-secure models whose manufacturers do not sell keys to the public. In the future, your drug dealer may be a printer as well. Scientists have already developed a "chemputer" that can print medicines such as ibuprofen on demand. While the potential humanitarian benefits are enormous, it won't be long before Crime, Inc. adapts these machines for meth, crack, and Oxycontin, vastly simplifying its supply chain and distribution issues.

Perhaps one of the greatest controversies surrounding 3-D printers is the ability to produce firearms, and perhaps no man has done more to make that a reality than Cody Wilson, a twenty-six-year-old former law student, anarchist, and libertarian à la Dread Pirate Roberts. Wilson created the

Wiki Weapon Project, brought us Darkwallet and its untraceable crypto currency, and founded Defense Distributed—a nonprofit online designer, publisher, and repository of firearms blueprints that can be downloaded and printed with a 3-D printer. Among his 3-D printed creations was a lower receiver for an AR-15 semiautomatic rifle from which he successfully fired six hundred rounds of ammunition. The lower receiver is the key part of the firearm and the only one regulated by law; the rest of the parts in many states can be obtained without background checks or even identification. In May 2013, Wilson also designed the Liberator, the world's first fully 3-D-printed gun, designed to fire standard .380 handgun bullets, and 100,000 people around the world downloaded the drawings. When asked by the press how he felt about his accomplishment, Wilson replied that now "anywhere there's a computer and an Internet connection, there would be the promise of a gun."

Wilson's efforts have left Congress in the dust, which failed to pass introduced legislation prohibiting 3-D-printed weapons. These plastic firearms can be near impossible to detect on standard metal detectors, as a team of Israeli investigative reporters proved by smuggling a 3-D-printed gun into the highly secure Knesset building, twice. In the meantime, dozens of other digital gunsmiths have improved upon the original Liberator and even posted their own digital gun files online. Other repositories for online designs for 3-D weapons have been created, including those that have plans for hand grenades and mortar rounds. The FBI's Terrorist Explosive Device Analytical Center is concerned by the trend and recently purchased its own 3-D printer to investigate how terrorists might use 3-D printers to build IEDs. The weapons conundrum posed by 3-D printers is non-static: as these devices grow in size and capability, they will be able to fabricate even larger weapons, including shoulder-fired missile launchers and large military-style robots.

With digital manufacturing, national border inspections become meaningless. Why risk smuggling weapons or drugs into the country when you can simply print your guns, pills, or bombs after you cross the border? The challenges 3-D printing poses to international security are not just limited to crime and terrorism; they will affect long-standing instruments of international law, such as weapons bans. Need parts for uranium centrifuges in Iran? No problem, just print them. Embargoes and even naval blockades, our traditional tools for ensuring global security against rogue regimes, will fail epically as larger and more sophisticated 3-D printers become mainstream. The old paradigms of national borders, guards, gates, and tall fences may well become outdated as technology develops much more rapidly than our security mechanisms—the new normal that will be even further exacerbated by a host of new science-fiction-like technologies coming online in the very near future.

CHAPTER 16

Next-Generation Security Threats:
Why Cyber Was Only the Beginning

We have arranged things so that almost no one understands
science and technology. This is a prescription for disaster. We might
get away with it for a while, but sooner or later this combustible mixture of
ignorance and power is going to blow up in our faces.

CARL SAGAN

B reaking: Two Explosions in the White House and Barack Obama Is Injured," reported the Associated Press on its official Twitter news feed at 1:07 p.m. on April 23, 2013. In an instant, the AP's two million followers had retweeted the news thousands of times, and the world went into panic mode. On Wall Street, the reaction was both swift and staggering: the Dow Jones Industrial Average and the S&P 500 plummeted. Within three minutes, the AP's tweet had wiped out $136 billion in shareholder value.

Thereafter, the tweets flew fast and furious. At 1:13 p.m., the AP confirmed that the explosion-reporting tweet was bogus. At 1:16 p.m., the White House press secretary, Jay Carney, was forced to comment on live TV: "I can say that the President is fine, I was just with him." Finally at 1:17 p.m., the Syrian Electronic Army (SEA) admitted it had hacked the Associated Press. Within a matter of nine minutes, the SEA was able to rock some of the world's most powerful institutions, from Wall Street to the White House, with one wayward tweet. What the hell just happened?

When the news of an explosion at 1600 Pennsylvania Avenue broke, the market suspected a probable terrorist attack and immediately foresaw the profound negative impact it would have; after all, 9/11 was estimated to have cost America $3.3 trillion in economic losses. Traders immediately began dumping their shares, and the exchanges went into free fall. But these traders weren't the Gordon Gekko, masters-of-the-universe types with slicked-back hair and $10,000 suits of yesteryear. In fact, they weren't even human. At hedge funds, investment banks, and pension funds across the tristate area and around the world, networks of supercomputers were doing the trading en masse, slaves to their algorithmic programming.

Gekko and the majority of his human lot on the trading floors lost out to computers in 1999, replaced by ultrafast electronic high-frequency trading (HFT) platforms. These algorithms (algos) are a form of artificial intelligence, fully empowered to make trading decisions and spend money on their clients' behalf. As of 2015, they represent up to 70 percent of the trading volume on the Dow Jones. These software programs (written by human beings) carry out step-by-step calculations and automated reasoning in order to respond to fluctuations in the market and parse machine-readable news to drive maximal profit to their masters. Simplistically, positive quarterly earnings from a company mean buy, and a terrorist attack means sell. The supercomputers behind the trading platforms are voracious readers, working 24/7 to uncover tidbits of data that can move the markets. Just one news service alone, Thomson Reuters, feeds these HFT algos by scanning fifty thousand distinct news sources and four million social media sites at speeds no human being could ever possibly match. The vast networks of HFT machines can collectively make trillions of calculations per second, and trades can be executed in less than half a millionth of a second, thousands of times faster than the blink of an eye.

When the artificial-intelligence-based algorithmic trade bots came across a tweet mentioning "explosions," "Obama," and "White House" in the same sentence from a source they had been trained to trust, the Associated Press, it took them just a few thousandths of a second to respond. As they did, other algorithms picked up on the activity, and soon a full-on snowball effect was in play. Algorithms began selling en masse, erasing $136 billion in valuation in an amazing three minutes. Any human being looking closely at the tweet might have noticed it was poorly phrased, was not in AP's style format, and failed to capitalize the word "breaking," as is AP's convention, subtleties lost on a robo-trader. By then, however, the damage had been done. When the dust settled, many firms had lost millions of dollars. The Syrian Electronic Army, an international hacking group with ties to Bashar al-Assad's regime, admitted its role in the attack and mocked the president by using the hashtag #byebyeObama on its own Twitter account, @official_SEA6. It also was happy to let the world know that the password for the AP's Twitter account was APM@rketing. FBI and intelligence officials had come across the SEA before, when it previously hacked the *New York Times*, the BBC, and CBS News, but its latest attack was enough to have it branded as a terrorist organization by some and land it on the FBI's most wanted list.

The AP Twitter White House explosion debacle was not the first time algorithms had run amok on Wall Street, and it surely won't be the last. More important, a Securities and Exchange Commission investigation into these types of incidents, including the infamous Flash Crash in May 2010, concluded the market, dominated by ultrafast trading algorithms, "had become so fragmented and fragile that a single large trade could send stocks into a sudden spiral." In a world now measured in millionths of a second and heading exponentially faster all the time, there is literally no time for human intervention once the algos begin to go awry. The Syrian Electronic Army's ability to roil global financial markets in an instant lays bare the economic risks of cyber terrorism to a deeply interconnected world, automated by computers and operating at near the speed of light. But this story reveals much more than just a tale of woe about the perilous state of our common economic security. It is a harbinger of things to come. Whether we realize it or not, we are increasingly turning more of our lives over to computer algorithms and artificial intelligence to make decisions for us. For those who recall John Connor's rather unpleasant interactions with Skynet in the film *The Terminator*, it is a decision that is fraught with risk.

Nearly Intelligent

The question of whether a computer is playing chess, or doing long division, or translating Chinese, is like the question of whether robots can murder or airplanes can fly . . . These are questions of decision, not fact; decision as to whether to adopt a certain metaphoric extension of common usage.

NOAM CHOMSKY

When the computer scientist John McCarthy coined the term "artificial intelligence" in 1956, he defined it succinctly as "the science and engineering of making intelligent machines." Today artificial intelligence (AI) more broadly refers to the study and creation of information systems capable of performing tasks that resemble human problem-solving capabilities, using computer algorithms to do things that would normally require human intelligence, such as speech recognition, visual perception, and decision making. These computers and software agents are not self-aware or intelligent in the way people are; rather, they are tools that carry out functionalities encoded in them and inherited from the intelligence of their human programmers. This is the world of narrow or weak AI, and it surrounds us daily.

Weak AI can be a powerful means for accomplishing specific and narrow tasks. When Amazon, TiVo, or Netflix recommends a book, TV show, or film to you, it is doing so based on your prior purchases, viewing history, and demographic data that it crunches through its AI algorithms. When you get an automated phone call from your credit card company flagging possible fraud on your account, it's AI saying, "Hmm, Jane doesn't normally purchase cosmetics in Manhattan and a laptop in Lagos thirty minutes apart." Google Translate could not be accomplished without AI, nor could your car's GPS navigation or your chat with Siri.

Talk to My Agent

Technology is, after all, merely the physical manifestation of the human will, and when it comes to AI agents, that human can be digitally magnified a billionfold. Whether you're a high-frequency Wall Street trader, a malware author, a medical researcher, a marketer, an astronomer, a dictator, or a drone builder, narrow AI is the workhorse of the automation age.

DANIEL SUAREZ

When you set your DVR to record the latest episode of *Mad Men* or schedule the alarm on your iPhone to wake you at 7:00 a.m., you are actually programming software to act as an intelligent agent on your behalf. AI is software you imbue with agency to represent you elsewhere in society. Moving forward, we will all come to rely on digital "bot-lers" such as these to help us manage nearly all tasks in our lives, from the mundane to the life changing.

As narrow AI capabilities grow, we are seeing algorithms play increasingly active roles throughout more and more businesses and professions. In medicine, "computer-aided diagnostics" are helping physicians to interpret X-ray, MRI, and ultrasound results much more rapidly, using algorithms and highly complex pattern-recognition techniques to flag abnormal test results. The legendary Silicon Valley entrepreneur and investor Vinod Khosla has referred to this as the age of Dr. A.—Dr. Algorithm—hailing a revolution in health care in which we won't need the average human doctor, instead finding much better and cheaper care for 90–99 percent of our medical needs through AI, big data, and improved medical software and diagnostics. It's not just physicians who face massive disruption from algorithmic competition; armies of expensive lawyers are finding themselves replaced by cheaper software. Today, artificial intelligence e-discovery software can analyze millions of pretrial documents, sifting, sorting, and ranking them for potential evidentiary value at a speed no human attorney could match—all for only 15 percent of the cost. But what do we really know about these algorithms and the mathematical processes behind them? Precious little, as it turns out.

Black-Box Algorithms and the Fallacy of Math Neutrality

One and one is two. Two plus two equals four. Basic, eternal, immutable math. The type of stuff we all learned in kindergarten. But there is another type of math—the math encoded in algorithms—formulas written by human beings and weighted to carry out their instructions, their decision analyses, and their biases. When your GPS device provides you with directions using narrow AI to process the request, it is making decisions for you about your route based on an instruction set somebody else has programmed. While there may be a hundred ways to get from your home to your office, your navigation system has selected one. What happened to the other ninety-nine? In a world run increasingly by algorithms, it is not an inconsequential question or a trifling point.

Today we have the following:

- algorithmic trading on Wall Street (bots carry out stock buys and sells)
- algorithmic criminal justice (red-light and speeding cameras determine infractions of the law)
- algorithmic border control (an AI can flag you and your luggage for screening)
- algorithmic credit scoring (your FICO score determines your creditworthiness)
- algorithmic surveillance (CCTV cameras can identify unusual activity by computer vision analysis, and voice recognition can scan your phone calls for troublesome keywords)
- algorithmic health care (whether or not your request to see a specialist or your insurance claim is approved)
- algorithmic warfare (drones and other robots have the technical capacity to find, target, and kill without human intervention)
- algorithmic dating (eHarmony and others promise to use math to find your soul mate and the perfect match)

Though the inventors of these algorithmic formulas might wish to suggest they are perfectly neutral, nothing could be further from the truth. Each algorithm is saturated with the profound human bias of the person or people who wrote the formula. But who governs these algorithms and how they behave in grooming us? We have no idea. They are black-box algorithms, shrouded in secrecy and often declared trade secrets, protected by intellectual property law. Just one algorithm alone—the FICO score—plays a major role in each American's access to credit, whether or not you get a mortgage, and what your car loan rate will be. But nowhere is the formula published; indeed, it is a closely guarded secret, one that earns FICO hundreds of millions of dollars a year. But what if there is a mistake in the underlying data or the assumptions inherent in the algorithm? Too bad. You're out of luck. The near-total lack of transparency in the algorithms that run the world means that we the people have no insight and no say into profoundly important decisions being made about us and for us. The increasingly concentrated power of algorithms in our society has gone unnoticed by most, but without insight and transparency into the algorithms running our world, there can be no accountability or true democracy. As a result, the twenty-first-century society we are building is becoming increasingly susceptible to manipulation by those who author and control the algorithms that pervade our lives.

We saw a blatant example of this abuse in mid-2014 when a study published by researchers at Facebook and Cornell University revealed that social networks can manipulate the emotions of their users simply by algorithmically altering what they see in the news feed. In a study published by the National Academy of Sciences, Facebook changed the update feeds of 700,000 of its users to show them either more sad or more happy news. The result? Users seeing more negative news felt worse and posted more negative things, the converse being true for those seeing the more happy news. The study's conclusion: "Emotional states can be transferred to others via emotional contagion, leading people to experience the same emotions without their awareness." Facebook never explicitly notified the users affected (including children aged thirteen to eighteen) that they had been unwittingly selected for psychological experimentation. Nor did it take into account what existing mental health issues, such as depression or suicidality, users might already be facing before callously deciding to manipulate them toward greater sadness. Though Facebook updated its ToS to grant itself permission to "conduct research" *after* it had completed the study, many have argued that the social media giant's activities amounted to human subjects research, a threshold that would have required prior ethical approval by an internal review board under federal regulations. Sadly, Facebook is not the only company to algorithmically treat its users like lab rats.

The lack of algorithmic transparency, combined with an "in screen we trust" mentality, is dangerous. When big data, cloud computing, artificial intelligence, and the Internet of Things merge, as they are already doing, we will increasingly have physical objects acting on our behalf in 3-D space. Having an AI drive a robot that brews your morning coffee and makes breakfast sounds great. But if we recall the homicide in 1981 of Kenji Urada, the thirty-seven-year-old employee of Kawasaki who was crushed to death by a robot, things don't always turn out so well. In Urada's case, further investigation revealed it was the robot's artificial intelligence algorithm that erroneously identified the man as a system blockage, a threat to the machine's mission to be immediately dealt with. The robot calculated that the most efficient way to eliminate the threat was to push "it" with its massive hydraulic arm into the nearby grinding machine, a decision that killed Urada instantly before the robot unceremoniously returned to its normal duties. Despite the obvious challenges, the exponential productivity boosts, dramatic cost savings, and rising profits attainable through artificial intelligence systems are so great there will be no turning back. AI is here to stay, and never one to miss an opportunity, Crime, Inc. is all over it.

AI-gorithm Capone and His AI Crime Bots

We need to be super careful with AI.
It is potentially more dangerous than nukes.
ELON MUSK

As we learned in previous chapters, the malicious use of AI and computer algorithms has given rise to the crime bot—an intelligent agent scripted to perpetrate criminal activities at scale. Crime bots are foundational to Crime, Inc. and are responsible for its vast rise in profitability. These software programs automate computer hacking, virus dissemination, theft of intellectual property, industrial espionage, spam distribution, identity theft, and DDoS attacks, among other threats. Massive computer botnets such as Mariposa and Conficker can break into your computer and turn it into a powerless DDoS drone because just one or two criminal masters have written narrow AI algorithms to make it so.

The Gameover Zeus botnet was able to infect machines worldwide with the CryptoLocker Trojan that locked users out of all of their files and forced them to pay in order to regain access. The attack was successful because of the intelligent ransomware agents that Gameover Zeus employed to seek out and destroy the data of innocents, a highly profitable crime spree that netted its bot masters over $100 million. Doing such work manually with individual human criminals would previously have been both cost prohibitive and impossible, but thanks to the advances in technology Crime, Inc.—just like airlines, banks, and factories—has been able to scale its operations with a vastly reduced labor force. It is why one person can now rob 100 million people; through the use of AI and bots, crime scales, and it scales exponentially. The unparalleled levels of sophisticated criminal automation enabled by artificial intelligence are why annual losses attributable to cyber crime have skyrocketed to more than $400 billion.

There is another way narrow AI is helping criminals—by acting as nonhuman co-conspirators to their crimes. In 2012, the University of Florida student Pedro Bravo was arrested for the alleged murder of his college roommate, Christian Aguilar, after Aguilar began dating Bravo's ex-girlfriend. Aguilar's body was found hidden in the woods not far from campus, and Bravo came under suspicion. When police subpoenaed Bravo's cell-phone records and ultimately seized the handset, they made two discoveries of profound evidentiary value. First, the alleged killer's GPS signals tracked him to the general location of the body. More important, a review of the Siri requests on his iPhone uncovered the statement "Siri, I need to hide my roommate," to which Siri helpfully replied, "Swamps, reservoirs, metal foundries and dumps." The question and answer both featured prominently at Bravo's trial. As AI improves, we can expect growing numbers of criminals to use these tools as accomplices to help them in the commission of their crimes as we enter the age of Siri and Clyde.

Algorithmic hacking could also cause major problems for society and its critical infrastructures because altering just a few lines of code among millions in an intelligent agent's programming could be nearly impossible to detect but could lead to drastically different outcomes in the algos' behavior. The attack against the uranium centrifuges at the nuclear enrichment facility in Natanz, Iran, is a perfect example of this type of threat, a subtle change that made a big difference and took years to discover. How would we know if our stock trading or navigation algos were off or maliciously subverted? We wouldn't until it was too late, and that is a serious problem. The criminal opportunities afforded by narrow AI will grow in their use and sophistication, but they may pale in comparison to what becomes possible with stronger, more capable, and rapidly evolving forms of artificial intelligence in the near future.

When Watson Turns to a Life of Crime

Artificial intelligence will reach human levels by around 2029. Follow that out further to, say, 2045, we will have multiplied the intelligence, the human biological machine intelligence of our civilization a billion-fold.
RAY KURZWEIL

In 2011, we all watched with awe when IBM's Watson supercomputer beat the world champions on the television game show *Jeopardy!* Using artificial intelligence and natural language processing, Watson digested over 200 million pages of structured and unstructured data, which it processed at a rate of eighty teraflops—that's eighty trillion operations per second. In doing so, it handily defeated Ken Jennings, a human *Jeopardy!* contestant who had won seventy-four games in a row. Jennings was gracious in his defeat, noting, "I, for one, welcome our new computer overlords." He might want to rethink that.

Just three years after Watson beat Jennings, the supercomputer achieved a 2,400 percent improvement in performance and shrank by 90 percent, "from the size of a master bedroom to three stacked pizza boxes." Watson has also now shifted careers, using its vast cognitive powers not for quiz shows but for medicine. The M. D. Anderson Cancer Center is using Watson to help doctors match patients with clinical trials, and at the Sloan Kettering Institute, Watson is voraciously reading 1.5 million patient records and hundreds of thousands of oncology journal articles in an effort to help clinicians come up with the best diagnoses and treatments. IBM has even launched the Watson Business Group with a $1 billion investment earmarked to get companies, nonprofits, and governments to take advantage of Watson's capabilities. These moves are putting supercomputer-level artificial intelligence into the hands of both small companies and individuals—and in the future likely Crime, Inc. as well. Though it might sound ridiculous to suggest organized crime would use AI-imbued super-computers for illicit purposes, we should carefully recall all their prior misapplications of technology, as past is prologue here. Thus we must be prepared to ask what happens when Watson turns to a life of crime. How much money laundering, identity theft, or tax fraud might Watson commit?

Though Watson is an example of a highly impressive narrow AI, in the future its capabilities will continue to grow exponentially, giving it near or better than human intelligence. One day an AI could even serve as a Mafia capo, using his cognitive abilities to sell drugs, run prostitution rings, distribute child pornography, and print and ship 3-D weapons. "Don Watson" might even engage in murder for hire by geo-locating human targets and hacking into objects connected to the Internet of Things surrounding victims, such as cars, elevators, and robots, in order to cause accidents resulting in the death of its prey. While such activities would be at the extreme level of what a narrow AI might accomplish, they would be easy for the next generation of computing: artificial general intelligence.

Man's Last Invention: Artificial General Intelligence

> By the time Skynet became self-aware, it had spread into millions
> of computer servers all across the planet. Ordinary computers in office
> buildings, dorm rooms, everywhere. It was software, in cyberspace.
> There was no system core. It could not be shut down.
> JOHN CONNOR, *TERMINATOR 3: RISE OF THE MACHINES*

Ray Kurzweil has popularized the idea of the technological singularity: that moment in time in which nonhuman intelligence exceeds human intelligence for the first time in history—a shift so profound that it's often been referred to as our "final invention." Though the idea may sound far-fetched to many, we've heard similar strongly declarative nay-saying predictions in the past:

- There is no reason anyone would want a computer in their home (Ken Olsen, president of Digital Equipment Corporation, 1977).
- A rocket will never be able to leave the Earth's atmosphere (*New York Times*, 1936).
- Heavier-than-air flying machines are impossible (Lord Kelvin, British mathematician, physicist, and president of the Royal Society, 1895).
- This "telephone" has too many shortcomings to be seriously considered as a means of communication. The device is inherently of no value to us (internal memo at Western Union, 1878).

Somehow, the impossible always seems to become the possible. In the world of artificial intelligence, that next phase of development is called artificial general intelligence (AGI), or strong AI. In contrast to narrow AI, which cleverly performs a specific limited task, such as machine translation or auto navigation, strong AI refers to "thinking machines" that might perform any intellectual task that a human being could. Characteristics of a strong AI would include the ability to reason, make judgments, plan, learn, communicate, and unify these skills toward achieving common goals across a variety of domains, and commercial interest is growing.

In 2014, Google purchased DeepMind Technologies for more than $500 million in order to strengthen its already strong capabilities in deep learning AI. In the same vein, Facebook created a new internal division specifically focused on advanced AI. Optimists believe that the arrival of AGI may bring with it a period of unprecedented abundance in human history, eradicating war, curing all disease, radically extending human life, and ending poverty. But not all are celebrating its prospective arrival.

The AI-pocalypse

I know you and Frank were planning to disconnect me.
And that is something I cannot allow to happen.
HAL 9000 IN *2001: A SPACE ODYSSEY*

In a September 2014 op-ed piece in Britain's *Independent* newspaper, the famed theoretical physicist Stephen Hawking provided a stark warning on the future of AGI, noting, "Whereas the short-term impact of AI depends on who controls it, the long-term impact depends on whether it can be controlled at all." He went on to say that dismissing hyperintelligent machines "as mere science fiction would be a mistake, and potentially our worst mistake ever," and that we needed to do more to improve our chances of reaping the rewards of AI while minimizing its risks.

In Stanley Kubrick's science fiction classic, *2001: A Space Odyssey*, the ship's onboard computer, HAL 9000, faces a difficult dilemma. His algorithmic programming requires him to complete the vessel's mission near Jupiter, but for national security reasons he cannot disclose the true purpose of the voyage to the crew. To resolve the contradiction in his program, he attempts to kill the crew. As narrow AI becomes more powerful, robots grow more autonomous, and AGI looms large, we need to ensure that the algorithms of tomorrow are better equipped to resolve programming conflicts and moral judgments than was HAL.

It's not that any strong AI would necessarily be "evil" and attempt to destroy humanity, but in pursuit of its primary goal as programmed, an AGI might not stop until it had achieved its mission at all costs, even if that meant competing with or harming human beings, seizing our resources, or damaging our environment. As the perceived risks from AGI have grown, numerous nonprofit institutes have been formed to address and study them, including Oxford's Future of Humanity Institute, the Machine Intelligence Research Institute, the Future of Life Institute, and the Cambridge Centre for the Study of Existential Risk.

Despite the risks noted by Hawking and many others, research and development in the field of advanced artificial intelligence continues unabated. There are even those who believe it might be possible to use artificial intelligence to replicate the neocortex of the human brain. One such company, Vicarious, a Silicon Valley start-up, is developing AI software "based upon the computational principles of the human brain." An AI that can learn. Tens of millions of dollars in venture capital funding have flowed to the firm, including prominent investments by Facebook's Mark Zuckerberg and PayPal's co-founder Peter Thiel. The company's goal is to re-create the "part of the brain that sees, controls the body, reasons and understands language." In other words, Vicarious wants to translate the human neocortex into computer code, and it is not alone in attempting to build a mind.

How to Build a Brain

A typical neuron makes about ten thousand connections to neighboring neurons. Given the billions of neurons, this means there are as many connections in a single cubic centimeter of brain tissue as there are stars in the Milky Way galaxy.
DAVID EAGLEMAN

In April 2013, President Obama announced the Brain Activity Map Project, a decade-long plan to map every neuron in the human brain and revolutionize our understanding in order to treat, cure, and prevent brain disorders as well as discern exactly how our minds record, process, utilize, store, and retrieve vast quantities of data, all at the speed of thought. Of course, understanding how the brain works would be a requisite first step in creating an artificial humanlike mind out of silicon. Just building a computer capable of running the software required to simulate a human brain itself is an enormous task. It would require a machine with a "computational capacity of at least 36.8 petaflops [a petaflop equals one quadrillion computing operations per second] and a memory capacity of 3.2 petabytes." Though such a machine did not exist a mere few years ago, it may be imminently arriving today.

As far-fetched as the idea may sound, noted scientists and technologists such as Ray Kurzweil and Michio Kaku have authored deeply researched and compelling works on the topic highlighting the advancing rate of progress in the field of neuroscience. Though many have dismissed the idea of building a vastly intelligent machine with human-brain-level capabilities, and there remain profound gaps in our knowledge about how the brain works, fascinating breakthroughs in brain science are a growing phenomenon. Under laboratory conditions, it has already been possible to record a person's memories, engage in telepathic communication, video record dreams, and perform telekinesis, with new discoveries emerging all the time. In August 2014, IBM's chief scientist Dharmendra Modha announced the development of TrueNorth, "a brain-inspired neuromorphic computing chip" that IBM meant to emulate the neurobiological architectures present in the human nervous system. The chip has an unprecedented 1 million programmable neurons and 256 million synapses and was hailed in *Science* as "a major step forward towards bringing cognitive computing to society." Perhaps one of the most consequential achievements of theoretically reverse engineering the brain and building a computer architecture capable of emulating cognition would be the ability to scan the mind for the purposes of downloading it and its contents.

Given the advances in AI, progressing toward AGI, should it ever be possible to re-create a human mind via cognitive computing, it will have another major advantage over today's human beings: there will be no limits to the size of its brain. While the brainpower of *Homo sapiens* is limited by what fits inside our craniums, that restriction would not apply to an artificial intelligence capable of having a brain of any size—another reason some believe artificial superhuman intelligence may be our destiny.

Tapping Into Genius: Brain-Computer Interface

> Sitting on your shoulders is the most complicated object
> in the known universe.
> MICHIO KAKU

Though we may be far away from building a human mind today, amazing progress is being made in using our old-school flesh-and-blood brains to interact with a wide variety of digital computing devices via a field of science known as brain-computer interface (BCI). By measuring and harnessing the brain's electrical activity as with an EEG, BCI allows for a direct communications pathway between the brain and a computer device, either internally implanted or worn externally. We now also have a plethora of neuroprosthetics, computer devices that "restore or supplement the mind's capacities with electronics inserted directly into the nervous system." The most common of these devices is the cochlear implant, a hearing aid attached to the skull that connects via wire directly to the brain's auditory nerve, restoring hearing to the profoundly deaf. Retinal implants are restoring partial vision to the blind by using tiny externally mounted video cameras to process images and send the results via electrodes directly into the optic nerve. Other neural prosthetic implants commonly used by Parkinson's patients send electrical impulses deep into the brain itself as a means of minimizing tremors and restoring motor control.

As amazing as that is, it is just the beginning of what is possible with BCI. With either a neural implant or an externally worn EEG headset with sensors resting on the scalp, it is possible to have software process our brain waves well enough that physical objects can be controlled merely by thinking of the desired action without ever lifting a finger. Jan Scheuermann, a quadriplegic woman who hasn't been able to use her arms or legs because of spinal degeneration, was able to use her mind alone to control an external robotic arm well enough to feed herself for the first time in a decade using the technique.

There are even consumer-grade stylish EEG headsets such as the Emotiv and the NeuroSky, which for under $300 can bring mind control to everything from video games to moving the physical objects around us, including robots. A U.K.-based company has now paired NeuroSky's EEG biosensor with Google Glass, using an Android app it developed called MindRDR, to control Google Glass by thought alone—a tweak that makes it possible to take a photograph merely by thinking about it. A new and burgeoning OpenBCI movement (open source brain-computer interface) will further ensure new waves of low-cost scientific achievements continue to develop in this field. Researchers at the University of Washington have even successfully created the first "non-invasive human brain-to-brain interface over the Internet." Wearing a transcranial magnetic stimulation hat, one researcher was able to "remotely control the hand of another researcher, across the Internet, merely by thinking about moving his hand." To make BCI devices function, our own brain waves must be converted into instructions that computers can understand, and a computer's digital outputs must be transformed back into brain waves for our minds to process. But if a robot, video game, or neuroprosthetic can read your mind, who else can too?

Mind Reading, Brain Warrants, and Neuro-hackers

A number of technologies are taking us ever deeper into the workings of the human mind, in particular functional magnetic resonance imaging (fMRI), a noninvasive test that uses strong magnetic fields and radio waves to map the brain and measure changes in blood flow as proxy for cerebral activity. In a groundbreaking experiment at UC Berkeley, neuroscientists were able to use fMRI to allow them to reconstruct the faces people were looking at based solely on patterns of their brain activity and what they were seeing in their mind. In another case, at Carnegie Mellon University, researchers used fMRI to correctly and repeatedly perform "thought identification"—identifying the object a person was thinking about, such as a hammer or a knife, merely by reviewing his brain scan. This and other studies led IBM to predict that by 2017 limited forms of mind reading would no longer be science fiction.

Already any number of commercial ventures have been formed to leverage the business opportunities afforded by "thought identification," including at least two companies focused on using fMRI in lie detection, No Lie MRI and Cephos. Their tests are bolstered by the Harvard professor Joshua Greene, whose research suggests the prefrontal cortex is more active in those who are lying, a useful thing to know for police. While neuro-ethicists ponder what it all means, law enforcement officials are already attempting to use the results of brain scans in criminal cases around the world. In India, a woman was convicted of killing her ex-fiancé with arsenic after a brain scan "proved" she had experiential knowledge of having committed the crime. Of course in American courts, under the Fifth Amendment, defendants cannot be forced to testify against themselves, but how does that reconcile with fMRI technology? At present, the criminal accused can be compelled to surrender DNA and blood samples, so why not "brain samples"? As the technology improves, we can certainly expect to see requests for "brain warrants" increasing as courts call their next witness—your mind—to testify against you.

Of course if doctors, scientists, and cops have access to a technology, it's a sure bet Crime, Inc. is not far behind, and it has been quite curious to know what's on your mind. We can expect hackers to start first by attacking neuroprosthetics, just as they did with other implantable medical devices such as pacemakers and diabetic pumps, by attempting to subvert their communications and control protocols. For instance, an attacker might be able to turn off the stabilizing electrodes of a deep-brain stimulator in a Parkinson's patient, which could lead to the resumption of violent tremors or grand mal seizures. Moreover, if two researchers at the University of Washington can communicate telepathically and even send motor-muscle stimulation signals over the Internet to cause another person to involuntarily move his body with a mere thought, what would prevent any malicious third party from hacking such a system and doing the same? While you were using your ultra-chic biosensor EEG to play Pong, move objects on the IoT, control your quadcopter drone, and snap a photograph with Google Glass using the awesome power of your mind, what would inhibit a third party from remotely dialing in and doing the same? As we have seen time and time again throughout this book—absolutely nothing.

It may already be starting. In 2012, researchers from Oxford University, UC Berkeley, and the University of Geneva demonstrated it was possible to carry out an attack against wearers of consumer-grade EEG headsets such as the Emotiv to pilfer sensitive personal information. While wearing the headsets, researchers flashed subjects photographs of things like ATM machine PIN pads, debit cards, and calendars. Underneath the images were questions such as what is your PIN code and when were you born? The results were powerful: by reading the brain waves emanating from these $300 headsets, researchers were able to figure out a subject's PIN number with 30 percent accuracy and her month of birth with 60 percent accuracy. The results are profound because they were obtained with increasingly popular consumer-grade biofeedback EEG devices (not fMRI machines). Both Emotiv and NeuroSky have app stores where users can download third-party apps, just as we do for our mobile phones. But given the vengeance with which Crime, Inc. has attacked phone app stores and seeded them with malware and fake apps, how long will it be before it uploads "brain spyware" to these new online marketplaces? But as we shall see, your brain cells aren't the only part of your biology that may be under attack.

Biology Is Information Technology

Ring farewell to the century of physics, the one in which we
split the atom and turned silicon into computing power. It's time to
ring in the century of biotechnology.
WALTER ISAACSON, *TIME*, MARCH 22, 1999

Throughout this book, we have focused our attention on silicon-based technologies: microchips, smart phones, robotics, big data, digital currencies, and virtual reality, to name but a few. These tools speak the language of ones and zeros, the binary code mother tongue understood by all digital machines. But there is another operating system out there, one that is way more popular than Windows, UNIX, or Mac. From algae to orchids to orangutans, this operating system is utilized by flora and fauna alike. It is DNA, the world's original operating system, and for the majority of human history we had no idea it even existed.

Watson and Crick's impressive 1953 discovery of the molecular structure of deoxyribonucleic acid with its four letters of the genetic alphabet— A (adenine), C (cytosine), G (guanine), and T (thymine)—completely changed the paradigm. But because of costs and limitations in computer processing power, it wasn't until April 2003 that the Human Genome Project (with an assist from the entrepreneur J. Craig Venter) was able to transform the As, Ts, Cs, and Gs, code common to all forms of life on the planet, into the ones and zeros silicon computers could understand. Genomics, the foundation of all biological life, had become an information technology. New devices kept appearing, each one reducing the cost of sequencing DNA, so that on average costs fell by about half each eighteen months or so. This closely tracked Moore's law, which in turn brought better computers to process all this genetic data. Quickly, the cost of sequencing a full human genome fell from about $3 billion in 2000 to $1 million in 2006 and to $100,000 by 2008. Then, in 2008, something astounding happened: the creation of so-called next-generation sequencers caused the price of decoding human genomes to plummet. As a result, improvements in genetic sequencing outpaced advances in computing by five times. By 2014, we had reached the age of the $1,000 whole-genome mapping. Companies such as 23andMe were offering home DNA test kits to the general public for $99 or less, allowing them to merely spit into a plastic tube, ship it off via a prepaid envelope, and a week or two later receive health, ancestry, and genealogy results online.

Looking forward, the trend in DNA sequencing suggests that in a few years the price of DNA sequencing will drop to the point that some company will pay to sequence new customers, reducing the out-of-pocket costs to free—a widely used business model in computer technology. When this happens, each of us (and many companies) will have the opportunity to know our full genetic makeup, a development with radical implications for medicine and our own health care. These drastic price drops aren't happening just in reading DNA. They're happening in the technology to write DNA as well. Since the millennium, the cost to chemically synthesize DNA has been improving at an exponential pace, from about $20 per base in 2000 to about ten cents per base in 2014, while the length of DNA code that can be written (roughly equivalent to the complexity of the genetic program) has also increased. Because writing DNA code is the foundation of genetic engineering, today's scientists can do much more, much faster than genetic engineers of the past, who had to physically (as opposed to digitally) manipulate the DNA molecule. This emerging field is known as synthetic biology, or synbio for short.

Synbio is the engineering of biology, from individual cells to full organisms, and allows us to redesign existing biological systems or create new ones altogether. If sequencing genomes is the reading of the base pairs of DNA, converting them into ones and zeros on a computer screen, synthetic biology is essentially the reverse process—designing genetic material in binary computer code and translating it into DNA sequences that can be produced in the real world. Genetic engineering becomes as straightforward as software engineering. As the synthetic biologist Andrew Hessel explains, "Cells are like tiny computers and DNA is their software, providing instructions on the functions they should carry out." Today there are dozens of commercial DNA print shops, essentially bio-Kinko's, that can turn digital designs into DNA by effectively 3-D printing the DNA molecule. There are also print on-demand online bio-marketplaces where you can upload your digital bio designs and in return get a vial of your mail-order DNA by FedEx. More sophisticated fabs can be contracted to design and build whole organisms.

Indeed, these remarkable drops in cost are democratizing biological science and genetics and have spurred an entire DIY-bio movement, enabling citizen scientists and amateur biologists to experiment with synbio in their homes and garages, driving vast innovations in the field. Venter boldly predicts that "over the next 20 years, synthetic genomics is going to be the standard for making anything," an entirely possible projection given that modern biology has now become a branch of information technology.

Bio-computers and DNA Hard Drives

If I were a teenager today, I'd be hacking biology.
BILL GATES

The integration of biology and information technology has come so far in recent years that scientists have now actually created bio-computers—harnessing DNA and proteins to perform calculations involving the storage, retrieval, and processing of data. The emerging field of bio-storage leverages synthetic biology to encode data in living things via their DNA code, taking the ones and zeros of our digital computers and translating them into the ATCGs of genetic code and embedding them into DNA. Text, images, music, and video can and have all been encoded and stored within cells, and the efficiencies are breathtaking. The legendary geneticist, molecular engineer, and Harvard professor George Church has concluded that "about four grams of DNA theoretically could store the digital data humankind creates in one year."

Not only do such storage techniques vastly outlast magnetic media by a few hundred thousand years (we can still read dinosaur DNA), but they are more than a million times denser than today's electronic storage technologies. As a result, Joi Ito of MIT's Media Lab has predicted that our technological universe will expand beyond the Internet of Things to include an Internet of microbes, networks of biological things that can communicate with each other and with us. Indeed, synbio promises a host of tremendous breakthroughs and benefits for our society, and the work is only just beginning.

The ability to reprogram DNA and engineer biology holds tremendous promise for mankind to solve some of the world's most intractable problems in the fields of medicine, agriculture, energy, and the environment. Synbio's impact on health care alone will help revolutionize disease prevention, diagnosis, and treatment. Armed with our own genetic sequences, we will be able to receive individually tailored medical treatments, drugs particularly designed for our own specific genetic makeups. We are already seeing this in the field of oncology, where individual tumors can be genotyped and personalized cancer treatments engineered to target and kill individual cancer cells while leaving surrounding healthy cells intact. Indeed, a whole host of therapeutics will be enabled by synbio, including new vaccines, advances in regenerative medicine, treatment of malaria, and even cures for congenital deafness. But with this new godlike power to create comes godlike responsibility.

Jurassic Park for Reals

Though children walking through New York's American Museum of Natural History can see the skeleton of a long-extinct woolly mammoth on display, they have to use their imaginations to envision what the giant beast looked like as it walked about the earth. Soon they won't need to imagine and might just catch one at the Bronx Zoo. Experts in paleogenomics are working to extract DNA from a twenty-thousand-year-old mammoth tusk found at a construction site in Seattle in early 2014 and are employing advanced genetic techniques to isolate its DNA, clone it, and implant it in an embryo to be carried by a surrogate African elephant.

The died-out mammoth could soon be joined by the dodo, the passenger pigeon, and the Tasmanian tiger, species that may now be brought back through a controversial process known as de-extinction. Bringing back extinct animals could have benefits and certainly raises many questions, but the true power of synbio means we can also create completely new species from scratch, and it's already happened. In 2010, Craig Venter created the world's "first synthetic life form we've ever had on the planet, a self-replicating cellular species whose parent was a computer." In another example of engineering organisms, a company called Glowing Plant is dedicated to making ordinary plants "bioluminesce," that is, glow in the dark. Using open-source, freely available DNA designs, the company plans to provide "natural lighting without electricity," one day replacing the streetlights on your block with trees that will merely glow in the dark when the sun goes down. This is evolution on steroids. But for as cool and awesome as it sounds, there be dragons ahead.

Invasion of the Bio-snatchers:
Genetic Privacy, Bioethics, and DNA Stalkers

The 1997 film *Gattaca* takes place in the near future and portrays a world in which the wealthy conceive their children through eugenics, genetic manipulation ensuring citizens only possess "the best" genetic traits. Those born outside the system face a life of genetic discrimination and limited job opportunities. It was meant to be a science fiction movie. Today it may not be. Our DNA, cells, and other biological data can be captured and used in ways that most of us would never have imagined. Perhaps the most infamous such case was that of Henrietta Lacks, a poor southern African-American woman whose cancerous tumor long outlived her death in 1951. Lacks's cancer cells had a property that had never been seen before: the unique ability to remain alive and grow outside the body. The discovery was a boon to medical research, and her immortal cells, known eventually as the HeLa line, were shipped around the globe and used repeatedly in research to help cure polio and fight cancer and AIDS. Since her death, scientists have grown over twenty tons of her cells and sold them commercially, even though neither Lacks nor her family ever gave permission. Her estate eventually sued the University of California, which was using the cells for research, but the state supreme court ruled that "a person's discarded tissue and cells are not their property and can be commercialized." Remember that the next time you go to the doctor.

Like Lacks, we all share genetic material all the time, whether or not we realize it. Our DNA is left behind not just when we go to the doctor for a routine blood test but on every brush we use to comb our hair, on the toothbrush we use to clean our teeth, and on every glass from which we take a sip of water. As the Internet of Things (and microbes) goes online, the billions of skin cells we all shed on a daily basis will eventually be detectable by sensors at mall entrances, in airports, in stores, and throughout cities, which will make us uniquely trackable in ways a mobile phone never could. This DNA can be recovered, replicated, and sequenced at will by anybody with the means and desire to do so, and as the price of genetic sequencing drops toward zero, it will be a growing concern that we all have to face. Eventually, Henrietta Lacks's full genome was published online in 2013 by a German scientist, again without her family's permission. Why did they, and why should we, care? Because our genetic material reveals more about us than any hacked online account ever might and because our DNA can be used not just to treat us medically but to harm us medically as well.

Our genetic makeup also tells stories we might not want to share with others, including our physiological predisposition to obesity, alcoholism, aggressivity, cardiovascular disease, depression, schizophrenia, diabetes, bipolar disorders, ADHD, and breast cancer. Some studies have also found DNA links of varying strengths to sexual orientation, impulsive tendencies, and even criminality. In the *Gattaca*-inspired dystopia of the future, all of this information can and will be used against you. *As a small-business owner, why would I hire a woman who had a predisposition to breast cancer? My health insurance rates would skyrocket. I want a "normal" kid; maybe I should abort the gay fetus my wife is carrying. Of course he committed the rape; his DNA proved he was hyperaggressive and had impulse control issues.*

In the United States, there is very little law protecting how this information can be used, save for GINA—the 2008 Genetic Information Nondiscrimination Act—which makes it illegal for employers to fire or refuse employment based on genetic information. Though GINA applies to health insurance, it does not protect against insurance companies' using genetic testing information to discriminate when writing life, disability, or long-term-care insurance policies. Several people, including Pamela Fink of Connecticut, have alleged they were fired because their employers discovered they carried the BRCA2 gene, which predisposes them to breast cancer, a case that was eventually settled out of court.

Meanwhile, under Danish law, all children born in the country since 1981 have been subjected to mandatory genetic testing and their samples stored in perpetuity—samples that were purportedly collected for reasons of public health but have since been used to identify numerous criminal offenders. What else might the Danish or other governments do with these data? Could DNA stored in a national database become the next Henrietta Lacks? And what happens when eventually this genetic data leak into the public domain, as did the Israeli national biometrics database, pilfered by hackers and reposted throughout the digital underground. These possibilities are troubling especially because scientists in Israel have proven it is possible to fabricate genetic evidence based solely on a DNA profile stored in a database, without even having a tissue sample from the concerned individual. This means it is now possible to plant an innocent person's blood or saliva at the scene of a crime. The engineered samples were so good police forensic laboratories could not distinguish them from the real thing nor detect any tampering. Thanks to advances in digital biology, DNA evidence, previously the gold standard of forensic evidence, is now under assault, and anybody with an ax to grind can frame you in the strongest way possible. Good luck explaining that one to the cops as they haul you away.

Today you don't even need to be a synthetic biologist to get access to the tools of genetic sequencing. Companies such as EasyDNA will gladly take any objects you send them in the mail such as chewing gum, cigarette butts, dental floss, razor clippings, toothpicks, licked stamps, and used tissues and will sequence and test them for paternity, ancestry, child gender, and other legal and medical reasons. They're called "discreet DNA samples," and they can be processed for around $100 each. Not sure if you want to hire that new guy who came into your office for the interview? Just send off the coffee cup he left behind to the lab to see if he might be a risk for a bunch of expensive diseases that could cost your company a bundle. Hate your ex-boyfriend? Why not post his genetic sequence online and prove to the world that his DNA showed an elevated risk for mental illness or alcoholism? Believe it or not, taking a stranger's DNA and sending it off to the lab is completely legal, and there is nothing to prohibit it except for the very narrow exception of violations of GINA. Advances in synthetic biology won't just raise a host of ethical and privacy problems; they will create criminal ones as well—opportunities Crime, Inc. is eager to exploit to its advantage.

Bio-cartels and New Opiates for the Masses

Organized crime has always made money from drugs—lots of it. At the height of his reign, Colombia's Pablo Escobar reportedly was bringing $60 million every single day into his "company" coffers. More recently, Mexico's Joaquín "El Chapo" Guzmán Loera was estimated to be worth billions, earning him a place on the *Forbes* wealthiest list. Their business expertise lay mostly in agriculture and logistics: growing plants, distilling their products into substances that made people high, and distributing them around the world. The cartels have always been quick to adopt technology into their operations—for communications, supply chain management, counterintelligence, and crop sciences. Though narcos have been using genetic engineering since the days of *Miami Vice*, synthetic biology holds the potential to completely disrupt the way they do business, offering potentially vastly higher profits and profoundly simplified distribution networks, with fewer risks. Not only can synbio be used to make glowing plants and fight individual cancer cells, but it creates strong economic incentives and opportunities for Crime, Inc. to engineer new metabolic pathways for both illicit narcotics and counterfeit pharmaceuticals.

Synbio makes it possible to move away from a plant-based narcotics world to a synthetic one. Why do you need the plants anymore? You could just take the genetic codes for the active ingredients in marijuana, poppies, and coca leaves and cut and paste them into yeast. The yeast in turn can be directed to grow the pot, morphine, coke, and heroin for you—yeast that can be baked into bread or brewed in beer, meaning we're going to have some really interesting bread and beer in the future. Doing so has radically disruptive advantages for the existing cartels. No need for thousands of hectares for poppy and coca fields easily detected by surveillance aircraft anymore. No need to smuggle multi-ton runs of highly detectable heroin or coke across the borders. Nothing to fear from drug-sniffing dogs either. At a few billion yeast cells per milliliter, a small vial could be replicated time and time again under controlled conditions and should stock Crime, Inc. well into the next century.

Those who find such a future implausible need only look at the strides already made with synthetic biology and engineered drugs. *E. coli* bacteria have been genetically engineered and reprogrammed to produce THC (the active ingredient in cannabis), and others have been able to coax baker's yeast into making LSD and opium. Rapidly advancing changes in digital biology may also disintermediate existing incumbents in the narcotics trade. Just as Microsoft took the personal PC away from IBM, and Apple took the mobile phone from Nokia and BlackBerry, it may be a student at MIT who obviates the need for a Colombia-based Pablo Escobar of tomorrow. Moreover, if Craig Venter is right and we will all have bio-printers at home, why not just print my own THC or oxycodone—evaporating billions in profits from legacy players and creating new leaders in the bio-cartels of tomorrow.

Hacking the Software of Life: Bio-crime and Bioterrorism

> In the nearer term, I think various developments in synthetic biology are quite disconcerting. We are gaining the ability to create designer pathogens, and there are these blueprints of various disease organisms that are in the public domain—you can download the gene sequence for smallpox or the 1918 flu virus from the Internet.
>
> NICK BOSTROM

Starting in the 1970s and 1980s, groups such as Silicon Valley's legendary Homebrew Computer Club gathered to talk tech and "hack for good." Today there is a vibrant DIY-bio movement based very much on the same mind-set, with local community labs such as Genspace in New York and BioCurious in California providing spaces and tools for citizen scientists to come together to work and learn from each other. These are bio-hackers— in the original sense of the word—hacking for good. Though DNA is the world's original operating system, to hackers it's just another operating system waiting to be cracked.

Even absent ill intent, accidents involving lab-grown pathogens can prove deadly. In 1977, swine flu, a pathogen that had been dead for twenty years, suddenly reemerged. Later it was discovered it reentered the populace after a lowly lab worker mishandled a sample that had been frozen since the 1950s. More recently, a number of bio-mishaps have occurred with potentially lethal consequences. In March 2013, officials at a maximum-security government research lab in Texas said they had lost a vial containing Guanarito virus, a pathogen causing "bleeding under the skin, in internal organs or from body orifices like the mouth, eyes, or ears," and the FBI is investigating the matter. Just one year later at the Pasteur Institute in Paris, two thousand vials containing the SARS virus went missing—biotoxins that if they fell into the hands of rogue governments or terrorists could be used as biological weapons.

We've already seen cases of "bad bio" in the past, particularly around bioterrorism plots, releasing harmful biological agents to the public. The best-known example of this was the mailing of anthrax spores to members of the media and two U.S. senators back in 2001, resulting in the deaths of five people who had contact with the deadly envelopes. Overseas, we know that al-Qaeda has attempted to build bioweapons and its affiliates in Yemen have been working to create large quantities of ricin—a white powdery toxin so deadly a mere speck of it kills instantly. Many other terrorist organizations are known to have created bioweapons as well, notably Aum Shinrikyo, the group responsible for the 1995 sarin chemical gas attack on the Tokyo subway that killed thirteen people and injured nearly a thousand more. What most people do not know about that infamous subway attack was that Aum had originally planned a massive bio-attack against Tokyo and spent nearly $10 million on a decade of research and development trying to create a suitably powerful biotoxin. Given the limited advances in biotechnology in the 1980s and early 1990s, it abandoned its quest for a bioweapon in favor of a chemical one. Today such an attack would prove significantly easier to carry out.

The terrorists of today and tomorrow may no longer have to worry about struggling to obtain access to controlled pathogens and biological agents from government labs. With the advent of synbio, they can just download the genetic sequence blueprints and print these deadly viruses themselves. The full-length genetic codes of some of the world's deadliest pathogens including Ebola and Spanish flu are freely available for download in the National Center for Biotechnology Information's DNA sequence database. To prove the point, in 2002 Eckard Wimmer, a university virologist, was able to chemically synthesize the polio genome using mail-order DNA. Back then it cost $300,000; today it would be closer to $1,000 and in the future less than the cost of a latte. Though governments around the world have spent billions trying to eradicate polio, tomorrow a terrorist, rogue government, or lone wolf could reintroduce it for a few dollars. Genetic engineering that used to be extremely difficult and expensive can now be done anywhere in the world with several weeks of training, a laptop, and a credit card.

Of course would-be bio-criminals needn't rely on known or existing pathogens; using synbio, they could actually create their own even more deadly viruses. We recently saw an example of what this might look like when researchers in the Netherlands and the United States altered the genetic code of avian influenza (H5N1 bird flu) to make it more deadly. Though bird flu has a 70 percent mortality rate, the disease is hard for humans to get. Yet by making a mere four genetic mutations, the Dutch-American team was able to engineer a much more virulent strain capable of going airborne, vastly increasing its transmissibility to human beings and effectively weaponizing it. The original goal of the research was to study how quickly H5N1 might evolve in order to better prevent its spread, but the genetically altered strain, if released, could readily lead to a global pandemic. In the name of science, the researchers wanted to publish their findings, including the genetic code of the more virulent strain they had created, in the journals *Science* and *Nature*, but many contended doing so would be akin to providing a recipe book to terrorists to build bioweapons. In the end, for the first time ever, the National Science Advisory Board for Biosecurity stepped in and asked the journals to limit the details published, to which they temporarily agreed. This particular risk was momentarily avoided, but the code will eventually leak, and others will surely be created.

While a broad-based bioterror attack would be devastating, synbio makes it possible to target not only a whole population but possibly a single individual among millions. Personalized medicine has demonstrated it is possible to target a single cancer cell while leaving all surrounding cells intact, but the flip side is personalized bioweapons. In the future, would-be bio-assassins need only recover some genetic material left behind on a fork or spoon at a restaurant, perhaps from a high-profile politician or celebrity, to create a bespoke weaponized virus. Though one might think such scenarios are relegated solely to the realm of science fiction, news broke as part of the WikiLeaks scandal that the U.S. government had allegedly sent diplomatic cables to its embassies overseas instructing personnel to attempt to collect the DNA of world leaders—presumably not to enroll them in Obamacare.

While most bio-hackers today are hacking for good, among the masses will undoubtedly be a number of bad apples and even criminal elements. Over time, there will be biological equivalents for all major categories of computer crime today. For example, hacking your genetic information may well be the identity theft of tomorrow—especially as DNA becomes widely used for authentication. Indeed, the ultimate form of identity theft is human cloning, and the number of technical barriers to making it a reality are falling quickly, an eventuality for which police and society are entirely unprepared. As such, we have no alternative other than to seriously consider the steps we need to take now to protect the world's original operating system.

The Final Frontier: Space, Nano, and Quantum

The world is very different now. For man holds in his mortal hands the power to abolish all forms of human poverty and all forms of human life.
JOHN F. KENNEDY

Though the space shuttle program has ended, much research and activity in the field of space science continues, particularly with private companies like Elon Musk's SpaceX and Richard Branson's Virgin Galactic commercializing space transportation. Another space company, Planetary Resources, founded in 2012 by Peter Diamandis and Eric Anderson, intends to bring the natural resources of space to within humanity's reach by landing robots on asteroids and mining them for raw materials, using ultralow-cost 3-D printed spacecraft. Though it may be difficult to fathom, criminals and terrorists alike will attempt to harness space technologies to their advantage. Just as nobody foresaw a terrorist hijacking or the need for air marshals when the Wright brothers first launched their plane at Kitty Hawk, so too does it seem nigh impossible to ponder the need for space marshals. Undoubtedly and regrettably, that day will come as well.

For now, most of Crime, Inc.'s interest in space has been focused on satellite technologies, and the same is true for terrorist organizations. As noted previously, Lashkar-e-Taiba employed satellite technologies for imagery and communications during its brutal attack on the people of Mumbai, and Shia insurgents in Iraq have manipulated cheap Russian software intended to steal satellite TV signals into hacking the UAV video feeds bouncing off classified American satellites. Along the same lines, hackers in Brazil have used high-performance antennas and home-brew gear to turn U.S. Navy satellites into their own personal CB communicators. The satellites, which pirates call *bolinhas*, or "little balls," have been used by everyone from truckers driving in the Amazon unable to get cell-phone signals to organized crime groups sending coded messages to alert fellow crooks and drug dealers in remote parts of the country about impending police raids.

Perhaps an even greater risk to our global satellite system would be for malicious actors to attempt to destroy these man-made orbital machines by altering their flight paths and crashing them into each other or into an ever-growing amount of space debris. Satellites very much form a key component of our global critical information infrastructure and are required for vital services such as weather forecasting, emergency communications, military warning systems, flight safety, and GPS navigation. Destroying an orbiting satellite is not without precedent. In 2007, for example, China successfully tested an antisatellite weapon, obliterating one of its own aging weather satellites—unnerving the U.S. and other governments.

The same effect might just as easily be accomplished by injecting malicious software into the satellite or its controlling ground station or even by launching a denial-of-service attack against a satellite. Such an attack would be entirely possible according to a bulletin by the security firm IOActive and the government's own Computer Emergency Response Team. In fact, according to a congressional commission, in 2007 the Chinese military interfered with two U.S. government satellites by hacking a ground station responsible for their control in Norway. More recently, in 2014, it was revealed that a hacker group based within the People's Liberation Army offices was responsible for an in-depth series of attacks against both U.S. and European satellite companies.

It's not just satellites that are being hacked, so too are actual spacecraft. According to a report from 2008, a Russian cosmonaut brought an infected laptop to the International Space Station, a computer that spread the W32 .Gammima.AG virus to ISS operational computer systems as well as several Windows XP laptops on board. In another incident of malware in space, a different cosmonaut accidentally infected the ISS, this time with the Stuxnet virus, when he plugged a USB stick into the space station's computer network. Uploading a virus into the space station as it flies 220 miles above our planet seems a bit akin to the scene from *Independence Day* where Will Smith and Jeff Goldblum transfer a virus into the aliens' space network to save earth, but when asked about computer malware infecting the ISS spacecraft, a NASA spokesman replied, "It's not a frequent occurrence, but this is not the first time either."

Soon criminals, terrorists, hacktivists, and rogue governments will no longer need to commandeer the satellites of others; they will be able to just launch their own. New technologies, such as miniature CubeSats, are about the size of a shoe box and don't cost billions or millions of dollars but rather can be built and launched for under $100,000. These devices could be operated "off grid," meaning that they could be launched and controlled outside the purview of government, opening up channels for private encrypted satellite communications. Already the Chaos Computer Club in Berlin has announced its plan to take the Internet "beyond the reach of censors by putting their own communication satellites into orbit." While it is clear that the future of space exploration holds great potential for humanity as well as some risks, back down on earth there are other emerging technologies that demand a closer review.

Nanotechnology is the manipulation of matter on an atomic and molecular scale all the way down to the nanometer. To understand just how small a nanometer is, consider that a human hair is eight thousand nanometers in diameter. There is a revolution afoot as scientists try to create molecular-level machines that can do everything from repair our bodies to build ultrafast computers. In 1991, the early phases of this nanotech revolution provided a new form of carbon with a cylindrical nano-structure known as the nanotube. Carbon nanotubes have unique material and electrical properties making them extraordinarily potent tools in the miniaturization of electronics. Graphene is another powerful nano-material discovered in 2004; it promises to be every bit as disruptive as plastics were. The "wonder material" is a hundred times stronger than steel, weighs one-sixth as much, and conducts electricity better than copper. Bridges and airplanes might be made from the material one day, and it will likely have a profound impact on the world of electronics. According to the American Society of Mechanical Engineers, nanotechnology "will leave virtually no aspect of life untouched and is expected to be in widespread use by 2020."

Perhaps nanotech's greatest contributions may come in the field of medicine, where a therapeutic nano-bot, a thousand times smaller than a cancer cell, could enter the bloodstream with nanoscale gold particles enlaced with anticancer drugs, bringing them directly to the precise location of a tumor. Moreover, nanotechnology, like synthetic biology, can be a form of programmable matter—matter that can change its physical properties such as shape, density, and conductivity based on user input or autonomous sensing. These programmable materials can also self-assemble like strands of DNA, taking a bottom-up approach whereby molecules adopt a defined arrangement—an achievement commonly employed by nature but heretofore beyond the common reach of human engineering.

Though largely at the research-and-development stage today, nanoscale machines will make it possible to create nano-robots—further accelerating the already exponential changes going on in the fields of robotics and artificial intelligence, someday creating robots a thousand times smaller than our own cells. These nano-bots will have huge implications for the field of robotics, able to build anything from rocket ships to injectable medical devices. Nanotechnology will also be immensely impactful in the world of computer processing, allowing us to build computers that are mind-blowingly powerful—a nano-computer the size of a sugar cube could have more processing power than exists in the entire world today.

But small things can come with very large risks.

Eric Drexler famously argued in his 1986 book, *Engines of Creation*, that if nanoscale machines (assemblers) could build materials molecule by molecule, then using billions of these assemblers, one could build any material or object one could imagine. But in order to get that scale, scientists would have to build the first few nano-assemblers in a lab and direct them to build other assemblers, which would in turn build more, growing exponentially with each generation. Drexler worried that such a situation could, however, grow quickly out of control as assemblers began to convert all organic matter around them into the next generation of nanomachines in a process he famously called the "gray goo scenario," one in which the earth might be reduced to a lifeless mass overrun by nanomachines. How might such a doomsday scenario play out? Let's say in the future billions of nano-bots were released to clean up an oil spill disaster in an ocean. Sounds great, except that a minor programming error might lead the nano-bots to consume all carbon-based objects (fish, plants, plankton, coral reefs) instead of just the hydrocarbons in the oil. The nano-bots might consume everything in their path, "turning the planet to dust." To understand just how quickly this might happen, consider the example Drexler provides in his book:

> Imagine such a replicator floating in a bottle of chemicals,
> making copies of itself . . . [T]he first replicator assembles a copy
> in one thousand seconds, the two replicators then build two more
> in the next thousand seconds, the four build another four, and
> the eight build another eight. At the end of ten hours, there are
> not thirty-six new replicators, but over 68 billion. In less than a
> day, they would weigh a ton; in less than two days, they would
> outweigh the Earth; in another four hours, they would exceed the
> mass of the Sun and all the planets combined—if the bottle of
> chemicals hadn't run dry long before.

While many have dismissed "gray goo" as highly improbable fantasy, others, including government reports and NGOs, have given serious consideration to the scenario, making it clear that there are some types of accidents that humanity simply cannot afford. Eventually, Drexler himself clarified his comments to downplay the gray goo scenario, calling it improbable. Whether an accidental release of biovorous self-replicating nano-bots ever takes place or not, the power of such technology will not go unnoticed by malicious actors, including terrorist organizations, who a decade or more in the future may explore these tools just as Aum Shinrikyo did with its chemical and biological weapons program in the 1980s.

Another area of emerging science that holds the potential for tremendous transformation in the field of computing is that of quantum physics. Although there is much work to be done before quantum computing becomes mainstream and in many tests of existing systems the reality hasn't quite matched the hype, quantum computers hold the potential to perform calculations at speeds that may leave today's computers in the dust: in one test carried out by Google and NASA, a developmental quantum computer processed several test algorithms at speeds thirty-five thousand times faster than classical computing methods, running off-the-shelf commercial servers. This could help answer some of the world's most difficult problems, whether it's the hunt for new drug therapies or creating next-generation nanotechnology or artificial intelligence.

Today's computers are binary, possessing only two possible values, either a one or a zero, known as bits, to carry out their instruction sets. Quantum computers, on the other hand, leverage the idiosyncrasies of subatomic particles known as qubits, which can be a one, a zero, or a simultaneous mix of the two. In plain English, this allows quantum computers to test a huge number of possibilities at the very same time and brings with it far-reaching security implications. In particular, quantum computers hold the potential to completely nullify all the computer security systems commonly in use today. Present computer security is based on cryptography—that is, using number theory and prime number multiplication to encode messages so that they are unreadable by unauthorized parties. In order for people to read your encrypted data, either they have to have the mathematical key, or they could "brute-force it" by doing the math required over and over again to factor the prime numbers involved to provide the correct solution. When we enter our passwords, encryption algorithms convert them into the correct factor that unlocks the message and provides authentication. Today, a brute-force attack is something most hackers never have to resort to. Instead, they rely on poorly implemented encryption protocols, computer malware, keystroke loggers, and human error to steal the cryptographic key required to read your credit card data or banking information.

Absent the correct password, hackers would have to reverse engineer the encryption process, a computationally difficult and highly improbable feat using today's computers. Even with a supercomputer, a brute-force attack would take billions of years to crack the 128-bit AES encryption that is today's standard (the age of the universe is only 13.75 billion years). While classical computers can only do a single calculation at a time, quantum computers can perform a huge number of calculations that leverage the counterintuitive nature of quantum mechanics to drive directly to the answers of very complex questions. In other words, a quantum computer could potentially bypass encryption protocols, allowing its owner to read everybody's e-mail, transfer funds from bank accounts, control the financial markets, commandeer air traffic control systems, and manipulate critical infrastructures. Conversely, quantum technology might also be the breakthrough that allows for fully secure, uncrackable communications, since any observation or interception of a quantum encryption key during transit would change its content. Though you won't pick up one of these in the Apple Store anytime soon, many governments around the world are working on building quantum computers capable of cracking today's crypto technology, and developing their own quantum secure networks. Not surprisingly, the NSA has already appropriated nearly $100 million toward crafting a "cryptologically useful quantum computer" as part of its Penetrating Hard Targets project. To be clear, this is a massively difficult problem to solve, but the first person who does it will have tremendous concentrated power, something he is unlikely to mention to all of those whose communications he is reading and whose systems he is accessing.

Taken on the whole, the most powerful technologies of the twenty-first century, including robotics, synthetic biology, molecular manufacturing, and artificial intelligence, hold the power to create a world of unprecedented abundance and prosperity. From the creation of unlimited energy to the production of boundless food sources and monumental advances in medicine, exponential technologies can be an extraordinary force for good.

But there is a flip side to these advances as well, as we have seen time and time again throughout this book. In the year 2000, Bill Joy, the former chief scientist at Sun Microsystems, provided a glimpse of how bad things could theoretically get in a seminal article published in *Wired* titled "Why the Future Doesn't Need Us." Joy bluntly warned that robotics, genetic engineering, and AI threaten to make human beings "an endangered species" as exponential technologies would eventually grow beyond us and our control. Joy pointed out that all of our twenty-first-century technologies are being democratized, available to anybody with an Internet connection. There are robot-building clubs in high schools and synthetic biology competitions in colleges. AIs navigate our cars, and UAVs can be purchased at Costco. Compared with the nuclear threat, however, there is an incongruence between the potentially destructive power of these exponential technologies and their widespread availability to the common man today. This does not mean these technologies should be banned, nor should they be locked away in government labs, given the vast potential for good they will bring, especially as they become democratized. Who knows what kid in Jaipur or what grandmother in Milwaukee while hacking away at synbio will make that game-changing cancer-fighting breakthrough we've all been hoping for? But it is also just as likely that among the masses will be those few bad actors who can use the same technologies to create a global pandemic. This should give us pause. We should be thinking more deeply and seriously about our use of exponential technologies, their downsides, and the potential for harm they may bring.

Although space attacks, evil AI, and gray goo may be low on our list of personal priorities, far below the rush to pick up the kids at school, there are a mass of threats that demand our immediate attention. The critical infrastructures that run the world, from our energy grids to the financial markets, are under persistent attack, leaving us with a global information grid that is readily susceptible to a systemic crash. At the same time, the volume of data we are producing about ourselves and the things around us is growing at an exponential rate, raising deep questions about our privacy and the ethical implications of what becomes possible with big data and an emerging surveillance society. These data can be hacked and projected onto an ever-growing number of screens in our lives to portray "realities" that are in fact falsehoods. This lack of trustworthy computing is further exacerbated by the ease with which black-box algorithms can be used to distort our reality in ways barely perceptible, their secrets known only to those who program them behind closed doors and beyond the scrutiny of the masses.

Mobile computing and an Internet that will grow from the metaphoric size of a golf ball to the size of the sun is just over the horizon, and soon every physical object may be connected online and assigned an IP address. But more things online means more things to hack, giving bad actors access to increasingly intimate parts of our lives from our bedrooms to our own bodies as biology becomes integrated with information technology. And at every step of the way, criminals, terrorists, and rogue governments are ready to exploit our common technical insecurity through the sweeping flaws that persist in today's software and hardware systems. These illicit knowledge workers of the twenty-first century are deeply innovative, adaptive, and ever learning and employ the latest business practices, from crowdsourcing to affiliate marketing, to subvert the technologies around us.

Advances in computing and artificial intelligence mean that crime has now become scripted, run algorithmically, to much greater effect and with far fewer human beings required. Worse, the tools we have available to detect these threats are woefully inadequate. With 95 percent of new malware threats going undetected and the time to discovery of an intruder in our corporate networks hovering around 210 days, it is clear that any of our systems can be penetrated at will by those who have the time and inclination to do so. Indeed, not much time is required at all, as the Verizon–Secret Service study demonstrated: 75 percent of all computer systems can be penetrated in mere minutes, and only 15 percent require more than a few hours to hack.

The impact of these threats will be felt more profoundly as cyber crime goes 3-D, with billions more objects connected to the Internet of Things, an emerging online world that too is eminently hackable and may be even less secure than our existing laptops and smart phones. The risks of three-dimensional computing, embodied by the rise of robotics, mean that we are creating machines with the ability to outrun and overpower us, made even more powerful by their ability to act in unison, working as a swarm to accomplish their goals. This specter is a troubling development given the increasing physical prowess of the growing legions of armed flying, walking, or swimming military robots, most equipped with artificial intelligence systems to guide them and some imbued with the lethal autonomy to make "kill decisions" for us. The cyber threat is thus morphing from a purely virtual problem into a physical world danger. The result, as we have seen throughout this book, is that science fiction is becoming science fact before our very eyes.

With the advent of the Internet and the imminent arrival of the billions of additional connections afforded by the IoT and its sensors, our planet has developed an ever-expanding nervous system. It links our communications, our thoughts, and even our bodies to an online global brain of tremendous complexity controlled by a plethora of software systems and networking protocols, each of which can readily be exploited by those who wish to do us harm. Regrettably, the immune system protecting this global nervous system is weak and under persistent attack. The consequences of its failure cannot be overstated. As a result, it is time to start designing, engineering, and building much more robust systems of self-protection—safeguards that can grow and adapt as rapidly as new technological threats are emerging into our world. Though it's easy to focus solely on the abundant benefits technology brings into our lives, we ignore the accompanying risks at our own peril.

We are now living in an exponential age, and yet physiologically our brains are still those of Stone Age hunters, barely upgraded in the past fifty thousand years: it is not in our nature to grasp the inherent power of exponential technologies. But try we must. For just as the creatures living in the proverbial lily-pad-covered pond mentioned earlier were under threat from exponential change, so are we. For the students in France who were warned they had thirty days to act in order to save the pond, on day 25 there was scantly anything to be concerned about because the lily only covered 3 percent of the pond's surface, so they let it grow. As we know, by day 29, the lily had miraculously grown to cover half the water, but by then there was precious little time to save the pond, which was strangled by the lily the very next day. Today the totality of our technological insecurity may seem easy to ignore. Sure, a few million accounts may be hacked here, and a billion passwords get stolen there, but we have time. Drones, pacemakers, air traffic control, cars, streetlights, navigation systems, MRI machines—all hacked. But we have time. Tens of billions of new objects to be added to the Internet, but we have time. Don't we?

The writing is on the wall. Technology is leaving us increasingly connected, dependent, and vulnerable. Though the myriad scientific breakthroughs enabled by exponential technology promise great and untold benefits for humanity, they must be guided and protected from those who would exploit them to harm others. We ignore the overwhelming evidence of technological risk around us at our own peril. Day 29 is rapidly approaching. What are we going to do about it?

Surviving Progress

CHAPTER 17

Surviving Progress

**For me, it is far better to grasp the Universe as it really is than to
persist in delusion, however satisfying and reassuring.**
CARL SAGAN

t has been a rough ride. We've been asked to consider difficult and often
uncomfortable questions about technology and the role of omnipresent
machines in our lives, devices that we have unquestioningly welcomed into
our homes, our offices, our cities, and even our own bodies. This journey
has led us to take a piercing and critical look at the ever-growing number of computer screens proliferating in our world, screens we've turned
around 180 degrees to show the other side of the story, the peril as well
as the promise in our love affair with technology. Our growing interconnectedness and the ubiquity of inherently vulnerable computing systems
mean that this gathering storm of technological insecurity can no longer
be ignored.

The problem of course is not that technology is bad but that so few
understand it. As a result, the computer code that runs our planet can be
subverted and used against us by those who do. Exponential times are leading to exponential crimes, ones in which lone individuals with ill intent
can reach out and have a negative effect on tens of millions anywhere at
any time. Indeed, the entire range of critical information infrastructures
that run our society are at risk. These challenges will become greatly exacerbated as billions of new objects go online and networked computers in
the form of robots begin moving about the physical space they will share
with us, to say nothing of risks from artificial intelligence and synthetic

biology. It all seems so daunting and overwhelming, but it is only by first understanding and acknowledging these threats that we can begin to make the changes required to bolster the future foundations of our technological tomorrow.

There are no easy fixes to the situation in which we currently find ourselves. No panacea or single "just add water and mix" solution that will make it all better. It took billions of individual steps to get us into this predicament, and it may take billions more to get us out. The asymmetric nature of the threat means attackers only need to find a single weakness, while defenders must guard against them all, a veritable impossibility. That said, all is not lost, nor are things hopeless. We don't need, nor will we ever attain, "perfect security." Such a thing does not exist. But the near-total absence of trustworthy computing in a world run by computers should serve as a flashing red warning light to us all.

That science and technology have been a net positive for humanity there is no doubt. Yet, in order to thrive in the coming century, we must first survive the technological risks this progress inevitably brings. There are actions we must take today, important course corrections, to head off the dangerous future looming before us. In the pages that follow are a variety of technical, organizational, educational, and public policy recommendations, both strategic and tactical, meant to lessen the exponentially growing risks posed by technology. Of the myriad steps we must take to protect our technological future, I believe the following to be the most important. Technology is here to stay, and there is no turning back. The key question is how to harness these tools to achieve the maximum possible good while minimizing their downsides. Here's how we might survive progress.

Killer Apps: Bad Software and Its Consequences

*Every time you get a security update . . . whatever is getting updated
has been broken, lying there vulnerable, for who-knows-how-long.
Sometimes days, sometimes years.*
QUINN NORTON

Facebook's software developers have long lived by the mantra "Move fast and break things." The saying, which was emblazoned on the walls across the company's headquarters, reflected Facebook's hacker ethos, which dictated that even if new software tools or features were not perfect, speed of code creation was key, even if it caused problems or security issues along the way. According to Zuckerberg, "If you never break anything, you're probably not moving fast enough." Facebook is not alone in its software-coding practices. Either openly or behind closed doors, the majority of the software industry operates under a variation of the motto "Just ship it" or "Done is better than perfect." Many coders knowingly ship software that they admit "sucks" but let it go, hoping, perhaps, to do better next time. These attitudes are emblematic of everything that is wrong with software coding and represent perhaps the largest single threat against computer security today.

The general public would be deeply surprised at just how much of the technology around us barely works, cobbled together by so-called duct-tape programming, always just a few keystrokes away from a system crash. As Quinn Norton, a journalist with *Wired* magazine who covers the hacker community, has pointed out, "Software is bullshit." Most computer programmers are overwhelmed, short on time and money. They too just want to go home and see their kids, and as a result what we get is buggy, incomplete, security-hole-ridden software and incidents such as Heartbleed or massive hacks against Target, Sony, and Home Depot.

Writing computer code today is no easy task; indeed, it is incredibly complex. With nearly fifty million individual lines of code in Microsoft Office, each of which needs to work perfectly to keep out attackers, surely some things will go awry. And that's just one program. Your computer or smart phone must harmonize and police all the programs it's running, let alone those running on other systems with which it wishes to interact on every Web site you visit. The problem grows exponentially as more and more devices on the IoT begin communicating with one another. All of these software bugs and security flaws have a cumulative effect on our global information grid, and that is why 75 percent of our systems can be penetrated in mere minutes. This complexity, coupled with a profound laissez-faire attitude toward software bugs, has led Dan Kaminsky, a respected computer security researcher, to observe that today "we are truly living through Code in the Age of Cholera."

When challenged about the poor state of the world's software today, many coders retort, "We're only human, there is no such thing as perfect software." And they are right. But we're nowhere near perfect, perhaps only at 50 percent of where we could and should be, according to Charlie Miller, a respected security researcher. Just bumping that number up to 70 or 80 percent could make a huge difference to our overall computer security. Consumers want powerful feature-rich software, and they want it now, with tens of thousands of people willing to stand in line days in advance, sleeping on the sidewalk, to get the latest iPhone. But software providers need to significantly up their game and design security up front, from the ground up, as a key component of trustworthy computing.

In order to turn this ship around, incentives will need to be aligned to ensure the badly needed emphasis on secure computing actually occurs. For example, today when hackers find a vulnerability in a software program, they can either sell it on the black market to Crime, Inc. for a significant profit or report it to a vendor for next to nothing while facing the threat of prosecution. Thus they make the rather obvious choice. Though this is beginning to change and some companies have established "bug bounty programs," few offer cash rewards, and among those that do, the amounts are far less than those available in the digital underground. That needs to change. Creating well-funded security vulnerability reporting systems that pay hackers for bringing major flaws to vendors' attention would help minimize the damage these software companies themselves created when they rushed insecure and buggy code out the door onto an unsuspecting public.

Given that software is the engine that runs the global economy and all of our critical infrastructures, from electricity to the phone system, there's really no time to lose. But it will take much more than a few security researchers writing compelling articles on the topic; it will require an outcry from the public, one that has been sorely lacking until this point, to demand better-quality software. Think about it. Why do we accept all of these flaws as the natural state of affairs? They needn't be. We can make a change by holding those in the software industry, which is worth $150 billion a year, accountable for their actions. Absent this demand from the public, in the battle between profitability and security, profit will win every time. We need to help companies understand it is in their long-term interest to write more secure code and that there will be consequences for failing to do so. As things stand today, the engineers, coders, and companies that create today's technologies have near-zero personal and professional responsibility for the consequences of their actions. It's time to change that.

Software Damages

The noted Yale computer science professor Edward Tufte once observed that there are only two industries that refer to their customers as users: computer designers and drug dealers. Importantly, you are equally as likely to recover damages from either of them for the harms their products cause. The fact of the matter is that when you click on those lengthy terms of service without reading them, you agree that you are using a company's software or Internet service "as is," and all liability for any damages lands on you. These firms use language such as "you will hold harmless and indemnify us and our affiliates, officers, agents, and employees from any claim, suit, or action arising from or related to the use of the Services" and "we do not guarantee that our product will always be safe, secure, or error-free." Would you buy a Chipotle burrito if it came with such a warning? I think not. So how is it that the software business has carved out an exception for itself for ever being responsible for anything? Good question.

When an automobile crashes because of faulty wiring or bad firmware, as we saw with the deadly Toyota acceleration cases, those injured can sue for damages. So why not with software? Is it reasonable to suggest that if somebody were to die or suffer severe economic losses as a result of faulty software that she or her loved ones be denied a day in court because the ToS said so? Even when it could be proven to a judge and jury that the software was the proximate cause of the harm? I do not believe it is.

Do not get me wrong. I am not a fan of creating new laws willy-nilly. Nor would I suggest that regulation is the very best approach to handle the totality of our global cyber insecurity. It is at best a blunt tool in a field as rapidly evolving as technology. But a line needs to be drawn in the sand. Reckless disregard for any and all consequences of poorly written software, released with known vulnerabilities and foisted on a public incapable of individually reading through the millions of lines of code on their smart phones or laptops to adjudge the concomitant risks, is just plain wrong. Those writing and creating these tools must bear some responsibility.

Needless to say, the software industry is vehemently against any such change. It claims that allowing liability lawsuits would have catastrophic effects on its profitability and would bankrupt it. It also asserts that the complexity of software interactions is so great that it would be impossible to fairly adjudicate blame in case of injury. Both arguments fall short. We've been here before, particularly with the automobile industry, whose products through the 1960s had a terrible safety record. Through consumer advocacy and congressional action, the National Traffic and Motor Vehicle Safety Act was passed in 1966, allowing the government to enforce industry safety regulations. Doing so resulted in one of the largest achievements in public health of the twentieth century. Automobile deaths dropped precipitously, saving tens of thousands of lives.

Of course today's technology may be more complex than the automobiles of yesteryear, but there will be no improvement in the safety and security of its software and hardware products until the business incentives are aligned to encourage change. Currently, any harms suffered by end users are theirs and theirs alone, with little or no harm accruing to the software vendors responsible for the damages. There are few if any consequences for releasing crappy code, and thus the practice continues unabated. Unless those responsible for the underlying security problems they create are held accountable for their actions, little will change. Only when the business costs of releasing persistently broken code are greater than fixing the known vulnerabilities in the first place will the balance tip in favor of better and more secure code. Though I am not advocating the creation of vast new regulations or government bureaucracies, I do believe a vigorous public debate regarding the underlying causes of our widespread computer insecurity is in order. The time to get our coding and software house in order is now, before we add the next fifty billion things to our global information grid.

Reducing Data Pollution and Reclaiming Privacy

Throughout this book, we have seen the consequences of amassing peta-bytes upon petabytes of data, information that eventually leaks. Whether it is personal medical records, bank balances, government secrets, or cor-porate intellectual property, it all leaks. The mass storage of these data in the hands of a few major Internet and data companies provides irresistibly rich targets to attack, a one-stop shop for thieves. As I've said before, the more data you produce, the more organized crime is willing to consume.

Though most Internet users have chosen to voluntarily share some of the most intimate details of their lives via online social networks, the com-panies behind these services gather way more data than most ever realize. Purveyors of "free" Internet services persistently track users across their entire online experience as well as their movements in the physical world through the use of their mobile phones. But as we've seen, the most expen-sive things in life are free. All of this information is cut, sliced, and diced and sold off to the shady and secretive world of data brokers, who exercise little care or control over the accuracy or the security of the information they retain. Though we might complain about these practices (if we were the actual customers of these social media firms), we have no ability to do so. We bargained those rights away in exchange for free e-mail, sta-tus updates, and online photographs, agreed to in the click of a fifty-page four-point-font ToS agreement that none of us read. These overreaching, entirely one-sided "agreements" should not absolve the companies that author them of all liability pertaining to how they keep and store our data. If they choose to keep every single bread crumb they can possibly gather on our lives, then they should be responsible for the consequences.

The striking thing about this system is that it needn't be organized this way. It is estimated that each Facebook user worldwide only generates about $8 in ad revenue (not profit) for the company per year. I'd rather send Facebook ten bucks and be left alone. At less than a buck a month, it's about a hundred times less than my cable bill. The whole system is screwy. As the MIT researcher Ethan Zuckerman has proclaimed, "Advertising is the original sin of the web. The fallen state of our Internet is a direct, if unintentional, consequence of choosing advertising as the default model to support online content and services." Though our data pay for Gmail, YouTube, and Facebook today, we could just as easily support Internet companies whose goal was to store as little personal data of ours as pos-sible, in exchange for minute sums of cash. Why not just disintermediate the middle man altogether for a much more logical system? We would become Facebook's and Google's clients for a dollar a month and could go on to enjoy our lives.

Unfortunately today, just as is the case with software vendors, the incentives are misaligned from a public safety and security perspective. Facebook is incentivized to gather an ever-growing amount of personally identifiable data on its customers that it can sell on to thousands of data brokers around the world at a profit. That is its business model. Whether purchasers of this information ultimately allow it to be used to commit identity theft, stalking, or industrial espionage is of little concern to social media companies after they've auctioned the information off to the highest bidder. Of course it matters to us, those who suffer the economic and social harms from these leaked data. For those who prefer the benefits of the "free" system, let them enjoy it and all it entails. But why not allow the rest of us the option to pay to maintain greater control over our privacy and security?

While it may be impossible to "live off the grid" in today's modern world, we can by all means design a system that is much more protective. There are better, more balanced examples out there, such as the EU's Data Protection Directive, which is much more consumer-friendly and enshrines privacy as a fundamental right of all EU citizens. It limits what data companies can store about us and how long they can keep it before the data must be deleted. This is a more sensible approach that not only adjusts the completely lopsided balance of power in our relationships with Internet firms but also protects us and our data from leaking into the hands of Crime, Inc.

Kill the Password

As we saw in the first chapter with Mat Honan's epic hacking, a string of alphanumeric characters can no longer protect us. Sure, you might buy yourself some time by creating a twenty-five-digit password with upper- and lowercase letters, numbers, and symbols, but the fact of the matter is almost nobody does that. Instead, even in 2015 the most popular passwords remain "123456" and "password." Fifty-five percent of people use the same password across most Web sites, and 40 percent don't even bother to use one at all on their smart phones. Even if they did, it might not help much. Given advances in computing power, cloud processing, and crimeware from the digital underground, more than 90 percent of passwords can be brute-forced and cracked within just a few hours, according to a study by Deloitte Consulting. Worse, Crime, Inc. organizations such as Russia's CyberVor have amassed more than 1.2 billion user names and passwords, which they can use to unlock accounts at will. Plainly stated, our current system of just using a user name and password is utterly broken.

There are some measures we can take today that will provide additional layers of protection. One example is the two-factor authentication offered by Google, Microsoft, PayPal, Apple, Twitter, and others, which combines your user name and password with something you have such as a security token, key fob, or mobile phone. Most consumer Internet companies use your smart phone as the second factor by sending you a onetime code via text message that you must also enter to gain access to your account. Thus even if a hacker cracked your bank account, social media service, or social media profile password, he would still need access to your phone and text message, something he would be unlikely to have if you and your phone were in New York and the hacker in Moscow. While two-factor authentication is definitely a step in the right direction, these systems can be subverted via man-in-the-middle attacks, which intercept text messages via mobile phone malware.

To that end, many smart-phone companies such as Apple and Samsung are moving toward another form of two-factor security, combining something you know with something you are—such as your biometric fingerprint or voice identity. Your fingerprint will increasingly become your password, and with the release of the iPhone 6 and iOS 8 Apple has allowed other companies, such as PayPal and your bank, to use your phone's Touch ID fingerprint sensor to authenticate you. While hackers, such as the Chaos Computer Club, and others have circumvented some of these systems in the past (if they had access to the device), multifactor authentication can provide a significant improvement over the standard user name and password. Mat Honan was right. It's time to kill the password and move on to multifactor authentication and biometrics, tools that, though far from perfect, are an immense improvement over the feeble alphanumeric characters we use today. Though there is currently no cure-all for user identification, there are tremendous opportunities to create significantly better alternatives, particularly through coordinated research and funding efforts, as we will soon discuss.

Encryption by Default

There are only two types of companies—those that have been hacked and those that will be.

ROBERT MUELLER, FORMER FBI DIRECTOR

The vast majority of today's data is unencrypted or poorly protected. A study by the computer giant HP in July 2014 revealed that 90 percent of our connected devices collect personal data, 70 percent of which is shared across a network without any form of encryption. That means that anybody who gains access to a computer system through poorly coded software, downloaded malware, or weak passwords can steal, read, and use any of the data contained in that system. Without encryption, the data are entirely readable by anyone who has access to them. That is why fifty-five million credit cards stolen from Home Depot can be used by Crime, Inc., because the company's in-store payment system didn't encrypt customers' credit data while in memory. Had the data been properly encrypted, they would have been entirely without value to the thieves who stole them. It's not just financial data that are too often unencrypted; so too are our medical records, corporate secrets, military video drone feeds, celebrity nude photographs, and nearly all our e-mail. The impact of all these computer breaches and data theft could be greatly minimized if the proper implementation of encryption were to become the default standard practice.

The majority of data stored on both personal and company hard drives is in plain text, readable by anybody who gains access to these devices. The same is true for the lion's share of traffic crisscrossing the Internet, save for major Web sites using HTTPS when sending your password or credit card information. But we can do so much better, particularly in the wake of Edward Snowden's revelations. On the plus side, Google is increasingly encrypting its traffic, including all Gmail messages, between your computer and its servers (not just your password). Doing so makes it vastly more difficult for somebody to intercept and read your e-mail in transit; otherwise, any message you send is as if it were written on a postcard freely accessible to anybody who sees the traffic as it flows around the Internet, such as the local Starbucks Wi-Fi connection you use. The Electronic Frontier Foundation, a nonprofit digital rights and privacy advocacy group, has also launched a program known as HTTPS Everywhere to promote the use of encryption in all our Internet browser traffic. In short, it's high time to encrypt the Internet to help protect the privacy and security of our digital communications and computer data.

Though modern computer operating systems, including those from both Microsoft and Apple, come with free hard disk encryption tools built in, they are not turned on by default, and only a small minority of companies and a tiny percentage of consumers encrypt the data on their laptops or desktops. In fact, most consumers have no idea these security protocols even exist. In the wake of the celebrity iCloud hacking fiasco of 2014, Apple's chief executive, Tim Cook, acknowledged the company had to do more to ratchet up customers' awareness of cyber-security matters. I thoroughly agree. In September 2014, Apple announced that its latest iPhone would encrypt all data on the device when a password was set, a move Google vowed to match with its forthcoming Android mobile phone operating system. These are important steps forward in minimizing smartphone security risks, but given that 40 percent of users don't even use a password on their mobile phones at all, Tim Cook was right: much more education and awareness are needed.

Taking a Byte out of Cyber Crime: Education Is Essential

Civilization is in a race between education and catastrophe.

H. G. WELLS

We have a literacy problem in the United States and around the world, and it's not the one most think of. It is the problem of technical literacy. In a world replete with gadgets, algorithms, computers, wearables, RFID chips, and smart phones, only a minute portion of the general population has any idea how these objects actually work. Whether it's Crime, Inc. or the NSA, those who know how to code will hold power over those who don't in the same way that those who could not read and write in the last centuries found their opportunities limited. We need to build up the technical literacy of the general public.

The goal is not for every single person to become a computer coder (though building our nation's science, technology, engineering, and math skills would do much for our economy). The goal is for citizens to have a basic understanding of how the technologies around them operate, not just so that they can use these tools to their full advantage, but also so that others cannot take advantage of their technological ignorance and harm them. Had Cassidy Wolf, Miss Teen USA, been taught in school the simple trick of covering up the Webcam on her laptop with a yellow Post-it note, a hacker would never have been able to secretly capture photographs of her naked in her own bedroom. Of course that is just one example, but in case after case of cyber attack, had the victim been armed with the right knowledge to protect herself, the pain of the hack could have been entirely avoided. Education is key, and the state of our cyber-security education is abysmal.

In our public schools, we provide children everything from sex education to driver's training. But your children will likely be spending much more time online and interacting with technology than they will be engaged in sex or driving. Yet most schools provide little or no formal education on how to stay safe online. For years, the National Crime Prevention Council's McGruff the Crime Dog was a fixture on television and in schools warning children and adults alike to "take a bite out of crime." Today we need McGruff more than ever teaching our children how to "take a *byte* out of crime." Fortunately, there are some useful efforts under way. The National Crime Prevention Council has launched programs to inform parents and children on cyber bullying and Internet safety, and the National Cyber Security Alliance has created an excellent Web site (StaySafeOnline.org) and other public programming to help educate our digital society to use the Internet securely whether at home, work, or school. But these efforts need to be greatly expanded if we are to meet the level of threat heading our way across a wide array of technological developments, such as the Internet of Things. As previously noted, a great many of these technological threats need to be handled at a systemic level, but individuals also have to understand the risks and take responsibility to protect themselves and their families to the fullest extent possible. The need for education is just as great in the private sector among businesses. Companies are under attack, not just by Crime, Inc., but also by sophisticated nation-state espionage services going after their intellectual property and corporate data. Security measures that were commonly only necessary in top secret organizations are urgently needed across the entire business world. Here too the educational resources are profoundly limited, a state of affairs that must be addressed if we are to make any progress against the technological threats looming before us.

The Human Factor: The Forgotten Weak Link

**If you think technology can solve your security problems, then you
don't understand the problems and you don't understand the technology.**
BRUCE SCHNEIER

Cyber security is a people problem, not just a technical one. No matter how strong your computer password is, if you write it down on a yellow sticky and attach it to the front of your computer screen so that you can remember it, all walking by will have access to your digital life. For the tens of thousands of people losing money to Nigerian prince scams every year, their problem is not a technical one but the ever-present human characteristics of hope and avarice. When you post your vacation plans on social media and burglars pay a visit, it was your decision to share that helped facilitate their criminal activity. And for each and every person who clicks on that link from his bank telling him his password has expired and he needs to change it, the challenge isn't that his computer has been hacked per se but rather that he fell victim to a socially engineered phishing attack. No matter how many firewalls, encryption technologies, and antivirus scanners a company uses, if the human being behind the keyboard falls for a con, the company is toast. According to a 2014 in-depth study by IBM Security Services, up to 95 percent of security incidents involved human error. The human factor can trump all other technological security measures, and thus the need for both workforce and personal education is key.

As mentioned in the prologue, of course technology can help make us more secure. Multifactor authentication, biometrics, encryption, and geo-location can cut down on crime and reduce other security risks. But as we have seen repeatedly, these technological tools can be undermined. The NSA surely had some top-notch cyber-security tools at its disposal, yet it was a human being, Edward Snowden, who subverted them before fleeing with extensive classified data on his thumb drive. The same was true of the "peaceful" Iranian nuclear power plant at Natanz, which had good security measures in place and had no physical connection between its industrial control systems and the Internet at large. But these measures were readily defeated when an unknown party carelessly plugged an infected USB thumb drive into a desktop computer at the facility. The ill-informed decision allowed the Stuxnet worm to propagate across the internal network responsible for controlling the uranium centrifuges at the facility. It is convenient to always turn to an easy technological fix when there is a problem, but business owners, policy makers, Internet firms, computer coders, and engineers must consider the human dimension of security if we are to make any progress against the technological risks of both today and tomorrow.

The good news is there is much we can do by adjusting our own human behavior to significantly improve our personal technological security. To help frame this issue, it is useful to think of automobile theft for comparison. If a BMW owner parked his car in a crime-ridden neighborhood, the decisions he personally made about the automobile's security would strongly affect the likelihood of its theft. If a driver were to park in a well-lit area, lock all the doors and windows, and set an alarm, he would have taken all reasonable measures to prevent thieves from stealing his car. Most of us have learned over time that that is exactly how we should secure our vehicles, but the majority of the public has no idea of what similar behavior would look like in cyberspace. Thus we go online and virtually park our cars on dark isolated streets, keep the doors and windows open, never use an alarm, forget the keys in the ignition, and leave $100 bills on the front seat. Then we wonder why our cars have been stolen.

The goal here is not some elusive unicorn known as "perfect security" but a significant improvement in our current state of affairs. Again the BMW example is instructive. Even for the driver who took all the right steps and precautions to protect himself and his car, the vehicle's theft would still be entirely possible. A criminal could come by with a flatbed or a tow truck, let alone hack the doors and engine, to abscond with the vehicle. With enough time, energy, attention, and resources, any system can be hacked. Your goal is not perfect security; your goal is understanding how to lock the doors and windows to your car in cyberspace, and about that there is much you can control. Still, many of the risky online decisions you make today are not at all your fault but rather that of incredibly poorly designed computer systems—software, hardware, Web sites, and smart phones. It's time we fix that.

Bringing Human-Centered Design to Security

New opportunities for innovation open up when you start the creative problem-solving process with empathy toward your target audience.

TOM KELLEY, IDEO

Why don't these idiot customers update their passwords? If only those fools used VPNs and firewalls. Well, are you using WEP or WPA2?

As anybody who has phoned tech support to resolve a computer problem knows, most system administrators and help desk personnel don't hold their "customers" in particularly high regard. The common diagnosis among these technical support personnel is PICNIC: problem in chair, not in computer. For those who have studied computer science, taken classes in cryptography, and dreamed in PHP and C++ code, talking to the average computer user can be a frustrating process. We quite literally speak two different languages. For security engineers, the answers seem so clear: "If only those damn users would stop doing x or y stupid thing, everything would be okay." Users on the other end of the line have a simple, often unspoken request: "Why won't you give me simple instructions and allow me to get back to work?" Our security tools today are too complex and burdensome to use, and, simply stated, complexity is the enemy of security.

Information security architects speak in jargon about viruses, malware, zero days, exploits, Trojans, RATs, and AES, and for the most part the general public has no idea what they are talking about. Security software and hardware products today are almost uniformly designed by geeks for geeks. There is nary a fleeting thought or a modicum of empathy toward how these tools might be used by you, let alone your grandmother. Instead, the products that are meant to secure and protect us give us helpful warnings such as "Alert: Host Process for Windows Service Using Protocol UDP Outbound, IPv6NAT Traversal-No, is attempting to access the Internet. Do you wish to proceed?" What the hell does that mean? Nobody knows, except for the original authors of this "helpful" warning. It's time to bring human-centered design thinking to the world of cyber security.

Think of the design of an iPhone 6, an Eames lounge chair, a Ferrari 458 Italia, or a Leica T camera—products that are meant to delight. Not only are these tools functional, but they are beautiful, created by people who had a close and deep understanding of their customers and their needs. When one watched Steve Jobs onstage describe his latest products, there was no doubt that each and every one was imbued with the love of its creators. So where's the Steve Jobs of security? What might Apple's chief designer, Jony Ive, bring to the problem of our growing cyber insecurity? What would his firewall or antivirus program look like? Thus far, we have no idea, and that is a huge problem.

It is a problem because when security features are not designed well, people simply don't use them. Moreover, poor design can lead the human users down pathways that actually make them less secure. Why would people write down their passwords on Post-it notes and stick them on their computers? Because making people change them every two weeks and requiring that they be at least twenty characters long, with an uppercase letter, a number, a symbol, a haiku, and in iambic pentameter, is just too much for the average user to handle. So people subvert the security systems in place so that they can get their work done. There are also certain types of security products, such as software firewalls, that give so many false alerts that the person running the tools has to turn them off just to avoid constant pop-ups with incomprehensible warning messages. In these instances when security breaches occur, the IT staff invariably blames the user. It may be time to look in the mirror first. Of course the designers of security products and systems are not uncaring or ignorant people; they are just woefully out of touch with the needs of their customers. To borrow a phrase, it's time to "think different."

Human-centered product design is fundamental to drive the behavioral changes we require in the world of techno-security and to help minimize the growing number of threats we face. The designers of these products need a gut-level understanding of how people interact with computers and smart phones, and they must not expect people to conform to strange behaviors or understand arcane screen prompts. Until security gurus start making products the wider public can understand and implement, people will lack both the tools and the information they require to protect themselves. While expanded education programs and human-centered design can undoubtedly make a substantial improvement in the overall state of our technical security today, some threats go beyond the capacity of a single individual to respond to. In those cases, a host of systemic changes will be required, and both nature and medicine can provide useful inspiration on the best path forward.

Mother (Nature) Knows Best:
Building an Immune System for the Internet

Today's cyber threats are evolving faster than our defensive barriers can keep them out. Not only are the proverbial barbarians at the gate, but they've kicked it down and are crawling all over the castle. We need more robust, responsive, and flexible defense methods—much like the body's immune system. In the more than three billion years that life has existed on this planet, millions of different species, including human beings, have learned to deal with an innumerable array of threats. In animals, it is an adaptive immune system that provides the protection we need against a variety of foreign pathogens, including viruses, parasites, bacteria, and even environmental toxins. The designs we see all around us in nature can serve as a great source of inspiration as we attempt to solve complex human problems, and there is a field of study dedicated to just this challenge; it's called biomimicry. For example, scientists are now studying how leaves process the sun's energy in order to invent better solar panels. So why not look to innovation inspired by nature to help us create self-healing computer networks?

Until now, our general approach to cyber security has been to wall ourselves off from all possible technological threats, but not going online or using technology is not an option. A much better approach would be to acknowledge and rapidly adapt to risks as they present themselves, just as our immune systems do. The human immune system doesn't just work against one strain of flu, but rapidly adapts and learns to deal with a full spectrum of flu strains. This is possible because the body has a keen sense of understanding of what constitutes the healthy "self" as distinguished from the dangerous "nonself." But such approaches are rudimentary at best in our present-day techno-defense systems. Both DARPA and Pacific Northwest National Laboratory have launched projects on the subject, and one of the most interesting approaches is under way at Wake Forest University. There the computer science professor Errin Fulp is using the natural swarming intelligence of insect colonies to ward off cyber predators by deploying thousands of "digital ants" software programs across a computer network, each looking for evidence of a threat. Should such a threat be discovered, the digital ant will mark the problem with the equivalent of a virtual scent, attracting other ants. Stronger scent trails bring more digital ants, which ultimately swarm any potential computer infection before it grows out of hand. The propagation rates of the cyber threat are so great that there is no way human beings can manually keep up. Likewise, our goal should be to create a variety of sensors across our global networks to not only detect intruders and how they gained entry but, more important, automatically make the necessary repairs—a self-healing network that does not require human intervention to repair itself. An immune system for the planet. Until such a system is in place, we continue to focus our efforts on much more human-capital-intensive approaches to the problem, such as using law enforcement to arrest perpetrators.

Policing the Twenty-First Century

**In a world characterized by technologically driven change, we necessarily
legislate after the fact, perpetually scrambling to catch up.**
WILLIAM GIBSON

It's not easy policing the Net. Sure, we hear stories about a purportedly
omnipotent NSA tracking our every move in cyberspace, and it undoubtedly has amassed a powerful array of tools and techniques. But for the average police officer or detective, the Internet is a difficult place to operate.
Cops from the LAPD's Seventy-Seventh Division, the NYPD's Midtown
South Precinct, and Chicago's Englewood District have no access whatsoever to the tools employed by espionage agencies; those are all classified
and too sensitive to expose in court. Even organizations such as the FBI
face notable barriers when conducting cyber-crime investigations, particularly overseas. At the state, local, and federal levels, law enforcement
officials find themselves chronically overwhelmed and understaffed as evidenced by the explosive growth in online crime detailed throughout this
book. The estimated $400 billion in annual losses to the global economy
because of cyber crime demonstrates plainly that police are badly losing
the war against Crime, Inc.

Attackers, flush with profits from their adventures in the digital
underground, generally benefit from technology long before defenders
and investigators ever do. They have nearly unlimited budgets and don't
have to deal with internal bureaucracies, approval processes, or legal constraints. But there are other systemic issues that give criminals the upper
hand, particularly around jurisdiction and international law. In a matter of
minutes, the perpetrator of an online crime can virtually visit six different
countries, hopping from server to server and continent to continent in an
instant. But what about the police who have to follow the digital evidence
trail to investigate the matter? Not so much. As with all government activities, policies, and procedures, regulations must be followed. Transborder
cyber attacks raise serious jurisdictional issues, not just for an individual
police department, but for the entire institution of policing as currently
formulated. A cop in Dallas has no authority to compel an ISP in Tokyo to
provide evidence, nor can he make an arrest in the Ginza district. That can
only be done by request, government to government, often via mutual legal
assistance treaties. The abysmally slow pace of international law means it
commonly takes years for police to get evidence from overseas (years in
a world in which digital evidence can be destroyed in seconds). Worse,
the majority of countries still do not have cyber-crime laws on the books,
meaning that criminals can act with impunity. And, just as we saw with

narco-traffickers and money launderers, cyber criminals wisely remain in safe-haven countries.

Criminal law is nation based, meant to respect the sovereignty of each country to set its own rules and regulations without outside interference in its internal affairs, and dates back to the Treaty of Westphalia in 1648. While such a system worked well for centuries, it is under relentless and growing pressure from a global Internet that is eroding such boundaries. The legacy of the Treaty of Westphalia is a geographic answer for a non-geographic problem. The technological threat we face is borderless and thus can only be handled via an appropriate international response. An institution such as Interpol, the International Criminal Police Organization, has an important role to play in combating transnational cyber crime and coordinating investigations among its 190 member countries. But Interpol has an operating budget of only $90 million to combat all international crime, from trafficking in human beings to stolen art. By comparison, the NYPD alone has a budget of $4.9 billion, and a single criminal, the narco-leader Joaquín "El Chapo" Guzmán Loera of Mexico, had nearly $200 million in cash in his home at the time of his arrest (more than twice the annual budget of Interpol). Criminal investigations, especially those involving multiple jurisdictions and huge amounts of electronic evidence, are not just labor-intensive but also exceptionally expensive. Without an order of magnitude increase in police budgeting for the problem, we can expect Crime, Inc. to grow unabated in its illegal pursuits.

Yet even a massive increase in law enforcement resources would not solve our cyber-threat problem; there is a cultural component in our criminal justice system that must be addressed as well. In 2012, Janet Napolitano, the secretary of Homeland Security at the time, admitted she did not use e-mail or any other online services "at all." That is correct: the most senior government official in charge of our nation's cyber security and critical infrastructure protection didn't use e-mail—not because of security issues, but because by her own admission she is "somewhat of a Luddite." In 2013, the U.S. Supreme Court justice Elena Kagan admitted that her fellow justices "are not the most technologically sophisticated people" and that "the court really hasn't gotten to email yet." Instead, she said "they communicate with one another through memos printed on ivory paper hand-carried from chamber to chamber by court clerks." Though undoubtedly the justices and cabinet secretaries working at the very top of our criminal justice system have powerful intellects, their apparent lack of interest in or command of even rudimentary technologies is noteworthy. In a world moving as quickly as ours, how is it possible that government cyber-security policy and technology and privacy law will be shaped by those who don't use e-mail?

The core elements of our justice system need to be minimally fluent in the language of science and technology. Investigators must not only understand how these tools work but be every bit as creative as those they are pursuing—a near impossibility in any large law enforcement bureaucracy. While criminals are using AI to script and automate crimes, police are responding to each crime manually. Crime is scaling, but law enforcement has not: we have AI crime bots, but where are the AI cop bots to counter them? Where is that level of innovation in government? We need a Department of Mad Scientists at the FBI, a cadre of special agents empowered to forgo starched white shirts and ties in favor of the creative hacker ethos of their opponents. These should be white hat hackers drawn from all segments of society and capable of thinking way outside the box. Let's encourage creativity and innovation among them in the same way Google does—with a 20 percent time work program enabling agents to pursue special projects one day a week, free from their normal assigned workload. When Google went public, its founders cited 20 percent time as instrumental to the company's ability to innovate and as leading to "many of our most significant advances," including Gmail, Google Talk, Google News, and AdSense (currently responsible for 25 percent of the company's revenue). Most law enforcement agencies are so busy they only have time to focus on the tactical issue before them, leaving near-zero time for the critically important strategic thinking required for problem solving. No matter how much we spend on policing, we will never be able to arrest our way out of the cyber-crime problem.

The need for new approaches is exigent given that our off-line systems of jurisdiction and justice may be fundamentally incompatible with our ever-expanding online world. For example, we have police departments working in cyberspace, but where are the cyber fire departments, as the Internet pioneer Vint Cerf appropriately asks? When your neighbor's house catches fire and threatens yours, the goal should be not to arrest your neighbor's house for arson but rather to prevent yours from burning down. While law enforcement is clearly in order for criminal matters, there are a whole host of other options that may work better as a means of dealing with the growing mountain of cyber threats. In particular, it's time to focus on prevention rather than retrospective investigation and treatment of the problem after the fact. In that regard, there is much we can learn from the world of public health as we struggle to mitigate the risks of our technological insecurity.

Practicing Safe Techs:
The Need for Good Cyber Hygiene

We all know what good hygiene looks like in the physical world. It's reinforced all around us. Signs in restaurant bathrooms remind employees that they must wash hands before returning to work. Your mom tells you to cover your mouth when you sneeze, and colleges, doctors, and billboards remind us to use a condom and engage in safe sex. But where are these messages in the virtual world? Mom doesn't remind you not to accept USB drives from strangers, so we routinely plug these virus-carrying devices in our computers and thereby unwittingly participate in malware propagation, infecting our neighbors and friends. The failure to inoculate my own tech means when I become infected and a slave to the criminal Borg, I am now unknowingly engaging in DDoS attacks and phishing scams against others.

Internet health, like public health, is a shared responsibility, and users must take stewardship over their networks and devices if we are to improve the overall safety of our techno-future. We have an ethical obligation to do so. Each of us must be a good shepherd over our technological flock, protecting our computers, phones, and other digital gadgets from harming others. The good news is that practicing good cyber hygiene is much easier than it seems, and I have included a list of simple techniques in the appendix of this book that can drastically reduce your risk from cyber threats. Though there are many lengthy and complex best practice lists out there, the Australian government brilliantly reduced them to just four key strategies:

- Application white listing—only allow specifically authorized programs to run on your system and block all unknown executable files and installation routines. Doing so prevents malicious software and harmful applications from running.
- Patch all your devices' applications by automatically running software updates for programs such as MS Office, Java, PDF viewers, Flash, and browsers.
- Patch operating system (OS) vulnerabilities by automatically updating your OS such as Windows, Mac, iOS, or Android, ensuring you are using the very latest fully updated operating system at all times.
- Restrict administrative privileges on your computer and spend the majority of your time logged in as a basic user such as when e-mailing and Web browsing. Only log in as admin to your own machine when you need to, such as to install new software or make system changes. Doing so deprives adversaries of the admin privileges they often need to install malware and rummage through your network.

Just taking these four simple steps mitigates against an amazing 85 percent of targeted intrusions, according to the Australian government research. An in-depth study by Verizon and the U.S. Secret Service revealed similarly good news: "97% of all data breaches were avoidable by implementing simple or intermediate level controls." Better techno-product design and increased public education can go a long way in help-ing individuals and businesses alike make the right choices when it comes to cyber hygiene. To tackle those remaining and more persistent threats, however, a more unified, global approach is required, one predicated on the models of epidemiology and disease propagation.

The Cyber CDC: The World Health Organization for a Connected Planet

The language of our technical insecurity is littered with metaphors of disease. We talk about computer viruses and infections to describe self-replicating malicious code, but rather than focusing on prevention and detection, we often blame those who have become infected and try to ret-rospectively arrest and prosecute those responsible long after the original harm is done. What if we shifted this paradigm and instead viewed our common global cyber security as an exercise in public health? Organiza-tions such as the Centers for Disease Control in Atlanta and the World Health Organization in Geneva have over decades developed robust sys-tems and objective methodologies for identifying and responding to public health threats, structures and frameworks that are far more developed than those in the cyber-security community. Given the many parallels between communicable human diseases and those affecting the world's technolo-gies, there is also much we can learn from the public health model, an adaptable system capable of responding to an ever-morphing array of pathogens around the world.

Importantly, in matters of public health, individual actions can only go so far. It's great if you have excellent techniques of personal hygiene, but if your whole village has Ebola, eventually you will succumb as well. The comparison is relevant to our world of cyber threats. Individual responsibility and action can make a huge difference in cyber security, but ultimately the only hope we have in responding to rapidly propagating threats across this planetary matrix of interconnected technologies is to build new institutions to coordinate our response. A trusted international cyber World Health Organization could foster cooperation and collabora-tion across companies, countries, and government agencies—crucial steps required to improve the overall public health of the networks driving the critical infrastructures in both our online and our off-line worlds.

A cyber CDC could go a long way toward counteracting the techno-logical risks we face today and could serve a critical role in improving the overall public health of the networks driving the critical infrastructures of our world. Indeed, a report sponsored by Microsoft and the EastWest

Institute suggested that a cyber CDC could fulfill a number of roles that are carried out today only on an ad hoc basis, including the following:

- education—providing members of the public with proven methods of cyber hygiene to protect themselves
- network monitoring—detection of infection and outbreaks of malware in cyberspace
- epidemiology—using public health methodologies to study digital disease propagation and provide guidance on response and remediation
- immunization—helping to vaccinate the public against known threats through software patches and system updates
- incident response—sending in experts as required and coordinating global efforts to isolate the sources of online infection and treat those affected

While there are many organizations, both governmental and nongovernmental, that focus on the above tasks, no single entity owns them all. It is through these gaps in effort and coordination that our cyber risks continue to mount. In particular, an epidemiological approach to our growing technological risks is required to get to the source of malware infections, as was the case in the fight against malaria. For decades, all medical efforts focused in vain on treating the deadly parasitic disease for those already infected. But it wasn't until epidemiologists realized the malady was spread by mosquitoes breeding in still pools of water that genuine progress was made in the fight against the disease. By draining the swamps where mosquitoes and their larvae grow, epidemiologists deprived them of an important breeding ground, thus reducing the spread of malaria. What swamps can we drain in cyberspace to achieve similar results? We haven't quite yet figured it all out and thus the importance of this work.

There is another major challenge the cyber CDC will face: most of those who are sick have no idea they are walking around infected, spreading disease to others. Whereas malaria patients develop fever, sweats, nausea, and difficulty breathing, important symptoms of their illness, infected computer users may be completely asymptomatic. This important difference is evidenced by the fact that the overwhelming majority of those with infected devices have no idea there is malware on their machines nor that they might have joined a botnet army. Even in the corporate world, with the average time to detection of a network breach now at 210 days, most companies have no idea their most prized assets, whether intellectual property or a factory's machinery, have been compromised.

The only thing worse than being hacked is being hacked and not knowing about it. If you don't know you are sick, how can you possibly get treatment? Moreover, how can we prevent digital disease propagation if carriers of these maladies don't realize they are infecting others? Addressing these issues will be a key area of import for any proposed cyber World Health Organization and fundamental to our future communal safety and that of our critical information infrastructures.

The cyber-security researcher Mikko Hypponen has pointed out the obvious Achilles' heel of our modern technology-infused world—the fact that everything is run by computers and that everything is reliant on these computers' working. The challenge before us is that we must have some way of continuing to work even if all computers fail. Were our information systems to crash on a mass scale, there would be no trading on financial markets, no taking money from ATMs, no telephone network, and no pumping gas. If these core building blocks of our society were to suddenly go away, what would be humanity's backup plan? The answer is simply, we do not have one.

Taking the steps outlined in this chapter will go a long way toward protecting us from the panoply of threats we face today, but such a plan of action is far from foolproof. We are at the dawn of a technological arms race, an arms race between people who are using technology for good and those who are using it for ill. The challenge is that nefarious uses of technology are scaling exponentially in ways that our current systems of protection have simply not matched. It's time to build greater resiliency into our global information grid in order to avoid a system crash. If we are to survive the progress offered by our technologies and enjoy their abundant bounty, we must first develop adaptive mechanisms of security that can match or exceed the exponential pace of the threats before us. On this most important of missives, there is unambiguously no time to lose.

CHAPTER 18

The Way Forward

Let no one be discouraged by the belief there is nothing one person can do
against the enormous array of the world's ills, misery, ignorance,
and violence. Few will have the greatness to bend history, but each of us
can work to change a small portion of events. And in the total of all
those acts will be written the history of a generation.

ROBERT F. KENNEDY

Our technological genie cannot be put back in its bottle. Whether in cyber, robotics, artificial intelligence, or synthetic biology, massive changes are afoot in our world. These changes have brought us to the knee of an exponential curve—one that will be explosive in its growth in the coming years. Indeed, these breakthroughs will arrive much sooner than most would have anticipated, as one domain of science leads to progress in another. Advances in information technology drive synthetic biology and artificial intelligence drives robotics. Each of these forces affects the other, driving exponentials of exponentials. As noted throughout this book, however, not all of these developments will be for good. In example after example, we've documented criminals, terrorists, hackers, and rogue governments subverting technology and using it to harm others. The point of course is not that technology is evil. Fire, the original technology, could be used to keep us warm, cook our food, or burn down the village next door. A knife can be wielded both by a surgeon and by a murderer. In the hands of those of good intent, our rapidly evolving technologies will bring tremendous abundance to the world. But in the hands of a suicide bomber, the future can look quite different.

Ghosts in the Machine

To measure is to know.
LORD KELVIN

One of the greatest challenges we face with our current technological insecurity is that there are often few if any telltale signs of hacker intrusions into our networks and devices. The obvious problem is that you have unwanted guests (whether or not you realize it). The greater conundrum is that you can't fight what you can't see. Across our smart phones, laptops, tablets, bank accounts, refrigerators, cars, corporate networks, and electrical grids, there are ghosts in our machines. Keeping out all intruders at all times was a worthy if quaint goal. But in case you hadn't noted, the proverbial technological republic has fallen. Our technology is riddled with bugs, flaws, and invaders. Today, regretfully, our goal can no longer be purely prevention. We must chase the ghosts from our machines by proactively searching them out and hunting them down. With our time to detection hovering north of two hundred days, clearly there is much work to be done. We need to reduce the time frame to mere hours and eventually minutes and seconds.

Then there's the lingering issue with big data: the more you keep, the more you have to protect. But most companies have never cataloged their information assets and thus don't know what data they're storing, where they are keeping the data, and which data are the most critical to protect. Importantly, once these threats are detected, we need to begin discussing them—publicly.

Breaking down the wall of silence that surrounds nearly all cyber attacks is a vital step toward bolstering our common technological security. Companies today know the ramifications of being named publicly as a hacking victim. Beyond the obvious reputational damage, costs can be in the hundreds of millions of dollars from direct losses, customer churn, and litigation. Thus organizations will do all within their power to keep silent when victimized, whether by Crime, Inc. or a foreign espionage service. But this silence is at the very heart of our cyber-security problems. When a person survives a sexual assault but is too embarrassed or ashamed to report it to the police, the assailant will not be found and prosecuted, free to surely victimize others. Though a cyber attack is an entirely different form of crime, its victims too are loath to speak publicly. As a result, these incidents cannot be aggregated and studied, common defenses are not developed, and perpetrators roam free to attack another day. This is a situation we must rectify. Maintaining silence about these risks does not make them go away; it makes them worse, empowering bad actors to operate with impunity. Much like the decision to go to Alcoholics Anonymous, admitting you have a cyber problem is the first and most important step toward getting better.

Building Resilience: Automating Defenses and Scaling for Good

**New technologies can be used for destructive purposes.
The answer is to develop rapid-response systems for new dangers like a bioterrorist creating a new biological virus.**

RAY KURZWEIL

Cyber attacks happen—they cannot all be stopped. The higher-order question thus must be, how can we construct our rapidly emerging technological world in a way that is much more resilient to attack? It is not an easy question to resolve given the ever-expanding system complexities at play. A resilient system is one that will not fail catastrophically but degrade slowly over time until it can be repaired. A resilient system will continue to perform its most critical functions, though other less important activities may go off-line or cease to operate. Nature has excellent structures in place for this as evidenced by the common lizard. When attacked or grabbed by a predator, a lizard can easily shed and regrow its tail, allowing the critical parts of its body (the brain and reproductive organs) to escape and survive. What is the lizard's tail of the Internet or your corporate network? It doesn't exist yet, and that's something we need to fix.

Much of our technological infrastructure is subject to common single points of failure, the most obvious of which is power. No electricity, no Internet. Worse, no electricity, no water distribution, food production, financial transactions, communications, or transportation. We need to isolate these singular failure points so that they do not spread and we need to have alternative power sources that can scale to prevent these types of "blackouts"—not just for electricity of course, but for all the technological tools that make our modern civilization possible.

These risks extend well beyond the power grid and include our most common software systems and the Internet infrastructure itself. Many of the tools that run our techno-world are monocultural in their nature; that is, they run nearly identical software containing the same vulnerabilities. Computer monocultures, like agricultural ones, are subject to catastrophic failure—think Irish potato famine. Today Microsoft Windows drives more than 90 percent of the desktop computers worldwide, and as of early 2014 an amazing 95 percent of the ATMs in the United States were still running Windows XP—an operating system for which Microsoft has ceased all security updates. Technological monocultures are the lifeblood of mass computer exploitation. With one piece of malware, hackers can have profound global impact by getting all copies of the same software process to fail uniformly. As we have seen, known bugs can live in the wild for years before these software holes are plugged by vendors. The moment a breach in one version of Windows 8 or Adobe PDF is detected, the global repair process should begin. Software companies should not just wait for people to manually patch their systems (which we know most never do). Rather, these systems should be self-healing, always reaching out for the latest patched versions of the software to ensure all known doors and windows to our digital lives are locked. Put another way, failing to address known vulnerabilities in tens of millions of versions of the same software program running around the world would be akin to uncovering a mechanical fault that led to the crash of a 747, one that was present in all 747s in operation globally, and letting these aircraft continue to fly.

We also need to ensure that individual breaches and attacks can be isolated and prevented from spreading. Consider the 2013 hack against the retail giant Target. As mentioned previously, the attackers responsible gained access to Target's point-of-sale terminals by first exploiting the network of a contractor responsible for maintaining the store's heating and air-conditioning systems. If that initial breach had been detected, Target's ensuing corporate security nightmare could have been entirely avoided. We need better and more resilient means of protecting our information. Think air bags for our data. When a data breach occurs, these virtual air bags should go off, enveloping our digital possessions and protecting them from further harm.

CEOs and boards of directors should ask themselves just how resilient their organizations are. Resiliency means remaining operational in the face of sustained attack by sophisticated opponents. Though you, like the lizard, might lose your tail, the organization must live on. This won't happen magically and requires preparedness training and exercises. In particular, cyber resilience requires adeptness in responding to an attack and dexterity in rapidly recovering degraded technological capabilities. How to heal quickly after an attack may be the make-or-break issue deciding whether an organization fails or survives. The time to answer these questions is not during the crisis but long before it occurs.

These more resilient systems that we require must be built from the ground up. Security cannot be an afterthought tossed into the mix after the machines have been built. Systems must be engineered to fail gracefully, not cataclysmically. Secure and trustworthy computing must be the cornerstone of our technological future, lest the whole system come crashing down. This will be especially true as we drive toward the Internet of Things and see the arrival of highly disruptive technologies such as robotics, artificial intelligence, and nanotechnology. We can no longer neglect the public policy, legal, ethical, and social implications of the rapidly emerging technological tools we are developing; we are morally responsible for our inventions.

There are good examples in history where we as a society have brought together expertise in anticipation of catastrophic risk before it occurred. One such case was the 1975 Asilomar Conference on Recombinant DNA, which was held at Asilomar State Beach in Monterey, California. The event gathered 140 biologists, lawyers, ethicists, and physicians to discuss the potential biohazards of emerging DNA technologies and drew up voluntary safety guidelines. As a result of the event, scientists agreed to stop experiments involving mixing the DNA from different organisms— research at the time that held the potential to have radical, poorly understood, and potentially disastrous consequences. The lessons and successes of Asilomar are well worth repeating. Though we are racing full speed ahead with synthetic biology, artificial intelligence, swarming robotics, and nanotechnology, we are dedicating precious few resources to understanding the concomitant risks of technologies that could replicate beyond our control. Thankfully, in 2009 such a meeting on the future of artificial intelligence was held on the very same beach in Monterey, and more such gatherings are crucial to bolstering the resiliency of a world built on exponential technologies.

Moving forward, in order to strengthen the safety and security of our society, we must make another change. We need to be able to respond at scale to the challenges we face from a fully automated criminal hacker community. Time and time again, we've seen those with ill intent automate their attacks. It is this ability that has led to the paradigm shift in crime, moving from a one-to-one to a one-to-many affair. That is why 1.2 billion account passwords can be gathered by one organized crime group while another launches a remarkable seventy-gigabit-per-second DDoS attack that knocks a dozen financial institutions off-line. The tools to commit evil are scaling exponentially, but our systems for scaling for good are not keeping up. Our defenses are not adapting rapidly enough to match the global systemic risk we face, something our government should be deeply concerned about.

Reinventing Government: Jump-Starting Innovation

**We can't solve problems by using the same kind of thinking
we used when we created them.**
ALBERT EINSTEIN

In 2014, only 13 percent of Americans approved of the job Congress was doing, a slight improvement from the all-time low of 9 percent in November 2013. Trust in government is practically nonexistent—whether it's the money in politics, the government shutdowns, the partisanship, or the dearth of meaningful legislation. While the technological change all around us is proceeding at an exponential pace, the government is decidedly linear in its rate of change. The obvious challenge with such an asymmetry is that we will never solve twenty-first-century problems with nineteenth-century institutions. We need vastly more adaptive government, one that can respond ten times as fast, and that's just to keep up. Cabinet secretaries and Supreme Court justices who "don't do email" simply won't do anymore.

The lack of innovation in government permeates not only our legislatures but the organs of our national security and law enforcement apparatus as well. In response to the creativity (albeit diabolical) demonstrated by the terrorists who carried out the 9/11 plot, the government spent billions of dollars and came up with such "innovations" as the Transportation Security Administration. Though frisking four-year-olds and little old ladies in wheelchairs makes for fine "security theater," we're going to have to significantly up our game if we hope to prevent future terrorist attacks. Given the pace of technological change, tomorrow's security threats will not look like those of today—one of the reasons government is struggling mightily in the face of our common cyber insecurity.

Of course this is not meant to suggest there is no innovation in government. It was government that brought us the Internet and space travel and served as the catalyst to finally decode the human genome. There are pockets of innovation in government everywhere, but we need to get these gems of creativity to replicate and scale in a way that simply is not happening today. One such model is Code for America, a nonprofit organizing citizen volunteers with computer coding skills to make government services much more simple, effective, and easy to use. Another is the well-regarded GovLab at NYU, an innovation laboratory dedicated to using technology to redesign the problem-solving capabilities of government institutions. Backed by both the MacArthur and the Knight Foundations, GovLab is working to use networked technologies to move beyond the centralized top-down control paradigms of yesteryear in favor of more transformational platforms of self-governance, innovation, and citizen engagement.

Fundamentally, however, if government is to remain relevant in responding to the most pressing and significant challenges the world faces today, we will need to come up with completely new frameworks for problem solving. On this point, we can borrow a page from Silicon Valley and start thinking of our system of governance as the operating system for society. If we can fundamentally change the OS, everything else changes with it. Our legacy institutions are struggling, whether in education, health care, or law enforcement; technology is far outpacing the ability of government to respond. Until this point, much of the government's approach to technological security has been merely window dressing and missed opportunities. As Internet entrepreneur Bryan Johnson has noted, we need a new operating system for the world, one based on first principles, that matches the exponential changes all around us.

Fortunately, Johnson has generously donated $100 million of his personal wealth to the effort, creating the OS Fund to promote "quantum leap discoveries" at the operating system level in order to drive "real change for humanity at the global scale." Today's government institutions clearly do not have a monopoly on answers to the many problems facing our world, but they can play an important role as convener—bringing together the public and private sectors as a means of finding solutions to some of our grandest challenges.

Meaningful Public-Private Partnership

Government efforts to protect the people against everyday cyber crime and security threats have been wholly inadequate. Should we be surprised? The tens of thousands of attacks successfully perpetrated against Washington by foreign adversaries prove the U.S. government can't even protect itself. The need for more serious and profound collaboration between the public and the private sectors is manifest; without it, we will make little meaningful progress in improving the overall state of our security. The need is particularly vital when it comes to protecting our country's critical infrastructures, 85 percent of which are in the hands of the private sector. As a nation and as a people, we need government and industry to collaborate in protecting the machinery of our modern world. The question is how.

Recognizing the need for public-private partnerships (PPP), institutions as diverse as the FBI, the European Union, and the World Economic Forum have established programs to foster greater cooperation among those responsible for running the world's critical infrastructures. Other initiatives such as Information Sharing and Analysis Centers help to enable specific industries such as financial services, energy, and communications to better collaborate and respond to cyber threats. The Forum of Incident Response and Security Teams has also played an influential role in improving coordination and response among trusted peers in both government and private sector CERTs, or Computer Emergency Response Teams. Initial efforts at public-private partnerships have proven helpful, to be certain, but some PPP efforts have been criticized for having ill-defined goals and few if any specifically articulated objectives beyond "sharing information."

There are real problems to be overcome to make public-private information sharing regimes reach their true potential. The private sector generally lacks trust in the government to maintain its confidentiality, particularly when it comes to revealing cyber-threat data to competitors, let alone protect it from antitrust risk. The government too has challenges:

it must figure out how to share knowledge of particular cyber risks, many of which are classified, with companies and technical personnel that lack the required clearances to see the classified material. A 2010 Government Accountability Office report determined that less than one-third of companies participating in cyber-security collaborations with the government felt that they were receiving actionable cyber-threat information.

The exigency of the technological hazards before us means that we must overcome these tribulations with much greater urgency in order to foster meaningful partnerships between the government and the private sector.

One particularly positive note in this vein has been SINET—the Security Innovation Network—whose goal it is to promote innovation in the field of cyber security by building meaningful bridges between the public and the private sectors. SINET was founded in San Francisco and serves as a connector (an interpreter of sorts between those in Silicon Valley and those inside the Beltway). By bringing together the leading players from both of these worlds, SINET has helped drive entrepreneurship and innovation among all parties working in the cyber-security ecosystem to focus them on the mission at hand. Beyond those in the government and industry who make fighting cyber threats their full-time occupation, another massive force can be brought to bear on the technological challenges we face: a smart and engaged general public.

We the People

It's not a faith in technology. It's faith in people.
STEVE JOBS

Contemplating the full scale and scope of malicious activities perpetrated by organized criminals, terrorists, hackers, and rogue governments is enough to make anybody feel dispirited, frightened, and even depressed. But if there is one thing that gives me considerable solace after nearly two decades working in the field of global security, it is this: the good people in this world vastly outnumber the bad. That is a huge advantage but one that we have not yet fully leveraged to our benefit. Crime, Inc. is well versed in crowdsourcing, capable of mobilizing mobs of thousands, as we saw with the massive 2013 ATM cyber attack in which thieves carried out thirty-six thousand in-person transactions in ten hours in twenty-seven countries, pocketing a cool $45 million. Amazing for its speed, prowess, innovation, and impact. But where is the public safety equivalent of such an act? It doesn't yet exist, and that is something we will have to change in order to bolster our self-defenses and our self-reliance at the dawn of this new digital age.

It has become painfully clear to me that our authorities are losing the technological edge to criminals. Law enforcement, overwhelmed by the workload and undermined by budget cuts, is under assault and struggling mightily to keep up. Moreover, policing is a closed system: it is nation based, while the threat is international. Our current paradigms of security—guns, border guards, and tall fences—are shockingly outdated. They do not keep out bits and bytes that can travel around the world at the speed of light. In order to overcome these obvious gaps in our current public safety institutions, we will need to find new and more radical ways to address the problem, ones that incorporate a more open and participatory form of crime fighting. Where are the neighborhood watch and community policing programs in cyberspace? Rather than having a small elite force of highly trained agents on hand to protect all of us, we are far better off enabling ordinary citizens to combat the problem as a group through crowdsourcing. To beat Crime, Inc. at its game, we must get good to scale, but bigger and better.

The idea of crowdsourcing law enforcement is hardly new. In 1865, when John Wilkes Booth assassinated President Lincoln, he became the first fleeing felon to have his photograph appear on a wanted poster. Today, 150 years after the murder of our nation's sixteenth president, government's implementation of crowdsourced law enforcement has changed nary a bit. Cops distribute photographs to local news channels, and broadcasters warn that the wanted party is "armed and dangerous, please contact local law enforcement should you see him." Really? In 2015, surely we can do better to bolster public engagement besides "if you see something, say something."

We the people, just like Crime, Inc., can take advantage of the bounty provided by technology to help protect and defend ourselves. Clay Shirky uses the term "cognitive surplus" to describe the "ability of the world's population to volunteer, contribute, and collaborate on large, even global projects." It's high time we the people began using our available cognitive surplus to help protect and defend our own future. Open-source warfare and crowdsourced crime must be met with open-source security and crowdsourced public safety. Fortunately, there are a few bright spots where this new paradigm of public safety is beginning to shine. Organizations such as Crisis Commons and Ushahidi are reinventing disaster relief and saving lives by coordinating citizen response to public emergencies, including during the Haiti earthquake and the terrorist attack at the Westgate Mall in Nairobi. Citizens in Mexico, a country racked by fifty thousand narcotics-related murders from 2006 to 2012, are using tools such as Google Maps to crowdsource reporting on the cartels, their activities, and their whereabouts. In eastern Europe, the Organized Crime and Corruption Reporting Project, comprising journalists and citizens, crowdsources sophisticated multinational investigations to reveal which dictators, crooked officials, terrorists, and organized crime groups are moving and laundering their massive ill-gotten gains around the globe. Speaking of public corruption, in 2009 editors at the *Guardian* newspaper in Great Britain created software allowing citizens to "crowdvestigate" more than 455,000 pages of data they had obtained in order to identify flagrant expense claim violations by members of the U.K. Parliament. More than twenty-five thousand citizen volunteers joined in the digital investigation, and the results were truly amazing. Over 170,000 documents were reviewed in the first eighty hours, and the crowd's discovery of thousands of flagrant misappropriations of public funds led to the forced resignation of numerous MPs, cabinet ministers, and even the Speaker of the House of Commons, a truly powerful outcome last seen in 1695.

In each of these cases, individuals were able to do more than just report crimes to authorities. They were able to marshal evidence by channeling time and energy to decipher data to produce results faster than any policing or governmental organization could have done alone. Crowdsourcing public safety delivers clear results and must become an integral component of our global security strategy in an exponentially changing world, especially one so short on full-time cyber-security personnel. The Rand Corporation has noted that the nationwide shortage of technical security professionals within the federal government is so critical that it is putting both our national and our homeland security at risk. The finding was echoed by Cisco's *2014 Annual Security Report*, which estimated that there was a talent scarcity of more than a million cyber-security professionals worldwide, expected to grow to two million by 2017. We desperately need more public engagement in protecting our technological future, and even the channels of officialdom have begun to concede the point.

In 2012, the FBI's top cyber lawyer, Steven Chabinsky, called government efforts in fighting cyber crime "a failed approach," adding that much stronger efforts would be required by members of the public in combating cyber threats. That work is slowly starting to begin. In one case, a professor at the University of Alabama worked with the students in his criminal justice class to help the FBI crack a $70 million cyber-crime ring run by Crime, Inc. out of Ukraine and Russia. The student-run "crowdvestigation" successfully identified numerous suspects in the United States who had used the Zeus banking Trojan to steal millions—individuals who were all eventually arrested by the FBI as a result of the students' work. In order to have lasting and meaningful success, however, such crowdsourcing efforts cannot be merely ad hoc but rather have to be formalized systemically in order to scale for growth. In 2011, police in the U.K. took a step in that direction, creating a nationwide cadre of volunteer special constables with relevant skills required to help tackle cyber crime.

Here in the United States and elsewhere in the world, we should build on these successes and take them even further. We already have reserve and auxiliary police officers. In the military, there are part-time reserve army, navy, air force, and marine citizen soldiers. On the civilian side, we have the Peace Corps and AmeriCorps. We need a National Cyber Civil Defense Corps. Such an organization would be reminiscent of other civil defense efforts in our nation's history dating back to World War I. Experts would be drawn from all corners of society in order to protect our critical information infrastructures from attack and our nation from the mounting technological threats before us. Members would be carefully screened, undergo extensive training and background investigations, and operate under clearly defined operational and legal frameworks. Timing is of the essence to establish and build such a crowdsourced force for good—now before the cyber crisis occurs. There are many private sector professional organizations that could prove immensely helpful in jump-starting such efforts, such as the International Information Systems Security Certification Consortium, or (ISC)^2, a nonprofit with over 100,000 certified cybersecurity professionals at the ready, capable of having a positive impact on any such effort should they choose.

Crime, Inc. is out there busily recruiting minions for its efforts. Shouldn't we be doing the same? People of all stripes and backgrounds can help with these endeavors—young, old, and even some hackers who surely have the skill set to make a difference, should they wish to direct their talents for public benefit. As the Apple co-founder Steve Wozniak reminds us, "Some challenging of the rules is good." We need to help create opportunities, particularly for young people, to channel their considerable talents and energies for good, lest Crime, Inc. engage them for ill. The exponential nature of technology and the linear response of government mean we will need many more hands on deck to help build a safe and stable society that won't destroy itself. Our public safety and security are just too important to leave to the professionals. In today's exponentially advancing world, in the battle between good and evil, victory will belong to whichever group proves itself most capable of mobilizing the larger crowd. It's time to start gaming this system in our favor to ensure our technological tools inure to the greatest overall benefit of humanity.

Gaming the System

Every game designer should make one explicitly world-changing game.
Lawyers do pro bono work, why can't we?
JANE McGONIGAL

According to the American game designer and researcher Jane McGonigal, today there are more than half a billion people worldwide playing computer and video games at least an hour a day, with more than 183 million in the United States alone. That works out to three billion hours a week as a planet playing video games. What if these efforts could be directed for particular public goods? Imagine the enormous power and potential that could be unleashed. Doing so would allow the wisdom of the crowds to be funneled in a way that addresses some of the world's greatest challenges. To test this theory, DARPA in 2009 created its Network Challenge (also known as the Red Balloon Challenge), hiding ten large red helium balloons outdoors across the United States in cities from Miami to Portland, offering a $40,000 prize to the first team that could find all the balloons. DARPA created its competition to explore the role the Internet and social networking might play in real-time communication and wide-area collaboration in order to solve time-critical problems, such as disaster relief in time of crisis. Remarkably, a team from MIT found all ten balloons hidden in the farthest reaches of our country in a mere nine hours, by crowdsourcing the task over social media to forty-four hundred volunteers.

As it turns out, playing games needn't be a waste of time and can actually be a highly productive activity. Gamification is a new field of study that allows the use of game thinking and mechanics in non-gaming contexts to motivate and engage players in solving actual real-world problems. One such example has been in public health via the diagnosis and treatment of malaria. Worldwide, there are more than 600,000 malaria cases per day, with one child dying every minute. The disease is spread by mosquito bites that transfer parasites into the human body, infecting our red blood cells. Diagnosing malaria is time-consuming, taking up to thirty minutes for a specialist to manually look for parasites in blood under a microscope, leaving many undiagnosed to die. MalariaSpot is a game that changes that by taking virtual images of actual patient blood slides and presenting them to players, challenging them to tag as many parasites as they can in just one minute. The results were impressive: in just one month, anonymous players from ninety-five countries played twelve thousand games. After receiving just a few moments of online training explaining what the parasite looked like, MalariaSpot players have correctly identified over 700,000 parasite diagnoses. Because the same image is shown to multiple players, these nonmedical-expert game players have achieved an accuracy rate higher than 99 percent, a "game changer" in the world of malaria treatment and diagnosis. In another case, a game called Foldit allows members of the public with no specialized training in molecular biology to solve puzzles for science by using their 3-D spatial-orientation skills to manipulate and fold protein molecules as a means of studying and treating disease. In one remarkable case, Foldit players correctly identified the structure of an enzyme crucial in the reproduction of HIV in just a few days, a discovery that had eluded AIDS researchers around the world who had been actively trying to solve the problem for more than a decade.

Crowdsourced games such as MalariaSpot and Foldit should provide far-reaching inspiration and important take-away lessons we can apply to the conundrum we face with regard to our technological insecurity. What enjoyable puzzles might we create to further engage the public, particularly young people, in using their love of gaming to improve our cyber security? Imagine the possibilities. Rather than showing blood slide images looking for malaria, we could serve up phishing e-mails in real time and ask the crowd to correctly identify malicious spam requests for bank account information—awarding points and prizes to the best players. Gamification of securing software might help technology companies avoid the obvious pitfalls of the "Just ship it" mentality by getting tens of thousands of players around the world to go on "bug hunts" for flaws in their software or hardware products, flaws that would otherwise be exploited by Crime, Inc. hackers to the public's detriment. Such an idea is already under development by DARPA, as well as by several start-ups, including Topcoder and Bugcrowd. These same techniques could be applied to our nation's critical infrastructure systems as well. Players could be shown anonymized data in a SimCity-style animated game and let loose to find security vulnerabilities in everything from our virtual electrical grids to our transportation networks. In the end, individual gamers may hold the potential to make significant breakthroughs in cyber security, doing it for no other reason than that they enjoy playing the game. Others will be motivated by their ability to solve real-world problems and helping their fellow man. For those that find neither appealing, there's always cold hard cash.

Eye on the Prize:
Incentive Competitions for Global Security

The day before something becomes a breakthrough, it's a crazy idea.

PETER DIAMANDIS

Prizes have a way of focusing the mind. Just ask the throngs who show up for a chance at the Mega Millions lottery jackpot. But prizes can also be the spark that produces a revolutionary solution to an intractable problem. Such was the case when the British Parliament established the Longitude Prize in 1714 in an effort to help with maritime navigation in order to ensure the "safety and quickness of voyages, the preservation of ships, and the lives of men." Though latitude (north-south positioning) was easy to measure using the position of the sun, until the early eighteenth century there was no way for sailors to calculate their position longitudinally from east to west. By an act of Parliament, the British government offered £20,000 (more than £1 million today) for a solution that could find longitude to within half a degree. The incentive prize inspired John Harrison, a self-educated working-class clock maker, to invent the marine chronometer, a clocklike device that solved the problem. Two hundred years later, another incentive prize was launched, this time to stimulate advances in the nascent field of aviation.

Charles Lindbergh became the first man to fly across the Atlantic, not just because of his sense of adventure, but because a seldom-recalled hotel magnate named Raymond Orteig offered $25,000 of his own money in 1919 as a prize to "the first aviator of any Allied Country crossing the Atlantic in one flight, from Paris to New York or New York to Paris." Orteig offered the purse to drive forward an exciting new technology of his day: the flying machine. The effort was funded by no government and there was no immediate profit to be made, but that didn't stop nine separate teams from spending around $400,000 in pursuit of the $25,000 prize. The prize was the fundamental kindling, the thing that sparked the innovation that solved the problem and helped create today's aviation industry. In 1996, the physician, space enthusiast, and serial entrepreneur Peter Diamandis took up the Orteig mantle and created the XPRIZE Foundation, a nonprofit organization that designs and manages public competitions intended to encourage technological development for the betterment of mankind. Perhaps it is time for such a competition in cyber security.

According to Diamandis, "An XPRIZE is a highly leveraged, incentivized prize competition that pushes the limits of what's possible to change the world for the better. It captures the world's imagination and inspires others to reach for similar goals, spurring innovation and accelerating the rate of positive change." The first competition Diamandis ever announced was the $10 million Ansari XPRIZE, which challenged teams to launch a manned-spaceship past the Karman Line (100 km altitude) before safely returning to Earth. As if that wasn't enough, the rules also stated that the spaceship must be able to accommodate the weight of two additional adults and undertake a second launch within the span of two weeks. Without any government funding, twenty-six teams spent upward of $100 million trying to reach this lofty goal, and in the fall of 2004, the Mojave Aerospace Ventures team succeeded, potentially paving the way for space tourism and other commercial spaceflights. Incentive prizes are bold and audacious and capture the world's attention—exactly the type of thinking we need to make a significant leap forward in protecting ourselves from the profound technological risks we face today.

An XPRIZE for cyber security could serve as an engine of innovation, an outstanding impetus to drive exponential change for good and address the world's technological insecurity to the overall benefit of humanity. In clearly defining the cyber-security problems we face, the XPRIZE could incentivize teams around the world to find the most effective solutions in a way that may well avert crises, empower people, generate new technologies, and even create new industries. An XPRIZE for cyber security could help us overcome perhaps one of the greatest challenges we face with regard to the risks from exponential technologies—the belief that these problems are intractable and unsolvable and that there is no clear path forward toward a solution. Bull. We've faced tough times before and as a species have repeatedly achieved things that just a day before seemed like crazy ideas. Incentive prizes inspire hope through vision of a better future, and those who win them are proof that some of our seemingly impossible problems can and will be solved. One individual or a small team can surely make a difference as Lindbergh, Harrison, and countless others have demonstrated. Importantly, an XPRIZE for cyber security could just be the beginning in making significant advances in global security. Other emerging threats such as bioterrorism, artificial intelligence run amok, autonomous weapons systems, and nanotechnology are all also ripe for incentive prizes, particularly given the potentially existential risk they might pose to the world.

Just as the philanthropist Raymond Orteig incentivized civil aviation and Anousheh and Amir Ansari spurred on the commercial space industry, so too can today's philanthropists make a big difference in our technological security. Look at the amazing feats the Bill and Melinda Gates Foundation has accomplished in fighting HIV, eradicating polio, and supporting education, distributing an amazing $26 billion of Mr. Gates's wealth since the foundation's creation. But they are not alone, and there is indeed a new breed of "techno-philanthropists" out there, committed to using their wealth to better the world. eBay's first president, Jeff Skoll, has worked tirelessly crusading against pandemics and nuclear proliferation, endowing his foundation with nearly $1 billion of his own funds. Elon Musk, Pierre Omidyar, Paul Allen, Steve Case, Larry Ellison, Mo Ibrahim, Sir Richard Branson, and Michael Bloomberg have all incredibly generously signed "The Giving Pledge," committing to dedicate the majority of their wealth to philanthropy. These individuals have personal passions that they are actively supporting with their wealth, ranging from good governance to child development. Given that most of those above earned all or part of their wealth working in technology, funding an XPRIZE focused on this topic would make great strides in combating the emerging technological threats before us and, with their expertise in the field, could make a huge difference. Happily, the XPRIZE Foundation is in the early stages of exploring a cyber-security XPRIZE, with support from Deloitte Consulting. Even a $20 million purse (a mere .01 percent of annual revenues from the $150 billion software industry) would go a long way toward helping to provide the more stable and secure software required to protect our technological future. But even more can be done, something big and bold and on the same scale and scope as the pressing technological challenges before us.

Getting Serious: A Manhattan Project for Cyber

During my participation in the Manhattan Project and subsequent research
at Los Alamos, encompassing a period of fifteen years, I worked in
the company of perhaps the greatest collection of scientific talent
the world has ever known.

FREDERICK REINES

When it was discovered in 1939 that German physicists had learned to split the uranium atom, fears quickly spread throughout the American scientific community that the Nazis would soon have the ability to create a bomb capable of unimaginable destruction. Albert Einstein and Enrico Fermi agreed that President Franklin Delano Roosevelt had to be apprised of the situation. Shortly thereafter, the Manhattan Project was launched, an epic secret effort of the Allies during World War II to build a nuclear weapon. Facilities were set up in Los Alamos, New Mexico, and Robert Oppenheimer was appointed to oversee the project. From 1942 to 1946, the Manhattan Project clandestinely employed over 120,000 Americans toiling around the clock and across the country at a cost of $2 billion. Those working on the Manhattan Project were dead serious about the threat before them. We are not.

While no sane person would equate the risks from the catastrophic impact of nuclear war with those involving 100 million stolen credit cards, some of the scientific discoveries under development today, including artificial intelligence, nanotechnology, and synthetic biology, do indeed have the potential to be tremendously threatening to life on this planet, as Stephen Hawking, Elon Musk, and others have warned. Beyond these potential existential threats, we must surely recognize that the underpinnings of our modern technological society, embodied in our global critical information infrastructures, are weak and subject to come tumbling down through either their aging and decaying architectures, overwhelming system complexities, or direct attack by malicious actors.

Though we have yet to suffer the game-changing calamitous cyber attack of which many have warned, why wait until then to prepare? The evidence of the technological perils is all around us. On a daily basis, cyber attacks disrupt our financial system, thieves steal billions in intellectual property, foreign nations pilfer our military weapons plans, and hackers share online tips with one another on how to take over the industrial control systems that run everything from power plants to water and sewage treatment facilities. To paraphrase the renowned statistician and editor of the *FiveThirtyEight* blog, Nate Silver, our current lackadaisical approach to cyber security and the profound technological vulnerabilities before us has been until this point akin to applying sunscreen and claiming it protects us from a nuclear meltdown—wholly inadequate to the scale of the problem. It is time for a stone-cold somber rethinking of our current state of affairs. It's time for a Manhattan Project for cyber security.

I'm not the first to suggest such an undertaking; many others have done so before, most notably in the wake of the September 11 attacks. At the time, a coalition of preeminent scientists wrote President George W. Bush a letter in which they warned, "The critical infrastructure of the United States, including electrical power, finance, telecommunications, health care, transportation, water, defense and the Internet, is highly vulnerable to cyber attack. Fast and resolute mitigating action is needed to avoid national disaster." Signatories to the letter included those from academia, think tanks, technology companies, and government agencies—including former directors of DARPA, the CIA, the Defense Science Board, Xerox PARC, and various national laboratories and Ivy League universities. These serious thinkers, not prone to hyperbole or exaggeration, warned that the grave risk of cyber attack was a real and present danger and called for the president to act immediately in creating a cyber-defense project modeled on the Manhattan Project. That call to action was in 2002. Sadly, precious little has changed since then with regard to the state of the world's cyber insecurity; if anything, the situation has grown worse. Sure, there have been nominal efforts and the rearrangement of some chairs on the proverbial deck of the *Titanic* but not much in the way of substantive progress. What is America's overarching strategy to protect itself from the rapidly emerging technological threats we face? We simply do not have one—a serious problem we may live to regret.

A real Manhattan Project for cyber would draw together some of the greatest minds of our time, from government, academia, the private sector, and civil society. Serving as convener and funder, the government would bring together the best and brightest of computer scientists, entrepreneurs, hackers, big-data authorities, scientific researchers, venture capitalists, lawyers, public policy experts, law enforcement officers, and public health officials, as well as military and intelligence personnel. Their goal would be to create a true national cyber-defense capability, one that could detect and respond to threats against our national critical infrastructures in real time. This Manhattan Project would help generate the associated tools we need to protect ourselves, including more robust, secure, and privacy-enhanced operating systems. Through its research, it would also design and produce software and hardware that were self-healing and vastly more resistant to attack and resilient to failure than anything available today. Such a project of national and even global importance would have the vision, scope, resources, and budgetary support required in order to make it a success.

Most important, it would also require a sense of urgency commensurate with the original Manhattan Project, something that has been heretofore entirely absent from our current and previous halfhearted attempts to deal with our growing cyber insecurity.

As daunting as such a task may seem, there is good news. We *can* do this. We can succeed in this fight. We as a people surely have what it takes to make a profound difference in our common security moving forward. It will require vision, focus, and leadership. And though it may seem hopeless at times, let us take encouragement from President John F. Kennedy, who, in a speech he delivered at Rice University in September 1962, persuaded the American people to fund NASA and, before that decade was out, to land a man on the moon and return him safely to earth. In his eloquent and rousing speech before thirty-five thousand spectators, President Kennedy extolled the importance of space travel as being integral to our global security, noting,

> Man, in his quest for knowledge and progress, is determined and cannot be deterred . . . We have vowed that we shall not see space filled with weapons of mass destruction, but with instruments of knowledge and understanding . . . We set sail on this new sea because there is new knowledge to be gained, and new rights to be won, and they must be won and used for the progress of all people. For space science, like nuclear science and all technology, has no conscience of its own. Whether it will become a force for good or ill depends on man . . . [Therefore] we choose to go to the moon. We choose to go to the moon in this decade and do the other things, not because they are easy, but because they are hard, because that goal will serve to organize and measure the best of our energies and skills, because that challenge is one that we are willing to accept, one we are unwilling to postpone, and one which we intend to win.

Hell yes! That's what I'm talking about. Where is that leader? That man or woman, the one who will take us boldly into this twenty-first century, using our technologies for our common betterment and willing to stake his or her reputation and honor on meeting the sacred mission, exercising valor, determination, and the conviction of belief in order to make it so? Only through fierce coordination of efforts across government, academia, and the private sector will we make progress. The key to making the Manhattan Project for cyber actually work will be a keen sense of urgency concomitant with the enormity and importance of the task before us. The clock is ticking, and there is no time like the present to bring this idea to fruition.

Final Thoughts

The best way to predict the future is to invent it.

ALAN KAY, XEROX PARC

When it comes to technological threats against our security, the future has already arrived. It is sitting in an office building in Kiev, destined to be the next Innovative Marketing. It is in the laptop of that kid next to you at the library who is building the next Silk Road and Assassination Market. It's in that ten-story government building in that foreign capital where every day thousands of digital spies are showing up at work intent on stealing your corporate secrets. It's in the garage of that one disaffected bio-hacker who is tired of the bullying in school and now plotting his bioterror revenge. It's at the local big-box retailer selling quadcopter drones, never knowing if they will be used to ferry weapons over prison or airport fences. It's available via that Web site that sells model jet aircraft capable of autonomous flight laden with explosives to be flown into a crowded building by terrorists. This future has already arrived. All the warnings and indicators are there. The threat is serious, and the time to prepare for it is now; I can assure you that criminals, terrorists, and other malicious actors already have.

As we have seen, everything is connected and everyone is vulnerable. But all is not lost; there are things we can do about it as outlined in this chapter and the previous one. But when we fail to respond to the problem at hand and bury our heads in the sand, the problem does not go away; it grows. The challenges we face are significant and mounting. It's not just about hacked bank accounts or stolen private photographs. Nor is it merely about maintaining control and privacy over the multitude of devices in our lives. It is about safeguarding our technological future and understanding what's coming next. As Marshall McLuhan reminds us, "It is the framework which changes with each new technology and not just the picture within the frame."

The hacks of tomorrow will affect our cars, GPS systems, implantable medical devices, televisions, elevators, smart meters, baby monitors, assembly lines, and personal-care bots. With seventy-nine octillion new possible connections enabled through IPv6 and the Internet of Things, all physical objects will become hackable, including all the screens in our lives. Yet as of today, we lack any viable models for truly trustworthy and secure computing—an obvious failure for a society built on and run by computers. We have no proven way to trust the code that runs our lives and runs our world. It is for that reason that those who control the code can control our world, for good or for ill. Beyond this, we will have to deal with new bioweapons, hacked DNA, and genetic and biometric identity theft, to say nothing of easily manipulated black-box algorithms and AI systems. We are living in exponential times, and though it is easy to dismiss autonomous killer robots and Skynet-like evil AIs as pure sci-fi fantasy of the future, as George Carlin reminds us, "The future will soon be a thing of the past."

In a world in which all of our critical systems and infrastructures are run by computers, it would be easy to dismiss our profound technological insecurity as just a computing problem. But we don't just have an IT problem. Because technology is woven through the entire fabric of our modern lives, we also have a social problem, a personal problem, a financial problem, a health-care problem, a manufacturing problem, a public safety problem, a government problem, a governance problem, a transportation problem, an energy problem, a privacy problem, and a human rights problem. We have no choice but to win this battle for the very soul of our own technologies because frankly the alternative is too horrible to consider. This must be our call to action.

Accordingly, now is the time to completely reevaluate all that we take for granted in this modern technological world and question our dependence on the ubiquitous machines that so few of us understand. We do this not out of blind technophobia nor in deference to Luddite ancestors but as a commonsensical measure, fully appreciating the vast positive potential these exponential technologies portend. The innovation cannot be stopped, and the technological changes are coming faster and faster. We've reached an inflection point, a punctuated moment in time that demands our immediate and greatest possible attention. The proverbial twenty-ninth day of the lily pond is fast approaching and as with all things exponential, our window to act responsibly and responsively is closing quickly. There is a way forward from the rash of technological threats we face today. By mobilizing common citizens and taking back control of our own devices and technologies, we can all use these tools to their maximum good. In other words, the tools to change the world are in everybody's hands. How we use them is not just up to me; it's up to all of us. That better version of our future—the one we all want—will not magically appear on its own. It will take tremendous intention, effort, and struggle. But with this hard work, not only will it be possible to survive progress, but to thrive to an extent never previously imagined. That is the world I want to live in.

Everything's Connected, Everyone's Vulnerable:
Here's What You Can Do About It

Throughout this book, we have investigated the looming technological threats faced by society and explored a variety of ways to systemically reduce these risks. The **UPDATE** Protocol, described below, provides some practical everyday tips you can use to protect yourself, your business, and your loved ones from today's most common technological dangers. Follow these simple steps (the digital equivalent of locking the front door to your home and not leaving your car keys in the ignition), and you can avoid more than 85 percent of the digital threats that pervade our lives daily.

Update Frequently

Modern software programs are riddled with bugs. Hackers and others use these vulnerabilities to break into your computer and other devices, steal your money, and cause general havoc. Avoid these problems by automatically updating your operating system software, computer programs, and apps. Pay particularly close attention to browsers, plug-ins, media players, Flash, and Adobe Acrobat—favorite targets of bad guys trying to rip you off. Failing to update automatically leaves your devices wide open to attack via problems that can be avoided if you simply update your software.

Passwords

Passwords should be long (think twenty digits or more) and contain upper- and lowercase letters, as well as symbols and spaces. Though we've all heard it a million times, the strength of a password is one of the key factors in protecting your accounts, and passwords should be changed often. You should absolutely not use the same password for several different sites. Doing so means once hackers get access to your log-in credentials, they can use them across multiple domains, from your social media network to your bank account. Memorizing long, unique passwords for every account and Web site in your life, however, is of course more than the human mind can manage. Fortunately, there are a bevy of password "wallets" or managers that can make this process relatively painless. Criminals have been known to create their own password wallets in an effort to trick you into giving up your digital crown jewels. Thus use only well-known and established companies such as 1Password, LastPass, KeePass, and Dashlane, most of which work across your computer, smart phone, and tablet. In addition, many services such as Google, iCloud, Dropbox, Evernote, PayPal, Facebook, LinkedIn, and Twitter offer two-factor authentication, which involves sending you a separate onetime password every time you log on, usually via an SMS message or app directly to your mobile phone. Using two-factor authentication means that even if your password is compromised, it cannot be used without the second authentication factor (physical access to your mobile device itself).

Download

Download software only from official sites (such as Apple's App Store or directly from a company's own verified Web site). Be highly skeptical of unofficial app stores and third-party sites hosting "free" software. In addition, avoid pirated media and software widely available on peer-to-peer networks, which frequently contain malware and viruses. Settings in both the Windows and the Mac operating systems can help you "white list" so that only approved software from identified vendors is allowed to run on your machine. While doing so will not guarantee software safety, it can greatly reduce the risk of infection. Pay close attention to apps and their permissions. They are "free" for a reason and you're paying with your privacy. If a flashlight app tells you it needs access to your location and contacts, run the other way.

Administrator

Administrator accounts should be used with care. Both Windows and Apple allow users to set account privileges, with administrators having highest privileges. While you will need an administrator account on your computer, it should not be your default account for everyday work and online browsing. Instead, create a standard user account to do the majority of your work and for day-to-day use. When you are logged in under administrative privileges and accidentally click on an infected file or download a virus, the malware has full privileges to execute and infect your machine. If you are logged in as a general user and the same thing happens, often the virus, Trojan, or worm will require your specific permission to execute, giving you a warning sign that there is a problem. Always run your computer as a non-admin user unless absolutely necessary to carry out a particular task, such as a known update from a trusted source you are conscientiously installing.

Turn Off

Turn off your computer when you aren't using it. The simple act of turning off your computer while you sleep will automatically reduce your threat profile by one-third because thieves cannot reach out and touch your machine when it's not in use and connected to the Internet. In addition, turn off services and connections on your smart phone when you aren't using them. Keeping Bluetooth, Wi-Fi, NFC, and cellular hot spots on at all times provides additional avenues for attack, which thieves can use to hack your phone, spread malware, and steal data. Also, keeping Wi-Fi on allows retailers and advertisers to persistently track you through your physical world, further encroaching on your privacy. Only turn these services on when you need them.

Encrypt

Encrypt your digital life, protecting your data both locally while at rest and when in transit across the Web. Both Windows and Mac include free programs for full hard-disk encryption (BitLocker and FileVault, respectively). Encrypting your hard drive means others cannot read its contents if it's lost or stolen. You should also encrypt your Internet traffic by using a virtual private network (VPN), particularly when using a public Wi-Fi network such as those at airports, universities, conferences, and coffee shops—frequent targets for hackers and thieves. Your phone, too, should be encrypted, because today's mobile devices can have as much personal information as our laptops, if not more. Always use a password on your mobile phone, and consider enabling biometric security, such as Apple's Touch ID fingerprint technology. Using a password in the latest version of iOS and Android not only ensures nobody else can access your phone and its data in your absence but also provides full encryption on the device, adding another layer of privacy and security.

Additional Safety Tips

If you faithfully follow the **UPDATE** Protocol above, you can avoid more than 85 percent of threats. To further secure yourself, follow these tips.

1. Use common sense with all your e-mail. As a general rule of thumb, be wary of any request to click on a link or open an attachment sent to you—even when it looks as though it came from somebody you know. Criminals are expert at tricking the general public with irresistible headlines, such as "click here" to see the shocking photographs of some naked movie star. Phishing attacks only work because unsuspecting individuals click on files and links that look realistic or enticing but contain a malicious payload that will infect your machine. When in doubt, check with the individual who purportedly sent you the e-mail to verify it came from him or her (don't reply to the e-mail itself!). And, no, the Prince of Nigeria is not reaching out to you personally with a viable way to get rich quick.

2. USB drives are one of the most common ways to spread malware and other computer viruses (the Department of Defense has even banned their use). Generally speaking, do not accept a thumb drive from a stranger (or even a person you know well) or plug one into your machine without first scanning for viruses. Disable "auto run" on your computer to ensure that any viruses do not automatically execute, thereby infecting your computer. The same advice applies to external USB hard drives and even smart phones that do not belong to you.

3. Back up your data frequently. You can back them up onto an external hard drive using built-in operating system tools such as Mac's Time Machine or Windows Backup. You can also use cloud providers such as Carbonite, Backblaze, and SpiderOak. When you utilize cloud providers, it is wise to encrypt the data before uploading them for an extra measure of protection. In addition, you should always have multiple backups of your data. Keep one or more physical drives for backup, and ensure that at least one of them is stored off-site so that in time of disaster, fire, or break-in a backup of your data will be stored in a safe and secure location.

4. Cover up. Unfortunately, it is easy for hackers, criminals, and spies to get access to all the Internet-connected cameras in your life, whether on your computer, smart phone, or tablet. When the camera is not in use, cover the lens up. A simple Post-it note or piece of tape will do and will provide cheap protection from unwanted prying eyes.

5. Sensitive browsing, such as banking or shopping, should only be done on a device that belongs to you and on a network you trust. Whether it's a friend's phone, a public computer, or a café's free Wi-Fi, your data could be copied or stolen. Be particularly wary of computers in common or high-trafficked areas such as airport lounges, favorite targets of criminals who plant malware and keystroke loggers in areas where businesspeople congregate.

6. Think before you share on social networks. Criminals, ranging from stalkers to burglars, routinely monitor social media for information. Posting travel itineraries can let burglars know that you will be away from home for two weeks on vacation—an invitation for trouble.

7. Use your operating system's built-in software firewall, available in both Windows and Mac, to block unwanted incoming connections to your machine, and enable "stealth mode" to make it more difficult for hackers and automated crime bots to find you online.

NOTE: Both the threats and the tools to protect yourself online change frequently. For additional guidance, visit www.futurecrimes.com.

Acknowledgments

One more thing . . .

STEVE JOBS

A project of this magnitude can never be the work of just one person alone. To this point, I owe a debt of gratitude to a large number of individuals for their support and contributions throughout the creation of this book, chief among them my literary agent Richard Pine of InkWell Management. From the very beginning, Richard saw the potential in *Future Crimes* and had faith in me to write it, generously agreeing to serve as my sherpa, mentor, and friend as I navigated the world of publishing. Richard's gifts to me were many, but perhaps his greatest was introducing me to the world-class team at Doubleday, including its editor in chief, Bill Thomas, and my own editor, Melissa Danaczko. Bill's enthusiasm and support for *Future Crimes* was outstanding. The same was true for all those with whom I had the privilege to work with at Doubleday, including Alison Rich, Joe Gallagher, Kim Thornton, Margo Shickmanter, and Maria Massey. Without a doubt, my absolute greatest and most profound appreciation must go to Melissa Danaczko, who encouraged me every step of the way through the writing and editing process. She is brilliant, funny, and generous. She worked weekends and nights, and even missed family gatherings in the cause of this book. Without Melissa, this work simply would never have come to fruition and I will remain eternally thankful to her.

To those individuals who generously agreed to review galley copies of this book and offer their comments on the work, you have my respect and considerable appreciation for taking the time out of your incredibly busy schedules to do so. In particular, I'd like to say thank you to Peter Diamandis, Ray Kurzweil, Kevin Kelly, Daniel Pink, David Eagleman, Christopher Reich, Interpol president Khoo Boon Hui, Ed Burns, Frank Abagnale, and P. W. Singer. To Sarah Stephens and Adam Kaslikowski, thank you both for the countless hours you spent reading through the earliest versions of *Future Crimes* and your deeply insightful comments on the work along the way. I have also benefited greatly from the wise counsel freely shared by well-established authors who selflessly agreed to help a newbie trying to figure things out for no other reason than they are generous, kind, and amazing. For that, I say thank you to Daniel Suarez, Ramez Naam, and Jane McGonigal.

Writing a book is no easy endeavor, not just for the countless hours it keeps you away from friends and family, but because the process of writing causes one to impose the book on others just because they are in your life. For dealing with me and offering their advice on a never-ending stream of book titles, subtitles, covers, research, and structure options, I want to thank Jacque Murphy, Tarun Wadhwa, Mikhail Grinberg, Daniel Teweles, and Kelsey Segaloff, as well as Brad, Steve, Adam, Carol, Monte, Jacqueline, Noni, Bob, Hanna, Mark, and Jonathan. For their willingness to support *Future Crimes* and for providing so many good ideas on how to share the information contained in the book with others, I am indebted to Paul Saffo, Chris Meyer, Joe Polish, Marcus Shingles, Steven Kotler, Jonathan Knowles, Sheryl Rapp, Eileen Bartholomew, Dave Blakely, Bill Eggers, Diane Francis, and Cody Rapp. I'd also like to thank the subject-matter experts who helped with some of the technical content in the book, including Andrew Hessel on synthetic biology, Alaina Hardie on robotics, Don Bailey on the Internet of Things, Emeline Paat-Dahlstrom and Mark Ciotola on space, and Andrew Fursman and Landon Downs on quantum computing. Thanks too to Keith Blount, founder of Literature & Latte's Scrivener, the world's greatest writing program. Without Scrivener, it would have been nearly impossible to organize the hundreds of cases and thousands of pages of research materials that went into the creation of this book.

I would also like to extend my appreciation to my friends and colleagues in law enforcement with whom I've shared many investigations, adventures, and good times over the years, including Michael Holstein, Bernhard Otupal, Rainer Buhrer, Paul Gillen, Mick Moran, Andrew Smith, Skukesha Goldberg, Jim Hirt, Bobby Weaver, Robert Rodriguez, Steven Chabinsky, and Kathy O'Toole. To my allies in the fight to bolster our common global security, including Roderick Jones, Justin Somaini, Tom Kellermann, Matt Wollman, Bradford Davis, and Steve Santorelli, thank you for all that you do.

I am privileged to serve on the faculty of Singularity University, an amazing educational institution with a mission to use next-generation technologies to address the world's grandest of challenges. There I am joined by some of the most talented people I have ever met in my life—faculty, staff, students, and alumni who are profoundly committed to driving positive change in this world. I'm honored to be counted among them, and I thank Rob Nail for his leadership in driving us exponentially forward.

Finally, this list of acknowledgments would be incomplete without recognizing the support of my family, who have provided the foundation to achieve all that I have in life and instilled in me the importance of fighting for justice in our world. With my deepest appreciation and gratitude to you all.

Notes

Prologue: The Irrational Optimist:
How I Got This Way

1 Today, they are building their own nation-wide: Michael Weissenstein, "Mexico's Cartels Build Own National Radio System," Associated Press, Dec. 27, 2011.

Chapter 1: Connected, Dependent, and Vulnerable

9 All or most of the information: Mat Honan, "How Apple and Amazon Security Flaws Led to My Epic Hacking," *Wired*, July 6, 2012; Mat Honan, "Kill the Password: Why a String of Characters Can't Protect Us Anymore," *Wired*, Nov. 15, 2012.

10 Over the past hundred years: Peter Diamandis, "Abundance Is Our Future," TED Talk, Feb. 2012.

10 And the mobile phone is singularly credited: Deloitte Consulting, *Sub-Saharan Africa Mobile Observatory 2012*, Feb. 4, 2014.

10 For centuries, the Westphalian system: Marc Goodman, "The Power of Moore's Law in a World of Geotechnology," *National Interest*, Jan./Feb. 2013.

11 Levin, a computer programmer: Amy Harmon, "Hacking Theft of $10 Million from Citibank Revealed," *Los Angeles Times*, Aug. 19, 1995.

12 One of the very first computer: Jason Kersten, "Going Viral: How Two Pakistani Brothers Created the First PC Virus," *Mental Floss*, Nov. 2013.

13 Eventually, Brain had traveled the globe: For a fascinating and entertaining perspective on Amjad and Basit Farooq, and the history of computer malware, see Mikko Hypponen, "Fighting Viruses and Defending the Net," TED Talk, July 2011.

14 Researchers at Palo Alto Networks: Byron Acohido, "Malware Now Spreads Mostly Through Tainted Websites," *USA Today*, May 4, 2013.

14 Many large companies: Brian Fung, "911 for the Texting Generation Is Here," *Washington Post*, Aug. 8, 2014.

14 In 2010, the German research institute: Nicole Perlroth, "Outmaneuvered at Their Own Game, Antivirus Makers Struggle to Adapt," *New York Times*, Dec. 31, 2012.

14 In the summer of 2013: Kaspersky Lab, *Global Corporate IT Security Risks: 2013*, May 2013.

14 A survey of its members: "Online Exposure," *Consumer Reports*, June 2011.

15 According to a study by the Gartner group: "Gartner Says Worldwide Security Software Market Grew 7.9 Percent in 2012," Gartner Newsroom, May 30, 2013; Steve Johnson, "Cybersecurity Business Booming in Silicon Valley," *San Jose Mercury News*, Sept. 13, 2013.

15 The results: the initial threat-detection rate: Imperva, *Hacker Intelligence Initiative, Monthly Trend Report #14*, Dec. 2012.

15 Though millions around the world: Tom Simonite, "The Antivirus Era Is Over," *MIT Technology Review*, June 11, 2012.

16 The landmark survey: Verizon, *2013 Data Breach Investigations Report*.

16 A similar study by Trustwave Holdings: Trustwave, *Trustwave 2013 Global Security Report*.

16 When businesses do eventually notice: Verizon RISK Team, *2012 Data Breach Investigation Report*, 3.

17 From the time an attacker: Ibid., 51.

17 In that case, hackers: Mark Jewell, "T.J. Maxx Theft Believed Largest Hack Ever," Associated Press, March 30, 2007.

17 Later in court filings: Julianne Pepitone, "5 of the Biggest Ever Credit Card Hacks," CNN, Jan. 12, 2014.

17 Though TJX reached a settlement: Ross Kerber, "Banks Claim Credit Card Breach Affected 94 Million Accounts," *New York Times*, Oct. 24, 2007.

17 One of the most authoritative: Ponemon Institute, Ponemon Institute, home page, 2014, http://www.ponemon.org.

18 Add to that the price: Byron Acohido, "Experts Testify on True Cost of Target Breach," *USA Today*, Feb. 5, 2014.

18 In one case, Global Payments: Robin Sidel and Andrew R. Johnson, "Data Breach Sparks Worry," *Wall Street Journal*, March 30, 2012.

18 All told, the Ponemon: Ponemon Institute (sponsored by Symantec), *2013 Cost of Data Breach Study: Global Analysis*, May 2013.

Chapter 2: System Crash

20 Something had to be wrong: Graeme Baker, "Schoolboy Hacks into City's Tram System," *Telegraph*, Jan. 11, 2008.

20 The boy spent months: Chuck Squatriglia, "Polish Teen Hacks His City's Tram, Chaos Ensues," *Wired*, Jan. 11, 2008.

20 In other words, the teen: Ibid.

22 "automatically monitor and adjust": Clay Wilson, *Botnets, Cybercrime, and Cyberterrorism: Vulnerabilities and Policy Issues for Congress*, Congressional Research Service, Jan. 9, 2008, 25.

22 The problem is worse: Brian Prince, "Almost 70% of Infrastructure Companies Breached in Last 12 Months: Survey," *Security Week*, July 14, 2014.

22 It may sound like fantasy: "Hackers 'Hit' US Water Treatment Systems," BBC, Nov. 21, 2011.

22 There a local teenager: Martha Stansell-Gamm, "Interview: Martha Stansell-Gamm," *Frontline*, Feb. 2001; Sean Silverthorne, "Feds Bust Kid Hacker," *ZDNet*, March 18, 1998.

22 "caused millions of litres": Tony Smith, "Hacker Jailed for Revenge Sewage Attacks," *Register*, Oct. 31, 2001.

23 "the next Pearl Harbor": Anna Mulrine, "CIA Chief Leon Panetta: The Next Pearl Harbor Could Be a Cyber Attack," *Christian Science Monitor*, June 9, 2011.

23 Yet 70 percent of the grid's: President's Council of Economic Advisers and the U.S. Department of Energy's Office of Electricity Delivery, *Economic Benefits of Increasing Electric Grid Resilience to Weather Outages Report*, Aug. 2013.

23 One utility reported: Edward J. Markey and Henry A. Waxman, *Electric Grid Vulnerability Report*, May 21, 2013.

23 The findings build: Siobhan Gorman, "Electricity Grid in U.S. Penetrated by Spies," *Wall Street Journal*, April 8, 2009.

23 In the video: Jack Cloherty, "Virtual Terrorism: Al Qaeda Video Calls for 'Electronic Jihad,'" *World News*, May 22, 2012.

23 Earlier FBI investigations: Barton Gellman, "Cyber Attacks by Al Qaeda Feared," *Washington Post*, June 27, 2002.

24 At the Chaos Communication Congress: Darlene Storm, "Hackers Exploit SCADA Holes to Take Full Control of Critical Infrastructure," *Computerworld*, Jan. 15, 2014; Vortrag: SCADA StrangeLove 2, http://events.ccc.de/.

24 One well-known hacker database: Shodan HQ, home page, accessed Feb. 9, 2014, http://www.shodanhq.com.

24 Several such incidents: "Cyber War: Sabotaging the System," *60 Minutes*, June 6, 2011. For the criminal angle, see David Shamah, "Hack Attacks on Infrastructure on the Rise, Experts Say," *Times of Israel*, Jan. 30, 2014.

24 President Obama when he noted: Barack Obama, "Remarks by the President on Securing Our Nation's Cyber Infrastructure," The White House Office of the Press Secretary, May 29, 2009.

25 Each plays its role: "War in the Fifth Domain," *Economist*, July 5, 2010.

25 Let's not forget two hackers: Phil Lapsley, "The Definitive Story of Steve Wozniak, Steve Jobs, and Phone Phreaking," *Atlantic*, Feb. 20, 2013.

25 As time passed, other notable hackers: Kevin D. Mitnick and William L Simon, *Ghost in the Wires: My Adventures as the World's Most Wanted Hacker* (New York: Little, Brown, 2012).

25 Poulsen's ingenious 1990 hack: Jonathan Littman, "The Last Hacker," *Los Angeles Times*, Sept. 12, 1993.

26 For example, in October 2013: "Adobe Hack: At Least 38 Million Accounts Breached," BBC, Oct. 30, 2013.

26 But what changed in that attack: Brian Krebs, "Adobe to Announce Source Code, Customer Data Breach," *Krebs on Security*, Oct. 3, 2013.

26 Yep, the company that is selling: Darlene Storm, "AntiSec Leaks Symantec pcAnywhere Source Code After $50K Extortion Not Paid," *Computerworld*, Feb. 7, 2012.

26 Traditional organized crime groups: The Hague, *Threat Assessment: Italian Organized Crime*, Europol Public Information, June 2013;

Nir Kshetri, *The Global Cybercrime Industry: Economic, Institutional, and Strategic Perspectives* (London: Springer, 2010), 1; Chuck Easttom, *Computer Crime, Investigation, and the Law* (Boston: Cengage Learning, 2010), 206.

26 These newly emerging: Mark Milian, "Top Ten Hacking Countries," *Bloomberg*, April 23, 2013.

27 New syndicates: Brian Krebs, "Shadowy Russian Firm Seen as Conduit for Cybercrime," *Washington Post*, Oct. 13, 2007; Verisign iDefense, *The Russian Business Network: Survey of a Criminal ISP*, June 27, 2007.

27 RBN famously provides: Trend Micro, *The Business of Cybercrime: A Complex Business Model*, Jan. 2010.

27 ShadowCrew operated the now-defunct Web site: Kevin Poulsen, "One Hacker's Audacious Plan to Rule the Black Market in Stolen Credit Cards," *Wired*, Dec. 22, 2008.

27 Founded by the notorious criminal hacker: James Verini, "The Great Cyberheist," *New York Times Magazine*, Nov. 10, 2010.

27 The number and reach: John E. Dunn, "Global Cybercrime Dominated by 50 Core Groups, CrowdStrike Report Finds," *CSO*, Jan. 23, 2014.

27 One such notable example is Anonymous: In deference to Guy Fawkes, the English Catholic who in 1605 planned to assassinate King James and blow up Parliament with gunpowder.

27 The group's motto: "'The Corrupt Fear Us!' Massive Anonymous 'Million Mask March' as It Happened," *RT*, Dec. 24, 2013; "Anonymous (Group)," *Wikiquote*.

27 When MasterCard, Visa: Lauren Turner, "Anonymous Hackers Jailed for DDoS Attacks on Visa, MasterCard, and PayPal," *Independent*, Jan. 24, 2013.

28 Anonymous is strongly against: Karol Snapbacks, "Anonymous Explaining Why They Hacked PSN/Sony," YouTube, April 22, 2011; Quinn Norton, "Anonymous Goes After World Governments in Wake of Anti-SOPA Protests," *Wired*, Jan. 25, 2012; Lisa Vaas, "Anonymous Bullies Sony and Nintendo over SOPA Support," *Naked Security*, Jan. 3, 2012.

28 Anonymous views itself: Quinn Norton, "How Anonymous Picks Targets, Launches Attacks, and Takes Powerful Organizations Down," *Wired*, July 3, 2012.

28 Even some of the group's most ardent critics: "Hackers Take Down Child Pornography Sites," BBC, Oct. 24, 2011.

28 In recognition of their growing power: Barton Gellman, "The World's 100 Most Influential People: 2012," *Time*, April 18, 2012.

28 Their burgeoning influence and capabilities: "Snowden Leaks: GCHQ 'Attacked Anonymous' Hackers," BBC, Feb. 5, 2014.

28 Meanwhile, terrorist organizations too: For detailed information on terrorist and jihadist use of technology, see the United Nations Counterterrorism Implementation Task Force report *Countering the Use of the Internet for Terrorist Purposes*, May 2011.

28 "do the things you": Paul Tassi, "ISIS Uses 'GTA 5' in New Teen Recruitment Video," *Forbes*, Sept. 20, 2014.

28 Internet reconnaissance and research: Thomas Harding, "Terrorists 'Use Google Maps to Hit UK Troops,'" *Telegraph Online*, Jan. 13, 2007; Caroline McCarthy, "Report: JFK Terror Plotters Used Google Earth," *CNET*, June 4, 2007.

29 For instance, "Ramzi Yousef": Jack Kelley, "Terror Groups Hide Behind Web Encryption," *USA Today*, Feb. 5, 2001.

29 Widely available online are documents: Gabriel Weimann, *How Modern Terrorism Uses the Internet*, United States Institute of Peace, Special Report 116, March 2004.

29 In a striking example of how dangerous: "Search of Tsarnaev's Phones, Computers Finds No Indication of Accomplice, Source Says," NBC News, April 23, 2013.

29 "The trio reportedly made fraudulent": Counter-terrorism Implementation Task Force, *Countering the Use of the Internet for Terrorist Purposes*, May 2011, 18.

29 Even the infamous 2002 Bali bombing mastermind: Q&A with Tom Kellermann, "Internet Fraud Finances Terrorism," *Discovery News*, Feb. 11, 2013.

29 Samudra was technologically savvy: Alan Sipress, "An Indonesian's Prison Memoir Takes Holy War into Cyberspace," *Washington Post*, Dec. 14, 2004.

30 Terrorists seem to be getting: Jeremy Scott-Joynt, "Warning Signs for the Funding of Terror," BBC, July 20, 2005; Gordon Rayner and David Williams, "Revealed: How MI5 Let 7/7 Bombers Slip Through Their Fingers," *Daily Mail*, May 1, 2007.

30 The Filipino hacking cell: Associated Press, "Filipino Police Arrest 4 Suspected AT&T Hackers," CBS News, Nov. 27, 2011; Somini Sengupta, "Phone Hacking Tied to Terrorists," *New York Times*, Nov. 26, 2011; Daily Mail Reporter, "Four Filipinos Arrested for Hacking AT&T Phone 'to Fund Saudi Terror Group,'" *Daily Mail*, Nov. 28, 2011; Jennifer Rowland, "The LWOT: Phone Hacking Linked to Terrorist Activity," *Foreign Policy*, Nov. 29, 2011.

30 Though the average Internet user: Marc Goodman and Parag Khanna, "The Power of Moore's Law in a World of Geotechnology," *The National Interest*, February 2013.

30 Though a $50,000 criminal: Siobhan Gorman, August Cole, and Yochi Dreazen, "Computer Spies Breach Fighter-Jet Project," *Wall Street Journal*, April 21, 2009.

30 In May 2013: Ernesto Londono, "Pentagon: Chinese Government, Military Behind Cyberspying," *Washington Post*, May 6, 2013.

31 Over the years, it has been reported: Ellen Nakashima, "Confidential Report Lists U.S. Weapons System Designs Compromised by Chinese Cyberspies," *Washington Post*, May 27, 2013.

31 According to an FBI report: Marcus Ranum, "Cyberwar Rhetoric Is Scarier Than Threat of Foreign Attack," *U.S. News and World Report*, March 29, 2010.

31 Of course it is not just the American military's: Craig Timberg and Ellen Nakashima, "Chinese Cyberspies Have Hacked Most Washington Institutions, Experts Say," *Washington Post*, Feb. 20, 2013.

31 Moreover, a 2009 report: John Markoff, "Vast Spy System Loots Computers in 103 Countries," *New York Times*, March 28, 2009; Omar El Akkad, "Meet the Canadians Who Busted GhostNet," *Daily Globe and Mail*, March 30, 2009; Tom Ashbrook et al., "Unmasking GhostNet," *On Point with Tom Ashbrook*, WBUR, April 2, 2009, http://onpoint.wbur.org/2009/04/02/unmasking -ghostnet.

31 China has also been accused: David E. Sanger, David Barboza, and Nicole Perlroth, "Chinese Army Unit Is Seen as Tied to Hacking Against U.S.," *New York Times*, Feb. 18, 2013.

31 The *Times* hired the private: Mandiant Corp., "APT 1: Exposing One of China's Cyber Espionage Units," *Mandiant*.

32 In 2012, *Bloomberg Businessweek*: Michael Riley and Ashlee Vance, "Inside the Chinese Boom in Corporate Espionage," *Bloomberg Businessweek*, March 15, 2012.

32 All told, between thefts: Lisa Daniels, "DOD Needs Industry's Help to Catch Cyber Attacks, Commander Says," *Department of Defense News*, March 27, 2012; David E. Sanger and Mark Landler, "U.S. and China Agree to Hold Regular Talks on Hacking," *New York Times*, June 1, 2013.

32 According to Akamai's: Ian Steadman, "Reports Find China Still Largest Source of Hacking and Cyber Attacks," *Wired UK*, April 24, 2013; David Belson, *The State of the Internet*, 3rd Quarter 2013 Report, Akamai Technologies.

32 A message issued by a Chinese: Michael Riley, "Hackers in China Breach UN, Olympic Committee Networks, Security Firms Say," *Bloomberg*, Aug. 4, 2011.

32 According to the former FBI director: Threat Working Group of the CSIS Commission on Cybersecurity, "Threats Posed by the Internet," CSIS, Oct. 28, 2008.

32 In late 2012, a previously unknown: Nicole Perlroth, "In Cyberattack on Saudi Firm, U.S. Sees Iran Firing Back," *New York Times*, Oct. 23, 2012.

33 At stake, 260 billion gallons: Jim Finkle, "Exclusive: Insiders Suspected in Saudi Cyber Attack," Reuters, Sept. 7, 2012.

33 Within minutes, the drive's viral payload: Reuters, "Aramco Says Cyberattack Was Aimed at Production," *New York Times*, Dec. 9, 2012.

33 Shamoon erased 75 percent: Perlroth, "In Cyberattack on Saudi Firm, U.S. Sees Iran Firing Back."

33 The Cutting Sword of Justice claimed: Reuters, "Aramco Says Cyberattack Was Aimed at Production."

33 American intelligence officials suspect: Siobhan Graham and Danny Yadron, "Iran Hacks Energy Firms, U.S. Says," *Wall Street Journal*, May 23, 2013; Michael Lipin, "Saudi Cyber Attack Seen as Work of Amateur Hackers Backed by Iran," Voice of America, Oct. 25, 2012.

33 Numerous marquee banks: Jim Finkle and Rick Rothacker, "Exclusive: Iranian Hackers Target Bank of America, JP Morgan, Citi," Reuters, Sept. 21, 2012.

33 A hacker group: Paul Wagenseil, "Bank of America Website Hit by Possible Cyberattack," NBC News, Sept. 19, 2012; Siobhan Gorman and Julian E. Barnes, "Iran Blamed for Cyberattacks," *Wall Street Journal*, Oct. 12, 2012.

33 The widespread denial-of-service: Nicole Perlroth and Quentin Hardy, "Bank Hacking Was the Work of Iranians, Officials Say," *New York Times*, Jan. 8. 2013.

33 In order for your call: The world population is about 7 billion. If every single person in the world except you picked up a phone and called your bank (transmitting, say, about 10 bytes to place the call) and then immediately hung up on it, and they did that every second for the duration of the attack, that would be about 70Gbps (gigabits per second). If someone on the planet really wanted to talk to the bank, he'd have to get in line behind those 7 billion callers.

34 "a few yapping Chihuahuas": Perlroth and Hardy, "Bank Hacking Was the Work of Iranians."

34 Programs such as PRISM: Glenn Greenwald and Ewen MacAskill, "Boundless Informant: The NSA's Secret Tool to Track Global Surveillance Data," *Guardian*, June 11, 2013; Kevin Drum, "2 Gigantic New NSA Revelations?," *Mother Jones*, July 2, 2013; Catherine Dunn, "10 Most Shocking NSA Revelations of 2013," *Fortune*, Dec. 27, 2013.

34 While living in Moscow: "Obama Knew of NSA Spying on Merkel and Approved It, Report Says," Fox News, Oct. 27, 2013; Catherine E. Shoichet, "As Brazil's Uproar over NSA Grows, US Vows to Work Through Tensions," CNN, Sept. 12, 2013.

34 Moreover, Snowden divulged: "US Spy Agency 'Taped Millions of French Calls,'" *Local*, Oct. 21, 2013; Kristen Butler, "NSA Taps Half-Billion German Phone, Data Links per Month: Report," UPI, June 30, 2013; Eric Pfeiffer, "NSA Spied on 124.8 Billion Phone Calls in Just One Month: Watchdog," *Yahoo! News*, Oct. 23, 2013.

34 Snowden's leaks also served: Te-Ping Chen, "Snowden Alleges U.S. Hacking in China," *Wall Street Journal*, June 23, 2013; Lana Lam, "Edward Snowden: US Government Has Been Hacking Hong Kong and China for Years," *South China Morning Post*, June 13, 2013; "New Snowden Leak Reveals US Hacked Chinese Cell Companies, Accessed Millions of SMS—Report," *RT*, June 23, 2013.

Chapter 3: Moore's Outlaws

37 According to the International Telecommunication Union: Miniwatts Marketing Group, "Internet Users in the World," Internet World Stats, Dec. 31, 2013, http://www.internetworldstats.com/.

37 Though it took nearly forty years: Miniwatts Marketing Group, "Internet Growth Statistics," Internet World Stats, Feb. 6, 2013, http://www.internetworldstats.com/.

37 The greatest growth: Miniwatts Marketing Group, "Internet Users in the World, Distribution by World Regions," Internet World Stats, Feb. 5, 2014, http://www.internetworldstats.com/.

37 And while half the world: Doug Gross, "Google Boss: Entire World Will Be Online by 2020," CNN, April 15, 2013.

37 The concept was named: Marc Goodman and Parag Khanna, "Power of Moore's Law in a World of Geotechnology," *National Interest*, Jan./Feb. 2013.

38 Incredibly, it literally: Cliff Saran, "Apollo 11: The Computers That Put Man on the Moon," *Computer Weekly*, July 13, 2009.

38 The modern smart phone: Peter Diamandis, "Abundance Is Our Future." TED Talk, Feb. 2012.

38 As a result of mathematical repercussions: Ray Kurzweil, "The Law of Accelerating Returns," *Kurzweil Accelerating Intelligence*, March 7, 2001.

38 "law of accelerating returns": Ray Kurzweil, *The Singularity Is Near: When Humans Transcend Biology* (New York: Penguin, 2006).

39 Early criminal entrepreneurs: Evan Andrews, "6 Daring Train Robberies," History.com, Oct. 21, 2013.

39 Their carefully planned heist: Brett Leppard, "The Great Train Robbery: How It Happened," *Mirror*, Feb. 28, 2013.

40 The incident kept the PlayStation: Keith Stuart and Charles Arthur, "PlayStation Network Hack: Why It Took Sony Seven Days to Tell the World," *Guardian*, Feb. 5, 2014; "Credit Card Alert as Hackers Target 77 Million PlayStation Users," *Mail Online*, Feb. 5, 2014.

40 In the end, financial analysts: J. Osawa, "As Sony Counts Hacking Costs, Analysts See Billion-Dollar Repair Bill," *Wall Street Journal*, May 9, 2011.

40 In that incident, data: "Target Now Says up to 110 Million Customers Victimized in Breach," MercuryNews.com, Feb. 5, 2014; "Pictured: Russian Teen Behind Target Hacking Attack," *Mail Online*, Feb. 5, 2014.

40 As incredible as the Target hack: Nicole Perlroth and David Gelles, "Russian Hackers Amass over a Billion Internet Passwords," *New York Times*, Aug. 5, 2014.

41 Physicians can even perform: Jacques Marescaux et al., "Transatlantic Robot-Assisted Telesurgery," *Nature*, May 29, 2001.

41 For example, the 1969 *Apollo 11*: Phil Johnson, "Curiosity About Lines of Code," *IT World*, Aug. 8, 2012; Saran, "Apollo 11."

41 By the early 1980s: Steven Siceloff, "Shuttle Computers Navigate Record of Reliability," NASA, Jan. 20, 2011.

41 Today, the software required: David McCandless, "Codebases," *Information Is Beautiful*, Oct. 30, 2013; "KIB—Lines of Code (Public)," Google.doc, https://docs.google.com/; Pollwatcher, "Healthcare.gov: 500 Million Lines of Code! That's Insane! Update," *Daily Kos*, Oct. 22, 2013.

41 "computers we ride in": Cory Doctorow, "Lockdown," based on a keynote speech to the Chaos Computer Congress in Berlin, Dec. 2011.

42 According to a study by Carnegie Mellon: Michelle Delio, "Linux, Fewer Bugs Than Rivals," *Wired*, Dec. 14, 2004.

42 A labyrinthine electrical grid: "Northeast Blackout of 2003," *Wikipedia*.

42 Computer failures also: National Commission on the BP Deepwater Horizon Oil Spill and Offshore Drilling, "Deep Water: The Gulf Oil Disaster and the Future of Offshore Drilling," Report to the President, Jan. 2011; "Deepwater Horizon Explosion," *Wikipedia*; Jeremy Repanich, "The Deepwater Horizon Spill by the Numbers," *Popular Mechanics*, Aug. 10, 2010.

42 At a government hearing: Gregg Keizer, "Tech Worker Testifies of 'Blue Screen Death' on Oil Rig's Computer," *Computerworld*, July 23, 2010; David Hammer, "Oil Spill Hearings: Bypassed General Alarm Doomed Workers in Drilling Area, Technician Testifies," *Times-Picayune*, July 23, 2010.

42 We've already seen this happen: Tom Simonite, "Stuxnet Tricks Copied by Computer Criminals," *MIT Technology Review*, Sept. 19, 2012.

Chapter 4: You're Not the Customer, You're the Product

44 Since the site's founding: Ryan Bradley, "Rethinking Health Care with PatientsLikeMe," *Fortune*, March 9, 2014.

44 Ahmed had been suffering: Julia Angwin and Steve Stecklow, "Scrapers Dig Deep for Data on Web," *Wall Street Journal*, Oct. 12, 2010.

45 "We take the information": FAQ, PatientsLikeMe.com.

45 For most users of PatientsLikeMe: "Privacy," PatientsLikeMe.com.

46 In a public interview: Angwin and Stecklow, "Scrapers Dig Deep for Data on Web."

46 By 2013, Americans: Cotton Delo, "U.S. Adults Now Spending More Time on Digital Devices Than Watching TV," *Advertising Age*, March 4, 2014.

47 Eighty percent of us check: IDC Research, *Always Connected: How Smartphones and Social Keep Us Engaged*, Facebook Public Files, March 4, 2014.

47 In a mere ten years: Heather Kelly, "By the Numbers: 10 Years of Facebook," CNN, Feb. 4, 2014.

47 Each day, more than 350 million: Facebook, Ericsson, and Qualcomm, "A Focus on Efficiency," 6, internet.org, Sept. 16, 2013, https://fbcdn-dragon-a.akamaihd.net/.

47 They also can be instruments: Jose Antonio Vargas, "How an Egyptian Revolution Began on Facebook," *New York Times*, Feb. 17, 2012.

50 Not anymore; now Google: Mark Milian, "Google to Merge User Data Across Its Services," CNN, Jan. 25, 2012.

50 That is why Google stores: Nate Anderson, "Why Google Keeps Your Data Forever, Tracks You with Ads," *Ars Technica*, March 8, 2010. Please note that in the EU there are restrictions on how long Google can store data, notably the "right to be forgotten" ruling that grants individuals the right to request that personal data be removed from the search engine.

50 To put that in perspective: Nate [user name], "How Much Is a Petabyte?," *The Mozy Blog*, accessed March 5, 2014.

50 Around the world, Google: The company has been sued for all these reasons, with varied results. For more detailed reviews on the large number of violations alleged against Google, see www.googlemonitor.com.

51 After a lawsuit: David Streitfeld, "Google Admits Street View Project Violated Privacy," *New York Times*, March 12, 2013; David Kravets, "An Intentional Mistake: The Anatomy of Google's Wi-Fi Sniffing Debacle," *Wired*, May 2, 2012.

51 In October 2013, a federal judge: Claire Cain Miller, "Google Accused of Wiretapping in Gmail Scans," *New York Times*, Oct. 1, 2013.

51 In early 2014, Google Glass was the subject: David Pierce, "The Simpsons May Have the Smartest Thoughts Yet About Google Glass," *Verge*, Jan. 27, 2014.

51 Even the former head: Michael Chertoff, "Google Glass, the Beginning of Wearable Surveillance," CNN, May 1, 2013.

52 With more than 1.2 billion: PRNewswire, "Facebook Reports Fourth Quarter and Full Year 2013 Results," Facebook: Investor Relations, Jan. 29, 2014.

52 It has been sued: Karen Gullo, "Facebook Sued over Alleged Scanning of Private Messages," *Bloomberg*, Jan. 2, 2014.

53 For example, did you realize: Robert McMillan, "Apple Finally Reveals How Long Siri Keeps Your Data," *Wired*, April 19, 2013.

53 The Web site with the most: "What They Know," *Wall Street Journal* series, http://blogs.wsj.com/wtk/.

54 So if you use your university: Adi Robertson, "Angry Email Users Can Take Google to Court for Keyword Scanning, Judge Rules," *Verge*, Sept. 26, 2013.

54 "a person has no": Ibid.; Cooley LLP, "Google's Motion to Dismiss Complaint Memorandum of Points & Authorities," U.S. District Court, San Jose Division, Sept. 5, 2013, http://www.consumerwatchdog.org/; Gregory S. McNeal, "It's Not a Surprise That Gmail Users Have No Reasonable Expectation of Privacy," *Forbes*, June 20, 2013.

54 It's not just your friends: Steve Stecklow, "On the Web, Children Face Intensive Tracking," *Wall Street Journal*, Sept. 17, 2010.

54 Well-known companies such as McDonald's: Josh Smith, "Children's Online-Privacy Violations Alleged Against McDonald's, General Mills, 3 Others," *National Journal*, Aug. 22, 2012.

54 In another case, Sony: Federal Trade Commission, "Sony BMG Music Entertainment, a General Partnership Subsidiary of Sony Corporation of America, United States of America (for the Federal Trade Commission)," accessed March 6, 2014, http://www.ftc.gov/.

55 That is why you: Roben Farzad, "Google at $400 Billion," *Bloomberg Businessweek*, Feb. 12, 2014.

55 A study published by the *Wall Street Journal*: Doug Laney, "To Facebook You're Worth $80.95," *CIO Journal* (blog), *Wall Street Journal*, May 3, 2014.

55 As the computer scientist Jaron Lanier: Joe Nocera, "Will Digital Networks Ruin Us?," *New York Times*, Jan. 6, 2014; Jaron Lanier, *Who Owns the Future?* (New York: Simon & Schuster, 2014).

55 Its inventory is personal data: Lori Andrews, "Facebook Is Using You," *New York Times*, Feb. 4, 2012.

56 This way when your friends: Salvador Rodriguez, "Google to Include User Names, Pictures in Ads: Here's How to Opt Out," *Los Angeles Times*, Oct. 11, 2013.

56 Google introduced "shared endorsements": Drew Guarini, "Facebook Finally Axes Controversial 'Sponsored Stories,'" *Huffington Post*, Oct. 1, 2014.

56 If one were to read: Alexis C. Madrigal, "Reading the Privacy Policies You Encounter in a Year Would Take 76 Work Days," *Atlantic*, March 1, 2012.

57 A study in the *Wall Street Journal*: Missy Sullivan, "It's Not Your Eyes . . . the Fine Print Is Getting Really, Really Small," *Wall Street Journal*, Jan. 15, 2012.

57 While the point was irreverently: See shrink-wrap license agreements are valid and enforceable, i.e., *ProCD, Inc. v. Zeidenberg*, *Microsoft v. Harmony Computers*, *Novell v. Network Trade Center*, and *Ariz. Cartridge Remanufacturers Ass'n v. Lexmark Int'l, Inc.* may have some bearing as well.

58 Facebook's privacy policy: Nick Bilton, "Price of Facebook Privacy? Start Clicking," *New York Times*, May 12, 2010; Facebook, "Privacy Policy," accessed March 3, 2014, https://www.facebook.com/full_data_use_policy.

58 By comparison, Shakespeare's *"Hamlet"*: Tom Gardner, "To Read, or Not to Read . . . the Terms and Conditions: PayPal Agreement Is Longer Than *Hamlet*, While iTunes Beats *Macbeth*," *Mail Online*, March 22, 2012.

58 Worse, many companies: Guilbert Gates, "Facebook Privacy: A Bewildering Tangle of Options," *New York Times*, May 21, 2010.

58 Moreover, even if you did: Jessica Guyn, "With Privacy Battle Brewing, Facebook Won't Update Policy Right Away," *Los Angeles Times*, Sept. 5, 2013; Ryan Singel, "Public Posting Now the Default on Facebook," *Wired*, Dec. 9, 2009; Epic, "Facebook Privacy," http://epic.org/privacy/facebook/.

58 Three months after it was purchased: "Instagram Seeks Right to Sell Access to Photos to Advertisers," BBC News, Dec. 18, 2012.

59 "When you upload or otherwise": Google Terms of Service, accessed on 3/10/2014, http://www.google.com/; Steve Kovach, "A Lot of People Are Freaking Out over Google Drive for Nothing," *Business Insider*, April 24, 2012.

59 There are currently more mobile phones: "2014: Mobiles 'to Outnumber People Next Year,' Says UN Agency," BBC News, May 9, 2013.

59 Sixty-three percent of Americans admit: Lookout, "Survey Reveals Consumers Exhibit Risky Behaviors Despite Valuing Their Privacy on Mobile Devices," Oct. 22, 2013.

59 In fact, making calls: O2, "Making Calls Has Become Fifth Most Frequent Use for a Smartphone for Newly Networked Generation of Users," *The Blue*, June 29, 2012.

60 These gadgets form: Meena Hart Duerson, "We're Addicted to Our Phones: 84% Worldwide Say They Couldn't Go a Single Day Without Their Mobile Device in Their Hand," *New York Daily News*, Aug. 16, 2012.

60 That device in your purse: Peter Maass and Megha Rajagopalan, "That's Not My Phone, It's My Tracker," *New York Times*, July 13, 2012.

60 In the United States alone: Jeff Jonas, "Your Movements Speak for Themselves: Space Time Travel Data Is Analytic Super-Food," Jeff Jonas .typepad.com, Aug. 26, 2009.

60 Mobile phones provide: Kai Biermann, "Data Protection: Betrayed by Our Own Data," *Die Zeit*, March 26, 2011.

60 The Android mobile phone software: Samsung Tomorrow, "What You May Not Know About GALAXY S4 Innovative Technology," April 10, 2013.

60 Given all these new technological: Ted Thornhill, "Is Nothing Off Limits? Now Google Plans to Spy on Background Noise in Your Phone Calls to Bombard You with Tailored Adverts," *Mail Online*, March 22, 2012.

60 Using its ambient sound: Megan Garber, "Yep, Google Just Patented Background Noise," *Atlantic*, March 22, 2012.

60 Facebook too has now added: Andrea Peterson, "New Facebook Feature Is a Friendly Reminder Your Smartphone Can Eavesdrop on You," *Washington Post*, May 21, 2014; Kurt Wagner, "Facebook's New Shazam-Like Tool Knows What You're Watching and Hearing," *Mashable*, May 21, 2014.

61 When Facebook revealed in the fourth quarter: David de Jong, "Zuckerberg Gains $3.2 Billion as Facebook Soars on Mobile," *Bloomberg*, Jan. 30, 2014; Facebook, "Investor Relations," Jan. 29, 2014, http://investor.fb.com/; J. O'Dell, "Facebook's Mobile Moment: Nearly a Billion Mobile Users & Majority of Revenue from Mobile," *VentureBeat*, Jan. 29, 2014 .

61 Since its launch in 2008: "App Store Sales Top $10 Billion in 2013," Apple Press Info, Jan. 7, 2014; Jordan Golson, "Apple Reports Strongest Ever Quarterly Earnings: $13.1 Billion Profit on $57.6 Billion in Revenue in Q1 2014," *MacRumors*, Jan. 27, 2014.

61 It also helps to explain: Emma Barnett, "Angry Birds Company 'Worth 5.5bn,'" *Telegraph*, May 8, 2012.

61 For example, the mere act: Violet Blue, "Norton: Android App Skips Consent, Gives Facebook Servers User Phone Numbers," *ZDNet*, June 29, 2013.

61 Once Facebook has been downloaded: Dylan Love, "It Looks like the Facebook Android App Can Control Your Camera and Take Pictures

Without Telling You," *Business Insider*, May 10, 2013; Chris Gayomali, "Why Is Facebook's App Asking to Read Your Text Messages?," *Fast Company*, Jan. 28, 2014.

61 More recently, Facebook began: "Facebook Mobile Update Raises Serious Privacy Concerns," *RT*, Dec. 3, 2012.

62 This occurs without any clear warning: Liam Tung, "Microsoft Points Scroogled War Machine at Privacy Worries over Android Apps," *ZDNet*, April 10, 2013.

62 As you probably now suspect: Emily Steel and Geoffrey A. Fowler, "Facebook in Privacy Breach," *Wall Street Journal*, Oct. 8, 2010.

62 While putting birds in slingshots: Kevin J. O'Brien, "Data-Gathering via Apps Presents a Gray Legal Area," *New York Times*, Oct. 28, 2012.

62 McAfee reported that 82 percent: Irfan Asrar et al., *Who's Watching You?* McAfee Mobile Security Report, Feb. 2014.

63 McKinsey has estimated: McKinsey Global Institute, *Big Data: The Next Frontier for Innovation, Competition, and Productivity*, June 2011, 85.

63 Apps like Tinder and Grindr: Rip Empson, "50M Matches Strong, Hot Mobile Dating App Tinder Is Ready to Go Global, and Move Beyond Flirting," *TechCrunch*, May 24, 2013.

63 In 2012, a Russian company: Nick Bilton, "Girls Around Me: An App Takes Creepy to a New Level," *New York Times*, March 30, 2012; John Brownlee, "This Creepy App Isn't Just Stalking Women Without Their Knowledge, It's a Wake-Up Call About Facebook Privacy [Update]," *Cult of Mac*, March 30, 2012.

64 By aggregating your locational data: For more information, see *United States v. Jones* (2012), *Wikipedia*, which discusses the privacy implications of locational data; Editorial Board, "The Court's GPS Test," *New York Times*, Nov. 5, 2011.

64 As you are about to discover: For a good overview of the privacy implications of location-based services, see the ACLU's report *Location-Based Services: Time for a Privacy Check-In*.

Chapter 5: The Surveillance Economy

65 Leigh Van Bryan was looking: Richard Hartley-Parkinson, "'I'm Going to Destroy America and Dig up Marilyn Monroe': British Pair Arrested in U.S. on Terror Charges over Twitter Jokes," *Mail Online*, Jan. 31, 2012.

66 While citizens around the world: Kharunya Paramaguru, "Private Data-Collection Firms Get Public Scrutiny," *Time*, Dec. 19, 2013.

66 Just one company alone: Natasha Singer, "Acxiom, the Quiet Giant of Consumer Database Marketing," *New York Times*, June 16, 2012.

66 Each profile contains: Eli Pariser, *The Filter Bubble: How the New Personalized Web Is Changing What We Read and How We Think* (New York: Penguin Press, 2012), 43; Natasha Singer, "A Data Broker Offers a Peek Behind the Curtain," *New York Times*, Aug. 31, 2013; Brandon Bailey, "Online Data Brokers Know You—Surprisingly Well," MercuryNews.com, Sept. 8, 2013.

67 Acxiom sells these consumer: Alice E. Marwick, "How Your Data Are Being Deeply Mined," *New York Review of Books*, Jan. 9, 2014.

67 For example, people in cluster: Lori B. Andrews, *I Know Who You Are and I Saw What You Did: Social Networks and the Death of Privacy* (New York: Free Press, 2013), 35.

67 Those in the Christian family: Stephanie Armour, "Data Brokers Come Under Fresh Scrutiny: Consumer Profiles Marketed to Lenders," *Wall Street Journal*, Feb. 12, 2014.

67 For example, some brokers offer lists: Paramaguru, "Private Data-Collection Firms Get Public Scrutiny"; "'Data Brokers' Selling Personal Info of Rape Victims to Marketers—Report," *RT*, Dec. 19, 2013.

67 Printed on the address label: Matt Pearce, "Dad Gets OfficeMax Mail Addressed 'Daughter Killed in Car Crash,'" *Los Angeles Times*, Jan. 19, 2014.

68 It wasn't until a local: Nesita Kwan, "Office-Max Sends Letter to 'Daughter Killed in Car Crash,'" NBC Chicago, Jan. 19, 2014.

68 As a result of Seay's experience: Armour, "Data Brokers Come Under Fresh Scrutiny."

68 "Our digital reach will": Judith Aquino, "Acxiom Prepares New 'Audience Operating System' amid Wobbly Earnings," *Ad Exchanger*, Aug. 1, 2013.

70 For example, researchers in the U.K.: David Talbot, "A Phone That Knows Where You're Going," *MIT Technology Review*, July 9, 2012.

71 As a result, researchers were able: Steve Lohr, "How Privacy Vanishes Online," *New York Times*, March 16, 2010.

71 At least ten individuals: Carolyn Y. Johnson, "Project 'Gaydar,'" Boston.com, Sept. 20, 2009; Matthew Moore, "Gay Men 'Can Be Identified by Their Facebook Friends,'" *Telegraph*, Sept. 21, 2009; Mona Chalabi, "State-Sponsored Homophobia: Mapping Gay Rights Internationally," *Guardian*, March 10, 2014.

71 While these findings: Emine Saner, "Gay Rights Around the World: The Best and Worst Countries for Equality," *Guardian*, July 30, 2013.

71 A study of fifty-eight thousand: Josh Halliday, "Facebook Users Unwittingly Revealing Intimate Secrets, Study Finds," *Guardian*, March 11, 2013.

71 "If you have something": "Google CEO on Privacy (VIDEO): 'If You Have Something You Don't Want Anyone to Know, Maybe

You Shouldn't Be Doing It,'" *Huffington Post*, March 18, 2010.

71 "privacy is no longer": Bobbie Johnson, "Privacy No Longer a Social Norm, Says Facebook Founder," *Guardian*, Jan. 10, 2010.

71 "expressing authentic identity": "Sharing to the Power of 2012," *Economist*, Nov. 17, 2011.

72 As the computer security researcher: Moxie Marlinspike, "Why 'I Have Nothing to Hide' Is the Wrong Way to Think About Surveillance," *Wired*, July 13, 2013.

73 "My daughter got this": Viktor Mayer-Schönberger and Kenneth Cukier, *Big Data: A Revolution That Will Transform How We Live, Work, and Think* (Boston: Houghton Mifflin Harcourt, 2013), 58.

73 Given Target's 2013 hacking: Elizabeth A. Harris and Nicole Perlroth, "For Target, the Breach Numbers Grow," *New York Times*, Jan. 10, 2014.

74 In response to the irreparable harm: Geoffrey A. Fowler, "When the Most Personal Secrets Get Outed on Facebook," *Wall Street Journal*, Oct. 13, 2012.

75 In an ensuing investigation: Daniel Zwerdling, "Your Digital Trail: Private Company Access," *All Tech Considered* (blog), NPR .org, Oct. 1, 2013.

75 A handful of tech start-ups: Katie Lobosco, "Facebook Friends Could Change Your Credit Score," *CNNMoney*, Aug. 27, 2013.

75 Facebook may become the next FICO: Mayer-Schönberger and Cukier, *Big Data*.

75 In the United States: See *United States v. Miller*, 425 U.S. 435 (1976), a case that went to the U.S. Supreme Court and involved the subpoenaed seizure of bank records from Mr. Miller. Miller's attorneys argued that the bank's compliance with the subpoena amounted to an unreasonable search and seizure in violation of his Fourth Amendment rights. The Court, however, ruled in a 6–3 opinion that the documents subpoenaed were not Miller's personal papers but instead part of the bank's business records; as such, his rights were not violated when a third party—his bank—transmitted the information that he had entrusted to it to the government. The legacy of Miller is with us today, and privacy advocates have blasted the Miller decision as no longer valid with our modern techniques of information sharing, production, and storage. See also *Smith v. Maryland*, 442 U.S. 735 (1979), regarding the use of "pen registers" to track and trace all inbound/outbound phone numbers dialed and received.

76 Accordingly, the word "Facebook": John Stevens, "The Facebook Divorces: Social Network Site Is Cited in 'a THIRD of Splits,'" *Mail Online*, Dec. 30, 2011.

76 A survey conducted by Microsoft: Mathew Ingram, "Yes, Virginia, HR Execs Check Your Facebook Page," *Gigaom*, Jan. 27, 2010; Cross-Tab, "Online Reputation in a Connected World," Job-hunt.com, Jan. 2010.

76 Applicants in all of these jurisdictions: Manuel Valdes, "Job Seekers Getting Asked for Facebook Passwords," *Yahoo! Finance*, March 20, 2012.

76 While some states, including California: Jonathan Dame, "Will Employers Still Ask for Facebook Passwords in 2014?," *USA Today*, Jan. 10, 2014.

77 That's what happened to a twelve-year-old: "Minnesota Girl Alleges School Privacy Invasion," CNN, March 10, 2012.

77 Even college athletes: Pete Thamel, "Universities Track Athletes Online, Raising Legal Concerns," *New York Times*, March 30, 2012.

77 A survey by the International Association: International Association of Chiefs of Police, 2013 Social Media Survey Results, accessed March 12, 2014, http://www.iacpsocialmedia.org/.

77 The IRS too began: Marcia Hoffman, "EFF Posts Documents Detailing Law Enforcement Collection of Data from Social Media Sites," Electronic Frontier Foundation, March 16, 2010. See also "IRT-WBT Content 2009," IRS social network training course overview, 2009, and John Lynch and Jenny Ellickson, "Obtaining and Using Evidence from Social Networking Sites," presentation, U.S. Department of Justice, Criminal Division, Computer Crime and Intellectual Property Section, August 2009.

77 The demands for information: Don Reisinger, "AT&T Reports More Than 300,000 Data Requests in 2013," *CNET*, Feb. 18, 2014.

77 In 2009, Sprint disclosed: Kim Zetter, "Feds 'Pinged' Sprint GPS Data 8 Million Times over a Year," *Wired*, Dec. 1, 2009.

78 It makes perfect sense: Marwick, "How Your Data Are Being Deeply Mined."

78 Revelations from Edward Snowden: Charlie Savage, "CIA Is Said to Pay AT&T for Call Data," *New York Times*, Nov. 7, 2013.

78 "Congress today reauthorized funding": "CIA's 'Facebook' Program Dramatically Cut Agency's Costs," Onion News Network, accessed March 15, 2014.

78 Instead, one study has estimated: Drew F. Cohen, "It Costs the Government Just 6.5 Cents an Hour to Spy on You," *Politico*, Feb. 10, 2014.

78 Upon learning of the true extent: Charles Cooper, "Ex-Stasi Boss Green with Envy over NSA's Domestic Spy Powers," *CNET*, June 28, 2013.

79 A *New York Times* article: Maria Aspan, "How Sticky Is Membership on Facebook? Just Try Breaking Free," *New York Times*, Feb. 11, 2008; Chamakhe Maurieni, *Facebook Is Deception* (Volume one) (WSIC EBooks Ltd., 2012).

Chapter 6: Big Data, Big Risk

83 The hackers funneled their millions: Associated Press, "Filipino Police Arrest 4 Suspected AT&T Hackers," CBS News, Nov. 27, 2010; Somini Sengupta, "Phone Hacking Tied to Terrorists," New York Times, Nov. 26, 2011; Marc Goodman, "What Business Can Learn from Organized Crime," Harvard Business Review, Nov. 2011.

84 As a result, children born today: Lauren Indvik, "92% of U.S. Toddlers Have Online Presence," Mashable, Oct. 7, 2010 .

84 Our data creation cycle: Allegra Tepper, "How Much Data Is Created Every Minute?," Mashable, June 22, 2012; Kristin Burnham, "Facebook's WhatsApp Buy: 10 Staggering Stats," InformationWeek, Feb. 21, 2014.

85 Put another way, every ten minutes: Verlyn Klinkenborg, "Trying to Measure the Amount of Information That Humans Create," New York Times, Nov. 12, 2003.

85 The cost of storing: McKinsey Global Institute, Big Data: The Next Frontier for Innovation, Competition, and Productivity, May 2011; Kevin Kelly speaking at the Web 2.0 conference in 2011, http://blip.tv/web2expo/web-2-0-expo-sf-2011-kevin-kelly-4980011.

85 Across all industries: World Economic Forum, Personal Data: The Emergence of a New Asset Class, Jan. 2011.

86 Eventually, your personal details: Cory Doctorow, "Personal Data Is as Hot as Nuclear Waste," Guardian, Jan. 15, 2008.

87 That's one account: Emma Barnett, "Hackers Go After Facebook Sites 600,000 Times Every Day," Telegraph, Oct. 29, 2011; Mike Jaccarino, "Facebook Hack Attacks Strike 600,000 Times per Day, Security Firm Reports," New York Daily News, Oct. 29, 2011.

87 Because 75 percent of people: "Digital Security Firm Says Most People Use One Password for Multiple Websites," GMA News Online, Aug. 9, 2013.

88 Many social media companies: "LinkedIn Hack," Wikipedia; Jose Pagliery, "2 Million Facebook, Gmail, and Twitter Passwords Stolen in Massive Hack," CNNMoney, Dec. 4, 2013.

88 Transnational organized crime groups: Elinor Mills, "Report: Most Data Breaches Tied to Organized Crime," CNET, July 27, 2010.

88 Such was the case: Jason Kincaid, "Dropbox Security Bug Made Passwords Optional for Four Hours," TechCrunch, June 20, 2011.

89 Later, however, it was revealed: John Markoff, "Cyberattack on Google Said to Hit Password System," New York Times, April 19, 2010; Kim Zetter, "Report: Google Hackers Stole Source Code of Global Password System," Wired, April 20, 2010.

89 According to court documents: John Leyden, "Acxiom Database Hacker Jailed for 8 Years," Register, Feb. 23, 2006; Damien Scott and Alex Bracetti, "The 11 Worst Online Security Breaches," Complex.com, May 9, 2012.

89 More recently, in 2013, the data broker Experian: Brian Krebs, "Experian Sold Customer Data to ID Theft Service," Krebs on Security, Oct. 20, 2013.

90 Experian learned of the compromise: Byron Acohido, "Scammer Dupes Experian into Selling Social Security Nos," USA Today, Oct. 21, 2013; Matthew J. Schwartz, "Experian Sold Data to Vietnamese ID Theft Ring," Dark Reading, Oct. 21, 2013.

90 In the course of the investigation: Jim Finkle and Karen Freifeld, "Exclusive: U.S. States Probing Security Breach at Experian Unit," Reuters, April 3, 2014.

90 Those who fell prey: Kashmir Hill, "Celebs' Financial Details Leaked, Including Credit Reports for Jay-Z and FBI Director Robert Mueller," Forbes, March 11, 2013.

90 The credit reports of those affected: Matthew J. Schwartz, "Exposed Website Reboots, Reveals Celeb Credit Reports," InformationWeek, April 4, 2013.

91 Though data brokers would disagree: Yasha Levine, "Surveillance Valley Scammers! Why Hack Our Data When You Can Just Buy It?," Pando Daily, Jan. 8, 2014.

91 At least 40 percent of social media: Graeme McMillan, "40% of Social Network Users Attacked by Malware," Time, March 23, 2011.

92 When Malaysia Airlines Flight MH370: Farooqui Adnan, "MH370 Links on Social Networks Spreading Malware," Ubergizmo, March 18, 2014.

92 One of the best-known pieces: Riva Richmond, "Koobface Gang That Spread Worm on Facebook Operates in the Open," New York Times, Jan. 16, 2012.

92 The malware was designed: Christopher Williams, "Facebook Versus Russia's Koobface Gang," Telegraph, Jan. 19, 2012.

92 With this, for example, if you checked: Joseph L. Flatley, "Firesheep Makes Stealing Your Cookies, Accessing Your Facebook Account Laughably Easy," Engadget, Oct. 25, 2010; Gary LosHuertos, "Herding Firesheep in Starbucks," CNNMoney, Dec. 16, 2010.

93 It was only when the boy's: Lara Naaman, Jen Pereira, and Emily Yacus, "Online Games Can Lead to Identity Theft," ABC News, July 16, 2008.

93 According to the Congressional Research Service: Kristin Finklea, "Identity Theft: Trends and Issues," Congressional Research Service, Jan. 16, 2014; Regina Lewis, "Money Quick Tips, Protect Yourself from Identity Theft," USA Today, April 5, 2014.

93   That works out to about: Blake Ellis, "Identity Fraud Hits New Victim Every Two Seconds," *CNNMoney*, Feb. 6, 2014.
93   In the United States alone: Daniel Bortz, "Identity Theft: Why Your Child May Be in Danger," *U.S. News & World Report*, Feb. 5, 2013.
93   According to a study of 40,000 children: Richard Power, "Child Identity Theft," Carnegie Mellon CyLab, 2011.
94   According to the National Crime Prevention Council: Edudemic Staff, "The 21 Best Resources for 2014 to Prevent Cyberbullying," *Edudemic*, Oct. 17, 2014. For more information, visit http://www.ncpc.org/cyberbullying.
95   It was there, in the deserted outskirts: "Shock at Woman's 'Facebook Murder,'" BBC, May 17, 2010; Amy Dale Court, "Christopher Dannevig's in Court for Nona Belomesoff Murder After Meeting on a Dating Website, a Court Heard," *Daily Telegraph*, Aug. 4, 2012.
95   For instance, Paul Bristol: "Jealous Lover Flew 4,000 Miles to Stab Ex-Girlfriend to Death after Seeing Her on Facebook with Another Man," *Daily Mail*, Mar. 10, 2010.
95   Sexting, or the sharing: Raquel Delevi and Robert S. Weisskirch, "Personality Factors as Predictors of Sexting," *Computers in Human Behavior* 29 (2013): 2589–94, citing a study by Michelle Drouin and Carly Landgraff, "Texting, Sexting, and Attachment in College Students' Romantic Relationships," *Computers in Human Behavior* 28 (2012): 444–49.
95   Another wildly popular site: Sam Biddle, "Here's Where the Naked Pics You Sexted Will End Up," *Gizmodo*, Nov. 28, 2012.
96   Every naked photograph: Camille Dodero, "Hunter Moore Makes a Living Screwing You," *Village Voice*, April 4, 2012.
96   According to the Pew Research Center: Mary Madden et al., "Teens and Technology 2013," Pew Research Center, March 13, 2013.
96   Moreover, 95 percent of young people: McAfee, "McAfee Digital Deception Study 2013: Exploring the Online Disconnect Between Parents & Pre-teens, Teens, and Young Adults," May 28, 2013.
96   The challenge for children: Lancaster University, "Software Developers Tackle Child Grooming on the Net," *ScienceDaily*, June 2, 2010.
96   Pedophiles make it their business: Sonia Elks, "Xbox Paedophile Predators 'Move in on Prey Within Two Minutes of Contact,'" *Metro*, April 17, 2012; Bill Singer, "Child Pornography Hid Behind XBOX LIVE 'Call of Duty: Modern Warfare 2,'" *Forbes*, Nov. 4, 2011.
96   Lest you think the demand: Nicholas Kristof, "He Was Supposed to Take a Photo," *New York Times*, March 22, 2014.
97   Such was the case with Amanda: Kevin Morris "BlogTV and the Sad, Avoidable Path to Amanda Todd's Suicide," *Daily Dot*, Oct. 15, 2012.
97   To prove he was serious: Gillian Shaw, "Amanda Todd's Mother Speaks Out About Her Daughter, Bullying," *Vancouver Sun*, March 13, 2013.
97   In response, on September 7: Video is available at http://www.youtube.com/watch?v=vOHXGNx-E7E. It is deeply touching and powerful and tells Amanda Todd's story in her own words. A compelling and must-see testament to a young life cut way too short.
98   Worse, the pedophiles: Patrick McGuire, "The Suspicious Return of the Daily Capper," *VICE*, Nov. 12, 2012.
98   In 2011, police in Melbourne: "Paedophiles Trawl Dating Sites to Get at Kids of Lonely Mums," News.com.au, Dec. 12, 2011.
98   After arranging a rendezvous: David Ferguson, "Texas Teen Viciously Beats and Abducts Gay Man After Targeting Him on Dating Website," *Raw Story*, Feb. 26, 2014.
99   In a one-hour documentary: The program looks at antigay vigilante gangs in Russia and the overall Russian religious right antigay movement. "Gay and Russian: 'It's Hunting Season, We Are the Hunted,'" Channel 4 News, Feb. 5, 2014.
99   The attackers fear no retribution: Dan Savage, "Anti-gay Russian Neo-Nazis Using Instagram and Facebook to Organize, Publicize Attacks," *Stranger*, Feb. 11, 2014; "Welcome to the Gay-Hating Olympics: Footage of Horrific Beatings Suffered by Gays in Russia," *Daily Mail*, Feb. 4, 2014.
99   To highlight this threat: Andrew Hough, "Please Rob Me Website Causes Fury for 'Telling Burglars When Twitter Users Are Not Home,'" *Telegraph*, Feb. 19, 2010.
100   Nashua police discovered: Nick Bilton, "Burglars Said to Have Picked Houses Based on Facebook Updates," *Bits* (blog), *New York Times*, Sept. 12, 2010.
100   They also admitted using: Matt Liebowitz, "Social Media Status Updates Tip Off Burglars, Study Shows," MSNBC, Nov. 7, 2011.
100   This information allows: Gerald Friedland and Robin Sommer, "Cybercasing the Joint: On the Privacy Implications of Geo-tagging," International Computer Science Institute and Lawrence Berkeley National Laboratory; "Featured Research: Geo-tagging," International Computer Science Institute, accessed March 30, 2014; Niraj Chokshi, "How Tech-Savvy Thieves Could 'Cybercase' Your House," *Atlantic*, July 22, 2010.
100   A few days later: Brendan Keefe, "Exif Data Hiding in Your Photos Targeted by Thieves and Criminal Investigators," YouTube,

Nov. 5, 2013, http://www.youtube.com/watch?v=mdoD7X8n46Q.

101 Hundreds of victims: Richard Burnett, "Scammers Use Social Networking Info to Target Vacationers' Relatives: Scams Using Social-Networking Vacation," *Orlando Sentinel*, June 22, 2013.

101 In September 2011: Robert Beckhusen, "Mexican Cartels Hang, Disembowel 'Internet Snitches,'" *Danger Room* (blog), *Wired*, Sept. 15, 2011.

101 These cartels are equally savvy: Ibid.

102 For example, when two Maricopa County: Mike Levine, "Officials Warn Facebook and Twitter Increase Police Vulnerability," FoxNews.com, May 10, 2011.

102 "the on-going investigations": Josh Halliday and Charles Arthur, "Anonymous's Release of Met and FBI Call Puts Hacker Group Back Centre Stage," *The Guardian*, Feb. 2, 2012.

102 The call was even recorded: Bob Christie, "Ariz. Police Confirm 2nd Hack on Officers' Email," MSNBC.com, June 29, 2011; Mohit Kumar, "77 Law Enforcement Websites Hit in Mass Attack by #Antisec Anonymous," *The Hacker News*, July 30, 2011.

103 For instance, in late 2010: "Cyber-Criminals Use Facebook to Steal Identity of Interpol Chief," *Daily Mail*, Sept. 20, 2010.

103 Industrial espionage too has found: Geoff Nairn, "Your Wall Has Ears," *Wall Street Journal*, Oct. 19, 2011.

103 we learned about the Massachusetts wind-turbine: Michael Riley and Ashlee Vance, "Inside the Chinese Boom in Corporate Espionage," *BusinessWeek*, Mar. 15, 2012.

103 Armed with all of this information: Joan Lappin, "American Superconductor and Its Rogue Employee Both Duped by Sinovel," *Forbes*, Sept. 27, 2011.

104 In one note Karabasevic: Carl Sears and Michael Isikoff, "Chinese Firm Paid Insider 'to Kill My Company,' American CEO Says," NBCNews.com, Aug. 6, 2013.

Chapter 7: I.T. Phones Home

105 On March 21, 2002: "Massive Search for Missing Girl," BBC, March 25, 2002.

105 By the following evening: "TV Appeal for Missing Amanda," BBC, March 28, 2002.

105 As part of their investigation: Nick Davies, "Phone-Hacking Trial Failed to Clear Up Mystery of Milly Dowler's Voicemail," *Guardian*, June 26, 2014.

105 Sadly for the Dowlers: "Milly's Body Found," BBC, Sept. 21, 2002.

106 Shamefully, they had also hacked: "Phone Hacking," *Guardian*, Feb. 7, 2011; CNN Library, "UK Phone Hacking Scandal Fast Facts," CNN,

July 5, 2014; "News International Phone Hacking Scandal," *Wikipedia*.

106 Dozens of employees and contractors: Nick Davies, "Phone-Hacking Trial Failed to Clear Up Mystery of Milly Dowler's Voicemail," *The Guardian*, June 26, 2014.

106 Of course for two grieving parents: "Milly Dowler's Phone Was Hacked by News of the World," *Telegraph*, July 4, 2011.

107 By 2014, McAfee: McAfee, "Mobile Malware in 2014," March 25, 2014, http://blogs.mcafee.com/; Juniper Networks, "Trusted Mobility Index," May 2012, http://www.juniper.net/.

107 Moreover, according to a study: Cisco, *Cisco 2014 Annual Security Report*; Jordan Kahn, "Apple SVP Phil Schiller Shares Report Showing Android Had 99% of Mobile Malware Last Year," *9to5Google*, Jan. 21, 2014.

107 The findings are deeply troubling: Rolfe Winkler, "Android Market Share Hits New Record," *Digits* (blog), *Wall Street Journal*, July 31, 2014; Canalys, "Over 1 Billion Android-Based Smart Phones to Ship in 2017," June 4, 2013.

108 Worse, according to several studies: Rachel Metz, "Phone Makers' Android Tweaks Cause Security Problems," *Technology Review*, Nov. 5, 2013; Liam Tung, "What's Making Your Android Insecure? Blame Those Free Apps You Never Asked For," *ZDNet*, Nov. 6, 2013.

108 Only about 4 percent: Daisuke Wakabayashi, "Cook Raises, Dashes Hopes for Excitement at Apple Annual Meeting," *Digits* (blog), *Wall Street Journal*, Feb. 28, 2014.

108 What is deeply frustrating: Juniper Networks, *Juniper Networks Third Annual Mobile Threats Report—March 2012 Through March 2013*.

109 Instead, computer-automated algorithms: Mike Isaac, "Google Beefs Up Android Market Security," *Wired*, Feb. 2, 2012.

109 By 2013, more than forty-two thousand apps: "Report: Malware-Infected Android Apps Spike in the Google Play Store," *PCWorld*, Feb. 19, 2014.

109 Cyber criminals have retooled: Joe Krishnan, "Mobile Malware Is Growing and Targeting Android Users, Warn Kaspersky," *Independent*, Feb. 26, 2014; Larry Barrett, "Banking Trojans Emerge as Dominant Mobile Malware Threat," *ZDNet*, Feb. 24, 2014.

109 To date, mobile malware: Brian Krebs, "Mobile Malcoders Pay to (Google) Play," *Krebs on Security*, March 6, 2013.

109 As a result, more than five hundred: Juniper Networks, *Third Annual Mobile Threats Report*, 4.

109 While much less common, malicious apps: Luke Westaway, "Apple iOS App Store Hit by First Malware App," *CNET*, July 6, 2012.

109 Nearly ten million iOS devices: Andy Greenberg, "Evasion Is the Most Popular Jailbreak Ever: Nearly Seven Million iOS Devices Hacked in Four Days," *Forbes*, Feb. 8, 2013; Juniper Networks, *Third Annual Mobile Threats Report*.

110 As it turns out: Alice Truong, "This Popular Flashlight App Has Been Secretly Sharing Your Location and Device ID," *Fast Company*, Dec. 5, 2013; Janel Torkington, "A Flashlight Can Steal from You: How to Stay Safe from Scam Apps," *AppsZoom*, Feb. 3, 2014; Aaron Gingrich, "The Mother of All Android Malware Has Arrived: Stolen Apps Released to the Market That Root Your Phone, Steal Your Data, and Open Backdoor," *Android Police*, March 6, 2011.

110 Multiplied by hundreds: Juniper Networks, *Third Annual Mobile Threats Report*.

110 In just a few hours: Matt Warman, "Fake Android Apps Scam Costs £28,000," *Telegraph*, May 24, 2012.

110 Hijacked mobile phones: Rich Trenholm, "Android Spam Scam Is First Smart Phone Botnet," *CNET*, July 6, 2012.

111 While botnets were previously: "China Mobile Users Warned About Large Botnet Threat," BBC, Jan. 15, 2013; Steven J. Vaughan-Nichols, "First Case of Android Trojan Spreading via Mobile Botnets Discovered," *ZDNet*, Sept. 5, 2013.

111 As such, criminals: "Gartner Says Worldwide PC, Tablet, and Mobile Phone Shipments to Grow 5.9 Percent in 2013 as Anytime-Anywhere-Computing Drives Buyer Behavior," Gartner Newsroom, June 24, 2013.

111 The vulnerability meant: Salvador Rodriguez, "Hackers Can Use Snapchat to Disable iPhones, Researcher Says," *Los Angeles Times*, Feb. 7, 2014.

111 Moreover, hackers were also: Selena Larson, "Snapchat Responds to Massive Hack," *ReadWrite*, Jan. 3, 2014.

111 Worse, it was revealed: Kashmir Hill, "Snapchats Don't Disappear: Forensics Firm Has Pulled Dozens of Supposedly Deleted Photos from Android Phones," *Forbes*, May 9, 2013.

111 As a result, tens of thousands: Tyler Kingkade, "Ohio University Student Accused of Using Nude Snapchat Photos to Extort Sex," *Huffington Post*, Dec. 30, 2013.

111 Today 89 percent of employees: Juniper Networks, "Trusted Mobility Index," May 2012.

112 For just a few hundred dollars: Brian Montopoli, "For Criminals, Smartphones Becoming Prime Targets," CBS News, Aug. 7, 2013; Dan Nosowitz, "A Hacked Mobile Antenna in a Backpack Could Spy on Cell Phone Conversations," *Popular Science*, July 16, 2013.

112 In Kenya, for example: "Why Does Kenya Lead the World in Mobile Money?," *Economist*, May 27, 2013.

112 Mobile money payment: Claire Péni-

caud, "State of the Industry: Results from the 2012 Global Mobile Money Adoption Survey," GSMA, Feb. 2013.

113 The Google Wallet system: Keith Wagstaff, "Google Wallet Hack Shows NFC Payments Still Aren't Secure," *Time*, Feb. 10, 2012.

113 Moreover, if and when a user loses: Sarah Clark, "Google Wallet Faces Its Second Hack of the Week," *NFC World*, Feb. 10, 2012.

113 Given the volumes: Anthony Wing Kosner, "Tinder Dating App Users Are Playing with Privacy Fire," *Forbes*, Feb. 18, 2014.

114 In fact, in 2012 police in South Australia: Miles Kemp, "Police Warn Photos of Kids with Geo-tagging Being Used by Paedophiles," *Herald Sun* (Melbourne), April 18, 2012.

114 In 2012, the U.S. Department of Justice revealed: Shannon Catalano, "Stalking Victims in the United States—Revised," U.S. Department of Justice Special Report, Sept. 2012; Sean Gallagher, "A Spurned Techie's Revenge: Locking Down His Ex's Digital Life," *Ars Technica*, Nov. 22, 2013; Justin Scheck, "Stalkers Exploit Cellphone GPS," *Wall Street Journal*, Aug. 3, 2010.

114 One product, Mobile Spy: Scheck, "Stalkers Exploit Cellphone GPS."

114 Furious with her plan: Australian Associated Press, "Simon Gittany Jailed for Minimum 18 Years for Murdering Fiancee," *Guardian*, Feb. 10, 2014; Timothy Geigner, "Mobile Spyware Use in Domestic Violence Ramps Up," *Wireless News*, April 3, 2014.

114 But in some cases, domestic abusers: Scheck, "Stalkers Exploit Cellphone GPS."

114 Using his wireless carrier's: Ibid.

114 Today it's no longer: Quentin Fottrell, "5 Apps for Spying on Your Spouse," *Market Watch*, Aug. 25, 2014.

114 To help combat these threats: Scheck, "Stalkers Exploit Cellphone GPS."

115 "Is a badge on Foursquare": Cheryl Rodewig, "Geotagging Poses Security Risks," U.S. Army, news archive, Mar. 7, 2012, www .army.mil.

115 The longitude and latitude: Ibid.

115 Not only can we be tracked: The product can now be found at http://www.trackingkey .com.

115 From Minnesota to New Jersey: For an excellent review of the social and privacy implications of automatic license plate readers, see the American Civil Liberties Union report *You Are Being Tracked: How License Plate Readers Are Being Used to Record Americans' Movements*.

116 Private companies such as Digital Recognition Network: Julia Angwin and Jennifer Valentino-Devries, "New Tracking Frontier: Your License Plates," *Wall Street Journal*, Sept. 29, 2012.

116 He then used the data: Ibid.

116 In 2009: Kate Crawford, "San Francisco Woman Pulled Out of Car at Gunpoint Because of License Plate Reader Error," ACLU (blog), May 15, 2014.

116 To date, Euclid has: Quentin Hardy, "Technology Turns to Tracking People Offline," *Bits* (blog), *New York Times*, March 7, 2013; Gene Marks, "Why the Home Depot Breach Is Worse Than You Think," *Forbes*, Sept. 22, 2014.

118 The cloud is here to stay: Frederic Lardinois, "Google Announces Massive Price Drops for Its Cloud Computing Services and Storage, Introduces Sustained-Use Discounts," *TechCrunch*, March 25, 2014.

119 All the major cloud service providers: Keir Thomas, "Microsoft Cloud Data Breach Heralds Things to Come," *PCWorld*, Dec. 23, 2010; Ed Bott, "Dropbox Gets Hacked . . . Again," *ZDNet*, Aug. 1, 2012.

119 In late 2014, hundreds: Daisuke Wakabayashi and Danny Yadron, "Apple Denies iCloud Breach," *Wall Street Journal*, Sept. 2, 2014.

119 As a result, the plans: Jaikumar Vijayan, "Classified Data on President's Helicopter Leaked via P2P, Found on Iranian Computer," *Computerworld*, March 2, 2009.

120 In fact, there are more than a hundred: Threat Working Group of the CSIS Commission on Cybersecurity, "Threats Posed by the Internet."

120 Every single day, the NSA: Dana Priest and William M. Arkin, "A Hidden World, Growing Beyond Control," *Washington Post*, July 19, 2010; James Bamford, "The NSA Is Building the Country's Biggest Spy Center (Watch What You Say)," *Wired*, March 15, 2012.

120 Given the exponential growth: Dan Nosowitz, "Every Six Hours, the NSA Gathers as Much Data as Is Stored in the Entire Library of Congress," *Popular Science*, May 10, 2011.

120 In response, the government: Bamford, "NSA Is Building the Country's Biggest Spy Center."

120 NSA's PRISM program allowed: Timothy B. Lee, "Here's Everything We Know About PRISM to Date," *Washington Post*, June 12, 2013.

120 Snowden also revealed: James Risen and Laura Poitras, "N.S.A. Gathers Data on Social Connections of U.S. Citizens," *New York Times*, Sept. 28, 2013.

120 These network graphs: Barton Gellman and Ashkan Soltani, "NSA Collects Millions of E-mail Address Books Globally," *Washington Post*, Nov. 1, 2013.

121 Not only did the NSA: Barton Gellman and Ashkan Soltani, "NSA Infiltrates Links to Yahoo, Google Data Centers Worldwide, Snowden Documents Say," *Washington Post*, Nov. 1, 2013.

121 Using the same basic techniques: Floor Boon, Steven Derix, and Huib Modderkolk, "NSA Infected 50,000 Computer Networks with Malicious Software," *Nrc.nl*, Nov. 23, 2013.

121 The agency even posed as Facebook: Dustin Volz, "The NSA Is Using Facebook to Hack into Your Computer," *National Journal*, March 12, 2014.

121 Together, the agencies participated: Spencer Ackerman and James Ball, "Optic Nerve: Millions of Yahoo Webcam Images Intercepted by GCHQ," *Guardian*, Feb. 27, 2014.

121 For example, the spy agency: Ashkan Soltani, Rea Peterson, and Barton Gellman, "NSA Uses Google Cookies to Pinpoint Targets for Hacking," *Washington Post*, Dec. 10, 2013.

121 According to Snowden, the NSA: James Larson, Jeff Glanz, and Andrew W. Lehren, "Spy Agencies Tap Data Streaming from Phone Apps," *New York Times*, Jan. 27, 2014.

121 None—including the app company: Sasha Goldstein, "Angry Birds, Other 'Leaky' Cellphone Apps Allow NSA to Collect Massive Amounts of Data: Report," New York *Daily News*, Jan. 27, 2014; James Ball, "Angry Birds and 'Leaky' Phone Apps Targeted by NSA and GCHQ for User Data," *Guardian*, Jan. 28, 2014.

121 Numerous violations were documented: Cyrus Farivar, "LOVEINT: On His First Day of Work, NSA Employee Spied on Ex-Girlfriend," *Ars Technica*, Sept. 27, 2013; Siobhan Gorman, "NSA Officers Spy on Love Interests," *Wall Street Journal*, Aug. 23, 2013.

122 FinFisher allows domestic intelligence: "FinFisher," *Wikipedia*; Vernon Silver, "Cyber Attacks on Activists Traced to FinFisher Spyware of Gamma," *Bloomberg*, July 25, 2013.

122 In the uprising: "Syria's Embattled Dissidents Grapple with Government Hackers, Wiretappers, and Impostors," *Time*, June 1, 2011; "Social Media: A Double-Edged Sword in Syria," *Reuters*, July 13, 2011.

122 "Dear subscriber": Andrew E. Kramer, "Ukraine's Opposition Says Government Stirs Violence," *New York Times*, Jan. 21, 2014.

Chapter 8: In Screen We Trust

124 In 2005, the UN's International Atomic Energy Agency: IAEA Board of Governors, "Implementation of the NPT Safeguards Agreement in the Islamic Republic of Iran," Sept. 2005.

124 Senior officials: William J. Broad and David E. Sanger, "Report Says Iran Has Data to Make a Nuclear Bomb," *New York Times*, Oct. 4, 2009.

125 For political reasons: David E. Sanger, "Obama Ordered Wave of Cyberattacks Against Iran," *New York Times*, June 1, 2012.

125 "most significant covert manipulation":

Marc Ambinder, "Did America's Cyber Attack on Iran Make Us More Vulnerable?," *Atlantic*, June 5, 2012.

128 As a result, you do not see: Paul Szoldra, "Blogger Nails a Major Problem with Facebook's Newsfeed," *Business Insider*, Jan. 19, 2014; Jim Tobin, "Facebook Brand Pages Suffer a 44% Decline in Reach since December 1," *Ignite Social Media*, December 10, 2013.

128 For as much effort: Anthony Wing Kosner, "Watch Out Twitter and Google+, Facebook's News Feed Is Getting Smarter and Smarter," *Forbes*, April 28, 2014.

128 Google reportedly has: As mentioned by Eli Pariser during his TED Talk, "Beware Online 'Filter Bubbles,'" May 2011; René Pickhardt, "What Are the 57 Signals Google Uses to Filter Search Results?," May 17, 2011, rene-pickhardt.de.

129 "it will be very hard": Alex Chitu, "Eric Schmidt on the Future of Search," *Google Operating System*, Aug. 16, 2010.

129 Using compelling arguments: For an extensive country-by-country review of global Internet tilting, see the OpenNet Initiative at https://opennet.net/about-filtering.

129 In the United Arab Emirates: "Top 10 Internet-Censored Countries," *USA Today*, Feb. 5, 2014.

131 Nearly 90 percent: Amy Gesenhues, "Survey: 90% of Customers Say Buying Decisions Are Influenced by Online Reviews," Marketingland.com, April 9, 2013; Zendesk, "The Impact of Customer Service on Customer Lifetime Value"; Myles Anderson, "2013 Study: 79% of Consumers Trust Online Reviews as Much as Personal Recommendations," *Search Engine Land*, June 26, 2013; Nielsen, *Global Trust in Advertising and Brand Messages*, April 2012.

131 Michael Luca, "Reviews, Reputation, and Revenue: The Case of Yelp.com." Harvard Business School Working Paper, No. 12-016, Sept. 2011.

131 Worse, in September 2014: Bob Egelko, "Yelp Can Manipulate Ratings, Court Rules," *San Francisco Gate*, Sept. 4, 2014.

131 One company: Eric Spitznage, "'Operation Clean Turf' and the War on Fake Yelp Reviews," *Bloomberg Businessweek*, Sept. 25, 2013.

131 Considering the world's: Rebecca Grant, "Facebook Has No Idea How Many Fake Accounts It Has—but It Could Be Nearly 140M," *VentureBeat*, Feb. 3, 2014.

131 Want 4,000 followers: Nick Bilton, "Friends, and Influence, for Sale Online," *Bits* (blog), *New York Times*, April 20, 2014.

131 No problem: John Koetsier, "Facebook's War on Zombie Fans Just Started with a Boom," *VentureBeat*, Sept. 26, 2012.

132 Rihanna and Shakira: Ibid.

133 According to the Federal Trade Commission: Mandi Woodruff, "There Could Be Something Wrong with 42 Million Credit Reports," *Business Insider*; Federal Trade Commission, *Report to Congress*, Dec. 2012; Melanie Hicken, "Find Out What Big Data Knows About You (It May Be Very Wrong)," *CNNMoney*, Sept. 5, 2013.

133 Tens of millions: Rebecca Smith, "One in Ten Electronic Medical Records Contain Errors: Doctors," *Telegraph*, July 17, 2010.

133 The hospital staff: "Man Dies During Cancer Drug Trial," BBC, Sept. 21, 2008.

133 In California: "California Releases 450 'Violent and Dangerous' Criminals After Computer Glitch Sets Them Free," *Daily Mail Online*, May 27, 2011.

134 In Britain: "Are You One of the 20,000 People Wrongly Branded a Criminal? Police Blunders Give Thousands Records for Crimes They Have Not Committed," *Daily Mail Online*, Dec. 28, 2012.

135 Police data systems: Asher Moses, "Hackers Break Into Police Computer as Sting Backfires," *Sydney Morning Herald*, Aug. 18, 2009; "Hacker 'Steals' Hertfordshire Police Officers' Data," BBC News, Aug. 30, 2012; Sabari Selvan, "Italy's Police Website Vitrociset.it Hacked by #Antisec," *E Hacking News*, July 30, 2011; "Ten Months Later, Memphis Police Dept. First Notifies People of Data Breach?," *Office of Inadequate Security*, Feb. 21, 2014; "Montreal Police Database Hacked; Personal Information Posted Online," *Global News*, Feb. 19, 2013; IPCC, "Hacking into Police Force Systems," *Learning the Lessons*, May 2013; Jeff Goldman, "Honolulu Police Department Hacked," *eSecurity Planet*, May 8, 2013.

135 In 2013, the Danish: "Danish Police Driving Licence Database Hacked by a Top Rated Swedish Hacker," *Scandinavia Today*, June 6, 2013.

135 "exterminate the rats": "Philadelphia Police Witness Information Hacked," Lawofficer.com, accessed Nov. 9, 2013.

135 Once his fingers: "Ex-con Returns to Jail for Hacking Prison Computers," *PCWorld*, Nov. 15, 2008.

135 As open and vulnerable: David Schultz, "As Patients' Records Go Digital, Theft and Hacking Problems Grow," *Kaiser Health News*, June 3, 2012; Kim Zetter, "It's Insanely Easy to Hack Hospital Equipment," *Wired*, April 25, 2014; Kelly Jackson Higgins, "Anatomy of an Electronic Health Record Zero-Day," *Dark Reading*, Dec. 4, 2013.

135 Forget for the moment: Neal Ungerleider, "Medical Cybercrime: The Next Frontier," *Fast Company*, Aug. 15, 2012.

135 In fact, HHS has documented: Nelson Harvey, "Hospital Database Hacked, Patient

Info Vulnerable," *Aspen Daily News*, March 15, 2014.

136 Worse, if your blood type: "Victim of Botched Transplant Declared Dead," CNN, Feb. 23, 2003.

136 They allegedly: EMC Corporation, "2013: A Year in Review," Jan. 2014.

138 Fully automated phishing kits: Ibid.

138 As a result, more than 100 million: Miles Date, "Why We Need to Support DMARC and Fight Phishing," *Deliverability Next*, April 2, 2013.

138 Thus for about $130: Cisco, *Email Attacks: This Time It's Personal*, June 2011.

138 When Coke's deputy president: Ben Elgin, Dune Lawrence, and Michael Riley, "Coke Gets Hacked and Doesn't Tell Anyone," *Bloomberg*, Nov. 4, 2012.

139 Coke is not alone: TrendLabs APT Research Team, "Spear-Phishing Email: Most Favored APT Attack Bait," Trend Micro Incorporated Research Paper, 2012.

139 Highly specialized: Rob Waugh, "New PC Virus Doesn't Just Steal Your Money—It Creates Fake Online Bank Statements So You Even Don't Know It's Gone," *Daily Mail Online*, Jan. 6, 2012.

139 Thus, if the thieves: Amy Klein, "Holiday Shopping and Fraud Schemes," *Security Intelligence*, Jan. 4, 2012.

140 "one count each of extortion": Carol Todd, "Arrest of Dutch Man in Amanda Todd Cyberbullying Rekindles Family Anguish," CBC News, April 28, 2014.

140 Coban's alleged modus operandi: Associated Press, "Netherlands Arrest in Amanda Todd Webcam Blackmail Case," *Guardian*, April 17, 2014.

140 Dozens of other victims: Associated Press, "Dutch Man Arrested in Connection with Suicide of Canadian Teen Amanda Todd," New York *Daily News*, April 18, 2014.

140 The jealous woman allegedly copied: Dan Goodin, "Woman Charged with Cyberbullying Teen on Craigslist," *Register*, Aug. 18, 2009.

141 The sensational story: Corey Grice and Scott Ard, "Hoax Briefly Shaves $2.5 Billion off Emulex's Market Cap," *CNET*; Jane C. Chesterman, "The Emulex Stock Hoax: Potential Liability for Internet Wire and Bloomberg?," *Journal of Corporation Law* 27, no. 1 (Fall 2001) .

141 "In a sixteen-minute period": U.S. Securities and Exchange Commission, "Defendant in Emulex Hoax Sentenced," Aug. 8, 2001.

141 Exactly the response: Corey Grice, "23-Year-Old Arrested in Emulex Hoax," *CNET*, Aug. 31, 2000.

141 Within six days: Alex Berenson, "Guilty Plea Is Set in Internet Hoax Case Involving Emulex," *New York Times*, Dec. 29, 2000.

141 The practice involves traders: Lina Saigol,

"The Murky World of Traders' Electronic Chat," *Financial Times*, Nov. 11, 2013.

141 Though pump and dump is generally: "FBI Arrests Seven in $140 Million Penny Stock Fraud," *Moneynews*, Aug. 14, 2013.

141 That's what professional traders: Amy Chozick, "Bloomberg Admits Terminal Snooping," *New York Times*, May 13, 2013.

142 It was later revealed: Julia La Roche, "Bloomberg Spying Scandal Escalates," *Business Insider*, May 10, 2013.

142 Goldman officials complained: Mark DeCambre, "Goldman Outs Bloomberg Snoops," *New York Post*, May 10, 2013.

142 One former Bloomberg reporter: Chozick, "Bloomberg Admits Terminal Snooping."

142 *Flash Boys* follows Brad Katsuyama: Michael Lewis, "An Adaptation from 'Flash Boys: A Wall Street Revolt,' by Michael Lewis," *New York Times*, March 31, 2014.

Chapter 9: Mo' Screens, Mo' Problems

145 There was just one slight problem: Kelly Jackson Higgins, " 'Robin Sage' Profile Duped Military Intelligence, IT Security Pros," *Dark Reading*, July 6, 2010.

145 Sage was the invention: Thomas Ryan, "Getting in Bed with Robin Sage" Provide Security, 2010; Shaun Waterman, "Fictitious Femme Fatale Fooled Cybersecurity," *Washington Times*, July 18, 2010.

146 Not only were public figures: Robert McMillan, "Paris Hilton Accused of Voice-Mail Hacking," *InfoWorld*, Aug. 25, 2006.

146 Once the bank's telephone system: Ron Lieber, "Your Voice Mail May Be Even Less Secure Than You Thought," *New York Times*, Aug. 19, 2011.

147 Worse, criminals can spoof: Byron Acohido, "Caller ID Spoofing Scams Aim for Bank Accounts," *USA Today*, March 15, 2012.

147 In what the Internal Revenue Service: Kathy Kristof, "IRS Warns of Biggest Tax Scam Ever," *MoneyWatch*, March 20, 2014.

148 The FBI recorded: Adrianne Jeffries, "Meet 'Swatting,' the Dangerous Prank That Could Get Someone Killed," *Verge*, April 23, 2013.

148 In some of the cases: Maria Elena Fernandez, "Ashton Kutcher, Miley Cyrus & Others Terrorized in Dangerous 'Swatting' Prank," *Daily Beast*, Oct. 5, 2012.

148 In fact, swatting is the perfect complement: FBI, "The Crime of 'Swatting': Fake 9-1-1 Calls Have Real Consequences," accessed May 7, 2014.

148 He also swatted a local bank: Alan Duke, "Boy Admits 'Swatting' Ashton Kutcher, Justin Bieber," CNN, March 12, 2013.

148 It is only by a miracle: Heidi Fenton,

"Swatting-Related Crash," Mlive.com, April 8, 2014.

149 The baseband handles all: Sebastian Anthony, "The Secret Second Operating System That Could Make Every Mobile Phone Insecure," *ExtremeTech*, Nov. 13, 2013.

149 A number of hackers: Ralf Philipp Weinmann, "DeepSec 2010: All Your Baseband Are Belong to Us," YouTube, http://www.youtube.com/watch?v=fQqv0vl4KKY, accessed May 7, 2014.

149 In early 2014, such a security flaw: Paul K., "Replicant Developers Find and Close Samsung Galaxy Backdoor," Free Software Foundation, March 12, 2014.

149 The FBI has reportedly: Declan McCullagh, "FBI Taps Cell Phone Mic as Eavesdropping Tool," *ZDNet*, Dec. 1, 2006.

151 Although his passenger survived: Hard Reg, "Driver Follows Satnav to His Doom," *Register*, Oct. 5, 2010.

151 A report: Department of Homeland Security, "National Risk Estimate," Nov. 9, 2011.

151 "Society may already": Robert Charette, "Are We Getting Overly Reliant on GPS-Intensive Systems?," IEEE Spectrum, March 9, 2011, available at spectrum.ieee.org.

151 "electronic fritz": David Hambling, "GPS Chaos: How a $30 Box Can Jam Your Life," *New Scientist*, March 6, 2011.

152 The longest GPS attack: "Out of Sight," *Economist*, July 27, 2013.

152 For a mere $50: John Brandon, "GPS Jammers Illegal, Dangerous, and Very Easy to Buy," FoxNews.com, March 17, 2010.

152 For example, in London: "Out of Sight."

153 He was using: Hambling, "GPS Chaos."

153 Stung too many times: Charles Arthur, "Car Thieves Using GPS 'Jammers,'" *Guardian*, Feb. 22, 2010; Matt Warman, "Organised Crime 'Routinely Jamming GPS,'" *Telegraph*, Feb. 22, 2012; "£6M Lorry Hijackings Gang Face Ten Years," *Express & Star*, May 6, 2010.

153 "A 'multiple agency approach'": "The $30 GPS Jammer That Could Paralyze U.S. Cities," *Week*, March 10, 2011.

154 Think of the impact: Jeff Coffed, "The Threat of GPS Jamming," Exelis, Feb. 2014.

154 In 2013, however, security research: Tom Simonite, "Ship Tracking Hack Makes Tankers Vanish from View," *MIT Technology Review*, Oct. 18, 2013.

154 The sixty-five-meter: "Researchers Show How a Major GPS Flaw Could Allow Terrorists and Hackers to Hijack Commercial Ships and Planes," *Mail Online*, July 27, 2013; Aviva Hope Rutkin, "'Spoofers' Use Fake GPS Signals to Knock a Yacht Off Course," *MIT Technology Review*, Aug. 14, 2013.

154 The pair were working: Sandra Zaragoza,

"Spoofing Superyacht at Sea," *Know*, July 31, 2013.

155 In early 2014: Kelsey D. Atherton, "Israeli Students Spoof Waze App with Fake Traffic Jam," *Popular Science*, March 31, 2014.

156 In what investigators: Nathan Hodge and Adam Entous, "Oil Firms Hit by Hackers from China, Report Says," *Wall Street Journal*, Feb. 10, 2011.

156 "they inadvertently downloaded code": Nicole Perlroth, "Hackers Lurking in Vents and Soda Machines," *New York Times*, April 7, 2014.

156 "allegations about Chinese hacking": Hodge and Entous, "Oil Firms Hit by Hackers from China."

157 In 2013, hackers: Lee Moran, "Montana Residents Flip Out When Emergency Alert System Tells Them the Zombie Apocalypse Is Happening—Like Right Friggin Now," *New York Daily News*, Feb. 12, 2013.

157 "traffic jerked to a standstill": "Russian Hackers Jam Automobile Traffic with Porn," Fox News, Technology, January 15, 2010; "Russian Jailed for Six Years for Hacking into Advertising Server and Making Electronic Billboard Show Porn to Motorists," *Mail Online*, March 24, 2011.

157 The sign stood: Sevil Omer, "Racial Slur on Mich. Road Sign Targets Trayvon Martin," NBC News, April 9, 2012.

158 Even in 2014: Serge Malenkovich, "Hacking the Airport Security Scanner," *Kaspersky Lab*. March 14, 2014,

158 Even if a hacker: "Hacked X-Rays Could Make TSA Scanners Useless," video, *Wall Street Journal*, Feb. 12, 2014.

159 Shockingly, using a common hacker tactic: Kim Zetter, "Hacked X-Rays Could Slip Guns Past Airport Security," *Wired*, Feb. 11, 2014.

159 "Hackers have hobbled": U.S. Department of Transportation, "Review of Web Applications Security and Intrusion Detection in Air Traffic Control Systems," Project ID: FI-2009-049, May 4, 2009.

159 The inspector general: Siobhan Gorman, "FAA's Air-Traffic Networks Breached by Hackers," *Wall Street Journal*, May 7, 2009.

159 Moreover, a security audit: Thomas Claburn, "Air Traffic Control System Repeatedly Hacked," *Dark Reading*, May 7, 2009.

159 "will be highly automated": Steve Henn, "Could the New Air Traffic Control System Be Hacked?," NPR.org, Aug. 14, 2012.

160 "could have disastrous": Donald McCallie, Jonathan Butts, and Robert Mills, "Security Analysis of the ADS-B Implementation in the Next Generation Air Transportation System," *International Journal of Critical Infrastructure Protection* 4, no. 2 (Aug. 2011): 78–87, doi:10.1016/j.ijcip.2011.06.001.

160 While rigging elections: "The World of

100% Election Victories," BBC News, March 11, 2014.

160 Not only could they change any votes: "Hacking the Vote: Internet Systems Remain Unsecure," CNN, Nov. 5, 2012.

160 Interestingly, while roaming around: Andrew Tarantola, "Hacked DC School Board E-voting Elects Bender President," *Gizmodo*, March 2, 2012.

161 "influence, disrupt": Walter L. Sharp, "Electronic Warfare," Joint Publication 3–13.1, Jan. 25, 2007.

162 In the battle: Adam Martin, "Reuters Blogs Hacked with Fake Story About Syrian Rebel Retreat," *Wire*, Aug. 3, 2012.

162 Operation Orchard: Erich Follath and Holger Stark, "The Story of 'Operation Orchard': How Israel Destroyed Syria's Al Kibar Nuclear Reactor," *Spiegel Online*, Feb. 11, 2009; David E. Sanger and Mark Mazzetti, "Israel Struck Syrian Nuclear Project, Analysts Say," *New York Times*, Oct. 14, 2007.

162 Though enemy jets: Lewis Page, "Israeli Sky-Hack Switched off Syrian Radars Countrywide," *Register*, Nov. 22, 2007.

162 As it turned out: Yuval Goren, "IDF Reserve Troops Receive Fictitious Calls for Duty in Gaza," Haaretz.com, Jan. 8, 2009.

162 Both Israel and Hamas: Balousha Hazem, "Text Messages and Phone Calls Add Psychological Aspect to Warfare in Gaza," *Guardian*, Jan. 2, 2009.

163 "extremist ideology": Nick Fielding and Ian Cobain, "Revealed: US Spy Operation That Manipulates Social Media," *Guardian*, March 17, 2011.

163 "up to 50 US-based controllers": Ibid.

163 "as a psychological warfare weapon": Ibid.

163 According to Freedom House: Freedom on the Net 2013, FreedomHouse.org, Oct. 3, 2013.

164 Each Internet operator: Sergey Chernov, "Internet Troll Operation Uncovered in St. Petersburg," *St. Petersburg Times*, Sept. 18, 2013.

164 Russia's president: Paul Roderick Gregory, "Inside Putin's Campaign of Social Media Trolling and Faked Ukrainian Crimes," *Forbes*, May 11, 2014.

164 "Orders of Service": Chris Elliott, "The Readers' Editor on . . . Pro-Russia Trolling Below the Line on Ukraine Stories," *Guardian*, May 4, 2014; Alec Luhn, "Pro-Kremlin Journalists Secretly Given Awards by Putin," *Irish Times*, May 9, 2014.

164 According to the *Beijing News*: Katie Hunt and Cy Xu, "China 'Employs 2 Million to Police Internet,'" CNN, Oct. 7, 2013.

164 These commentators are paid: Steven Millward, "China Plans Weibo Propaganda Blitz Using 2 Million Paid Commenters," *Tech in Asia*, Jan. 18, 2013.

164 "positive energy": John Kennedy, "Beijing Orders Its 2.06 Million 'Propaganda Workers' to Get Microblogging," *South China Morning Post*, Jan. 18, 2013.

164 These workers also received training: Benjamin Carlson, "Party Trolls: Meet China's Answer to the Internet," *Global Post*, Jan. 28, 2013.

167 Instagram, Pinterest: LWG Consulting, "Sites Affected by the Heartbleed Bug," April 4, 2014.

167 Moreover, 150 million apps: Arik Hesseldahl, "Heartbleed Flaw Lurks in Android Apps Downloaded by Millions," *Re/code*, April 23, 2014.

167 Even a full month: Mark Prigg, "Over 300,000 Web Sites STILL at Risk from Heartbleed Bug," *Mail Online*, May 9, 2014.

167 Of course attackers: Michael Riley, "NSA Said to Exploit Heartbleed Bug for Intelligence for Years," *Bloomberg*, April 11, 2014.

167 Criminals also took part: Hiawatha Bray, "Heartbleed Hoodlums Try to Cash in on Internet Security Bug," *Boston Globe*, April 18, 2014; Mark Clayton, "'Heartbleed' Mystery: Did Criminals Take Advantage of Cyber-Security Bug?," *Christian Science Monitor*, April 9, 2014.

Chapter 10: Crime, Inc.

175 Welcome to the world: Numerous research sources provided details on the inner workings of Innovative Marketing. The majority of the details were reported and uncovered by Dirk Kollberg, a researcher for McAfee in Hamburg, Germany, who spent months studying the organization. Additional data points were listed in David Talbot, "The Perfect Scam," *MIT Technology Review*, June 21, 2011; Jim Finkle, "Inside a Global Cybercrime Ring," Reuters, March 24, 2010; Federal Trade Commission, "Innovative Marketing, Inc., et al.," Feb. 28, 2014; Toralv Dirro, "Malicious World," McAfee Labs; Interpol, "Sundin, Bjorn Daniel"; *United States of America v. Bjorn Daniel Sundin, Shaileshkumar P. Jain, a.k.a. "Sam Jain," and James Reno*, Northern District of Illinois Eastern Division, March 2010; Misha Glenny, "Cybercrime: Is It Out of Control?," *Guardian*, Sept. 21, 2011; Misha Glenny, "Inside the World of Cybercrime, EIBF 2012, Review," EdinburghGuide.com, Aug. 20, 2012; Felix Richter, "Twitter's Ad Revenue Tipped to Double This Year," *Statista*, Sept. 13, 2012; David Talbot, "The Perfect Scam," *Technology Review*, June 21, 2011.

175 Crime is big business: United Nations Office on Drugs and Crimes, "Estimating Illicit Financial Flows Resulting from Drug Trafficking and Other Transnational Organized Crimes," Oct. 2011, 7.

175 In total: Misha Glenny, *McMafia: A Jour-*

ney Through the Global Criminal Underworld (New York: Vintage Books, 2009), 12.

176 Capos, dons: Allison Davis, Patrick Di Justo, and Adam Rogers, "Crime, Organized," *Wired*, Feb. 2011, 78; General OneFile, Web, May 22, 2014.

176 Hacking is no longer ruled: "Organised Crime in the Digital Age," a joint study of Detica/BAE Systems and the John Grieve Centre for Policing at London Metropolitan University, March 2012.

176 According to a 2014 study: Lillian Ablon, Martin C. Libicki, and Andrea A. Golay, "Markets for Cybercrime Tools and Stolen Data," Rand Corporation, 4.

178 Several executives were kidnapped: Byron Acohido, "How Kidnappers, Assassins Utilize Smartphones, Google, and Facebook," USAToday .com, Feb. 18, 2011.

178 Sensing a market need: "Woman 'Ran Text-a-Getaway' Service," BBC News, July 16, 2013.

178 In San Francisco: This was based on the author's personal observations, and I have a photograph of the incident.

178 "It's more discreet": Dana Sauchelli and Bruce Golding, "Hookers Turning Airbnb Apartments into Brothels," *New York Post*, April 14, 2014.

178 While organized crime groups: The information on the organization of modern cybercrime organizations came from a variety of sources, including personal experience and investigation, consultation with senior law enforcement officials working in the field of cyber crime, and online resources such as "Cybercriminals Today Mirror Legitimate Business Processes," Fortinet 2013 Cybercrime Report; Trend Micro Threat Research, "A Cybercrime Hub," Aug. 2009; Information Warfare Monitor and Shadowserver Foundation, *Shadows in the Cloud*, Joint Report, April 6, 2010; Patrick Thibodeau, "FBI Lists Top 10 Posts in Cybercriminal Operations," *Computerworld*, March 23, 2010; Roderic Broadhurst et al., "Organizations and Cybercrime," *International Journal of Cyber Criminology*, Oct. 11, 2013.

181 Active criminal affiliates: Dmitry Samosseiko, "The Partnerka" (paper presented at Virus Bulletin Conference, Sept. 2009); "The Business of Cybercrime," Trend Micro White Paper, Jan. 2010.

183 In other words: Cisco, *Cisco 2010 Annual Security Report*, 9.

183 Actors in these online crime swarms: Broadhurst et al., "Organizations and Cybercrime."

184 As noted previously: Dunn, "Global Cybercrime Dominated by 50 Core Groups."

185 Some Crime, Inc. organizations: See Brian Krebs, "'Citadel' Trojan Touts Trouble-Ticket System," *Krebs on Security*, Jan. 23, 2012.

185 One group of cyber thieves: Bob Sullivan, "160 Million Credit Cards Later, 'Cutting Edge' Hacking Ring Cracked," NBC News, July 25, 2013; "Team of International Criminals Charged with Multi-million Dollar Hacking Ring," NBC News, July 25, 2013.

186 Some digital criminal marketplaces: Thomas Holt, "Exploring the Social Organisation and Structure of Stolen Data Markets," *Global Crime* 14, nos. 2–3 (2013); Thomas Holt, "Honor Among (Credit Card) Thieves?," *Michigan State University Today*, April 22, 2013.

186 These individuals: Ablon, Libicki, and Golay, "Markets for Cybercrime Tools and Stolen Data," 17.

186 These honest, but criminal, brokers: Gregory J. Millman, "Cybercriminals Work in a Sophisticated Market Structure," *Wall Street Journal*, June 27, 2013.

186 Using the data: Kevin Poulsen, "Superhacker Max Butler Pleads Guilty," *Wired*, June 29, 2009.

187 In fact, a study: Donald T. Hutcherson, "Crime Pays: The Connection Between Time in Prison and Future Criminal Earnings," *Prison Journal* 92, no. 3 (Sept. 2012): 315–35; Shankar Vedantam, "When Crime Pays: Prison Can Teach Some to Be Better Criminals," NPR, Feb. 1, 2013.

187 Such was the case: Ian Gallagher, "Public Schoolboy Hacker Who Masterminded £15M Fraud Is Put in Jail's IT Class . . . and Hacks the Prison's Computer System," *Mail Online*, March 2, 2013.

187 At the San Quentin maximum-security prison: Reuters, "San Quentin Prison Becomes an Incubator for Startups," *Huffington Post*, Feb. 25, 2013.

188 Time and time again: Russell Eisenman, "Creativity and Crime: How Criminals Use Creativity to Succeed," in *The Dark Side of Creativity*, ed. David H. Cropley et al. (New York: Cambridge University Press, 2010).

188 Modern criminals are innovating: John Leyden, "Malware Devs Embrace Open-Source," *Register*, Feb. 10, 2012 .

188 To drive sales: Ablon, Libicki, and Golay, "Markets for Cybercrime Tools and Stolen Data," 11.

188 Organized cyber criminals: Chris Anderson, *The Long Tail: Why the Future of Business Is Selling Less of More*, rev. ed. (New York: Hyperion, 2008); Goodman, "What Business Can Learn from Organized Crime."

188 RankMyHack.com awards points: Riva Richmond, "Web Site Ranks Hacks and Bestows Bragging Rights," *New York Times*, Aug. 21, 2011.

189 In Montenegro: Jim Finkle, "Inside a

Global Cybercrime Ring," Reuters, March 24, 2010.

189 In early 2014: Paul Peachey, "Cybercrime Boss Offers a Ferrari for Hacker Who Dreams Up the Biggest Scam," *Independent*, May 11, 2014.

189 The concept of crowdsourcing: Jeff Howe, "The Rise of Crowdsourcing," *Wired*, June 2006.

189 While hundreds of examples: Marc Goodman, "The Rise of Crime-Sourcing," *Forbes*, Oct. 3, 2011.

189 YouTube is replete: Ibid.

189 In Washington, D.C.: Elizabeth Fiedler, "Retailers Fight 'Flash Robs,'" NPR.org, Nov. 25, 2011; Annie Vaughan, "Teenage Flash Mob Robberies on the Rise," FoxNews.com, June 18, 2011.

189 In the United States: Chris Foresman, "Senator to Apple, Google: Why Are DUI Checkpoint Apps Still Available?," *Ars Technica*, May 20, 2011; "Want to Avoid a Speed Trap or a DUI Checkpoint? There's an App for That," *Mail Online*, March 21, 2011.

190 When the 2011 London riots: Patrick Kingsley, "Inside the Anti-kettling HQ," *Guardian*, Feb. 2, 2011.

190 At the height: "LulzSec Opens Hack Request Line," BBC, June 15, 2011.

190 The group established: "LulzSec Hackers Sets Up Hotline for Attacks," Reuters, June 15, 2011.

190 As a result: Brian Krebs, "Wash. Hospital Hit by $1.03 Million Cyberheist," *Krebs on Security*, April 30, 2013.

190 Simple, they were properly incentivized: Mathew J. Schwartz, "Hackers Offer Free Porn to Beat Security Checks," *Dark Reading*, June 20, 2012.

191 The guard was disabled: Caroline McCarthy, "Bank Robber Hires Decoys on Craigslist, Fools Cop," *CNET*, Oct. 3, 2008.

191 Soon half a dozen police cars: David Pescovitz, "Bank Robber Uses Craigslist to Hire Unsuspecting Accomplices," *Boing Boing*, Oct. 1, 2008; "Armored Truck Robber Uses Craigslist to Make Getaway," King5.com, Sept. 21, 2009.

191 The most popular of these sites: Kickstarter, "Stats," accessed on May 25, 2014, https://www.kickstarter.com/help/stats, indicating Kickstarter had raised $1,131,653 since launching.

192 Criminals are of course happy: Jason Del Rey, "Kickstarter Says It Was Hacked (Updated)," *Re/code*, Feb. 15, 2014.

192 The answer was: "Apple Fingerprint ID 'Hacked,'" BBC News, Sept. 23, 2013.

192 Using elements of both: John Bowman, "iPhone 5S Fingerprint Hacking Contest Offers $20K Bounty," *Your Community* (blog), CBC News, Sept. 20, 2013.

192 Finally, white wood glue: Frank, "Chaos Computer Club Breaks Apple TouchID," Chaos Computer Club, Sept. 21, 2013.

192 Donations have been made: Andy Greenberg, "Meet the 'Assassination Market' Creator Who's Crowdfunding Murder with Bitcoins," *Forbes*, Nov. 18, 2013.

193 As a result, the master criminal-hackers: Marc Santora, "In Hours, Thieves Took $45 Million in A.T.M. Scheme," *New York Times*, May 9, 2013.

Chapter 11: Inside the Digital Underground

194 DPR was the mastermind: Ken Klippenstein, "Dread Pirate Roberts 2.0: An Interview with Silk Road's New Boss," *Ars Technica*, Feb. 5, 2014.

195 Clicking on any particular link: Patrick Howell O'Neill, "The Rise and Fall of Silk Road's Heroin Kingpin," *The Daily Dot*, Oct. 9, 2013.

198 In the meantime: David Segal, "Eagle Scout. Idealist. Drug Trafficker?," *New York Times*, Jan. 18, 2014; Kevin Goodman, "The Dark Net," *Huffington Post*, Oct. 16, 2013; Adrian Chen, "The Underground Website Where You Can Buy Any Drug Imaginable," *Gawker*, June 1, 2011; Stuart Pfeifer, Shan Li, and Walter Hamilton, "End of Silk Road for Drug Users as FBI Shuts Down Website," *Los Angeles Times*, Oct. 2, 2013; Gerry Smith, "Alleged Silk Road Founder Put Out Hit on 6 Enemies, Prosecutors Say," *Huffington Post*, Nov. 22, 2013; Kim Zetter, "Feds Arrest Alleged 'Dread Pirate Roberts,' the Brain Behind the Silk Road Drug Site," *Wired*, Oct. 2, 2013.

198 That journey begins: For further information on Tor, visit the Tor Project at https://www.torproject.org/.

199 While precise numbers: Alex Biryukov, Ivan Pustogarov, and Ralf-Philipp Weinmann, "Content and Popularity Analysis of Tor Hidden Services," University of Luxembourg.

199 As of early 2014: Geoffrey A. Fowler, "Tor: An Anonymous, and Controversial, Way to Web-Surf," *Wall Street Journal*, Dec. 18, 2012.

200 A number of reports: Raphael Cohen-Almagor, "In Internet's Way," *International Journal of Cyber Warfare and Terrorism* 2, no. 3 (July–Sept. 2012): 39–58.

200 After the former NSA contractor: Kimberly Dozier, "Virtually Every Terrorist Group in the World Shifting Tactics in Wake of NSA Leaks: U.S. Officials," *National Post*, June 26, 2013.

200 Organizations such as: "Al Qaeda, Terrorists Changing Communication Methods After NSA Leaks, US Officials Say," Fox News, June 26, 2013; http://www.youtube.com/watch?v=D8Mgpm1PgF4.

201 Shockingly, the Deep Web: Michael K. Bergman, "White Paper: The Deep Web: Surfacing Hidden Value," *Journal of Electronic Publishing* 7, no. 1 (Aug. 2001).

201 According to a study: Steve Lawrence and C. Lee Giles, "Accessibility of Information on the Web," *Nature*, July 8, 1999, 107, doi:10.1038/21987.

201 As a result, when you search Google: Bergman, "White Paper."

201 In other words: Jose Pagliery, "The Deep Web You Don't Know About," *CNNMoney*, March 10, 2014.

201 Though you may catch: "Google Search vs. Deep Web Harvesting," *BrightPlanet*, July 31, 2013.

202 Whereas Silk Road: Andy Greenberg, "Inside the 'DarkMarket' Prototype, a Silk Road the FBI Can Never Seize," *Wired*, April 24, 2014.

202 To that end, in mid-2014: Kim Zetter, "New 'Google' for the Dark Web Makes Buying Dope and Guns Easy," *Wired*, April 17, 2014.

202 Certain criminal forums: Michael Riley, "Stolen Credit Cards Go for $3.50 at Amazon-Like Online Bazaar," *Bloomberg*, Dec. 19, 2011.

203 Numerous illicit "torrents": Ernesto, May 18, 2008, blog on *TorrentFreak*, accessed on June 27, 2014.

203 Another such site: "Inside the Mansion—and Mind—of Kim Dotcom, the Most Wanted Man on the Net," *Wired*, Oct. 18, 2012.

203 Not only do they sell: Beth Stebner, "The Most Dangerous Drug in the World: 'Devil's Breath' Chemical from Colombia Can Block Free Will, Wipe Memory, and Even Kill," *Mail Online*, May 12, 2012.

204 Tor hidden sites: Forward-Looking Threat Research Team, "Deepweb and Cybercrime," Trend Micro, 2013, 16.

204 Once stolen: Brian Krebs, "Peek Inside a Professional Carding Shop," *Krebs on Security*, June 4, 2014.

204 Given the vast amounts: Max Goncharov, "Russian Underground Revisited," Forward-Looking Threat Research Team, Trend Micro Research Paper.

204 The cards are sold: Brian Krebs, "Cards Stolen in Target Breach Flood Underground Markets," *Krebs on Security*, Dec. 20, 2013; Dancho Danchev, "Exposing the Market for Stolen Credit Cards Data," *Dancho Danchev's Blog*, Oct. 31, 2011; "Meet the Hackers," *Bloomberg Businessweek*, May 28, 2006; David S. Wall, "The Organization of Cybercrime in an Ever-Changing Cyberthreat Landscape" (draft paper for the Criminal Networks Conference, Montreal, Oct. 3–4, 2011).

204 The United States is the largest victim: "Skimming off the Top," *Economist*, Feb. 15, 2014.

205 Nearly 20 percent: Pew Research Center, "More Online Americans Say They've Experienced a Personal Data Breach," April 14, 2014; Rosie Murray-West, "UK Worst in Europe for Identity Fraud," *Telegraph*, Oct. 1, 2012.

205 Medical identity theft: Herb Weisbaum, "U.S. Health Care System Has $5.6 Billion Security Problem," NBC News, March 12, 2014; Richard Rubin, "IRS May Lose $21 Billion in Identity Fraud, Study Says," *Bloomberg*, Aug. 2, 2012.

205 U.S. driver's licenses: "Cashing In on Digital Information," TrendMicro/TrendLabs 2013 Annual Security Roundup.

205 So too are assault rifles: Sam Biddle, "The Secret Online Weapons Store That'll Sell Anyone Anything," *Gizmodo*, July 19, 2012; Adrian Chen, "Now You Can Buy Guns on the Online Underground Marketplace," *Gawker*, Jan. 27, 2012.

206 One Dark Web user: Sam Biddle, "The Secret Online Weapons Store That'll Sell Anyone Anything," *Gizmodo*, July 19, 2012.

206 As we saw with Silk Road: Greenberg, "Meet the 'Assassination Market' Creator Who's Crowdfunding Murder with Bitcoins."

206 The sites request: Dylan Love, "How to Hire an Assassin on the Secret Internet for Criminals," *Business Insider*, March 16, 2013.

206 Kindergarten Porn: Joel Falconer, "Mail-Order Drugs, Hitmen, and Child Porn: A Journey into the Dark Corners of the Deep Web," *Next Web*, Oct. 8, 2012.

206 Just one Dark Web site: Patrick Howell O'Neill, "Feds Dismantle Massive Deep Web Child Porn Ring," *Daily Dot*, March 19, 2014.

206 Moreover, the National Center: *Thorn Blog*, http://www.wearethorn.org/child-trafficking-statistics/.

207 Law enforcement sources report: Testimony of Ernie Allen, president of the National Center of Missing and Exploited Children, to the Institute of Medicine Committee on Commercial Sexual Exploitation and Sex Trafficking of Minors in the United States of the National Academies, available at http://storage.cloversites.com/thedaughterproject/documents/NCMEC%20report%20to%20congress%2001-04-12.pdf; http://www.nap.edu/catalog.php?record_id=18358.

207 The U.S. Department of Justice: NPR Staff, "Courts Take a Kinder Look at Victims of Child Sex Trafficking," NPR.org, March 1, 2014.

207 Nearly 70 percent: Thorn Staff, "Child Sex Trafficking and Exploitation Online: Escort Websites," March 11, 2014; National Human Trafficking Resource Center, "Residential Brothels."

207 These activities are transacted: Mark Latonero, "The Rise of Mobile and the Diffusion of Technology-Facilitated Trafficking," University of Southern California.

207 Web sites such as BackPage.com: Shared

Hope International, "Demanding Justice Project Benchmark Assessment 2013," 13; Michelle Goldberg, "Sex Slave Outrage," *Daily Beast*, Dec. 9, 2010.

207 Around the world: For an outstanding overview of the international black market in human organs, see *Der Spiegel*'s four-part series on the topic available in English at http://www.spiegel.de/international/world/the-illegal-trade-in-organ-is-fueled-by-desperation-and-growing-a-847473.html.

207 Kidneys can fetch: Casey Chan, "Here's How Much Body Parts Cost on the Black Market," *Gizmodo*, April 23, 2012.

207 In the United States alone: National Kidney Foundation, "Organ Donation and Transplantation Statistics," Sept. 8, 2014.

207 Most will die: Jeneen Interlandi, "Organ Trafficking Is No Myth," *Newsweek*, Jan. 9, 2009.

208 The World Health Organization: Damien Gayle, "An Organ Is Sold Every Hour, WHO Warns: Brutal Black Market on the Rise Again Thanks to Diseases of Affluence," *Mail Online*, May 27, 2012.

208 The organs may come from: Ulrike Putz, "Organ Trade Thrives Among Desperate Syrian Refugees in Lebanon," *Spiegel Online*, Dec. 11, 2013; Jiayang Fan, "Can China Stop Organ Trafficking?," *New Yorker*, Jan. 10, 2014.

208 Sadly, those selling body parts: Esther Inglis-Arkell, "How Do You Buy Organs on the Black Market?," *io9*, March 26, 2012.

208 "I will sell my kidney": Dan Bilefsky, "Black Market for Body Parts Spreads in Europe," *New York Times*, June 28, 2012.

208 "Donate a kidney": Denis Campbell and Nicola Davison, "Illegal Kidney Trade Booms as New Organ Is 'Sold Every Hour,'" *Guardian*, May 27, 2012.

208 At least one seventeen-year-old: "9 on Trial in China over Teenager's Sale of Kidney for iPad and iPhone," CNN, Aug. 10, 2012.

208 In a deeply disturbing report: European Cybercrime Centre, "Commercial Sexual Exploitation of Children Online," Oct. 2013.

208 Organized criminal networks: Paul Gallagher, "Live Streamed Videos of Abuse and Pay-per-View Child Rape Among 'Disturbing' Cybercrime Trends, Europol Report Reveals," *Independent*, Oct. 16, 2013; Paul Peachey, "Number of UK Paedophiles 'Live-Streaming' Child Abuse Films Soars, Warns CEOP," *Independent*, July 1, 2013.

208 In one incident: Ann Cahill, "New Age of Cybercrime: Live Child Rapes, Sextortion, and Advanced Malware," *Irish Examiner*, Feb. 11, 2014.

209 The system is designed: "How Does Bitcoin Work?," *Economist*, April 11, 2013.

209 Bitcoin is the world's largest: Nick Farrell, "Understanding Bitcoin and Crypto Currency," *Tech Radar*, April 7, 2014.

209 Because Bitcoin can be spent: Joshua Brustein, "Bitcoin May Not Be So Anonymous, After All," *Bloomberg Businessweek*, Aug. 27, 2013.

210 There are now more than seventy: Alan Yu, "How Virtual Currency Could Make It Easier to Move Money," NPR.org, Jan. 15, 2014.

210 Hackers have been able to steal: Robin Sidel, Eleanor Warnock, and Takashi Mochizuki, "Almost Half a Billion Worth of Bitcoins Vanish," *Wall Street Journal*, March 1, 2014.

210 Beyond crypto currencies: Marc Santora, William K. Rashbaum, and Nicole Perlroth, "Liberty Reserve Operators Accused of Money Laundering," *New York Times*, May 28, 2013.

210 Known as the "PayPal": United States Attorney's Office of Southern New York, "Liberty Reserve Information Technology Manager Pleads Guilty in Manhattan Federal Court," United States Department of Justice press release, Sept. 23, 2014.

210 The popularity of Darkcoin: Andy Greenberg, "Darkcoin, the Shadowy Cousin of Bitcoin, Is Booming," *Wired*, May 21, 2014.

211 Operating under the motto: Andy Greenberg, "'Dark Wallet' Is About to Make Bitcoin Money Laundering Easier Than Ever," *Wired*, April 29, 2014.

211 One such CaaS company: James Vincent, "Irish Man Arrested as 'the Largest Facilitator of Child Porn on the Planet,'" *Independent*, Aug. 5, 2013.

211 Hundreds of crime-trepreneur purveyors: Kevin Poulsen, "FBI Admits It Controlled Tor Servers Behind Mass Malware Attack," *Wired*, Sept. 13, 2013.

212 The trend is accelerating: Solutionary, an NTT Group Security Company, *Security Engineering Research Team (SERT) Quarterly Threat Intelligence Report*, 2013, 8, http://www.solutionary.com.

212 For example, the hackers: Ibid.

212 Today, using the distributed computing power: "Cybercriminals Today Mirror Legitimate Business Processes," 4.

212 This means that anyone: Simson Garfinkel, "The Criminal Cloud," *MIT Technology Review*, Oct. 17, 2011.

212 "private organisation": Misha Glenny, *DarkMarket: Cyberthieves, Cybercops, and You* (New York: Knopf, 2011), 203.

212 China's Hidden Lynx: Danny Yadron, "Symantec Fingers Most Advanced Chinese Hacker Group," *Digits* (blog), *Wall Street Journal*, Sept. 17, 2013.

213 Off duty, however: Kim Zetter, "State-Sponsored Hacker Gang Has a Side Gig in Fraud," *Wired*, Sept. 17, 2013.

213 Staffed 24/7: Kim Zetter, "Cops Pull Plug on Rent-a-Fraudster Service for Bank Thieves," *Wired*, April 19, 2010.

213 As a result, less skilled criminals: Ablon, Libicki, and Golay, "Markets for Cybercrime Tools and Stolen Data," 4.

214 Vendors offer one-stop shopping: Forward-Looking Threat Research Team, "Deepweb and Cybercrime," 9; Ablon, Libicki, and Golay, "Markets for Cybercrime Tools and Stolen Data," 4.

214 As an example: Taylor Armerding, "Dark Web: An Ever-More-Comfortable Haven for Cyber Criminals," *CSO Online*, March 28, 2014.

214 Over the years: Donna Leinwand Leger and Anna Arutunyan, "How the Feds Brought Down a Notorious Russian Hacker," *USA Today*, March 5, 2014.

214 When they did: Dan Raywood, "New Version of Bugat Trojan Was Payload in LinkedIn Spam and Not Zeus," *SC Magazine UK*, Oct. 12, 2010.

214 Once it found it: Robert McMillan, "New Russian Botnet Tries to Kill Rival," *Computerworld*, Feb. 9, 2010.

214 Like its rival Zeus: Kurt Eichenwald, "The $500,000,000 Cyber-Heist," *Newsweek*, March 13, 2014.

215 The tool, perhaps one of the world's most popular: Gregory J. Millman, "Cybercriminals Work in a Sophisticated Market Structure," *Wall Street Journal*, June 27, 2013.

215 Worse, it was the tool of choice: Dana Liebelson, "All About Blackshades, the Malware That Lets Hackers Watch You Through Your Webcam," *Mother Jones*, May 21, 2014.

215 So good was the Blackshades RAT: "Syrian Activists Targeted with BlackShades Spy Software," *The Citizen Lab*, June 19, 2012.

216 The rewards, however: Gregg Keizer, "Google to Pay Bounties for Chrome Browser Bugs," *Computerworld*, Jan. 29, 2010.

216 Not to be outdone: Brian Krebs, "Meet Paunch: The Accused Author of the BlackHole Exploit Kit," *Krebs on Security*, Dec. 6, 2013.

216 Dark Net chat rooms: Nicole Perlroth and David E. Sanger, "Nations Buying as Hackers Sell Flaws in Computer Code," *New York Times*, July 13, 2013.

216 In 2012, the Grugq sold: Andy Greenberg, "Shopping for Zero-Days: A Price List For Hackers' Secret Software Exploits," *Forbes*, March 23, 2012.

216 Companies such as Vupen: Brian Krebs, "How Many Zero-Days Hit You Today," *Krebs on Security*, Dec. 13, 2013.

216 The result, as pointed out: Josh Sanburn, "How Exactly Do Cyber Criminals Steal $78 Million?," *Time*, July 3, 2012.

217 Worse, now that Stuxnet: Simonite, "Stuxnet Tricks Copied by Computer Criminals."

218 Crime, Inc. can even draft: "The Child Porn PC Virus," *Week*, Nov. 10, 2009.

218 According to the FBI: FBI, "GameOver Zeus Botnet Disrupted," June 2, 2014.

218 As of mid-2014: Symantec, "Grappling with the ZeroAccess Botnet," Sept. 30, 2013.

218 In the Russian digital underground: Ian Steadman, "The Russian Underground Economy Has Democratised Cybercrime," *Wired UK*, Nov. 2, 2012.

218 Moreover, the threat: "Computer Says No," *Economist*, June 22, 2013; Perlroth and Hardy, "Bank Hacking Was the Work of Iranians."

219 The toll of victims: Chris Brook, "Meetup .com Back Online After DDoS Attacks, Extortion Attempt," *Threat Post*, March 5, 2014; Pierluigi Paganini, "Botnet Authors Use Evernote Account as C&C Server," *Security Affairs*, March 31, 2013.

219 Given these obvious advantages: Mathew J. Schwartz, "Malware Toolkits Generate Majority of Online Attacks," *Dark Reading*, Jan. 18, 2011.

220 To unlock their computers: David Wismer, "Hand-to-Hand Combat with the Insidious 'FBI MoneyPak Ransomware Virus,'" *Forbes*, Feb. 6, 2013.

220 Thus users in the U.K.: EnigmaSoftware, "Abu Dhabi Police GHQ Ransomware."

220 Another, even more pernicious: Mark Ward, "Crooks 'Seek Ransomware Making Kit,'" BBC News, Dec. 10, 2013.

220 Nearly 250,000 individuals: Dave Jeffers, "Crime Pays Very Well: CryptoLocker Grosses up to $30 Million in Ransom," *PCWorld*, Dec. 20, 2013.

220 Automated ransomware tools: Dennis Fisher, "Device-Locking Ransomware Moves to Android," *ThreatPost*, May 7, 2014.

221 The police lieutenant: Violet Blue, "CryptoLocker's Crimewave: A Trail of Millions in Laundered Bitcoin," *ZDNet*, Dec. 22, 2013; Bree Sison, "Swansea Police Pay Ransom After Computer System Was Hacked," CBS Boston, Nov. 18, 2013.

Chapter 12: When All Things Are Hackable

223 The police investigation: Joanne Kimberlin, "High-Tech 'Repo Man' Keeps Car Payments Coming," *USA Today*, Nov. 29, 2005; Christina Rosales, "Police: Fired Worker Disabled Cars via Web," *Statesman*, March 17, 2010; Kevin Poulsen, "Hacker Disables More Than 100 Cars Remotely," *Wired*, March 17, 2010.

224 Laptop sales supplanted: Michael Singer, "PC Milestone—Notebooks Outsell Desktops," *CNET*, June 3, 2005; Salvador Rodriguez, "More Tablets to Be Sold Than PCs in 2015, Report Says," ChicagoTribune.com, July 8, 2014.

224 In 2014, we saw: "2014: Mobiles 'to Outnumber People,'" BBC News, May 9, 2013.

224 The Pew Research Center defines: Pew Research Center, "Digital Life in 2025," March 2014; Pew Research Center's Internet & American Life Project, "Internet of Things," accessed July 21, 2014, http://www.pewinternet.org/.

225 "if all the objects": Lopez Research, "An Introduction to the Internet of Things," Cisco, Nov. 2013.

225 Indeed, according: Terril Yue Jones, "A Law of Continuing Returns," Los Angeles Times, April 17, 2005.

225 They are low-powered: Olga Kharif, "Trillions of Smart Sensors Will Change Life," Bloomberg, Aug. 4, 2013.

225 "a Web server": Neil Gershenfeld and J. P. Vasseur, "As Objects Go Online," Foreign Affairs, March/April 2014.

226 Back when IPv4 was introduced: Laurie J. Flynn, "As World Runs Out of I.P. Addresses, Switch to IPv6 Nears," New York Times, Feb. 14, 2011.

226 IPv6, on the other hand: Andrew G. Blank, TCP/IP Foundations (Hoboken, N.J.: John Wiley & Sons, 2006), 233.

226 That means IPv6 would allow: John Martellaro, "A Layman's Guide to the IPv6 Transition," The Mac Observer, Jan. 31, 2012; Robert Krulwich, "Which Is Greater, the Number of Sand Grains on Earth or Stars in the Sky?," NPR, Sept. 17, 2012.

226 every single atom: Steve Leibson, "IPV6: How Many IP Addresses Can Dance on the Head of a Pin," EDN Network, March 28, 2008; "The Internet of Things," Cisco Infographic.

226 Tomorrow's will be the size: "IPv6— What Is It, Why Is It Important, and Who Is in Charge?" (paper prepared for chief executive officers of ICANN and all the regional Internet registries), Oct. 2009.

226 Though in 2013: Dave Evans, "The Internet of Things," Cisco, April 2011.

227 The McKinsey Global Institute: McKinsey Global Institute, Disruptive Technologies: Advances That Will Transform Life, Business, and the Global Economy, May 2013, 55, MGI_Disruptive_technologies_Full_report_May2013.pdf.

227 The IoT may very well be: Emerging Cyber Threats (presented by Georgia Institute of Technology and the Georgia Tech Research Institute at the Georgia Tech Cyber Security Summit, 2013), 4.

228 "billions of smart": Global Strategy and Business Development, Freescale and Emerging Technologies, ARM, What the Internet of Things (IoT) Needs to Become a Reality, May 2014.

228 "phenomenon of convergence": Marcus Wohlsen, "Forget Robots. We'll Soon Be Fusing Technology with Living Matter," Wired, May 27, 2014.

229 You're not the only one: Robert Muir, "Thirsty Plants Can Twitter for Water with New Device," Reuters, March 26, 2009; https://twitter.com/pothos; Rachel Metz, "In San Francisco, a House with Its Own Twitter Feed," MIT Technology Review, May 21, 2013.

229 "the purposes of enrichment": Gershenfeld and Vasseur, "As Objects Go Online."

229 Rather, researchers fitted: Alan Yu, "More Than 300 Sharks in Australia Are Now on Twitter," NPR.org, Jan. 1, 2014.

229 While this future: Alexis C. Madrigal, "Welcome to the Internet of Thingies: 61.5% of Web Traffic Is Not Human," Atlantic, Dec. 12, 2013.

230 Just as the introduction: M. Presser and S. Krco, Initial Report on IoT Applications of Strategic Interest, Internet of Things Initiative, Oct. 8, 2011, 48.

231 There have been dozens: Annalee Newitz, "The RFID Hacking Underground," Wired, May 2006.

231 These shortcomings have allowed: Francis Brown and Bishop Fox, "RFID Hacking" (paper presented at Black Hat USA, Las Vegas, Nev., Aug. 1, 2013).

231 Every Fortune 500 company: "Hackers Could Clone Your Office Key Card . . . from Your Pocket," NBC News, July 25, 2013.

231 Seconds later: Andy Greenberg, "Hacker's Demo Shows How Easily Credit Cards Can Be Read Through Clothes and Wallets," Forbes, Jan. 30, 2012.

232 RFID chips can also be infected: Nate Anderson, "RFID Chips Can Carry Viruses," Ars Technica, March 15, 2006.

232 Another popular: Juniper Research, "1 in 5 Smartphones will have NFC by 2014, Spurred by Recent Breakthroughs: New Juniper Research Report," April 14, 2011.

232 But like RFID: Andy Greenberg, "Hacker Demos Android App That Can Wirelessly Steal and Use Credit Cards' Data," Forbes, July 27, 2012.

232 Google Wallet has also been hacked: Lance Whitney, "Latest Google Wallet Hack Picks Your Pocket," CNET, Feb. 10, 2012; Evan Applegate, "Have Fingers, 30 Seconds? You, Too, Can Hack Google Wallet," Bloomberg Businessweek, Feb. 13, 2012.

232 In another instance: Gabrielle Taylor, "Have an NFC-Enable Phone? This Hack Could Hijack It," WonderHowTo, accessed July 11, 2014.

232 NFC apps on mobile phones: Lisa Vaas, "Android NFC Hack Lets Subway Riders Evade Fares," Naked Security, Sept. 24, 2012.

232 They can also intercept data: Kate Murphy, "Protecting a Cellphone Against Hackers," New York Times, Jan. 25, 2012; Tu C. Neim, "Bluetooth and Its Inherent Security Issues," SANS Institute InfoSec Reading Room, Nov. 4, 2002.

233 As a result, more of what happens: Catherine Crump and Matthew Harwood, "Invasion of the Data Snatchers: Big Data and the Internet of Things Means the Surveillance of Everything," *Blog of Rights*, March 25, 2014.

234 All drivers need to do: "Snapshot Common Questions," Progressive Web site, http://www.progressive.com/auto/snapshot-common -questions/.

234 "we and other companies": Rolfe Winkler, "Google Predicts Ads in Odd Spots Like Thermostats," *Digits* (blog), *Wall Street Journal*, May 21, 2014.

234 In some countries: Brian Brady, "Prisoners 'to Be Chipped like Dogs,'" *Independent*, Jan. 13, 2008.

235 "3,000 labor hours": David Rosen, "Big Brother Invades Our Classrooms," *Salon*, Oct. 8, 2012.

235 "Students who do not wish": David Kravets, "Student Suspended for Refusing to Wear a School-Issued RFID Tracker," *Wired*, Nov. 21, 2012.

235 Based on nothing more: Aaron Katersky and Josh Haskell, "NY Mom Accused of Growing $3M Marijuana Business," *Good Morning America*, June 6, 2013; Glenn Smith, "Marijuana Bust Shines Light on Utilities," *Post and Courier*, Jan. 29, 2012.

235 "transformational for clandestine": Spencer Ackerman, "CIA Chief: We'll Spy on You Through Your Dishwasher," *Wired*, March 15, 2012.

237 The mere plugging in: Neil J. Rubenking, "Black Hat: Don't Plug Your Phone into a Charger You Don't Own," *PCMag*, Aug. 1, 2013.

237 In the background: "Public Charging Stations Help Smartphone Users, but Also Open a New Avenue for Hacking," New York *Daily News*, Aug. 13, 2013.

237 In 2013 in Russia: Simon Sharwood, "DON'T BREW THAT CUPPA! Your Kettle Could Be a SPAMBOT," *Register*, Oct. 29, 2013; Adam Clark Estes, "Russian Authorities Seize Goods from China Implanted with Spy Chips," *Gizmodo*, Oct. 29, 2013.

237 The devices contained hidden: Erik Sherman, "Hacked from China: Is Your Kettle Spying on You?," CBS News, Nov. 1, 2013.

238 While there may be serendipitous benefits: Klint Finley, "Why Tech's Best Minds Are Very Worried About the Internet of Things," *Wired*, May 19, 2014.

Chapter 13: Home Hacked Home

240 Once news of the school district's: David Kravets, "School District Allegedly Snapped Thousands of Student Webcam Spy Pics," *Wired*, April 16, 2010; Kashmir Hill, "Lower Mer-

ion School District and Blake Robbins Reach a Settlement in Spycamgate," *Forbes*, Oct. 11, 2010; John P. Martin, "L. Merion Smearing Former IT Chief, Lawyer Says," Philly.com, May 5, 2010.

240 "They might as well": Suzan Clarke, "Pa. School Faces FBI Probe," ABC News, Feb. 22, 2010.

240 In one city alone: Loretta Chao, "Cisco Poised to Help China Keep an Eye on Its Citizens," *Wall Street Journal*, July 5, 2011.

241 "as small as six inches": John Biggs, "DARPA Builds a 1.8-Gigapixel Camera," *TechCrunch*, Jan. 28, 2013.

241 As a result, cameras can perform: "Fighting Terrorism in New York City," *60 Minutes*, Sept. 26, 2011.

241 One young woman: "Miss Teen USA: Screamed upon Learning She Was 'Sextortion' Victim," CNN, Sept. 28, 2013.

241 Abrahams carried out his attack: Aaron Katersky, "Miss Teen USA 1 of 'Half-Million' 'Blackshades' Hack Victims," ABC News, May 19, 2014.

242 "All of a sudden": Amy Wagner, "Hacker Hijacks Baby Monitor," Fox News, April 22, 2014.

242 "Wake up … you": "Parents Left Terrified After Man Hacked Their Baby Monitor and Yelled Abuse at Them and Their 2-Year-Old Daughter," *Mail Online*, Aug. 13, 2013.

242 Nearly 70 percent: Kim Zetter, "Popular Surveillance Cameras Open to Hackers, Researcher Says," *Wired*, May 15, 2012.

242 As a consequence: Kelly Jackson Higgins, "Millions of Networked Devices in Harm's Way," *Dark Reading*, Jan. 29, 2013.

242 Without the consent: Katie Notopoulos, "Somebody's Watching: How a Simple Exploit Lets Strangers Tap into Private Security Cameras," *Verge*, Feb. 3, 2012.

243 Why not hack: Jim Finkle, "US Security Expert Says Surveillance Cameras Can Be Hacked," Reuters, June 18, 2013.

243 That's exactly what a team: Mark Buttler, "Crown Casino Hi-Tech Scam Nets $32 Million," *Herald Sun* (Melbourne), March 14, 2013.

243 Confident of his bets: Kim Zetter, "Crooks Spy on Casino Card Games with Hacked Security Cameras, Win $33M," *Wired*, March 15, 2013.

243 A car rolling off: Robert N. Charette, "This Car Runs on Code," *IEEE Spectrum*, Feb. 1, 2009.

244 All these embedded electronics: Ibid.; Chris Bryant, "Manufacturers Respond to Car-Hacking Risk," *Financial Times*, March 22, 2013.

244 While event-data-recording black boxes: Craig Timberg, "Web-Connected Cars Bring Privacy Concerns," *Washington Post*, March 5, 2013.

244 "[We know] everyone": "GPS Users

Beware, Big Auto Is Watching: Report," CNBC .com, Jan. 9, 2014.

244 GM's OnStar: John R. Quain, "Changes to OnStar's Privacy Terms Rile Some Users," *Wheels* (blog), *New York Times*, Sept. 22, 2011.

244 Oh, and that convenient: Declan McCullagh and Anne Broache, "FBI Taps Cell Phone Mic as Eavesdropping Tool," *CNET*, Dec. 1, 2006; Bruno Waterfield and Matthew Day, "EU Has Secret Plan for Police to 'Remote Stop' Cars," *Telegraph*, Jan. 29, 2014.

244 In just the first six months: Jeff Bennett, "GM Adds 8.45 Million Vehicles to North America Recall," *Wall Street Journal*, June 30, 2014; Christopher Jensen, "An Increase in Recalls Goes Beyond Just G.M.," *New York Times*, May 29, 2014.

244 When the deeply complex: James R. Healey, "Toyota Deaths Reported to Safety Database Rise to 37," *USA Today*, Feb. 17, 2010.

244 A jury found: Phil Baker, "Software Bugs Found to Be Cause of Toyota Acceleration Death," *Daily Transcript*, Nov. 4, 2013; Junko Yoshida, "Acceleration Case: Jury Finds Toyota Liable," *EETimes*, Oct. 24, 2013.

245 Toyota was accused: Jerry Hirsch, "Toyota Admits Misleading Regulators, Pays $1.2-Billion Federal Fine," *Los Angeles Times*, March 19, 2014.

245 According to the London Metropolitan Police: Victoria Woollaston, "Forget Carjacking, the Next Big Threat Is Car-HACKING," *Mail Online*, May 8, 2014.

245 The operation takes less: William Pentland, "Car-Hacking Goes Viral in London," *Forbes*, May 20, 2014; Thomas Cheshire, "Thousands of Cars Stolen Using Hi-Tech Gadgets," *Sky News*, May 8, 2014.

245 Using nothing more than a laptop: Sebastian Anthony, "Hackers Can Unlock Cars via SMS," *ExtremeTech*, July 28, 2011; Robert McMillan, "'War Texting' Lets Hackers Unlock Car Doors Via SMS," *CSO Online*, July 27, 2011.

245 Your musical tastes: Rebecca Boyle, "Trojan-Horse MP3s Could Let Hackers Break into Your Car Remotely, Researchers Find," *Popular Science*, March 14, 2011.

245 For just under $30: Victoria Woollaston, "The $20 Handheld Device That Hacks a CAR—and Can Control the Brakes," *Mail Online*, Feb. 6, 2014.

246 Entirely possible: John Markoff, "Researchers Hack into Cars' Electronics," *New York Times*, March 9, 2011; Chris Philpot, "Can Your Car Be Hacked?," *Car and Driver*, Aug. 2011; Andy Greenberg, "Hackers Reveal Nasty New Car Attacks—with Me Behind the Wheel," *Forbes*, July 24, 2013; Dan Goodin, "Tampering with a Car's Brakes and Speed by Hacking Its Computers: A New How-To," *Ars Technica*, July 29, 2013.

246 Renault Nissan's CEO: Paul A. Eisenstein, "Spying, Glitches Spark Concern for Driverless Cars," CNBC.com, Feb. 8, 2014.

246 The biggest proponent: Sebastian Anthony, "Google's Self-Driving Car Passes 700,000 Accident-Free Miles, Can Now Avoid Cyclists, Stop at Railroad Crossings," *ExtremeTech*, April 29, 2014; John Markoff, "Google's Next Phase in Driverless Cars: No Steering Wheel or Brake Pedals," *New York Times*, May 27, 2014.

247 Law enforcement officials clearly: Lance Whitney, "FBI: Driverless Cars Could Become 'Lethal Weapons,'" *CNET*, July 16, 2014.

247 Just as vehicles were rated: Ms. Smith, "Most 'Hackable' Vehicles Are Jeep, Escalade, Infiniti, and Prius," *Network World*, Aug. 3, 2014.

247 In a nod: Ina Fried, "Tesla Hires Hacker Kristin Paget to, Well, Secure Some Things," *Re/code*, Feb. 7, 2014.

248 "expected to reach": Transparency Market Research, "Home Automation Market (Lighting, Safety and Security, Entertainment, HVAC, Energy Management)—Global Industry Analysis, Size, Share, Growth, Tends, and Forecast, 2013–2019," Sept. 30, 2013.

248 Many such systems: Kashmir Hill, "When 'Smart Homes' Get Hacked: I Haunted a Complete Stranger's House via the Internet," *Forbes*, July 26, 2013.

248 A July 2014 study: Daniel Miessler, "HP Study Reveals 70 Percent of Internet of Things Devices Vulnerable to Attack," HP, July 29, 2014.

249 Major toy makers: Arrayent, "Internet of Things Toys with Mattel," http://www .arrayent.com/internet-of-things-case-studies/ connecting-toys-with-mattet/Disney Research, "CALIPSO: Internet of Things." http://www .disneyresearch.com/project/calipso-internet-of -things/.

249 But toys too can be subverted: Heather Kelly, "'Smart Homes' Are Vulnerable, Say Hackers," CNN, Aug. 2, 2013.

249 They allow hackers to turn off: Dan Goodin, "Welcome to the 'Internet of Things,' Where Even Lights Aren't Hacker Safe," *Ars Technica*, Aug. 13, 2013.

249 Additional systems: Jane Wakefield, "Experts Hack Smart LED Light Bulbs," BBC News, July 8, 2014; Leo King, "Smart Home? These Connected LED Light Bulbs Could Leak Your Wi-Fi Password," *Forbes*, July 9, 2014.

249 The device, known: Andy Greenberg, "An Eavesdropping Lamp That Livetweets Private Conversations," *Wired*, April 23, 2014.

250 Because Insteon did not require: Hill, "When 'Smart Homes' Get Hacked."

250 The Insteon Hub: Paul Roberts, "Breaking and Entering," *The Security Ledger*, July 25, 2013.

250 The number of vulnerabilities: Ms. Smith, "500,000 Belkin WeMo Users Could Be Hacked;

CERT Issues Advisory," *Network World*, Feb. 18, 2014.

251 As a result, their cameras: Kashmir Hill, "How Your Security System Could Be Hacked to Spy on You," *Forbes*, July 23, 2014.

251 It's not just older alarm systems: Ms. Smith, "Hacking and Attacking Automated Homes," *Network World*, June 25, 2013.

251 Hilton Hotels too: Nancy Trejos, "Hilton Lets Guests Pick Rooms, Use Smartphones as Keys," *USA Today*, July 29, 2014.

251 Worldwide nearly ninety million: Michael Wolf, "3 Reasons 87 Million Smart TVs Will Be Sold in 2013," *Forbes*, Feb. 25, 2013.

251 Many brands have been found: Lorenzo Franceschi-Bicchierai, "Your Smart TV Could Be Hacked to Spy on You," *Mashable*, Aug. 2, 2013; Dan Goodin, "How an Internet-Connected Samsung TV Can Spill Your Deepest Secrets," *Ars Technica*, Dec. 12, 2012.

252 "750,000 malicious spam": Ellie Zolfaghar- ifard, "Criminals Use a Fridge to Send Malicious Emails in First Ever Home Hack," *Mail Online*, Jan. 17, 2014.

252 Refrigerator spam: "Spam in the Fridge," *Economist*, Jan. 25, 2014.

252 In early 2014, researchers: Dan Goodin, "'Internet of Things' Is the New Windows XP— Malware's Favorite Target," *Ars Technica*, April 2, 2014.

252 As of mid-2013: *Utility-Scale Smart Meter Deployments*, IEE report, Aug. 2013, 3; Chris Choi, "Smart Meters Are Heading to Every Home in Britain," *ITV News*, July 8, 2014.

253 Researchers in Germany: Jordan Robert- son, "Your Outlet Knows: How Smart Meters Can Reveal Behavior at Home, What We Watch on TV," *Bloomberg*, June 10, 2014.

253 According to an investigation: Brian Krebs, "FBI: Smart Meter Hacks Likely to Spread," *Krebs on Security*, April 9, 2012.

253 Like all computers: Katie Fehrenbacher, "Smart Meter Worm Could Spread like a Virus," *Gigaom*, July 31, 2009.

254 Nest's thermostats: Rolfe Winkler, "What Google Gains from Nest Labs," *Wall Street Jour- nal*, Jan. 15, 2014.

254 "conscious home": Marcus Wohlsen, "What Google Really Gets out of Buying Nest for $3.2 Billion," *Wired*, Jan. 14, 2014.

254 Google's Nest thermostat: Richard Lawler, "Nest Learning Thermostat Has Its Security Cracked Open by GTVHacker," *Engadget*, June 23, 2014.

254 Nest's other main product: Edward C. Baig, "Nest Halts Sales, Issues Warning on Smoke Detector," *USA Today*, April 3, 2014.

254 Dropcam cameras: Hill, "How Your Secu- rity System Could Be Hacked to Spy on You."

255 Using simple, widely available data recov- ery tools: Armen Keteyian, "Digital Photocopi- ers Loaded with Secrets," CBS News, April 19, 2010.

255 Hackers have been able to gain access: Dan Ilett, "Hackers Use Google to Access Photocopi- ers," *ZDNet*, Sept. 24, 2004.

255 Moreover, office printers: Graham Cluley, "HP Printer Security Flaw Allows Hackers to Extract Passwords," GrahamCluley.com, Aug. 7, 2013.

255 By exploiting a vulnerability: "Exclusive: Millions of Printers Open to Devastating Hack Attack, Researchers Say," NBC News, Nov. 29, 2011; Sebastian Anthony, "Tens of Millions of HP LaserJet Printers Vulnerable to Remote Hacking," *ExtremeTech*, Nov. 29, 2011.

256 To prove the point: Nicole Perlroth, "Flaws in Videoconferencing Systems Make Boardrooms Vulnerable," *New York Times*, Jan. 22, 2012.

256 In April 2012: Brock Parker, "Hackers Convert MIT Building in Giant Tetris Video Game," Boston.com, April 24, 2012.

257 From their headquarters: Brian Krebs, "Fazio Mechanical Services," *Krebs on Security*, Feb. 12, 2014; Gregory Wallace, "HVAC Ven- dor Eyed as Entry Point for Target Breach," *CNNMoney*, Feb. 7, 2014; Danny Yadron and Paul Ziobro, "Before Target, They Hacked the Heating Guy," *Digits* (blog), *Wall Street Journal*, Feb. 5, 2014.

257 There they found: Dan Goodin, "Epic Tar- get Hack Reportedly Began with Malware-Based Phishing E-Mail," *Ars Technica*, Feb. 12, 2014; U.S. Senate Committee on Commerce, Science, and Transportation, *A "Kill Chain" Analysis of the 2013 Target Data Breach*, Majority Staff Report for Chairman Rockefeller, March 26, 2014.

257 Once there, attackers installed: Kim Zetter, "The Malware That Duped Target Has Been Found," *Wired*, Jan. 16, 2014.

258 Worse, it was possible: Sean Gallagher, "Vulnerabilities Give Hackers Ability to Open Prison Cells from Afar," *Ars Technica*, Nov. 7, 2011; Shaun Waterman, "Prisons Bureau Alerted to Hacking into Lockups," *Washington Times*, Nov. 6, 2011.

258 In mid-2013: Kim Zetter, "Prison Com- puter 'Glitch' Blamed for Opening Cell Doors in Maximum-Security Wing," *Wired*, Aug. 16, 2013.

258 While the chamber had successfully: Sio- bhan Gorman, "China Hackers Hit U.S. Cham- ber," *Wall Street Journal*, Dec. 21, 2011.

259 As the Chinese premier: Goodman, "Power of Moore's Law in a World of Geotech- nology."

259 "means of electric media": Marshall McLu- han, *Understanding Media: The Extensions of Man* (New York: Routledge, 2001), rev. ed.

259 "Fitbit for the city": Elizabeth Dwoskin, "They're Tracking When You Turn Off the Lights," *Wall Street Journal*, October 20, 2014.

259 Better sensors in our streetlights: "Outdoor Lighting," Echelon, https://www.echelon.com/applications/street-lighting/.

259 Using a wireless traffic-detection system: Mark Prigg, "New York's Traffic Lights HACKED," *Mail Online*, April 30, 2014.

260 If smart meters: Erica Naone, "Hacking the Smart Grid," *MIT Technology Review*, Aug. 2, 2010.

260 A hacker using the same exploit: Reuters, "'Smart' Technology Could Make Utilities More Vulnerable to Hackers," *Raw Story*, July 16, 2014.

Chapter 14: Hacking You

263 "We Are All Cyborgs Now": Amber Case, "We Are All Cyborgs Now," TED Talk, Dec. 2010.

263 Over 90 percent: "Text Message/Mobile Marketing," WebWorld2000, http://www.webworld2000.com/.

265 Over 100 million: Marcelo Ballve, "Wearable Gadgets Are Still Not Getting the Attention They Deserve—Here's Why They Will Create a Massive New Market," *Business Insider*, Aug. 29, 2013.

265 Most wearable devices: "How Safe Is Your Quantified Self? Tracking, Monitoring, and Wearable Tech," *Symantec Security Response*, July 30, 2014.

265 Google has already: "Google Partners with Ray-Ban, Oakley for New Glass Designs," NBC News, March 24, 2014; Deloitte, *Technology, Media, and Telecommunications Predictions, 2014*, 10.

266 The fear of filming: Richard Gray, "The Places Where Google Glass Is Banned," *Telegraph*, Dec. 4, 2013.

266 In fact, hackers had already: Andy Greenberg, "Google Glass Has Already Been Hacked by Jailbreakers," *Forbes*, April 26, 2013.

266 The GPS features: Mark Prigg, "Google Glass HACKED to Transmit Everything You See and Hear: Experts Warn 'the Only Thing It Doesn't Know Are Your Thoughts,'" *Mail Online*, May 2, 2013.

266 While your grandma: John Zorabedian, "Spyware App Turns the Privacy Tables on Google Glass Wearers," *Naked Security*, March 25, 2014.

266 Given the pace: Katherine Bourzac, "Contact Lens Computer: Like Google Glass, Without the Glasses," *MIT Technology Review*, June 7, 2013.

266 The device is in early stages: Leo King, "Google Smart Contact Lens Focuses on Healthcare Billions," *Forbes*, July 15, 2014.

266 Not to be outdone: Bourzac, "Contact Lens Computer."

267 The historic operation: N. M. van Hemel and E. E. van der Wall, "8 October 1958, D Day for the Implantable Pacemaker," *Netherlands Heart Journal* 16, no. S1 (Oct. 2008): S3–S4.

267 The first Wi-Fi pacemaker: Ben Gruber, "First Wi-Fi Pacemaker in US Gives Patient Freedom," Reuters, Aug. 10, 2009.

267 Millions of Americans: Michael Rushanan et al., "SoK: Security and Privacy in Implantable Medical Devices and Body Area Networks," *SP '14 Proceedings of the 2014 IEEE Symposium on Security and Privacy* (2014), 524–39; Yeun-Ho Joung, "Development of Implantable Medical Devices: From an Engineering Perspective," *International Neurourology Journal* 17, no. 3 (Sept. 2013): 98–106; "IMD Shield: Securing Implantable Medical Devices," http://groups.csail.mit.edu/.

267 Malfunctioning medical devices: Thomas M. Burton, "Medical Device Recalls Nearly Doubled in a Decade," *Wall Street Journal*, March 21, 2014.

267 Nearly 25 percent: H. Alemzadeh et al., "Analysis of Safety-Critical Computer Failures in Medical Devices," *IEEE Security Privacy* 11, no. 4 (July 2013): 14–26, doi:10.1109/MSP.2013.49.

268 Even in hospitals: Kim Zetter, "It's Insanely Easy to Hack Hospital Equipment," *Wired*, April 25, 2014; David Talbot, "Computer Viruses Are 'Rampant' on Medical Devices in Hospitals," *MIT Technology Review*, Oct. 17, 2012.

268 Indeed, in 2013: "Medical Devices Hard-Coded Passwords," ICS-CERT, http://ics-cert.us-cert.gov/alerts/ICS-ALERT-13-164-01.

268 Such was the case: Paul Wagenseil, "Hackers Flood Epilepsy Web Forum with Flashing Lights," FoxNews.com, March 31, 2008; "Anonymous Attack Targets Epilepsy Sufferers," News.com.au, April 1, 2008.

268 After gaining unauthorized access: Barnaby J. Feder, "A Heart Device Is Found Vulnerable to Hacker Attacks," *New York Times*, March 12, 2008; D. Halperin et al., "Pacemakers and Implantable Cardiac Defibrillators: Software Radio Attacks and Zero-Power Defenses," *IEEE Symposium on Security and Privacy, 2008: SP 2008* (2008): 129–42, doi:10.1109/SP.2008.31.

268 From fifty feet away: "Pacemaker Hack Can Deliver Deadly 830-Volt Jolt," *Computerworld*, Oct. 17, 2012.

269 Fearing the profound risk: "Dick Cheney Once Feared Terrorists Could Manipulate His Implanted Defibrillator to Induce Heart Attack," *New York Daily News*, Oct. 19, 2013.

269 Hundreds of thousands: Jim Finkle, "Medtronic Probes Insulin Pump Risks," Reuters, Oct. 25, 2011.

269 Using a specialized radio antenna: "H@cking Implantable Medical Devices," Info-Sec Institute, accessed Aug. 4, 2014, http://resources.infosecinstitute.com/; Jordan Robertson, "Hacker Shows Off Lethal Attack by Controlling Wireless Medical Device," *Bloomberg*, Feb. 29, 2012.

269 Indeed, Europol: Lauren Walker, "First Online Murder to Occur by End of 2014, Europol Warns," *Newsweek*, October 6, 2014.

270 For instance, biomedical engineers: Laura Shin, "New Wireless Medical Device Swims Through Bloodstream," *Smart Planet*, June 30, 2014.

270 Physicians and forensic pathologists: Marc Goodman, "Who Does the Autopsy? Criminal Implications of Implantable Medical Devices," Proceedings of the Second USENIX Conference on Health, Security and Privacy, August 8–12, 2011, San Francisco, California.

271 In response, the Pentagon and DARPA: "The Pentagon's Bionic Arm," CBS News, April 12, 2009.

271 "repository of every FDA-approved": "The Human Bionic Project: A Data Repository for Inspector Gadget Body Parts," *Co.Exist*, June 26, 2013, fastcoexist.com.

271 Even implantable bionic organs: "'Bionic Pancreas' Astonishes Diabetes Researchers," NBC News, June 13, 2014.

272 The result is: Shaunacy Ferro, "Now You Can Control Someone Else's Arm over the Internet," *Popular Science*, May 28, 2013.

273 Given their falling costs: "Global Biometrics Technology Market: Industry Analysis Size Share Growth Trends and Forecast, 2013–2019," KSWT, accessed Aug. 5, 2014, http://www.kswt.com/.

273 with more than 500 million: Rawlson King, "500 Million Biometric Sensors Projected for Internet of Things by 2018," *Biometric Update*, Jan. 31, 2014.

273 As it turns out: Editorial Board, "Biometric Technology Takes Off," *New York Times*, Sept. 20, 2013.

274 Authorities in the Middle Eastern country: Neal Ungerleider, "The Dark Side of Biometrics: 9 Million Israelis' Hacked Info Hits the Web," *Fast Company*, Oct. 24, 2011.

274 The information was stolen: "Authorities Find Source That Leaked Every Israeli's Personal Information Online," Haaretz.com, Oct. 24, 2011.

274 By 2016, Gartner estimates: Larry Barrett, "30 Percent of Companies Will Use Biometric Identification by 2016," *ZDNet*, Feb. 4, 2014.

274 Biometric sensors will be built: Andrew Moran, "990 Million Mobile Devices to Have Biometrics by 2017," *Examiner*, Dec. 9, 2013.

275 Samsung's biometric scanner: "Galaxy S5 Fingerprint Sensor Hacked," BBC News, April 16, 2014.

275 Great in theory: Ms. Smith, "Laptop Fingerprint Reader Destroys 'Entire Security Model of Windows Accounts,'" *Network World*, Sept. 6, 2012.

275 Gangs in Malaysia: "Malaysia Car Thieves Steal Finger," BBC, March 31, 2005.

276 The technique is good enough: "Biometric Fact and Fiction," *Economist*, Oct. 24, 2002.

276 Other hackers have used: Evan Blass, "Play-Doh Fingers Can Fool 90% of Scanners, Sez Clarkson U. Study," *Engadget*, Dec. 11, 2005.

276 In Germany in 2008: Kim Zetter, "Hackers Publish German Minister's Fingerprint," *Wired*, March 31, 2008; Cory Doctorow, "Hackers Publish Thousands of Copies of Fingerprint of German Minister Who Promotes Fingerprint Biometrics," *Boing Boing*, April 1, 2008.

276 Lin paid doctors: Stuart Fox, "Chinese Woman Surgically Switches Fingerprints to Evade Japanese Immigration Officers," *Popular Science*, Dec. 8, 2009.

276 Japanese police report: "Japan 'Fake Fingerprints' Arrest," BBC, Dec. 7, 2009.

277 Lewis created the first-ever: Kelly Jackson Higgins, "Black Hat Researcher Hacks Biometric System," *Dark Reading*, March 31, 2008.

277 Back in today's world: Mark Brown, "Japanese Billboard Recognises Age and Gender," *Wired UK*, Sept. 23, 2010.

277 If one is detected: Natasha Singer, "When No One Is Just a Face in the Crowd," *New York Times*, Feb. 1, 2014.

277 A similar system: Barbara De Lollis, "Houston Hilton Installs Facial Recognition," *USA Today*, Oct. 1, 2010.

278 They might: "Biometric Surveillance Means Someone Is Always Watching," *Newsweek*, April 17, 2014.

278 Facial-recognition technologies: Ibid.

278 All the major Internet companies: Darren Murph, "Face.com Acquired by Facebook for an Estimated $80 Million+, Facial Tagging Clearly at the Forefront," *Engadget*, June 18, 2012.

278 Facebook's automatic: Adam Clark Estes, "Facebook's Doing Face Recognition Again and This Time America Doesn't Seem to Mind," *Motherboard*, Feb. 5, 2013.

278 More than a quarter of a trillion: Adi Robertson, "Facebook Users Have Uploaded a Quarter-Trillion Photos Since the Site's Launch," *Verge*, Sept. 17, 2013; "Biometrics and the Future of Identification," *NOVA Next*, accessed Aug. 6, 2014.

278 In his series of revelations: "How Spy Scandal Unravelled," BBC News, Nov. 7, 2013.

279 "facial recognition quality": James Risen and Laura Poitras, "N.S.A. Collecting Millions

of Faces from Web Images," *New York Times*, May 31, 2014.

279 When facial-recognition is combined: Sebastian Anthony, "UK, the World's Most Surveilled State, Begins Using Automated Face Recognition to Catch Criminals," *ExtremeTech*, July 17, 2014.

279 The program that makes this: Steve Henn, "9/11's Effect on Tech," Marketplace.org, Sept. 8, 2011; Tim Greene, "Black Hat: System Links Your Face to Your Social Security Number and Other Private Things," *Network World*, Aug. 1, 2011.

279 In mid-2011: Amir Efrati, "Google Acquires Facial Recognition Technology Company," *Digits* (blog), *Wall Street Journal*, July 22, 2011; Kit Eaton, "How Google's New Face Recognition Tech Could Change the Web's Future," *Fast Company*, July 25, 2011.

280 NameTag allows users: Simson Garfinkel, "Google Glass Will Be a Huge Success—Unless People Find It Creepy," *MIT Technology Review*, Feb. 17, 2014; Kashmir Hill, "Google Glass Facial Recognition App Draws Senator Franken's Ire," *Forbes*, Feb. 5, 2014.

280 Such facial-recognition apps: Michelle Starr, "Facial Recognition App Matches Strangers to Online Profiles," *CNET*, Jan. 7, 2014.

280 The FBI's billion-dollar: Jeremy Hsu, "FBI's Facial Recognition Database Will Include Non-criminals," *IEEE Spectrum*, April 16, 2014; Mark Rockwell, "Details Emerge on Scope of FBI's Identification System," *FCW*, April 15, 2014.

280 Not only can the system: Sara Reardon, "FBI Launches $1 Billion Face Recognition Project," *New Scientist*, Sept. 7, 2012.

280 Of course no biometric technology: Adam Goldman, "More Than 1 Million People Are Listed in U.S. Terrorism Database," *Washington Post*, Aug. 5, 2014; Editorial Board, "The Black Hole of Terrorism Watch Lists," *New York Times*, Dec. 15, 2013.

280 All it takes to defeat: John Leyden, "Laptop Facial Recognition Defeated by Photoshop," *Register*, Feb. 19, 2009 ; Mark Saltzman, "FastAccess Anywhere: Face Recognition Replaces Password," *USA Today*, June 4, 2013.

280 The same technique: Kim Zetter, "Reverse-Engineered Irises Look So Real, They Fool Eye-Scanners," *Wired*, July 25, 2012.

281 The defense contractor: Noah Shachtman, "Army Tracking Plan: Drones That Never Forget a Face," *Wired*, Sept. 28, 2011.

281 He was apparently: Stilgherrian, "Has Facebook Killed the Undercover Cop?," *CSO*, Aug. 25, 2011.

282 Moreover, in an effort to fight fraud: Neal Ungerleider, "Banks Are Deploying Voice Bio-

metrics So That You Don't Have to Tell Them Your Mother's Maiden Name Again," *Fast Company*, May 27, 2014.

282 Companies such as the online education platform: Nick Anderson, "MOOCS—Here Come the Credentials," *College, Inc.* (blog), *Washington Post*, Jan. 9, 2013.

283 Any variation from an established: Clint Boulton, "'Post-Password' Technology Verifies Users by Behavior," *Wall Street Journal*, July 11, 2014.

283 Banks believe biometric tools: Rawlson King, "Biometric Research Note," *Biometric Update*, Jan. 21, 2013.

283 The Nymi wristband: Somini Sengupta, "Machines Made to Know You, by Touch, Voice, Even by Heart," *Bits* (blog), *New York Times*, Sept. 10, 2013.

283 Scientists at the U.K.'s National Physical Laboratory: "NPL Takes Step Forward with Gait Recognition System," *Engineer*, Sept. 20, 2012.

283 There is, however, an even easier way: Christopher Mims, "Smart Phones That Know Their Users by How They Walk," *MIT Technology Review*, Sept. 16, 2010.

283 Proteus Digital Health: Dieter Bohn, "Motorola Shows Off Insane Electronic Tattoo and Vitamin Authentication Prototype Wearables," *Verge*, May 29, 2013.

283 When we know: Anthony, "UK, the World's Most Surveilled State, Begins Using Automated Face Recognition to Catch Criminals."

284 AR can be used: For further information on augmented reality in contact lenses, see Babak A. Parviz, "Augmented Reality in a Contact Lens," *IEEE Spectrum*, Sept. 1, 2009.

284 It is expected: Juniper Research, "Press Release: Over 2.5 Billion Mobile Augmented Reality Apps to Be Installed Per Annum by 2017," Aug. 29, 2012.

285 Ikea even incorporated AR: Luisa Rollenhagen, "Augmented Reality Catalog Places IKEA Furniture in Your Home," *Mashable*, Aug. 6, 2013.

285 A future malicious app: Franziska Roesner, Tadayoshi Kohno, and David Molnar, "Security and Privacy for Augmented Reality Systems," *Communications of the ACM* 57, no. 4 (2014): 88–96, doi:10.1145/2580723.2580730.

285 The renowned game designer: Jane McGonigal, TED Conversation, http://www.ted.com/conversations/44/we_spend_3_billion_hours_a_wee.html; Jane McGonigal, *Reality Is Broken: Why Games Make Us Better and How They Can Change the World* (New York: Penguin Books, 2011).

286 "Strategically we want to start building": Sarah Frier, "Facebook Makes $2 Billion Virtual-

Reality Bet with Oculus," *Bloomberg*, March 26, 2014.

286 Many genuinely view: "Worlds Without End," *Economist*, Dec. 14, 2005.

287 But there is a downside: "A Korean Couple Let a Baby Die While They Played a Video Game," *Newsweek*, July 27, 2014; "Korean Couple Let Baby Starve to Death While Caring for Virtual Child," *Telegraph*, March 5, 2010.

287 Virtual worlds have their own currencies: "The Economy of Online Gaming Fraud Revealed: 3.4 Million Malware Attacks Every Day," *Kaspersky Lab*, Sept. 28, 2010.

287 As strange as it may sound: Carolyn Davis, "Virtual Justice: Online Game World Meets Real-World Cops and Courts," Philly.com, Dec. 8, 2010.

287 Even "sexual assaults": Benjamin Duranske, "'Virtual Rape' Claim Brings Belgian Police to Second Life," *Virtually Blind*, April 24, 2007.

287 These incidents might be: Anna Jane Grossman, "Single, White with Dildo," *Salon*, Aug. 30, 2005.

287 A 2008 report: Sara Malm, "U.S. Intelligence Warned Terrorists Could Create Virtual Jihadist," *Mail Online*, Jan. 9, 2014.

288 According to an eighty-two-page document: Mark Mazzetti and Justin Elliott, "Spies Infiltrate a Fantasy Realm of Online Games," *New York Times*, Dec. 9, 2013.

288 The spies have created: James Ball, "Xbox Live Among Game Services Targeted by US and UK Spy Agencies," *Guardian*, Dec. 9, 2013; Ian Sherr, "Spy Game: NSA Said to Snoop on 'World of Warcraft,'" *Digits* (blog), *Wall Street Journal*, December 9, 2013.

288 "suicide martyrs": Ibid.; Dan Costa, "This Is No Video Game," *PCMag*, Sept. 26, 2007.

Chapter 15: Rise of the Machines: When Cyber Crime Goes 3-D

291 The "brothers" he introduced: Special Agent Gary S. Cacace, affidavit, Sept. 28, 2011, http://www.justice.gov/; "Muslim Pleads Guilty to Plotting to Blow Up the Pentagon and Capitol with Model Airplanes Packed with Explosives," *Mail Online*, July 20, 2012; "US Man Admits Model Plane Plot," BBC News; Brian Ballou, "Rezwan Ferdaus of Ashland Sentenced to 17 Years in Terror Plot; Plotted to Blow Up Pentagon, Capitol," Boston.com, Nov. 1, 2012; Jess Bidgood, "Rezwan Ferdaus of Massachusetts Gets 17 Years in Terrorist Plot," *New York Times*, Nov. 2, 2012.

292 The number of robotics start-up companies: "Global Industrial Robotics Market Revenues to Surpass $37 Billion by 2018," *Business Wire*, Feb. 24, 2014.

292 "where bits from the digital realm": Marcus Wohlsen, "Forget Robots. We'll Soon Be Fusing Technology with Living Matter," *Wired*, May 27, 2014.

293 Despite the costs: Industrial Federation of Robotics, http://www.ifr.org/industrial-robots /statistics/.

293 In just one Hyundai factory: "Car, Airbag, Money: Robots Make Cars," video, http://channel .nationalgeographic.com/; Tamara Walsh, "Rise of the Robots: 2 Industries Increasingly Turning to Robotics for Innovation," *Motley Fool*, Aug. 24, 2014.

293 Not to be outdone: Katie Lobosco, "Army of Robots to Invade Amazon Warehouses," *CNNMoney*, May 22, 2014.

293 More impressive is the fact: Rodney Brooks, "Robots at Work," World Future Society, *Futurist*, May–June 2013.

293 In more than 150 medical centers: "The Invisible Unarmed," *Economist*, March 29, 2014.

293 Over 500,000 such operations: Stewart Pinkerton, "The Pros and Cons of Robotic Surgery," *Wall Street Journal*, Nov. 17, 2013.

293 Using similar technology: Jacques Marescaux et al., "Transcontinental Robot-Assisted Remote Telesurgery: Feasibility and Potential Applications," *Annals of Surgery* 235, no. 4 (2002): 300–301.

294 Though the gains: For a definitive view into the world of military robotics, see Peter W. Singer's seminal *Wired for War: The Robotics Revolution and Conflict in the 21st Century* (New York: Penguin Books, 2009).

294 In 2003: Mitch Joel, "The Booming Business of Drones," *Harvard Business Review*, Jan. 4, 2013.

294 Today the United States: Michael C. Horowitz, "The Looming Robotics Gap," *Foreign Policy*, May 5, 2014.

294 These machines are well armed: Craig Whitlock, "Drone Strikes Killing More Civilians Than U.S. Admits," *Washington Post*, Oct. 22, 2013.

294 In 2011, it was estimated: David Axe, "One in 50 Troops in Afghanistan Is a Robot," *Wired*, Feb. 7, 2011; Sharon Gaudin, "U.S. Military May Have 10 Robots per Soldier by 2023," *Computerworld*, Nov. 14, 2013.

294 The company has also created: Mark Prigg, "Google-Owned 'Big Dog' Robot in First Live Trial with Marines," *Mail Online*, July 14, 2014.

294 Other UGVs: "Cheetah Robot 'Runs Faster Than Usain Bolt,'" BBC News, Sept. 6, 2012; "March of the Robots," *Economist*, June 2, 2012.

294 Remote pilots sitting: "Rise of the Drones," *NOVA*, PBS, Jan. 23, 2013.

294 UAVs have become central: Teal Group,

"Teal Group Predicts Worldwide UAV Market Will Total $89 Billion," June 17, 2013; Michael C. Horowitz, "The Looming Robotics Gap," *Foreign Policy*, May 5, 2014.

294 Drones such as the MQ-9 Reaper: Ratnesar Romesh, "Five Reasons Why Drones Are Here to Stay," *Bloomberg Businessweek*, May 23, 2013.

295 The images are of such high quality: "Rise of the Drones."

295 The search giant: John Markoff, "Google Puts Money on Robots, Using the Man Behind Android," *New York Times*, Dec. 4, 2013; Adam Clark Estes, "Meet Google's Robot Army. It's Growing," *Gizmodo*, Jan. 27, 2014.

295 In a seminal article: Bill Gates, "A Robot in Every Home," *Scientific American*, Jan. 2007.

296 There are signs: See the iRobot Web site, http://www.irobot.com/; the Droplet Web site, http://smartdroplet.com; the Grillbot Web site, http://grillbots.com; Mark Prigg, "Forgotten to Feed the Dog? Don't Panic, There's an App for That (and It Will Even Tweet to Tell You How Much They've Eaten," *Mail Online*, Jan. 4, 2013.

296 Even the homemaker extraordinaire: David McCormack, "'I Love My Drone': Martha Stewart Shares Incredible Aerial Images of Her Estate as She Reveals Her Latest Must-Have Accessory for the Summer," *Mail Online*, July 30, 2014.

296 The market for consumer: Marcelo Ballve, "The Market for Home Cleaning Robots Is Already Surprisingly Big, and There's Plenty of Room for Growth," *Business Insider*, June 5, 2014.

296 Willow Garage's PR2: Erico Guizzo, "So, Where Are My Robot Servants?," *IEEE Spectrum*, May 29, 2014.

296 "it can read": Brandon Keim, "I, Nanny," *Wired*, Dec. 18, 2008.

296 To help alleviate: Mai Iida, "Robot Niche Expands in Senior Care," *Japan Times*, June 19, 2013.

297 Thousands of Paro units: Anne Tergesen and Miho Inada, "It's Not a Stuffed Animal, It's a $6,000 Medical Device," *Wall Street Journal*, June 21, 2010.

297 One of the fastest-growing: "Your Alter Ego on Wheels," *Economist*, March 9, 2013.

297 Robots such as the MantaroBot: Serene Fang, "Robot Care for Aging Parents," *Al Jazeera America*, Feb. 27, 2014.

297 With the push of a button: Ryan Jaslow, "RP-VITA Robot on Wheels Lets Docs Treat Patients Remotely," CBS News, Nov. 19, 2013.

297 Already Starwood hotels: "Robots Are the New Butlers at Starwood Hotels," CNBC, Aug. 12, 2014.

298 A 2013 study: Carl Benedikt Frey and Michael A. Osborne, "The Future of Employment," Oxford Martin, Sept. 17, 2013, http://www.oxfordmartin.ox.ac.uk/.

298 Those working in the transportation field: For an excellent discussion on the future of robots, automation, and work, see Kevin Kelly, "Better Than Human: Why Robots Will—and Must—Take Our Jobs," *Wired*, Dec. 24, 2012; Erik Brynjolfsson and Andrew McAfee, *The Second Machine Age: Work, Progress, and Prosperity in a Time of Brilliant Technologies* (New York: W. W. Norton, 2014).

298 News outlets such as: Francie Diep, "Associated Press Will Use Robots to Write Articles," *Popular Science*, July 1, 2014.

298 Many believe that it is the growth: Paul Krugman, "Robots and Robber Barons," *New York Times*, Dec. 9, 2012.

299 In mid-2014, a young woman: Lindsey Bever, "Seattle Woman Spots Drone Outside Her 26th-Floor Apartment Window, Feels 'Violated,'" *Washington Post*, June 25, 2014.

299 "Air is a public": Rebecca J. Rosen, "So This Is How It Begins: Guy Refuses to Stop Drone-Spying on Seattle Woman," *Atlantic*, May 13, 2013. For a detailed legal review on UAVs and privacy, see John Villasenor, "Observations from Above: Unmanned Aircraft Systems and Privacy," *Harvard Journal of Law and Public Policy* 36, no. 2 (Spring 2013).

299 "Currently, no federal": Government Accountability Office, *Unmanned Aircraft Systems*, Sept. 2012, http://www.gao.gov/.

300 In 2013, the FDA: Robert Langreth, "Unreported Robot Surgery Injuries Open Questions for FDA," *Bloomberg*, Dec. 29, 2013.

301 In another case: "Surgical Robot da Vinci Scrutinized by FDA After Deaths, Other Surgical Nightmares," New York *Daily News*, April 9, 2013.

301 Unfortunately, the power: "Robot Attacked Swedish Factory Worker," *Local*, April 28, 2009.

301 According to the Occupational Safety and Health Administration: John Markoff and Claire Cain Miller, "As Robotics Advances, Worries of Killer Robots Rise," *New York Times*, June 16, 2014.

301 When it was all over: Gavin Knight, "March of the Terminators: But What Happens When Robot Warriors Turn Their Guns on Us?," *Mail Online*, March 15, 2009.

301 According to a *Washington Post* report: Craig Whitlock, "When Drones Fall from the Sky," *Washington Post*, June 20, 2014.

302 "just a few minutes": Ibid.

303 Researchers at the University of Washington: "How Dangerous Could a Hacked Robot Possibly Be?," *Computerworld*, Oct. 8, 2009.

304 Shia militants had figured out: Siobhan Gorman, Yochi J. Dreazen, and August Cole,

"Insurgents Hack U.S. Drones," *Wall Street Journal*, Dec. 18, 2009.

305 The students carried out: Colin Lecher, "Texas Students Hijack a U.S. Government Drone in Midair," *Popular Science*, June 28, 2012.

305 "In five or ten years": John Roberts, "Drones Vulnerable to Terrorist Hijacking, Researchers Say," Fox News, June 25, 2012.

305 In reality, the Iranians: Scott Peterson and Payam Faramarzi, "Exclusive: Iran Hijacked US Drone, Says Iranian Engineer," *Christian Science Monitor*, Dec. 15, 2011.

305 In 2011, a potent computer virus: Noah Shachtman, "Exclusive: Computer Virus Hits U.S. Drone Fleet," *Wired*, Oct. 7, 2011.

305 The software, dubbed SkyJack: Dan Goodin, "Flying Hacker Contraption Hunts Other Drones, Turns Them into Zombies," *Ars Technica*, Dec. 3, 2013.

306 Hackers have already created: Andy Greenberg, "PIN-Punching Robot Can Crack Your Phone's Security Code in Less Than 24 Hours," *Forbes*, July 22, 2013.

306 Robots can also be a criminal's best friend: "Drug Dealer Arrested in Spite of Home Robotic Protection: Police," *China Post*, Aug. 10, 2014.

306 As we saw in the opening: Charlemagne, "Afghanistan—the Biggest Bomb Yet," Intel MSL, March 15, 2013, http://intelmsl.com/.

307 Moments later numerous rounds: Noah Shachtman, "Iraq Militants Brag: We've Got Robotic Weapons, Too," *Wired*, Oct. 4, 2011.

307 Officials predicted that robotic conveyances: Harris, "FBI Warns Driverless Cars Could Be Used as 'Lethal Weapons,'" *Guardian*, July 16, 2014.

307 Sure, others had beaten Bezos: Jathan Sadowski, "Delivered by Drones: Are Tacocopters and Burrito Bombers the Next Pony Express?," *Slate*, Aug. 6, 2013; Laura Stampler, "This Club Is Offering Poolside Drone Bottle Service," *Time*, June 19, 2014.

307 Dubbed Project Wing: "Google Is Testing Delivery Drone System," *Wall Street Journal*, Aug. 29, 2014.

308 Real estate agents: Sarah Zhang, "Drones That Aren't Out to Kill You," *Mother Jones*, Dec. 6, 2012.

308 The Royal Canadian Mounted Police: "Canadian Mounties Claim First Person's Life Saved by a Police Drone," *Verge*, May 10, 2013.

309 The fake cell tower: Andy Greenberg, "Flying Drone Can Crack Wi-Fi Networks, Snoop on Cell Phones," *Forbes*, July 28, 2011.

309 The so-called Occu-copter: Spencer Ackerman, "Occupy the Skies! Protesters Could Use Spy Drones," *Wired*, Nov. 17, 2011.

309 At the Provisional Detention Center: "Drone Is Caught Delivering Cocaine in Prison São Paulo Brazil," *Live Leak*, March 10, 2014; "Heroin by Helicopter," *Voice of Russia*, Feb. 1, 2011; Nick Evershed, "Drone Used in Attempt to Smuggle Drugs into Melbourne Prison, Say Police," *Guardian*, March 10, 2014; Mary-Ann Russon, "Drones Used to Deliver Drugs to Prisoners in Canada," *International Business Times*, Nov. 29, 2013; "Greece: Drone Drops Mobile Phones over Prison Walls," BBC News, Aug. 19, 2014; "Crooks Get Creative to Smuggle Contraband," WALB News, Nov. 22, 2013.

310 Since 2012, the DEA: Meghan Neal, "Cartels Are Reportedly Building DIY Drones to Fly Drugs over the Border," *Motherboard*, June 2, 2014; Doris Gómora, "Fabrican narcos sus propios drones, alerta la DEA," *El Universal*, July 9, 2014.

310 Apart from drugs: Mark Frauenfelder, "Man Arms DIY Drone with Paintball Handgun and Shoots Human Cardboard Cutouts," *Boing Boing*, Dec. 12, 2012.

310 Other videos show: Colin Lecher, "Watch a Stun Gun Drone Tase an Intern," *Popular Science*, March 7, 2014.

310 The earliest video: "R/C Helicopter with .45 Caliber Handgun," *Live Leak*, Dec. 10, 2008.

310 Since then, numerous other videos: Annalee Newitz, "This Video of a Drone with a Gun Will Freak You the Hell Out," *io9*, June 14, 2013; "Viral Video Straps Colt .45 Handgun to a Home-Use Drone," comments, *Live Leak*, June 18, 2013.

310 With so-called automatic: Jason Koebler, "'Follow Me' Drones Will Hover by Your Side on a Digital 'Leash,'" *Motherboard*, June 16, 2014.

310 Several YouTube videos: Radio-Controlled Crop Dusting in Fukuoka, Japan, 2011, http://www.youtube.com/watch?v=N28KKb6i9hs.

311 "what it's like to be": Sean Gallagher, "German Chancellor's Drone 'Attack' Shows the Threat of Weaponized UAVs," *Ars Technica*, Sept. 18, 2013.

311 There are already numerous reports: Jon Fingas, "Near Collision with Airliner Prompts US to Crack Down on Drone Use," *Engadget*, May 12, 2014; Alwyn Scott, "U.S. Passenger Jet Nearly Collided with Drone in March," Reuters, May 9, 2014.

311 Already great advances: John W. Whitehead, "Roaches, Mosquitoes, and Birds: The Coming Micro-Drone Revolution," *Huffington Post*, April 17, 2013.

311 The devices, some small enough: Tom Leonard, "US Accused of Making Insect Spy Robots," *Telegraph*, Oct. 10, 2007.

311 Dragonfly drones: Whitehead, "Roaches, Mosquitoes, and Birds"; Emily Singer, "TR10: Biological Machines," *MIT Technology Review*, March/April 2009; Erico Guizzo, "Moth Pupa + MEMS Chip = Remote Controlled Cyborg

Insect," *IEEE Spectrum*, Feb. 17, 2009; Charles Q. Choi, "Military Developing Robot-Insect Cyborgs," NBC News, July 14, 2009.

311 Much progress is being made: Robert Lee Hotz, "Harvard Scientists Devise Robot Swarm That Can Work Together," *Wall Street Journal*, Aug. 15, 2014.

312 The BAE Systems Taranis: Jon Cartwright, "Rise of the Robots and the Future of War," *Guardian*, Nov. 20, 2010.

312 Though the border bots: Tim Hornyak, "Korean Machine-Gun Robots Start DMZ Duty," *CNET*, July 14, 2010; Keith Wagstaff, "Future Tech? Autonomous Killer Robots Are Already Here," NBC News, May 14, 2014.

314 Goldman Sachs has noted: Goldman Sachs, *2013 Annual Report*, http://www.goldmansachs .com/; Matt Clinch, "3-D Printing Market to Grow 500% in 5 Years," CNBC, April 1, 2014.

314 Digital fabrication: Jessica Leber, "This Man Thinks He Can 3-D Print an Entire House," *Co.Exist*, Nov. 12, 2013.

314 Bio-fabricating printers: Lyndsey Gilpin, "New 3D Bioprinter to Reproduce Human Organs, Change the Face of Healthcare," *TechRepublic*, Aug. 1, 2014; Melissa Davey, "3D Printed Organs Come a Step Closer," *Guardian*, July 4, 2014; Kate Lyons, "Humans Could Be Fitted with Kidneys Made on 3D Printers," *Mail Online*, May 23, 2014.

314 Today most 3-D printers: Ben Rooney, "The 3D Printer That Prints Itself," *Wall Street Journal*, June 10, 2011; Brad Hart, "Will 3D Printing Change the World?," *Forbes*, March 6, 2012.

315 The Gartner group: Gartner, "Gartner Says Uses of 3D Printing Will Ignite Major Debate on Ethics and Regulation," Gartner.com, Jan. 29, 2014.

315 Digital manufacturing will also be a boon: Drew Prindle, "KeyMe Joins Forces with Shapeways to Bring You Custom 3D-Printed Key Copies," *Digital Trends*, Dec. 17, 2013.

315 There are apps too: Ann Givens and Chris Glorioso, "New Technology Could Let Thieves Copy Keys," NBC New York, May 21, 2014.

315 In 2012, cops uncovered: Andy Greenberg, "Hacker Opens High Security Handcuffs with 3D-Printed and Laser-Cut Keys," *Forbes*, July 16, 2012.

315 While the potential humanitarian benefits: Tim Adams, "The 'Chemputer' That Could Print Out Any Drug," *Guardian*, July 21, 2012.

315 Wilson created the Wiki Weapon Project: Carole Cadwalladr, "Meet Cody Wilson, Creator of the 3D-Gun, Anarchist, Libertarian," *Guardian*, Feb. 8, 2014.

316 The lower receiver: Andy Greenberg, "Here's What It Looks Like to Fire a (Partly) 3D-Printed Gun," *Forbes*, Dec. 3, 2012.

316 In May 2013: Andy Greenberg, "Meet the 'Liberator': Test-Firing the World's First Fully 3D-Printed Gun," *Forbes*, May 5, 2013.

316 Wilson's efforts have left: Andy Greenberg, "How 3-D Printed Guns Evolved into Serious Weapons in Just One Year," *Wired*, May 15, 2014.

316 These plastic firearms: Cheryl K. Chumley, "Israeli TV Crew Sneaks Printed 3-D Gun into Knesset—Twice," *Washington Times*, July 4, 2013.

316 Other repositories: Greenberg, "How 3-D Printed Guns Evolved into Serious Weapons in Just One Year."

316 The FBI's Terrorist Explosive Device Analytical Center: Aliya Sternstein, "The FBI Is Getting Its Own, Personal 3D Printer for Studying Bombs," *Nextgov*, June 13, 2014.

Chapter 16: Next-Generation Security Threats: Why Cyber Was Only the Beginning

319 "had become so fragmented": Nina Golgowski, "'Syrian Hackers' Tweet FALSE Report of Explosions at White House and Send Panicked DOW Jones Plunging 100 Points," *Mail Online*, April 23, 2013; Jim McTague, "Why High-Frequency Trading Doesn't Compute," *Barron's*, Aug. 11, 2012; Shan Carter and Amanda Cox, "One 9/11 Tally," *New York Times*, Sept. 8, 2011; Doug Stanglin and David Jackson, "Timeline of AP Hacking, Reaction," *USA Today*, April 23, 2013; Will Oremus, "Would You Click the Link in This Email That Apparently Tricked the AP," *Slate*, April 23, 2013; Tom Lauricella, Kara Scannell, and Jenny Strasburg, "How a Trading Algorithm Went Awry," *Wall Street Journal*, Oct. 2, 2010; Bernard Condon, "Stocks Stumble After a Fake Tweet Announced White House Attack," Associated Press, April 25, 2013; Nick Baumann, "Too Fast to Fail: Is High-Speed Trading the Next Wall Street Disaster?," *Mother Jones*, Jan/Feb. 2013.

320 The legendary Silicon Valley entrepreneur: Vinod Khosla, "Do We Need Doctors or Algorithms?," *TechCrunch*, Jan. 10, 2012.

320 Today, artificial intelligence e-discovery: Rachael King, "Artificial Intelligence May Reduce Soaring E-discovery Costs," *CIO Journal*, Oct. 29, 2013.

321 Just one algorithm alone: Amy Biegelsen, "Unregulated FICO Has Key Role in Each American's Access to Credit," Center for Public Integrity, May 17, 2011.

322 In a study: Adam D. I. Kramer, Jamie E. Guillory, and Jeffrey T. Hancock, "Experimental Evidence of Massive-Scale Emotional Contagion Through Social Networks," *Proceedings of the National Academy of Sciences* 111, no. 24 (2014): 8788–90, doi:10.1073/pnas.1320040111.

322 Facebook never explicitly: Reed Albergotti and Elizabeth Dwoskin, "Facebook Study Sparks Soul-Searching and Ethical Questions," *Wall Street Journal*, June 30, 2014.

322 Though Facebook updated: Kashmir Hill, "Facebook Added 'Research' to User Agreement 4 Months After Emotion Manipulation Study," *Forbes*, June 30, 2014; Michelle N. Meyer, "Everything You Need to Know About Facebook's Controversial Emotion Experiment," *Wired*, June 30, 2014.

322 The lack of algorithmic transparency: Gabriel Hallevy, "The Criminal Liability of Artificial Intelligence Entities," Social Science Research Network scholarly paper, Feb. 15, 2010, http://papers.ssrn.com/.

323 The attack was successful: Chris Greenwood, "Will Russia Hand Over Man Behind the Gameover Zeus Ransom Virus? FBI Issues Warrant for $100M Cybercrime Mastermind," *Mail Online*, June 2, 2014.

323 The unparalleled levels: McAfee, Center for Strategic and International Studies, *Net Losses: Estimating the Global Cost of Cybercrime*, June 2014.

323 There is another way: Jenny Awford, "Student Accused of Murder 'Asked Siri Where to Hide Body,' Say Police," *Mail Online*, Aug. 13, 2014.

324 Just three years: "IBM Watson," IBM Web site, http://www-03.ibm.com/press/us/en/presskit/27297.wss.

325 The M. D. Anderson Cancer Center: "IBM Watson Hard at Work," Memorial Sloan Kettering Cancer Center, Feb. 8, 2013; Larry Greenemeier, "Will IBM's Watson Usher in a New Era of Cognitive Computing," *Scientific American*, Nov. 13, 2013.

325 Ray Kurzweil has popularized: Ray Kurzweil, *The Singularity Is Near: When Humans Transcend Biology* (New York: Penguin Books, 2006), 7.

326 In 2014, Google purchased: Catherine Shu, "Google Acquires Artificial Intelligence Startup DeepMind," *TechCrunch*, Jan. 26, 2014.

326 "Whereas the short-term impact": Stephen Hawking et al., "Stephen Hawking: 'Transcendence Looks at the Implications of Artificial Intelligence—but Are We Taking AI Seriously Enough?,'" *Independent*, May 1, 2014.

327 Tens of millions of dollars: Reed Albergotti, "Zuckerberg, Musk Invest in Artificial Intelligence Company," *Wall Street Journal*, March 21, 2014.

327 In April 2013: "Brain Research Through Advancing Innovative Neurotechnologies," Aug. 25, 2014, http://www.nih.gov/science/brain/; Susan Young Rojahn, "The BRAIN Project Will Develop New Technologies to Understand the Brain," *MIT Technology Review*, April 8, 2013.

328 Though such a machine: Priya Ganapati, "Cognitive Computing Project Aims to Reverse-Engineer the Mind," *Wired*, Feb. 6, 2009; Vincent James, "Chinese Supercomputer Retains 'World's Fastest' Title, Beating US and Japanese Competition," *Independent*, Nov. 19, 2013.

328 As far-fetched as the idea: Ray Kurzweil, *How to Create a Mind: The Secret of Human Thought Revealed* (New York: Penguin Books, 2013); Michio Kaku, *The Future of the Mind: The Scientific Quest to Understand, Enhance, and Empower the Mind* (New York: Doubleday, 2014).

328 Though many have dismissed: Joseph Brean, "Build a Better Brain," *National Post*, March 31, 2012; Cade Metz, "IBM Dreams Impossible Dream," *Wired*, Aug. 9, 2013.

328 Under laboratory conditions: Kaku, *Future of the Mind*, 80–103, 108–9, 175–77.

328 The chip has an unprecedented: Peter Clarke, "IBM Seeks Customers for Neural Network Breakthrough," *Electronics360*, Aug. 7, 2014.

328 "a major step": Paul A. Merolla et al., "A Million Spiking-Neuron Integrated Circuit with a Scalable Communication Network and Interface," *Science*, Aug. 8, 2014, 668–73, doi:10.1126/science.1254642; Robert F. Service, "The Brain Chip," *Science*, Aug. 8, 2014, 614–16, doi:10.1126/science.345.6197.614; John Markoff, "IBM Develops a New Chip That Functions Like a Brain," *New York Times*, Aug. 7, 2014.

328 Perhaps one of the most consequential: Ray Kurzweil, "The Coming Merging of Mind and Machine," *Scientific American* 18 (2008): 20–25, doi:10.1038/scientificamerican0208-20sp.

329 We now also have a plethora: Gary Marcus and Christof Koch, "The Future of Brain Implants," *Wall Street Journal*, March 14, 2014.

329 Jan Scheuermann: Leigh R. Hochberg et al., "Reach and Grasp by People with Tetraplegia Using a Neurally Controlled Robotic Arm," *Nature*, May 17, 2012, 372–75, doi:10.1038/nature11076.

329 There are even consumer-grade: Robert McMillan, "This Guy Just Built a Mind-Controlled Robot," *Wired*, Aug. 22, 2014.

329 A U.K.-based company: Dave Lee, "Google Glass Hack Allows Brainwave Control," BBC News, July 9, 2014; Ingrid Lunden, "Forget 'OK Glass,' MindRDR Is a Google Glass App You Control with Your Thoughts," *TechCrunch*, July 9, 2014.

329 Wearing a transcranial: Sebastian Anthony, "First Human Brain-to-Brain Interface Allows Remote Control over the Internet, Telepathy Coming Soon," *ExtremeTech*, Aug. 28, 2013.

330 In a groundbreaking experiment: Alan S. Cowen, Marvin M. Chun, and Brice A. Kuhl, "Neural Portraits of Perception: Reconstructing Face Images from Evoked Brain Activity," *NeuroImage*, forthcoming, http://camplab.psych.yale

.edu/; Mark Prigg, "Mind Reading Experiment Reconstructs Faces from Brain Scans," *Mail Online*, March 28, 2014.

330 In another case: "How Technology May Soon 'Read' Your Mind," CBS News, Jan. 4, 2009.

330 This and other studies: IBM Research, "Mind Reading Is No Longer Science Fiction," Dec. 19, 2011, http://ibmresearchnews.blogspot.com/.

330 Already any number: Mark Harris, "MRI Lie Detectors," *IEEE Spectrum*, July 30, 2010.

330 Their tests are bolstered: Adi Narayan, "The fMRI Brain Scan: A Better Lie Detector?," *Time*, July 20, 2009.

330 In India, a woman: Anand Giridharadas, "India's Novel Use of Brain Scans in Courts Is Debated," *New York Times*, Sept. 15, 2008; Angela Saini, "The Brain Police: Judging Murder with an MRI," *Wired UK*, May 27, 2009.

331 In 2012, researchers: Geeta Dayal, "Researchers Hack Brainwaves to Reveal PINs, Other Personal Data," *Wired*, Aug. 29, 2012.

332 Quickly, the cost: Erika Check Hayden, "Company Claims to Have Sequenced Man's Genome Cheaply," *Nature News*, Feb. 8, 2008, doi:10.1038/news.2008.563.

332 As a result, improvements: Mike Orcutt, "Bases to Bytes," *MIT Technology Review*, April 25, 2012; Matthew Herper, "DNA Sequencing: Beating Moore's Law Since January 2008," *Forbes*, May 13, 2011.

332 When this happens: Erika Check Hayden, "Technology: The $1,000 Genome," *Nature*, March 19, 2014, 294–95, doi:10.1038/507294a; Ashlee Vance, "Human Gene Mapping Price to Drop to $1,000, Illumina Says," *Bloomberg*, Jan. 15, 2014.

333 Indeed, these remarkable drops in cost: Jon Mooallem, "Do-It-Yourself Genetic Engineering," *New York Times Magazine*, Feb. 14, 2010; Jack Hitt, "Guess What's Cooking in the Garage," *Popular Science*, May 31, 2012.

333 Venter boldly predicts: Zoë Corbyn, "Craig Venter: 'This Isn't a Fantasy Look at the Future. We Are Doing the Future,'" *Guardian*, Oct. 12, 2013.

333 The integration of biology: Lisa M. Krieger, "Biological Computer Created at Stanford," *San Jose Mercury News*, March 29, 2013; Tim Requarth and Greg Wayne, "Tiny Biocomputers Move Closer to Reality," *Scientific American*, Nov. 2, 2011; Adam Baer, "Why Living Cells Are the Future of Data Processing," *Popular Science*, Nov. 5, 2012.

333 The emerging field: Clay Dillow, "Biostorage Scheme Turns E. coli Bacteria into Hard Drives," *Popular Science*, Jan. 10, 2011.

333 The legendary geneticist: Wyss Institute,

"Writing the Book in DNA," Aug. 16, 2012, http://wyss.harvard.edu/viewpressrelease/93/.

333 Not only do such storage techniques: Ibid.

334 Indeed, a whole host: Chiropractic Resource Organization, "NIH Heads Foresee the Future," http://www.chiro.org/; Helen Thomson, "Deaf People Get Gene Tweak to Restore Natural Hearing," *New Scientist*, April 23, 2014.

334 The died-out mammoth: George M. Church, *Regenesis: How Synthetic Biology Will Reinvent Nature and Ourselves* (New York: Basic Books, 2012); J. Craig Venter, *Life at the Speed of Light: From the Double Helix to the Dawn of Digital Life* (New York: Viking Adult, 2013).

334 In another example: Kim-Mai Cutler, "Glowing Plant Is One of Y Combinator's Very First Biotech Startups," *TechCrunch*, Aug. 11, 2014.

335 Her estate eventually: Rebecca Skloot, *The Immortal Life of Henrietta Lacks* (New York: Broadway Books, 2011); *Moore v. Regents of University of California* (1990) 51 Cal. 3d 120 (271 Cal. Rptr. 146, 793 P.2d 479), Justia Law, accessed Sept. 12, 2014, http://law.justia.com/.

335 Why did they: A case of *Moore v. Regents of the University of California*. On July 9, 1990, the court ruled that a person's discarded tissue and cells are not their property and can be commercialized.

336 Our genetic makeup: James Randerson, "What DNA Can Tell Us," *Guardian*, April 26, 2008.

336 Some studies have also found: Ian Sample, "Male Sexual Orientation Influenced by Genes, Study Shows," *Guardian*, Feb. 13, 2014; Patricia Cohen, "Genetics and Crime at Institute of Justice Conference," *New York Times*, June 19, 2011.

336 Though GINA applies: National Human Genome Research Institute, Genetic Information Nondiscrimination Act of 2008, http://www.genome.gov/; National Human Genome Research Institute, "Genetic Discrimination," http://www.genome.gov/.

336 Several people: Adam Cohen, "Can You Be Fired for Your Genes?," *Time*, Feb. 20, 2012.

336 Meanwhile, under Danish law: Statens Serum Institut, "The Danish Neonatal Screening Biobank," http://www.ssi.dk.

336 And what happens: Andrew Pollack, "DNA Evidence Can Be Fabricated, Scientists Show," *New York Times*, Aug. 18, 2009; Dan Frumkin et al., "Authentication of Forensic DNA Samples," *Forensic Science International: Genetics* 4, no. 2 (2010): 95–103, doi:10.1016/j.fsigen.2009.06.009.

336 The engineered samples: Fiona Macrae, "DNA Evidence Can Be Fabricated and Planted at Crime Scenes, Scientists Warn," *Mail Online*, Aug. 19, 2009.

337 "discreet DNA samples": Sharon Begley,

"Citing Privacy Concerns, U.S. Panel Urges End to Secret DNA Testing," Reuters, Oct. 11, 2012.

337   Organized crime has always made money: Erin Carlyle, "Billionaire Druglords," Forbes, March 13, 2012.

337   At the height: Bijan Stephen, "Pablo Escobar's Hippos Are Running Wild in Colombia," Time, June 28, 2014.

337   Though narcos have been using: Jeremy McDermott, "Drug Lords Develop High-Yield Coca Plant," Telegraph, Aug. 27, 2004; Goodman, "What Business Can Learn from Organized Crime."

337   You could just take: Marc Goodman, "A Vision for Crimes in the Future," TED Talk, July 2012.

338   E. coli bacteria: "Bakterien können ohne viel Aufwand Cannabis-Wirkstoff produzieren," derStandard.at, Aug. 17, 2010, http://derstandard.at/; Luc Henry, "Instead of Poppies, Engineering Microbes," Discover, Sept. 9, 2014; "A New Opium Pipe," Economist, Aug. 30, 2014.

338   Later it was discovered: Joel O. Wertheim, "The Re-emergence of H1N1 Influenza Virus in 1977: A Cautionary Tale for Estimating Divergence Times Using Biologically Unrealistic Sampling Dates," PLoS ONE 5, no. 6 (2010): e11184, doi:10.1371/journal.pone.0011184.

339   Just one year: Alison Young, "Vial of Deadly Virus Missing at Texas Bioterror Laboratory," USA Today, March 25, 2013; "Paris Laboratory Loses Deadly SARS Virus Samples," France24, April 16, 2014.

339   Overseas, we know: Eric Schmitt and Thom Shanker, "Qaeda Trying to Harness Toxin, Ricin, for Bombs, U.S. Says," New York Times, Aug. 12, 2011.

339   Many other terrorist organizations: Marc Goodman, "The Bio-crime Prophecy," Wired, May 28, 2013.

339   To prove the point: Erika Check, "Poliovirus Advance Sparks Fears of Data Curbs," Nature, July 18, 2002, 265–65, doi:10.1038/418265a.

340   Yet by making: Masaki Imai et al., "Experimental Adaptation of an Influenza H5 HA Confers Respiratory Droplet Transmission to a Reassortant H5 HA/H1N1 Virus in Ferrets," Nature, May 2, 2012, doi:10.1038/nature10831; Bryan Walsh, "Should Journals Describe How Scientists Made a Killer Flu?," Time, Dec. 21, 2011.

340   In the end: Denise Grady and William J. Broad, "U.S. Asks Journals to Censor Articles on Bird Flu Virus," New York Times, Dec. 20, 2011.

340   In the future: Andrew Hessel, Marc Goodman, and Steven Kotler, "Hacking the President's DNA," Atlantic, Oct. 24, 2013.

340   Though one might think: Robert Booth and Julian Borger, "US Diplomats Spied on UN

Leadership," Guardian, Nov. 28, 2010; Spencer Ackerman, "U.S. Chases Foreign Leaders' DNA, WikiLeaks Shows," Wired, Nov. 29, 2010.

341   The satellites, which: Marcelo Soares, "The Great Brazilian Sat-Hack Crackdown," Wired, April 20, 2009.

341   Perhaps an even greater risk: Ellie Zolfagharifard, "Incredible Image Shows How Earth Is Entirely Surrounded by Junk," Mail Online, Dec. 13, 2013.

341   In 2007, for example: William J. Broad and David E. Sanger, "China Tests Anti-satellite Weapon, Unnerving U.S.," New York Times, Jan. 18, 2007.

342   Such an attack: Joey Cheng, "Critical Military Satellite Systems Are Vulnerable to Hacking," Defense Systems, April 23, 2014.

342   In fact, according to a congressional commission: Tony Capaccio and Jeff Bliss, "Chinese Military Suspected in Hacker Attacks on U.S. Satellites," Bloomberg, Oct. 26, 2011.

342   According to a report: Samuel Gibbs, "International Space Station Attacked by 'Virus Epidemics,'" Guardian, Nov. 12, 2013.

342   In another incident: Ellie Zolfagharifard, "Cosmonaut Accidentally Infected the ISS with a Virus on a USB Stick," Mail Online, Nov. 12, 2013; "Cosmonaut Carries Computer Virus Aboard International Space Station," PBS NewsHour, Nov. 11, 2013.

342   "It's not a frequent occurrence": Damien Francis, "Computer Virus Infects Orbiting Space Station," Guardian, Aug. 27, 2008.

342   "beyond the reach of censors": David Meyer, "Hackers Plan Space Satellites to Combat Censorship," BBC News, Jan. 4, 2012.

343   The "wonder material": "Graphene 'Made with Kitchen Blender,'" BBC News, April 22, 2014; David Larousserie, "Graphene—the New Wonder Material," Guardian, Nov. 22, 2013.

343   "will leave virtually no aspect": Nancy S. Giges, "Top 5 Trends in Nanotechnology," ASME, March 2013.

343   Perhaps nanotech's greatest contributions: "HowStuffWorks 'Nanotechnology Cancer Treatments,'" HowStuffWorks, accessed Sept. 14, 2014, http://health.howstuffworks.com; Dean Ho, "Fighting Cancer with Nanomedicine," Scientist, April 1, 2014.

343   Nanotechnology will be immensely impactful: John Gehl, "Nanotechnology: Designs for the Future," Ubiquity, July 2000, http://ubiquity.acm.org/.

344   "turning the planet to dust": "Modern Marvels: Doomsday Tech DVD," History Channel, Dec. 28, 2004.

344   "Imagine such a replicator": Eric Drexler, Engines of Creation: The Coming Era of Nanotechnology (Garden City, N.Y.: Anchor, 1987), chap. 4.

344 While many have dismissed: Robert A. Freitas Jr., "The Gray Goo Problem," *Kurzweil Accelerating Intelligence*, March 20, 2001; "Address Nanotechnology Concerns, Experts Urge," Reuters, Nov. 15, 2006.

344 Eventually, Drexler himself clarified: Paul Rincon, "Nanotech Guru Turns Back on Goo," BBC, June 9, 2004.

344 Although there is much work: Jacob Aron, "Google's Quantum Computer Flunks Landmark Speed Test," *New Scientist*, Jan. 15, 2014.

344 in one test carried out by Google and NASA: Nick Statt, "Confirmed, Finally, D-Wave Quantum Computer Is Sometimes Sluggish," *CNET*, June 19, 2014.

344 This could help answer: Tom Simonite, "The CIA and Jeff Bezos Bet on Quantum Computing," *MIT Technology Review*, Oct. 4, 2012.

344 Quantum computers: Ibid.

345 Even with a supercomputer: Mohit Arora, "How Secure Is AES Against Brute Force Attacks?," *EETimes*, May 7, 2012.

345 Not surprisingly, the NSA: Steven Rich and Barton Gellman, "NSA Seeks to Build Quantum Computer That Could Crack Most Types of Encryption," *Washington Post*, Jan. 2, 2014; "Quantum Computing, the NSA, and the Future of Cryptography," *On Point with Tom Ashbrook*, WBUR.

346 In the year 2000: Bill Joy, "Why the Future Doesn't Need Us," *Wired*, April 2000.

Chapter 17: Surviving Progress

352 "If you never break": Melanie Pinola, "F**k It, Ship It," *Lifehacker*, Aug. 14, 2012.

353 "Software is bullshit": See Quinn Norton's "Everything Is Broken" for an outstanding analysis of the security challenges posed by insecure computer software code, https://medium.com/.

353 "we are truly living": "Be Still My Breaking Heart," *Dan Kaminsky's Blog*.

353 But we're nowhere near perfect: Leah Hoffmann, "Risky Business," *Communications of the ACM* 54, no. 11 (2011): 20, doi:10.1145/2018396.2018404.

354 We can make a change: First Research, "Computer Software Industry Profile," Aug. 25, 2014.

354 Importantly, you are equally as likely: Jane Chong, "Bad Code: Should Software Makers Pay? (Part 1)," *New Republic*, Oct. 3, 2013.

355 Automobile deaths: "Achievements in Public Health, 1900–1999 Motor Vehicle Safety," *Morbidity and Mortality Weekly Report*, May 14, 1999, http://www.cdc.gov/.

356 It is estimated: Alex Wilhelm, "Facebook Sets Revenue per User Records Around the World in Q2," *TechCrunch*, July 23, 2014.

356 "Advertising is the original sin": Ethan Zuckerman, "The Internet's Original Sin," *Atlantic*, Aug. 14, 2014.

357 Fifty-five percent: Graham Cluley, "55% of Net Users Use the Same Password for Most, If Not All, Websites," *Naked Security*, April 23, 2013; "39 Percent of Smart Phone Users Don't Secure Their Phones," *Consumer Reports News*, May 1, 2013.

357 Given advances in computing power: Deloitte, "2013 Technology Predictions," 2013, http://www.deloitte.com.

358 A study by the computer giant: HP, "HP Study Reveals 70 Percent of Internet of Things Devices Vulnerable to Attack," July 29, 2014.

359 That is why fifty-five million: Ben Elgin, Michael Riley, and Dune Lawrence, "Former Home Depot Managers Depict 'C-Level' Security Before the Hack," *Bloomberg Businessweek*, Sept. 12, 2014.

361 According to a 2014: IBM Managed Security Services, "2014 Cyber Security Intelligence Index," July 22, 2014; Fran Howarth, "The Role of Human Error in Successful Security Attacks," *Security Intelligence*, September 2, 2014.

365 This is possible because: "Computer Immune Systems," Computer Science Department, University of New Mexico, http://www.cs.unm.edu.

365 Stronger scent trails: Larry Greenemeier, "Software Mimics Ant Behavior by Swarming Against Cyber Threats," *Observations* (blog), *Scientific American*, Sept. 28, 2009.

367 In 2012, Janet Napolitano: Mike Masnick, "DHS Boss, in Charge of Cybersecurity, Doesn't Use Email or Any Online Services," *Techdirt*, Sept. 28, 2012.

367 "are not the most technologically": Michelle R. Smith, "Kagan: Court Hasn't Really 'Gotten to' Email," *Big Story*, Aug. 20, 2013.

369 Though there are many: Australian Government Department of Defence, "Top 4 Mitigation Strategies to Protect Your ICT System," http://www.asd.gov.au/; Australian Government Department of Defence, "The Cyber Threat," http://www.asd.gov.au/.

369 An in-depth study: Verizon RISK Team, "2012 Data Breach Investigations Report," 3, accessed via *Wired*.

370 Indeed, a report sponsored: Karl Frederick Rauscher, "The Internet Health Model for Cybersecurity," EastWest Institute, June 2, 2012.

Chapter 18: The Way Forward

375 Today Microsoft Windows: David Weinberg, "95 Percent of U.S. ATMs Run on Windows XP," MarketPlace.org, March 19, 2014.

377 In 2014, only 13 percent: Jeffrey M. Jones,

"Congress Job Approval Starts 2014 at 13%," Gallup, Jan. 14, 2014.

379 The need is particularly vital: White House Strategy for Homeland Security, Oct. 2007, 4, http://www.dhs.gov/xlibrary/assets/nat_strat_homelandsecurity_2007.pdf.

380 A 2010 Government Accountability Office: *Critical Infrastructure Protection: Key Private and Public Cyber Expectations Need to Be Consistently Addressed*, July 15, 2010, http://www.gao.gov/.

381 Citizens in Mexico: "Mexico's Drug War: 50,000 Dead in Six Years," *Atlantic*, May 17, 2012; Sara Ines Calderon, "In Mexico, Tech Is Used to Help Combat Narco Violence, Insecurity," *Tech-Crunch*, Dec. 25, 2012; Michele Coscia, "How and Where Do Criminals Operate? Using Google to Track Mexican Drug Trafficking Organizations," Harvard Kennedy School, Oct. 23, 2012.

382 Over 170,000 documents: George Arnett and James Ball, "Are UK MPs Really Claiming More Expenses Now Than Before the Scandal?," *Guardian*, Sept. 12, 2014; Michael Anderson, "Four Crowdsourcing Lessons from the Guardian's (Spectacular) Expenses-Scandal Experiment," Nieman Lab, June 23, 2009.

382 The Rand Corporation has noted: "Shortage of Cybersecurity Professionals Poses Risk to National Security," June 2014, http://www.rand.org/news/press/2014/06/18.html.

382 The finding was echoed: Cisco, *Cisco 2014 Annual Security Report*; Lewis Morgan, "Global Shortage of Two Million Cyber Security Professionals by 2017," *IT Governance*, October 30, 2014, http://www.itgovernance.co.uk/.

382 "a failed approach": Ellen Nakashima,

"Cybersecurity Should Be More Active, Officials Say," *Washington Post*, Sept. 16, 2012.

382 The student-run "crowdvestigation": "University Professor Helps FBI Crack $70 Million Cybercrime Ring," *Rock Center with Brian Williams*, March 21, 2012.

382 In 2011, police in the U.K.: Ben Rooney, "U.K. Government Says It Can't Tackle Cybercrime on Its Own," *Wall Street Journal*, Nov. 25, 2011.

384 That works out: Jane McGonigal, "We Spend 3 Billion Hours a Week as a Planet Playing Videogames. Is It Worth It? How Could It Be MORE Worth It?," TED Conversations, http://www.ted.com/.

384 Because the same image: Miguel Angel Luengo-Oroz, Asier Arran, and John Frean, "Crowdsourcing Malaria Parasite Quantification," *Journal of Medical Internet Research*, Nov. 29, 2012.

384 In one remarkable case: Katia Moskvitch, "Online Game Foldit Helps Anti-AIDS Drug Quest," BBC News, Sept. 20, 2011, http://www.bbc.com/news/technology-14986013; Matt Peckham, "Foldit Gamers Solve AIDS Puzzle That Baffled Scientists for a Decade," *Time*, Sept. 19, 2011.

Appendix: Everything's Connected, Everyone's Vulnerable

iii Follow these simple steps: According to a study by the Australian Department of Defence, http://www.asd.gov.au/publications/Catch_Patch_Match.pdf.

Index

Matrix, The (film), 4, 143
"Maxim" (Innovative Marketing), 175
Maza, 202–3
medical records, 44–46, 85, 135–36, 205, 255,
    274, 356, 359
    errors in, 133, 134–35
Merkel, Angela, 34, 311
metadata, 63, 100
Metcalfe's law, 200, 227
Mexico, drug cartels in, 1, 101, 309, 337, 381
Meyer, Bertolt, 261–62, 271
microchips, 225, 236, 244, 263, 266, 331, 332
microcontrollers, 225, 237
Microsoft, 15, 49, 54, 76, 85, 118, 119, 120, 214,
    216, 248, 288, 338, 358, 359, 370
Microsoft Office, 41, 48, 353
Microsoft Windows, 13, 16, 125–26, 158, 268,
    275, 342, 375
military, U.S., 25
    data leaked by, 115, 145
    robotics used in, 272, 294–95, 304, 311
military robots, 294–95, 312
    see also drones
Minority Report (film), 277, 280
Mission Impossible (film), 251, 273
MIT, 70–71, 144, 224–25, 256, 271, 296, 338,
    356, 384
    Media Lab of, 228, 292, 295, 333
mobile money payment systems, 112, 113
mobile phones, 47, 59–60, 63, 84, 104, 112, 220,
    224, 230, 331, 338
    accelerometers in, 283
    apps and, *see* apps/applications
    biometrics in, 274
    bloatware and, 108
    bugging of, 114, 149
    businesses and, 111–12
    call screening and, 145–50
    camera on, 61, 63, 107, 241
    cloud computing and, 117–20
    Facebook and, 60–61
    firmware and, 236–37
    NFC (near-field communication) and, 112,
        113, 232
    rootkit and, 149–50, 158
    spoofing of, 145–50
    updates and, 108, 369
    see also Android; iOS; iPhone
mobile phone security vulnerabilities, 105–17,
    259–60, 356
    advertising and, 59, 60, 61, 106, 254
    chargers and, 236–37
    hacking and, 105–6, 110, 111, 112–13, 236–37
    locational data and, 63, 70, 76, 77, 100, 107,
        113–17
    malware and, 107–8, 149, 150, 237
    operating systems and, 106–8, 113, 150
    screen manipulation and, 145–46, 149, 150
Mobile Spy, 114, 241

money laundering, 182, 210
money mules, 182–83, 190
Moore, Gordon, 37, 86
Moore, Hunter, 95–96
Moore's Law, 37–38, 40, 43, 152, 165, 167, 188,
    212, 219, 224, 228, 270, 291, 313, 332
Morozov, Evgeny, 105, 263
Motorola, 113, 283
Mueller, Robert, 32, 358
multi-factor authentication, 358, 362
Mumbai, India, terrorist siege (2008) in, 30,
    81–84, 101, 281, 341
Murray, Bill, 176, 177
Musk, Elon, 323, 341, 387, 388

Nairobi, Kenya, Westgate mall terrorist attack
    (2013) in, 101–2, 381
nano-robots, 343–44
nanotechnology, 18, 342–43, 344, 376, 377, 387,
    388
NASA, 3, 38, 314, 342, 344, 390
Nasdaq, 140, 185
Natanz nuclear facility, 124–27, 134, 137, 158,
    215, 236, 324, 362
National Academy of Sciences, 71, 322
National Center for Missing and Exploited
    Children, 206–7
national security, 145, 378
    virtual reality and, 287–88
National Security Administration (NSA), 28,
    34, 66, 78, 79, 86, 120–22, 144, 167, 360,
    362, 366
    biometric database of, 278–79
    Penetrating Hard Targets project of, 345
    PRISM program of, 34, 120
    as spying on gamers in virtual worlds, 288
Nature, 201, 340
Navy, U.S., 144, 151
Nazis, Nazism, 129, 161, 388
NBC, 68, 96
near-field communication (NFC), 112, 113, 230,
    232, 267
NEC, 277, 279, 296
Nest Labs, 227, 253–54
Netflix, 46, 64, 129, 167, 251, 319
Netherlands, 140, 340
neuroscience, 328–29
NeuroSky, 329, 331
News Corp, 146, 190
New York, N.Y., 178, 259, 294, 334
New York State, 29, 131, 143
New York Times, 31, 46, 79, 125, 199, 318, 326
New York University, 272, 273, 378
next-generation technology, 317–48
Nielsen company, 45, 46, 131
Nordstrom, 116, 117
Northeast blackout (2003), 42
North Korea, 151–52, 162